The Role of Plant Roots in Crop Production

NAND KUMAR FAGERIA

The Role of Plant Roots in Crop Production

CRC Press
Taylor & Francis Group
Boca Raton London New York

CRC Press is an imprint of the
Taylor & Francis Group, an **informa** business

CRC Press
Taylor & Francis Group
6000 Broken Sound Parkway NW, Suite 300
Boca Raton, FL 33487-2742

First issued in paperback 2019

ISBN-13: 978-1-4398-6737-2 (hbk)
ISBN-13: 978-0-367-38104-2 (pbk)

Visit the Taylor & Francis Web site at
http://www.taylorandfrancis.com

and the CRC Press Web site at
http://www.crcpress.com

To my grandchildren
Sofia, Maia, and Anjit
With great love and affection

Contents

Preface

A major challenge facing the agricultural scientific community today is to feed the rapidly increasing world population, especially in developing countries. The world population is projected to increase to more than 9 billion by 2050, a large portion of which would be constituted by people in the developing countries. Global food demand is likely to increase by 50% in the next 30 years. This large increase in demand will result primarily from a likely explosion in the world population growth at a time when cereal production on a per capita basis has been decreasing. Stagnation in grain yield of important food crops in recent years in the developed as well as in developing countries has contributed to the sharp increase in food prices. Hence, higher grain yield will be needed in the future to feed a burgeoning world population with a rising standard of living that requires more grain per capita. Diverting land to biofuel production will require even larger increases in yield to match the growth in demand. If this situation continues for a longer period of time, it may cause a serious food crisis worldwide. Hence, increasing food production worldwide has become a major issue in the twenty-first century. The decreasing food production and increasing food prices fuel the fears of agricultural scientists and economists that food production may fail to increase by 2.5% each year, the rate required to feed the growing population, raise incomes, and reduce malnutrition. To meet this challenge, technologies that will enhance productivity, ensure environmental safety, and conserve natural resources are required.

A growing population and rising dietary levels are driving global demand for increasing agricultural production while arable land is being lost to urbanization and degradation. Crop yields need to be increased by improving the use efficiency of resources such as water and nutrients while meeting the demand for diverse agricultural commodities. At present, about 1 billion people live in a state of absolute poverty, of which 800 million people live under uncertain food security, and yields of several crops have already reached a plateau in developed countries. Therefore, more of the productivity gains in the future will have to be achieved in developing countries through better natural resources management and crop improvement. Productivity gains are essential for long-term economic growth, but in the short term, these are even more important for maintaining adequate food supplies for the growing world population. The amount of land available for crop production is decreasing due to urbanization and land degradation, and the trend is expected to be much more dramatic in developing than in developed countries.

This book is a comprehensive presentation of our current knowledge of the role of environmental factors in root growth and development and their effect on the improvement of the yield of annual crops. I believe that this treatise will be useful to plant scientists worldwide and of interest to agronomists, horticulturists, plant physiologists, soil scientists, plant breeders, plant pathologists, and entomologists in its references to numerous annual crops. In addition, it brings together the issues and the state-of-the-art technologies that affect root growth, with comprehensive reviews to facilitate efficient, sustainable, economical, and environmentally responsible crop production. Much of my professional career has been spent at the National Rice and Bean Research Center of Empresa Brasileira de Pesquisa Agropecuária (EMBRAPA), Santo Antônio de Goiás, Brazil, where I was allowed the freedom to continue and build my studies on mineral nutrition of field crops. I have accumulated a lot of root data on field crops like upland and lowland rice, dry bean, corn, wheat, soybean, and legume cover crops of tropical origin. This has helped me significantly in writing many of the chapters. I sincerely thank all the staff members at the National Rice and Bean Research Center of EMBRAPA for extending their cooperation and providing me with an amiable academic environment in which to work and write this book. I would also like to sincerely thank the Brazilian Scientific and Technological Research Council (CNPq) for providing financial support to two of my research projects (grant number 300614/2010-1 and 471032/2010-7). Many results of these projects have been included in the chapters of this book. I sincerely thank Randy Leigh Brehm, and David Fausel, CRC Press (Taylor & Francis Group) for handling numerous issues excellently and for their dedication to producing a high-quality book.

Finally, I would like to thank my wife, Shanti; my daughter, Savita; my daughter-in-law, Neera; my sons, Rajesh and Satya Pal; and my grandchildren, Sofia, Maia, and Anjit, for their patience, encouragement, and understanding, without which I could not possibly have found the time and energy required for writing this book.

Nand Kumar Fageria

National Rice and Bean Research Center of Empresa Brasileira de Pesquisa
Agropecuária (EMBRAPA)
Santo Antônio de Goiás, Brazil

Author

Dr. Nand Kumar Fageria, doctor of science in agronomy, is a senior research soil scientist at the National Rice and Bean Research Center, Empresa Brasileira de Pesquisa Agropecuária (EMBRAPA), Santo Antônio de Goiás, Brazil, since 1975 and has also been research fellow and ad hoc consultant to the Brazilian Scientific and Technological Research Council (CNPq) since 1989. He is a well-recognized expert in the area of mineral nutrition of crop plants at the national and international level. Dr. Fageria was the first to identify zinc deficiency in upland rice grown on Brazilian Oxisols in 1975. He has developed crop genotype screening techniques for aluminum and salinity tolerance and for nitrogen, phosphorus, potassium, and zinc use efficiency. He has also established adequate soil acidity indices like pH; base saturation; Al saturation; and Ca, Mg, and K saturation for dry bean grown on Brazilian Oxisols in conservation or no-tillage systems. He has determined adequate and toxic levels of micronutrients in soil and plant tissues of upland rice, corn, soybean, dry bean, and wheat grown on Brazilian Oxisols and has also determined an adequate rate of N, P, and K for lowland and upland rice grown on Brazilian lowland soils, locally known as "varzea," and Oxisols of "Cerrado" region, respectively. He has screened a large number of tropical legume cover crops for acidity tolerance and for N, P, and micronutrient use efficiency. Dr. Fageria has characterized chemical and physical properties of varzea soils of several states of Brazil, which can be helpful in better fertility management of these soils for sustainable crop production. He has also determined adequate rate and sources of P and acidity indices for soybean grown on Brazilian Oxisols. Results of all these studies have been published in scientific papers, technical bulletins, book chapters, and congress or symposium proceedings.

Dr. Fageria is the author/coauthor of 11 books and more than 300 scientific journal articles, book chapters, review articles, and technical bulletins. His two books, *The Use of Nutrients in Crop Plants* (2009) and *Growth and Mineral Nutrition of Field Crops, Third Edition* (2011), have become best sellers for CRC Press. He has been invited several times by the editor of *Advances in Agronomy*, a well-established and highly regarded serial publication, to write review articles on nutrient management, enhancing nutrient use efficiency in crop plants, ameliorating soil acidity by liming on tropical acid soils for sustainable crop production, and the role of mineral nutrition in the root growth of crop plants. He has been an invited speaker to several national and international congresses, symposia, and workshops. He is a member of the editorial boards of the *Journal of Plant Nutrition* and the *Brazilian Journal of Plant Physiology*. He is also a member of the international steering committee of the symposium on plant–soil interactions at low pH since 1990.

1 Types of Plant Roots, Their Measurements, and Associations with Yield

1.1 INTRODUCTION

The root is an important organ that plays a significant role in the growth and development of plants. Roots not only supply water and nutrient to plants but also give mechanical support from seedling through maturity and development. In addition, growth hormones are also supplied by roots to the plants, which help in their better growth and development. Furthermore, roots improve the organic matter content of the soil, which is responsible for improving the physical, chemical, and biological properties of the soil, resulting in higher crop yields. Barber (1979) estimated that 8%–11% of corn stalk residues were transformed into soil organic carbon, while at least 18% of corn roots were transformed into soil organic carbon. Allmaras et al. (2004) showed that the measurement of the quantity of plant roots alone underestimates the role of roots in the formation of soil organic carbon. They noted that rhizodeposition in the field is a large contributor to the total carbon (C) cycle and must be included in the analysis at the field level. Johnson et al. (2006) summarized the contributions of different plant parts from different plant species to soil organic carbon and gave guidelines for including the contributions of plant roots and rhizodeposition to the total C cycle when analyzing changes in soil organic carbon. Belowground deposition of fixed C in structural root biomass, exudates, mucilage, and sloughed cells may be a major source for soil organic carbon accumulation (Bottner et al., 1999; Allmaras et al., 2004). Benjamin et al. (2010) reported that the contribution of the crop root system to the formation and increase of soil organic carbon is important when considering the selection of a crop rotation in a cropping system.

The influence of roots on soil organic carbon pools could be relatively greater than the influence of aboveground C inputs because of the continuous release of C from roots and the complex nature of the rhizosphere–soil interface (Michulmas et al., 1985; Boone, 1994; Norby and Cotrufo, 1998). Puget and Drinkwater (2001) studied the fate of root-derived C versus shoot-derived C of hairy vetch (*Vicia villosa* Roth) to identify the factors contributing to any differences in the retention of aboveground versus belowground C inputs. At the end of the growing season, nearly one-half of the root-derived C was still present in the soil, whereas only 13% of the shoot-derived C remained. The differences in decomposition rates for root- and shoot-derived litter were partly due to differences in decomposability. Hairy vetch shoot litter, with a lignin/N of 1:1, was mineralized more rapidly and to a greater extent than was root litter, which had a lignin/N slightly greater than 6:1 (Puget and Drinkwater, 2001). Roots influence aggregate stability directly by physically enmeshing soil particles and indirectly by stimulating microbial biomass that in turn synthesizes polymers that act as binding agents (Tisdall and Oades, 1979; Jastrow et al., 1998).

Soils are the largest pool of C in the terrestrial environment (Schlesinger, 1990; Jobbagy and Jackson, 2000; Entry et al., 2002). The amount of C stored in soils is twice the amount of C in the atmosphere and three times the amount of C stored in living plants (Schlesinger, 1990); therefore, a change in the size of the soil C pool could significantly alter the atmospheric CO_2 concentration (Wang et al., 1999). Several studies have reported that roots have relatively greater influence than shoots on soil C pools and have attributed the greater retention of root-derived C partly to physical protection within soil aggregates (Gale et al., 2000; Wander and Yang, 2000; Puget and Drinkwater, 2001; Kong and Six, 2010). In legume-based systems, Puget and Drinkwater (2001) suggested that the greater proportion of root-derived C associated with occluded particulate organic matter and silt and clay indicates a greater persistence of root C in soil. Gale et al. (2000) reported that root-derived intra-aggregate particulate organic matter was more important than residue-derived C in the stabilization of small macro-aggregates under no-till soil conditions. In addition, Wander and Yang (2000) reported that root-derived materials are more rapidly occluded by aggregates than aboveground materials in no-till and moldboard plowed soils. Kong and Six (2010) reported that at harvest of corn, about 52% of the root-derived C was still present in the soil, while only about 4% of shoot-derived C remained. Similarly, Balesdent and Balabane (1996) reported that although aboveground (345 g C m^{-1} year^{-1}) input was nearly 3.3 times higher than the belowground input (152 g C m^{-1} year^{-1}), the latter contributes more to the soil organic matter pool (57 g C m^{-1} year^{-1}) than the aboveground corn residues (36 g C m^{-1} year^{-1}).

Roots also help in sequestrating carbon, which is a hot topic in the twenty-first century due to climate change and its adverse impact on environment and, consequently, human and animal health. Soil organic carbon is considered a key component in removing CO_2 from the atmosphere to decrease greenhouse gas emissions and mitigate global climate change (Christopher et al., 2009). Hinsinger et al. (2009) reported that roots of higher plants anchor the aboveground diversity of terrestrial ecosystems, and provide much of the carbon to power the soil ecosystem. The production of ramified root system is important, especially under environmental stresses like drought and mineral nutrition (Fageria et al., 2006; Jaleel et al., 2009). Early growth of vigorous roots has proven to be a major factor in increasing N uptake in wheat (Liao et al., 2004; Noulas et al., 2010). Larger root systems provide greater root–soil contact, which is particularly important for the uptake of P (Gahoonia and Nielsen, 2004). Mobile nutrients like nitrate can be depleted at low rooting density, while for less mobile nutrients, like P and K, depletion is often closely related to root length (Atkinson, 1991).

For many years (until the middle of the last century), roots were considered the "hidden half" of plants (Waisel et al., 2002), with a significant scarcity of research results on this issue throughout the world (Otto et al., 2009). The reasons for this lack of data are historically explained by methodological difficulties, by the inaccessibility of the root system itself as an objective of experimentation, by its three-dimensional complexity, and by its notable spatial and temporal variability (Van Noordwijk, 1993). The study of root system is very laborious and time consuming. In addition, the time required for activities of quantification of the root system and the uncertainties of the results were strongly discouraging factors for research on roots (Zonta et al., 2006). Currently, there is consensus on the importance of the study of root system, varied types of plant roots, their measurements and their associations with yield that is critical for improving crop productivity. It could potentially lead to a more efficient use of nutrients and water by crops and to reduced negative impacts on the environment. Despite this potential, roots have received relatively little attention in such areas as architecture, nutrition, and hydrology. The aerial portions of plant species have received greater attention and study, probably because of their conspicuousness and relative ease of measurement, while the subterranean portions were neglected (Fageria et al., 2006). Part of this neglect sprang from the difficulty of accessing roots below the soil surface. An additional complication involves the rapid turnover of the root system. Roots continually grow, die, and decompose (Jones, 1985). As a result, information on root architecture and growth dynamics as a function of environmental factors in field crops is scarce and scattered. As early as 1926, Weaver (1926) reported of the difficulty of interpreting root data. Accurate measurement of the root system presents practical difficulties that many root scientists have struggled to overcome (Hoad et al., 2001). Although there are now well-established ways of measuring rooting using cores or permanent access tubes, the data collected may not be representative of the crop as a whole (Hoad et al., 2001). In addition, these methods (soil cores, pits, and trenches) are destructive in nature, labor intensive, and limited with respect to soil volume and surface area that can be assessed. Further, data obtained by these methods do not provide information on root distribution in the soil but on root biomass averages across plots or treatments (Butnor et al., 2003).

This chapter summarizes the current state of knowledge on roots of cereals, legumes, and grasses, root-to-shoot ratio, contribution of roots to the total plant weight, root length, specific root length (SRL), root dry weight, rooting depth and density, and root growth association with yield. Most of our discussion is supported by experimental results, which hopefully will be useful for those studying plant nutrition, water use, breeding, and plant physiology, and who may be interested in manipulating plant root systems for higher yields.

1.2 ROOTS OF CEREALS, LEGUMES, AND GRASSES

Cereal and legume seeds contain relatively large reserves of storage carbohydrates and nutrients, which allow the initial root system to grow rapidly to considerable depth (Marschner, 1998). Branching often begins before the leaves have unfolded, with the result that the plant establishes early contact with moist soil (Hoad et al., 2001). Generally, roots are classified into four groups: taproot, basal roots, lateral roots, and shoot born or adventitious roots (Zobel, 2005a). When plants produce secondary shoots (tillers) or shoot branches, which develop roots, these roots are commonly called adventitious roots. To indicate the true origin of these adventitious roots, the term "shoot born" is sometimes used (Zobel, 2005a). The primary function of the taproot, basal roots, and adventitious roots is to establish the most optimum framework from which to initiate small lateral roots to affect water and nutrient uptake (Zobel, 2005b). The taproot penetrates relatively deeply to ensure an adequate supply of soil water; the basal roots spread out laterally to ensure a structure for lateral roots that take up P and other nutrients that are less abundant in the lower levels of the soil profile (Zobel, 2005b), and to provide a degree of lodging resistance to the plant as it matures and produces seed (Stoffella et al., 1979; Barlow, 1986). For many grasses and other species in which root secondary thickening is not important, the shoot born roots take over the role of the basal roots. The shoot born roots continue to build the framework with larger and larger conducting roots as the plant increases in size (Zobel, 2005b). The basal and shoot born roots probably provide little direct uptake of nutrients and water (St. Aubin et al., 1986).

Monocots and dicots typically have different root system structures. Root systems of monocots are fibrous, whereas dicots often have taproots. The fibrous root systems of monocots consist of seminal, nodal, and lateral roots. Seminal roots develop from primordia within seeds, and nodal roots develop adventitiously from lower stem nodes. All adventitious roots of stem origin are called nodal roots to distinguish them from other adventitious roots that emerge from the mesocotyl or elsewhere on the plant. Nodal roots are identified by the node number from which they originate. Nodal roots may be functional or nonfunctional (Thomas and Kaspar, 1997). Functional nodal roots are defined as roots that have emerged from stem nodes, entered the soil, and developed lateral roots and/or root hairs. Nonfunctional nodal roots are defined as roots that have emerged from aboveground stem nodes and have not entered the soil or produced lateral roots (Thomas and Kaspar, 1997).

Initial seminal or nodal roots develop laterals that are classed as roots of the first order, roots that develop from the first-order roots are classed as second-order roots, and additional roots that develop from these laterals are classed as third-, fourth-order roots, etc. (Yamauchi et al., 1987a,b). Nodal roots are also known as adventitious, coronal, and/or crown roots. Roots of cereals such as rice include mesocotyl, radical (seminal), and nodal or adventitious roots (Yoshida, 1981). Mesocotyl roots emerge from the axis between the coleoptile node and the base of the radical, and they typically develop only when seeds are planted very deep or are treated with chemicals (Yoshida, 1981). Until adventitious roots develop, seedlings must rely on roots that initiate on the subcoleoptile internodes above the seed or seminal roots below the seed. Adventitious roots are important to seedling establishment because they can conduct more water than smaller diameter seminal roots. Adventitious roots may develop as early as 2 weeks after sowing. Seedling survival may increase when seeds are sown at greater soil depths, where greater soil water availability may increase adventitious root development (Fageria et al., 2006).

Tiller roots do not form in cereals until tillers have two to three leaves (Klepper et al., 1984), and until these roots have developed, parent culms must provide nutrients and water. Parent culms may also have to provide hormonal control, which is essential for tiller survival. Delayed root production by tillers may explain why late tillers often do not survive (Klepper et al., 1984). Figure 1.1 shows the radical and adventitious root system of upland rice (*Oryza sativa* L.) (cereal) and Figure 1.2 shows the taproot system of dry bean (*Phaseolus vulgaris* L.) (legume).

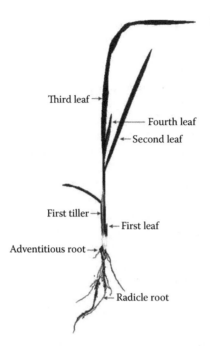

FIGURE 1.1   Root system of upland rice seedlings. (From Fageria, N.K., *J. Plant Nutr.*, 30, 843, 2007.)

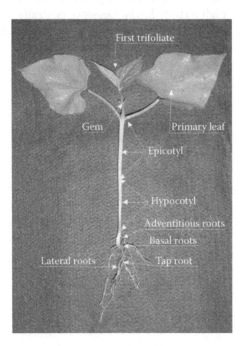

FIGURE 1.2   Root system of dry bean seedling. (From Fageria, N.K. and Santos, A.B., *J. Plant Nutr.*, 31, 983, 2007.)

Rooting patterns differ greatly between cereals and legumes, which could lead to more efficient exploration of soil volume by cereal–legume mixtures (Anil et al., 1998). However, when fertilizer N is limited, biological N fixation is the major source of N in legume–cereal mixed cropping systems (Fujita et al., 1992), but the amount of N_2 fixed by legumes generally declines with increasing soil N availability (Anil et al., 1998), and if the legumes are continuously shaded their ability to fix N_2 is further impaired (Willey, 1979; Szumigalski and Acker, 2006). The use of cereal grain companion crop is the most common legume establishment method in the north central United States, and has been used historically to establish 85% of the alfalfa fields in Iowa (Blaser et al., 2006). Spring-seeded oat is the companion crop of choice in this region. In addition, introducing a winter cereal grain/red clover intercrop into a corn–soybean rotation can provide produces with crop alternative that can diversify income (Exner and Cruse, 2001), improve yields of subsequent crops with reduced inputs (Singer and Cox, 1998), improve soil quality (Reicosky and Forcella, 1998), and disrupt pest cycles (Cook, 1988; Blaser et al., 2006). All these changes can bring improvement in root growth of crop plants grown in rotation, and intercropping consequently enhances the use of resources like nutrients and water.

Roots of grain legumes are less vigorous compared with cereals like rice and corn. Figure 1.3 shows root growth of dry bean at low and high fertility. Similarly, Figure 1.4 shows root growth of upland rice at low and high fertility levels. Root growth of both the species was more vigorous at high fertility level compared to low fertility level at physiological maturity. However, root growth of upland rice was more vigorous at low as well as high fertility levels. Legumes generally have low yield compared to cereals, and less vigorous root growth of legumes compared to cereals is one of the factors that contribute to lower grain yield of legumes. Another factor that

FIGURE 1.3 Root growth of dry bean cultivar Pérola at low and high fertility levels at physiological maturity.

FIGURE 1.4   Root growth of upland rice genotype BRA 02535 at low and high fertility levels.

is responsible for low grain yield of legumes is their high photorespiration compared to cereals (Fageria, 2009). In addition, legumes fix a large amount of atmospheric nitrogen, and a large part of energy is diverted to this process, which is also responsible for lower yield. The statement that legumes have lower root growth compared to cereals is only valid for grain legumes. It is not valid for tropical legume cover crops. Figure 1.5 shows that lablab, a tropical legume cover crop, has a very good root system at low as well as high soil fertility level.

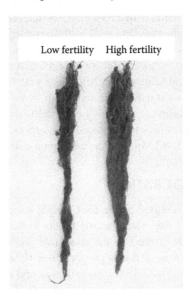

FIGURE 1.5   Root growth of lablab tropical cover crop species at low and high fertility levels.

In addition to their morphological differences, roots of cereals and legumes have different physiochemical properties. The surface of plant roots has a negative electric charge, mainly due to carboxyl groups in the pectin of the root cell walls. The density of this negative charge is defined as a cation exchange capacity (CEC) (Takenaga, 1995). The CEC of cereals such as rice, barley, and corn is typically lower than the CEC of legumes like dry bean and broad bean. Roots with high CEC absorb more divalent cations like Ca^{2+} and Mg^{2+} than monovalent cations such as K^+ and $NH_4{}^+$. On the other hand, roots with lower CEC absorb more monovalent than divalent cations. Hence, in grass–legume mixtures, legumes generally suffer from K^+ deficiency due to large uptake of this element by grasses. Essau (1977), Klepper (1992), Leskovarant and Stofella (1995), O'Toole and Bland (1987), Zobel (1991; 2005a,b), and Fageria et al. (2006) have dealt extensively with various types of monocotyledonous and dicotyledonous roots and root hairs, and their growth and morphology.

Grasses are an important forage source in pasture lands around the world. Globally, pasture lands are much higher compared to croplands. *Brachiaria* spp. are dominant forage grasses in Brazilian pasture lands. Forage grasses are important sources of cellulosic biomass for the developing bioenergy industry (Collins et al., 2010). Important bioenergy grasses are elephant grass (*Pennisetum purpureum* Schumach.), reed canarygrass (*Phalaris arundinacea* L.), miscanthus (*Miscanthus* spp.), big blue stem (*Andropogon gerardii* Vitman), Indian grass (*Sorghastrum nutans* L. NAS), guinea grass (*Panicum maximum*), and switgrass (*Panicum virgatum* L.). Perennial bioenergy grasses have been shown to improve soil quality, enhance nutrient cycling, improve wildlife habitat, and sequester C (Lemos and Lal, 2005; McLaughlin and Kszos, 2005). Root system of forage grasses is fibrous and volumes. Since grass roots are interminate and freely branching organs, they grow most rapidly where environmental conditions are most favorable (Jones, 1985). McLaughlin and Walsh (1998) reported that warm season grasses are characterized by well-developed root systems with a root biomass that is comparable to the annual aboveground biomass production. Ma et al. (2000) reported that >50% of the root distribution of switchgrass was concentrated at 30 cm of the soil surface after 6 years of growth. Zan et al. (2001) reported that switgrass has four to five times more root biomass than corn and could sequester 2.2 Mg C ha^{-1} year^{-1}. Tufekcioglu et al. (2003) found that mature switgrass stands contained six to seven times the C in their root systems than did corn. Collins et al. (2010) reported that total soil organic carbon derived from switgrass roots plus the C stored in the roots themselves represented a belowground C pool of 6.5–8.1 Mg ha^{-1}, depending on cultivars.

1.3   ROOT-TO-SHOOT RATIO

The root-to-shoot ratio is defined as the root weight divided by the shoot weight. Knowledge of this ratio is important in crop plants to have an idea about the distribution of photosynthetic product in the roots and shoots. In addition, the ratio of root-to-shoot biomass expressed as a proportion is needed to estimate and compare belowground biomass, including root exudates and other rhizodeposits, as a function of shoot biomass including grain (Allmaras et al., 2004). In addition, high root-to-shoot ratio is also used in breeding crop genotypes for drought tolerance

(Bonos et al., 2004; Karcher et al., 2008). Bonos and coworkers were able to achieve up to 81% and 130% gains in root-to-shoot ratios following two generations of selection of turf-type tall fescue (*Festuca arundinacea* Schreb.) and perennial ryegrass (*Lolium perenne* L.) varieties, respectively.

Functionally, roots absorb water and nutrients, and anchor the plant, while shoots photosynthesize and transpire, and are the site of sexual reproduction (Groff and Kaplan, 1988). Plants generally maintain tight coordination of biomass partitioning between roots and shoots (Davidson, 1969; Makela and Sievanen, 1987). Poorter and Nagel (2000) describe the allocation of biomass between shoots and roots in plants such that the shoot-to-root biomass ratio is very rapidly restored following the pruning of a large fraction of either roots or leaves.

Partitioning of dry matter in roots relative to shoots is high during the seedling stages of growth and steadily declines throughout development (Evans and Wardlaw, 1976). Overall, the root-to-shoot ratio decreased with the advancement of plant age, indicating that the major part of the photosynthesis is going to shoot. During early vegetative growth, up to 50% of assimilated carbon may be translocated to roots, of which as much as 50% is consumed in respiration or lost from the plant in exudation and rhizodeposition (Gregory, 1994). The principal direction of resource allocation changes markedly around flowering, with considerably less assimilates that transfer to the roots. After anthesis, the amount of assimilates translocated to the root system is very small, and the net size of the root system hardly increases and even declines as roots die (Gregory, 1994). Fageria (1992) reported that the root-to-shoot ratio of dry bran, rice, wheat, and cowpea decreased as plants advanced in age. Environmental factors like temperature, solar radiation, soil moisture content, soil fertility, soil salinity, soil texture, and structure influence the root-to-shoot ratio (Wardlaw, 1990). In addition, crop species and genotypes within species also influence the root-to-shoot ratio (Fageria et al., 2006).

The mineral nutrients P and N exert a pronounced influence on photosynthate and dry matter partitioning between shoots and roots (Costa et al., 2002). The P- and N-deficient plants have much more dry matter partitioned to roots than shoots, probably as a result of higher export rates of photosynthates to the roots. It is well known that leaf expansion is highly sensitive to low P concentrations in the tissue, leading to higher concentrations of sucrose and starch in P-deficient leaves because of reduced demand (Fredeen et al., 1989). Thus, roots become more competitive for photosynthates than shoots, which leads to higher export of carbohydrates to the roots with correspondingly lower shoot–root dry weight (Rufty et al., 1993). Cakmak et al. (1994) reported that dry matter partitioning between shoots and roots of common bean was very differently affected by low supply of P, K, and Mg. Although total dry matter production was more or less similar in P-, K-, and Mg-deficient plants, under P deficiency, a much greater proportion of dry matter was partitioned to roots, whereas in K-deficient and especially in Mg-deficient plants, shoot growth was enhanced at the expense of root growth (Cakmak et al., 1994). The shoot–root weight ratios were 1.8 in P-deficient, 4.9 in control, 6.9 in K-deficient, and 10.2 in Mg-deficient plants.

Yaseen and Malhi (2011) reported that wheat genotypes differed significantly in partitioning of their dry matter in roots and shoots. This difference partly can be attributed to their difference in genetic makeup. These authors also reported that the root-to-shoot ratio of wheat genotypes was significantly influenced by genotype, P level, and interaction of genotype × P level. This indicates the pattern of distribution of biomass between root and shoot. Significant differences in the root-to-shoot ratio were observed in stress P treatment. These differences were less pronounced and nonsignificant in adequate P treatment (Yaseen and Malhi, 2011). The root-to-shoot ratio was about fourfold higher at stress P level than that at adequate P level. This confirms the conclusion of higher root growth due to P-deficiency stress. The root-to-shoot ratio decreased with increasing P level, which may be due to more photosynthetic product translocated to shoot at higher P levels.

Environmental stresses like water and nitrogen increase relative weights of roots compared to shoots (Wilson, 1988; Eghball and Maranville, 1993). An increase in root growth due to water stress was reported in sunflower (Tahir et al., 2002) and *Catharanthus roseus* (Jaleel et al., 2009). An increase in the root-to-shoot ratio under drought conditions was related to abscisic acid content of roots and shoots (Sharp and LeNoble, 2002; Manivannan et al., 2007). The root-to-shoot ratio of 14 tropical legume crops as influenced by P fertilization is presented in Table 1.1. There was a

TABLE 1.1

Root-to-Shoot Ratio of 14 Tropical Legume Cover Crops as Influenced by Phosphorus Fertilization

Cover Crops	0 mg P kg^{-1}	100 mg P kg^{-1}	200 mg P kg^{-1}	Average
1. Crotalaria (*Crotalaria breviflora*)	0.25ab	0.12d	0.27ab	0.21a
2. Sunn hemp	0.14abc	0.25abcd	0.14bc	0.17ab
3. Crotalaria (*Crotalaria mucronata*)	0.04c	0.24abcd	0.30ab	0.18ab
4. Crotalaria (*Crotalaria spectabilis*)	0.13abc	0.33a	0.10c	0.18ab
5. Crotalaria (*Crotalaria ochroleuca*)	0.05c	0.29ab	0.21abc	0.18ab
6. Calopogonium	0.13abc	0.14cd	0.34a	0.20ab
7. Pueraria	0.07bc	0.21abcd	0.09c	0.12b
8. Pigeon pea (black)	0.28a	0.27abcd	0.07c	0.20ab
9. Pigeon pea (mixed color)	0.20abc	0.29abc	0.16bc	0.21a
10. Lablab	0.15abc	0.29abc	0.17abc	0.20ab
11. Mucuna bean ana	0.25ab	0.18bcd	0.15bc	0.19ab
12. Black mucuna bean	0.17abc	0.34a	0.20abc	0.23a
13. Gray mucuna bean	0.13abc	0.16bcd	0.20abc	0.16ab
14. White jack bean	0.13abc	0.13d	0.10c	0.12b
Average	0.15b	0.23a	0.18b	0.18

F-test

P level (P)	**			
Cover crops (C)	**			
P × C	**			

**Significant at the 1% probability level. Means followed by the same letter in the same column or in the same line under three P levels are statistically not significant at the 5% probability level by Tukey's test.

significant influence of P, cover crops, and P × cover crops interaction. Overall, the root-to-shoot ratio increased up to 100 mg P kg^{-1} and then decreased. This indicates that root weight increased up to 100 mg P kg^{-1} and then decreased. The decrease in root weight of crops at higher nutrient levels is reported by Fageria (2009) and Fageria and Moreira (2011). Across P levels, cover crops produced significant root-to-shoot ratio and varied from 0.12 to 0.23, with an average value of 0.18. Hence, it can be concluded that there was a significant variation in root as well as shoot dry weight among cover crops. Figures 1.6 and 1.7 show shoot growth of pigeon pea and lablab cover crops at three P levels. Both the crop species responded to P fertilization. However, shoot mass production was different at three P levels. Similarly, cover

FIGURE 1.6   Shoot growth of pigeon pea cover crops at three P levels.

FIGURE 1.7   Shoot growth of lablab cover crops at three P levels.

crops *Crotalaria ochroleuca*, calopogonium, and white jack bean root growth was different at three P levels (Figures 1.8 through 1.10). Cover crop *C. ochroleuca* produced minimum root growth compared to other two crop species at 0 mg P kg^{-1} soil. Fageria and Moreira (2011) reported significant differences in shoot and root growth of cover crops under different P levels.

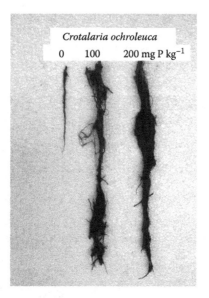

FIGURE 1.8   Root growth of *Crotalaria ochroleuca* cover crops at three P levels.

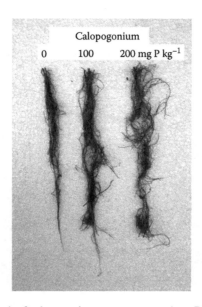

FIGURE 1.9   Root growth of calopogonium cover crops at three P levels.

White jack bean

0   100   200 mg P kg^{-1}

FIGURE 1.10   Root growth of white jack bean cover crops at three P levels.

The author studied the root-to-shoot ratios of upland rice, corn, wheat, soybean, and dry bean during the growth cycle of these crops grown on a Brazilian Oxisol (Table 1.2). The root-to-shoot ratios of five crops were significantly affected by plant age. Values were higher during the vegetative growth stage and decreased during reproductive or grain filling growth stages. The higher root-to-shoot ratio of five crop species in the beginning of crop growth (21–40 days growth period) indicates that more photosynthetic product was translocated to the roots. After this period, the shoot begins to gain dry mass more quickly than the root and the ratio of root-to-shoot dry mass begins to fall. The decrease in the root-to-shoot ratio during reproductive and grain filling growth stages may reflect that more photosynthetic product was translocated to aerial parts compared to root.

With only measured structural root biomass, Huggins and Fuchas (1997) calculated a root-to-shoot ratio of 0.25 for corn, while Buyanovsky and Wagner (1997) measured ratios of 0.28 for wheat, 0.21 for corn, and 0.23 for soybean at harvest. Buyanovsky and Wagner (1997) also reported that the root-to-shoot (vegetative + grain) ratios were 0.48, 0.35, and 0.38 for wheat, corn, and soybean, respectively, when the rhizodeposition was included. Bolinder et al. (1999) assembled a mean root-to-shoot ratio of 0.19 for corn when considering only plant biomass (vegetative + grain) and suggested a ratio of 0.38 when considering belowground soil carbon. The root-to-shoot ratio of drilled-planted, field-grown, flooded rice progressively decreased with plant development with root-to-shoot ratio of 0.23 during vegetative growth and 0.13 at 50% heading (Teo et al., 1995b). This was similar to the root-to-shoot ratio (0.14) of individual field-grown rice plants at physiological maturity (Slaton and Beyrouty, 1992). The root-to-shoot ratio is influenced by environmental factors and also controlled genetically (Fisher and Dunham, 1984). Hence, the differences in the root-to-shoot ratio reported by several authors between same crop species may be related to these factors.

TABLE 1.2

Root-to-Shoot Ratio of Upland Rice, Corn, Wheat, Soybean, and Dry Bean during Growth Cycle

Plant Age in Days	Upland Rice	Corn	Wheat	Soybean	Dry Bean
21	0.04	0.26	0.57	0.20	0.11
40	0.31	0.27	0.29	0.13	0.39
61	0.28	0.20	0.11	0.08	0.16
78	0.23	0.09	0.08	0.07	0.06
97	0.21	0.11	0.05	0.08	0.05
118	0.16	0.08			

Regression Analysis

Upland rice plant age vs. R/S ratio $(Y) = -0.0926 + 0.0106X - 0.000073X^2$, $R^2 = 0.5757**$

Corn plant age vs. R/S ratio $(Y) = 0.3482 - 0.0031X + 0.0000064X^2$, $R^2 = 0.7588**$

Wheat plant age vs. R/S ratio $(Y) = 0.9625 - 0.0215X + 0.00013X^2$, $R^2 = 0.9357**$

Soybean plant age vs. R/S ratio $(Y) = 0.3075 - 0.0060X + 0.000038X^2$, $R^2 = 0.9626**$

Dry bean age vs. R/S ratio $(Y) = 0.0460 + 0.0082X - 0.000089X^2$, $R^2 = 0.4018*$

Five experiments were conducted simultaneously and crops were harvested at physiological maturity. Upland rice and corn were harvested at 118 days; the remaining crops were harvested at 97 days after sowing.

*, **Significant at the 5% and 1% probability levels, respectively.

1.4   CONTRIBUTION OF ROOT, SHOOT, AND GRAIN IN THE TOTAL PLANT WEIGHT

The contribution of root, shoot, and grain in the total plant weight is an important photosynthetic distribution trait that determines economic yields of crop species. Large and vigorous root systems not only absorb more water and nutrients but also improve soil quality due to large biomass left in the soil after crop harvest. Similarly, higher shoot weight also improves soil quality if straw is incorporated in the soil after crop harvest. Data in Table 1.3 show contribution of root, shoot, and grain in the total plant weight of 20 upland rice genotypes grown on a Brazilian Oxisol. The contribution of roots varied from 11.59% to 30.46% at low N rate, with an average value of about 22%. At high N rate, the contribution of root varied from 2.48% to 21.28%, with an average value of about 14%. The contribution of shoot varied from 35.68% to 75.78%, with an average value of about 49% at low N rate and from 33.05% to 56.44% at high N rate, with an average value of about 39%. The contribution of grain in the total plant weight varied from 8.27% to 41.04%, with an average value of about 29% at low N rate. At high N rate, contribution of grain varied from 22.28% to 62.34%, with an average value of about 48%.

The N × genotype has significant interactions for root weight, shoot weight, and grain weight, indicating different responses of genotypes at two N rates. The decrease in root and shoot contribution at high N rate indicates that more photosynthetic product is translocated to grain with the improvement in N uptake. In addition, root as well as shoot weight was significantly higher at higher N rate compared to

TABLE 1.3

Contribution of Root, Shoot, and Grain in the Total Plant Weight of 20 Upland Rice Genotypes as Influenced by Nitrogen Fertilization

Genotype	Root Weight (%)		Shoot Weight (%)		Grain Weight (%)	
	N_0	N_{300}	N_0	N_{300}	N_0	N_{300}
BRA 01506	17.56abcd	2.48g	43.01e	35.17cd	39.44ab	62.34a
BRA 01596	18.08abcd	2.78g	46.18de	35.19cd	35.74abc	62.03a
BRA 01600	23.06abcd	5.87fg	42.93e	35.74cd	34.abcd	58.38ab
BRA 02535	24.63abcd	14.37cde	54.63cd	44.02bc	20.74def	41.61d
BRA 02601	22.76abcd	18.55abcd	45.34de	38.07cd	31.90abcd	43.38cd
BRA 032033	20.36abcd	15.29bcde	42.25e	36.28cd	37.40ab	48.43bcd
BRA 032039	14.29cd	20.55ab	73.23ab	51.97ab	12.49ef	27.48e
BRA 032048	11.59d	19.44abc	66.01abc	32.70d	22.40cdef	47.86bcd
BRA 032051	23.39abcd	14.05cde	45.70de	32.56d	30.92abcd	53.40abcd
BRA 042094	25.12abcd	12.28e	45.83de	40.40cd	29.05abcd	47.32bcd
BRA 042094	23.62abcd	14.66cde	45.21de	37.06cd	31.16abcd	48.28bcd
BRA 042160	27.54abc	14.59cde	46.35de	40.68cd	26.10bcde	44.72cd
BRA 052015	16.78abcd	13.34de	62.38bc	33.05d	20.83def	53.60abcd
BRA 052023	15.95bcd	21.28a	75.78a	56.44a	8.27f	22.28e
BRA 052033	22.28abcd	15.31bcde	35.68e	38.72cd	41.04a	45.97cd
BRA 052034	30.46a	12.33e	39.60e	37.51cd	29.94abcd	50.16abcd
BRA 052045	26.83abc	16.66abcde	39.69e	33.84d	33.48abcd	49.50bcd
BRA 052053	29.38ab	11.04ef	41.42e	33.41d	29.20abcd	55.54abc
BRS Primavera	27.54abc	12.23e	41.75e	34.62cd	30.71abcd	53.15abcd
BRS Sertaneja	25.63abc	13.63de	42.94e	43.97bc	31.43abcd	42.40d
Average	22.34a	13.54b	48.85a	38.57b	28.81b	47.89a
F-test						
N level (N)	*		*		**	
Genotype (G)	**		**		**	
N × G	**		**		**	
CV (%)	17.51		7.34		10.45	

N levels were 0 and 300 mg kg^{-1} of soil.

*, **Significant at the 5% and 1% probability levels, respectively. Means followed by the same letter in the same column or same line are not significantly different at the 5% probability levels by Tukey's test. Values of averages were compared in the same line for each parameter at low and high N levels.

lower N rate, which might have been also responsible for lower contribution. These findings are in agreement with those reported by Fageria and Baligar (2005) and Fageria (2009) that when N is absorbed in sufficient amount by crop plants, a major part of the photosynthesis goes to the grain. Figures 1.11 and 1.12 show shoot growth of upland rice genotypes BRA 02601 and BRA 032039 at low and high N rates. Similarly, Figures 1.13 through 1.15 show root growth of three upland rice genotypes as influenced by N fertilization. Root growth of three genotypes was more vigorous

FIGURE 1.11   Shoot growth of upland rice genotype BRA 02601 at two N levels.

FIGURE 1.12   Shoot growth of upland rice genotype BRA 032039 at two N levels.

at higher N rate compared with lower N rate. Overall, the contribution of upland rice roots in the total plant weight was 18%, the contribution of shoot was 44% in the total plant weight, and the contribution of grain was 38% (Table 1.3). Fageria (1989) reported that the contribution of old upland rice cultivars in the total plant weight was 14% by roots, 51% by shoot, and 35% by grain. In the modern upland rice cultivars, the contribution of roots was 10%, shoot 48%, and grain 42% (Fageria, 1989).

FIGURE 1.13   Root growth of upland rice genotype BRA 02601 at two N levels.

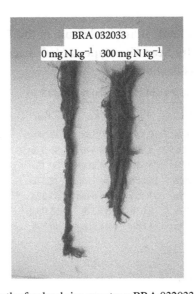

FIGURE 1.14   Root growth of upland rice genotype BRA 032033 at two N levels.

Hence, it can be concluded that there was an improvement in the root system of upland rice cultivars in Brazil, which is important because 1–2 week drought is very common in the central part of Brazil, locally known as the cerrado region, during the rainy season. Improvement in root system may improve water and nutrient uptake and, consequently, crop yield.

The contribution of root, shoot, and grain weight in the total plant weight of dry bean genotypes at two P levels is presented in Table 1.4. Root weight was

FIGURE 1.15 Root growth of upland rice genotype BRA 032048 at two N levels.

TABLE 1.4
Contribution of Root, Shoot, and Grain in the Total Plant Weight of Six Dry Bean Genotypes as Influenced by Phosphorus Fertilization

Genotype	Root Weight (%)		Shoot Weight (%)		Grain Weight (%)	
	P_{25}	P_{200}	P_{25}	P_{200}	P_{25}	P_{200}
CNFP 10103	18.09a	10.48ab	60.50a	38.58bc	21.41b	50.94ab
CNFP 10104	13.83a	9.90ab	38.44b	36.07bc	47.73a	54.03ab
CNFC 10429	17.09a	6.51b	44.28ab	29.22c	38.63ab	64.27a
CNFC 10431	17.42a	10.66ab	44.15ab	42.37bc	38.44ab	46.98ab
CNFP 10120	16.79a	13.65a	44.64ab	63.82a	38.57ab	22.54c
CNFC 10470	12.07a	8.63b	33.67b	53.06ab	54.25a	38.31bc
Average	15.42a	4.64b	44.28a	46.59a	41.86a	48.76a
F-test						
P level (P)	**		NS		NS	
Genotype (G)	**		**		**	
P × G	*		**		**	
CV (%)	15.22		15.60		19.49	

P_{25} is 25 mg P kg^{-1} and P_{200} is 200 mg P kg^{-1} soil.
NS, nonsignificant.
*, **Significant at the 5% and 1% probability levels, respectively. Means followed by the same letter in the same column or same line are not significantly different at the 5% probability levels by Tukey's test. Values of averages were compared in the same line for each parameter at low and high P levels.

significantly influenced by P and genotype treatments. Phosphorus × genotype interaction was also significant, indicating variation in root dry weight of genotypes with the variation in P levels. Overall, the contribution of root weight was higher at low P level compared to higher P level. This may be associated with higher yield of shoot and grain at higher P level. Overall, the contribution of root system in the total plant weight of dry bean was 10%; the contribution of shoot in the total plant weight was 45% and grain was also 45%. Figures 1.16 and 1.17 show shoot growth of dry bean genotype CNFC 10467 and CNFC 10432 at low P (25 mg P kg⁻¹) and high P (200 mg P kg⁻¹) levels. Shoot growth of genotype CNFC 10467 was much better at low P level compared with genotype CNFC 10432 at the same P level. This shows that there was a significant difference in dry bean genotype growth at different P levels. In conclusion, the root system of cereals and legumes were influenced not

FIGURE 1.16   Shoot growth of dry bean genotype CNFC 10467 at low and high P level.

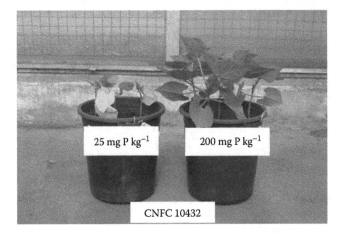

FIGURE 1.17   Shoot growth of dry bean genotype CNFC 10432 at low and high P level.

only by N and P fertilization but also by genotypes. Root growth of dry bean was less vigorous compared to upland rice.

Smucker (1993) reported that root production and maintenance costs 20%–47% of the photosynthates of the whole plant and that this consumption may be even greater when plants are grown in the field where they may be subjected to various environmental stresses. Partitioning of photosynthates and its effect on dry matter distribution between shoots and roots is influenced by several environmental factors, including low temperature, drought, and mineral nutrient deficiency (Wardlaw, 1990; Fageria and Moreira, 2011). The mineral nutrients P and N exert a pronounced influence on partitioning between shoots and roots (Fageria and Moreira, 2011). The P- and N-deficient plants have much more dry matter partitioned to roots than shoots, probably as a result of higher export rates of photosynthates to the roots. It is well known that leaf expansion is highly sensitive to low P concentrations in the tissue, leading to higher concentrations of sucrose and starch in P-deficient leaves because of reduced demand (Fredeen et al., 1989). Thus, roots become more competitive for photosynthates than shoots, which leads to higher export of carbohydrates to the roots with correspondingly lower shoot–root dry weight (Rufty et al., 1993; Fageria and Moreira, 2011).

1.5   ROOT LENGTH

Root length is one of the most important and widely used parameters for describing plant root systems and predicting their response to changes in the environment (Caldwell, 1987; Barber, 1995; Wang et al., 2011). Root length is a better parameter of root growth to correlate with absorption of water and nutrients. Root length and root diameter are the components determining the size of the root surface area, which determines water and nutrient uptake potential of the root system (Barber and Silberbush, 1984; Eissenstat, 1992; Noulas et al., 2010). Root acquisition of nutrients and water is based more upon root length and root surface area than mass since high root length density (RLD) is linked to a short travel distance of water and solutes (Andrew and Newman, 1970; Van Noordwijk and Brouwer, 1991). According to Mengel (1985), the following sequence represents parameters most important for P uptake: root length > P concentration in the bulk soil solution > root radius > P buffering power of the soil > diffusion coefficient. This emphasizes the importance of root length for the exploitation of phosphorus. Root length can be measured by Newman's technique. Newman (1966) developed a theory that root length can be estimated by the following equation:

$$R = \frac{\pi A N}{2H}$$

where
   R is the total length of roots
   N is the number of intercepts between the root and the straight lines on a grid
   A is the area of a square or rectangle of the grid
   H is the total length of the straight lines

Tennant (1975) proposed that Newman's formula can be simplified. He suggested that for a grid of indeterminate dimensions, the interaction counts can be converted to centimeter measurements using the equation

$$\text{Root length } (R) = 11/14 \times \text{Number of intersections } (N) \times \text{Grid unit}$$

As proposed by Tennant (1975), the 11/14 of the equation can be combined with the grid unit, and thus a length conversion factor is obtained. The factors for the 1, 2, and 5 cm grid squares are 0.786, 1.57, and 3.93, respectively.

It should be apparent that the method of counting the number of intersections between roots and grid lines influences the resulting root length. Historically, humans have hand-counted intersections, which inevitably introduce error and subjectivity. Different methods of washing and counting root intersections have led to widely different values of root length for similar crops grown under similar conditions (Fageria et al., 2006). The increasing use of scanning technology and recognition software has reduced, but not eliminated, subjectivity in determining root counts.

Root length per unit volume of soil, or RLD, is another widely used parameter in root studies. It can be calculated with the equation (Fageria et al., 2006)

$$\text{Root density (cm cm}^{-3}) = \frac{\text{Total root length in cm}}{\text{Soil volume where roots have been collected in cm}^3}$$

Most field methods used to study root growth are labor intensive and require at least partial destruction of the experimental site (Bohm, 1979). To overcome these disadvantages, methods were developed to observe root growth in situ. McMichael and Taylor (1987) provided a historical overview of the development of the various transparent wall methods.

Some research institutes have developed large rhizotron facilities to examine plant root systems. The word "rhizotron" (coined from the Greek word *rhizos* for root and *tron* for instrument) can be defined as a facility or building for viewing and measuring underground parts of plants through transparent surfaces (Klepper and Kaspar, 1994). These facilities allow an investigator simultaneous access to roots and shoots of plants growing in a field-like environment. The first rhizotrons were built in the early 1960s, and were used to monitor seasonal and diurnal changes in root system growth and function, cultivar differences in root growth parameters, and effects of soil treatments on root growth and water uptake (Klepper and Kaspar, 1994).

Other similar techniques have been developed in recent years, which include minirhizotrons and associated microvideo camera techniques to provide opportunities for quantifying in situ root systems of field-grown plants (Murphy et al., 1994). Linear regression of minirhizotron root count data, determined from minirhizotron images, on RLD, determined from destructive sampling, is often performed to calibrate minirhizotron root counts to RLD. These models use the general equation

$$\text{RLD} = C \times (\text{Root counts cm}^{-2} \text{ of tube})$$

where the conversion factor C has a value of 1.0 (Upchurch and Ritchie, 1983), 2.0 (Melhuish and Lang, 1968), or 3.3 (Upchurch, 1985). Bland and Dugas (1988) found that a value of $C = 2.0$ provided the best estimate of root length from minirhizotron root counts of cotton, whereas no simple relationship for sorghum was evident. Studies with wheat have reported C values of 2.8 (Bragg et al., 1983), 3.5 (Meyer and Baris, 1985), and 20 (Belford and Henderson, 1985); correlation coefficients (r) in these reports were 0.85, 0.90, and 0.95, respectively. Murphy et al. (1994) reported the empirically derived value of $C \geq 15$ for creeping bentgrass (*Agrostis palustris* Huds.) and annual bluegrass (*Poa annua* L.). Since there is so much disparity in the methods by which RLD is determined, it should not be surprising that there is also disparity among C values. Merrill (1992) presented detailed information about fabrication and usage of the minirhizotron systems for the observation of root growth in field experiments.

Root length is controlled genetically as well as influenced by environmental factors (Fageria, 1992). Table 1.5 shows the influence of genotypes and N rates on root length of 20 upland rice genotypes grown on a Brazilian Oxisol. Nitrogen × genotype interaction was significant for root length. Root length varied from 27 cm produced by genotype BRA 01600 to 43 cm produced by genotype BRA 032039, with an average value of 30.49 cm at the low N rate (0 mg N kg^{-1}). Similarly at high N rate (300 mg N kg^{-1}), root length ranged from 15.67 cm produced by genotype BRA 01596 to 40.33 cm produced by genotype BRS Sertaneja, with an average value of 31.99 cm. Overall, root length was 5% higher at the higher N rate compared to lower N rate. However, 35% genotypes produced lower root length at the higher N rate compared to lower N rate. Fageria (1992) reported higher root length of rice at low N rate compared to high N rate in nutrient solution. Fageria (1992) also reported that at nutrient-deficient levels, root length is higher compared to high nutrient levels because of the tendency of plants to tap nutrients from deeper soil layers. However, the results of present study show that root length in upland rice is determined by environmental as well as genetic factors. O'Toole and Bland (1987), in a review article, reported that there is ample evidence of genotype variation in the root characteristics of crop species. Russell (1977) also reported that the rate of root extension, like all characteristics of root systems, is subject to variation depending on genetic and soil factors. Similarly, Klepper (1990) reported that soil pH and soil fertility influence root growth of crop plants. Figures 1.18 and 1.19 show root length of upland rice cultivar Primavera and BRS Sertaneja at two N rates. Root length of both the cultivars was improved with the addition of N fertilizer. Root growth of two dry bean genotypes was significantly different at 0 mg K kg^{-1} soil grown in Brazilian Oxisol (Figure 1.20). Similarly, root length of 14 tropical legume cover crops varied significantly when grown on a Brazilian Oxisol (Figure 1.21). This shows genetic variation in root growth of genotypes of same species.

The influence of N, P, and K fertilization on root growth of upland rice grown on a Brazilian Oxisol is presented in Table 1.6. Root length was significantly improved with the addition of N and P but K did not influence root length significantly. Nitrogen significantly improved root length up to 150 mg N kg^{-1} soil and P improved root length up to 300 mg P kg^{-1} soil. These results suggest that P was the most important nutrient

TABLE 1.5

Root Length (cm) of Upland Rice Genotypes as Influenced by Nitrogen Fertilization

Genotype	0 mg N kg^{-1}	300 mg N kg^{-1}
BRA 01506	34.67ab	21.00cd
BRA 01596	30.00b	15.67d
BRA 01600	27.00ab	25.33bcd
BRA 02535	31.67ab	30.00abcd
BRA 02601	28.00b	32.67abc
BRA 032033	30.00a	28.00abcd
BRA 032039	43.00a	34.67abc
BRA 032048	28.50b	32.00abc
BRA 032051	30.67ab	35.67ab
BRA 042094	30.33b	38.00ab
BRA 042156	29.33b	33.00abc
BRA 042160	29.67b	32.50abc
BRA 052015	31.00ab	35.00abc
BRA 052023	29.67b	26.50abcd
BRA 052033	30.33b	31.33abc
BRA 052034	29.00b	37.67ab
BRA 052045	28.67b	37.67ab
BRA 052053	29.00b	36.50ab
BRS Primavera	29.33b	36.33ab
BRS Sertaneja	30.00b	40.33a
Average	30.49	31.99

F-test	
N rate (N)	NS
Genotype (G)	**
N × G	**
CV (%)	14.25

NS, nonsignificant.

**Significant at the 1% probability level. Means followed by the same letter in the same column are not significant at the 5% probability level by Tukey's test.

in improving root length of upland rice grown on a Brazilian Oxisol followed by N. Figure 1.22 shows that growth of upland rice was significantly increased with increasing P levels when N and K levels were constant. Similarly, grain yield of upland rice was significantly increased with the addition of N, P, and K in a Brazilian Oxisol (Figure 1.23). The author studied the root length of upland rice, corn, wheat, soybean, and dry bean during growth cycles of these crops (Table 1.7). Root length of five crop species was significantly influenced by age and follows a quadratic increase with the

FIGURE 1.18   Influence of N on root growth of upland rice cultivar Primavera.

FIGURE 1.19   Influence of N on root growth of upland rice cultivar BRS Sertaneja.

advancement of plant age. Upland rice and wheat produced minimum root length and corn produced maximum root length during different growth stages. Soybean and dry bean produced intermediate root lengths, respectively. Variation in the root length of crop species due to plant age was in the order of upland rice > corn > soybean > wheat > dry bean as indicated by R^2 values.

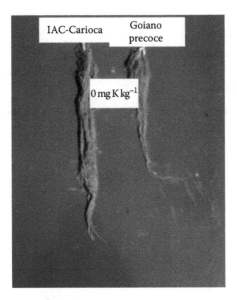

FIGURE 1.20   Root growth of two dry bean cultivars at low K level.

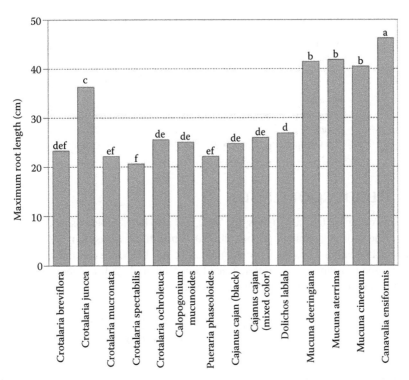

FIGURE 1.21   Maximum root length of 14 tropical legume cover crops grown on a Brazilian Oxisol.

TABLE 1.6
**Influence of Nitrogen, Phosphorus,
and Potassium on Root Length of Upland Rice
at Harvest**

Rate of N, P, and K (mg kg^{-1})	Maximum Root Length (cm)
N_0	29.56b
N_{150}	32.93a
N_{300}	34.63a
Average	32.37
F-test	**
CV (%)	3.8
P_0	24.70c
P_{100}	33.89b
P_{200}	38.52a
Average	32.37
F-test	**
CV (%)	2.5
K_0	32.37a
K_{100}	32.33a
K_{200}	32.41a
Average	32.37a
F-test	NS
CV (%)	4.4

NS, nonsignificant.

**Significant at the 1% probability level. Means followed by the
same letter in the same column under same nutrient levels are
not significantly different at 5% probability level by Tukey's test.

1.6   SPECIFIC ROOT LENGTH

The ratio of length to mass, that is, the SRL, has also been widely used as an indicator for fine root morphology (Zobel, 2005a). Since direct measurement of root surface area and volume has been problematic, SRL has been an excellent surrogate for surface area to volume ratio (Fitter, 1985; Eissenstat et al., 2000; Zobel, 2005). SRL can be calculated by using the following equation:

$$\text{SRL (cm g}^{-1}\text{ root dry wt.)} = \frac{\text{Root length in centimeter}}{\text{Root dry weight in gram}}$$

SRL is influenced by genotype and environmental factors. If the variation in total biomass reflects a variation in the amount of carbon invested for tissue synthesis, then plants that have lower cost of biomass to produce root length (high SRL) must be able to increase their total length root system more easily than plants with a low SRL.

FIGURE 1.22 Growth of upland rice at $N_{300}P_0K_{200}$, $N_{300}P_{100}K_{200}$, and $N_{300}P_{200}K_{200}$ levels.

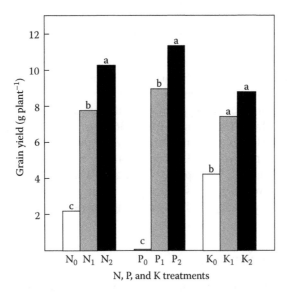

FIGURE 1.23 Grain yield of upland rice as influenced by N, P, and K fertilization.

These plants appear to absorb water and nutrients more efficiently (Eissenstat, 1992). The influence of N fertilization on SRL of 10 cover crops is presented in Table 1.8. The SRL was significantly affected by N, cover crops, and N × cover crops interaction. Maximum SRL was achieved at N_3 treatment (200 mg N kg^{-1} soil) and at this N level pueraria produced maximum SRL and black velvet bean produced

TABLE 1.7

Maximum Root Length (cm) of Five Crop Species during Growth Cycle

Plant Age in Days	Upland Rice	Corn	Wheat	Soybean	Dry Bean
21	9.67	26.00	14.67	16.67	23.00
40	19.33	41.00	19.67	32.00	31.00
61	24.00	43.67	19.67	31.33	32.67
78	24.00	40.00	17.67	27.33	27.33
97	26.33	39.33	14.67	36.33	29.00
118	24.67	48.67			

Regression Analysis

Upland rice plant age vs. root length (Y)
$\quad = -0.0744 + 0.5722X - 0.00309X^2$, $R^2 = 0.8985**$
Corn plant age vs. root length $(Y) = 21.9782 + 0.4145X + 0.0018X^2$, $R^2 = 0.5427**$
Wheat plant age vs. root length $(Y) = 8.0338 + 0.4113X - 0.0035X^2$, $R^2 = 0.5086**$
Soybean plant age vs. root length $(Y) = 9.5451 + 0.5395X - 0.0031X^2$, $R^2 = 0.5141**$
Dry bean age vs. root length $(Y) = 15.3092 + 0.4999X - 0.0038X^2$, $R^2 = 0.4835*$

Crops were harvested at physiological maturity. Upland rice and corn were matured at
    118 days. Wheat, soybean, and dry bean were matured at 97 days.
*, **Significant at the 5% and 1% probability levels, respectively.

minimum SRL. Overall, maximum SRL was also produced by gray velvet bean and minimum by pueraria. This may be due to maximum and minimum root dry weight produced by these two cover crops across N levels.

The SRL of upland rice genotypes was significantly influenced by P rate and genotype treatments (Table 1.9). It varied from 5.36 cm g^{-1} root dry weight produced by genotype BRA 042160 to 15.23 cm g^{-1} root dry weight produced by genotype BRA 032051, with an average value of 9.14 cm g^{-1} root dry weight. The SRL was higher at lower P level (25 mg P kg^{-1}) compared with high P level (200 mg P kg^{-1}). The lower SRL at higher P level may be associated with higher root weight at higher P level. Figure 1.24 shows that root growth was more vigorous at high P level compared with low P level.

1.7   ROOT DRY WEIGHT

Root weight is the most commonly used parameter for studies of root growth in response to environment. Generally, washed roots are dried and their weight determined. For plants grown in soil, the method of washing roots determines the diameter and percentage of fine roots that are lost and retained. Fresh weight is frequently recorded in plant pathology studies investigating nematodes and fungi on roots. To determine dry weight, the washed and cleaned roots are dried in an oven at 105°C for about 10–20 h, depending on the amount of roots (Bohm, 1979). Drying at 60°C–75°C is preferable for, although it takes longer, it has the advantage of preventing roots

TABLE 1.8

Specific Root Length of 10 Tropical Cover Crops as Influenced by Nitrogen Treatments

Cover Crops	N_0	N_1	N_2	N_3	Average
		Specific Root Length (cm g^{-1} Root Dry Weight)			
1. *Crotalaria*	161.11a	195.28a	130.42ab	145.15ab	157.99a
2. *Crotalaria*	108.01a	56.62b	38.19cd	91.63bc	73.61bc
3. *Crotalaria*	47.17b	43.76b	38.76cd	66.01cd	48.92cd
4. Calopogonium	99.14a	69.19b	81.25bc	118.25bc	94.21b
5. Pueraria	145.96a	172.43ab	158.06a	193.91a	167.59a
6. Pigeon pea (black)	43.41b	45.24b	74.17bcd	110.67bc	68.37bc
7. Lablab	35.76b	20.58b	18.87d	16.43d	22.91de
8. Black velvet bean	15.64b	11.54b	39.71cd	15.86d	20.69de
9. Gray velvet bean	13.74b	15.51b	16.67d	20.00d	16.48e
10. Jack bean	37.33b	25.56b	18.87d	21.28d	25.76de
Average	70.73ab	65.57ab	61.50b	79.92a	69.43

F-test	
N rate (N)	*
Cover crops (C)	**
N × C	**
CVN (%)	26.24
CVC (%)	32.13

*, **Significant at the 5% and 1% probability levels, respectively. Means in the same column followed by the same letter are not significantly different at the 5% probability level by Tukey's test. Average values were compared in the same line for significant differences among N rates. N_0 = 0 mg N kg^{-1}; N_1 = 0 mg N kg^{-1} + inoculant; N_2 = 100 mg N kg^{-1} + inoculant; and N_3 = 200 mg N kg.

from being pulverized or volatizing nitrogenous compounds. Root weight is a good parameter for characterizing the total mass of roots in a soil. It is also easy to measure in the experiments of controlled conditions.

Root dry weight is genetically controlled as well as influenced by environmental factors (Fageria, 1992). A global analysis of root measurements found that average root mass ranged from about 0.2 kg m^{-2} for croplands to about 5 kg m^{-2} (20–50 Mg ha^{-1}) for forests and sclerophyllous shrubs and trees (Jackson et al., 1996; Gregory, 2006a). Root mass in forest ecosystems ranged from 2 to 5 kg m^{-2}, while that in croplands, deserts, tundra, and grasslands was <1.5 kg m^{-2} (<15 Mg ha^{-1}). These figures belie the fact that there is a constant turnover of roots and root material, with exudates and other rhizodeposits contributing sustenance to the soil microbial and faunal populations (Gregory, 2006a).

Over the last decades, substantial amounts of data have been collected on root dry weight accumulation in different crop species at different growth stages (Greenwood

TABLE 1.9

Specific Root Length of Upland Rice Genotypes as Influenced by P Fertilization

Genotype	Specific Root Length (cm g⁻¹ Root Dry Weight)
BRA 01506	7.85bcd
BRA 01596	12.45abc
BRA 01600	12.74ab
BRA 025535	8.41bcd
BRA 02601	7.97bcd
BRA 032033	7.43bcd
BRA 032039	8.00bcd
BRA 032048	9.35bcd
BRA 032051	15.23a
BRA 042094	8.91bcd
BRA 042156	10.51abcd
BRA 042160	5.36d
BRA 052015	8.07bcd
BRA 052023	6.02d
BRA 052033	12.32abc
BRA 052034	8.11bcd
BRA 052045	10.50abcd
BRA 052053	8.78bcd
BRS Primavera	6.66cd
BRS Sertaneja	8.09bcd
Average	9.14
Low P level (25 mg kg⁻¹)	10.34a
High P level (200 mg kg⁻¹)	7.94b
F-test	
P level (P)	*
Genotype (G)	**
P × G	NS

Values are across two P levels.

NS, nonsignificant.

*, **Significant at the 5% and 1% probability levels, respectively. Means in the same column followed by the same letter are not significantly different at the 5% probability level by Tukey's test. Low and high P levels values were compared separately.

et al., 1982; Klepper et al., 1984; Gregory and Eastham, 1996; Watson et al., 2000; Gregory, 2006a,b; Fageria and Moreira, 2011). The author studied the influence of N on root dry weight upland rice genotypes (Table 1.10). Root dry weight varied from 0.87 g plant⁻¹ produced by genotype BRA 015966 to 1.78 g plant⁻¹ produced by genotype BRA 052034 and BRA 042094, with an average value of 1.38 g plant⁻¹ at the lower

FIGURE 1.24   Root growth of upland rice genotype BRA 01596 at low and high P levels. The P levels are in mg kg⁻¹.

N rate (0 mg N kg⁻¹). At the higher N rate (300 mg N kg⁻¹), root dry weight ranged from 0.40 g plant⁻¹ produced by genotype BRA 01506 to 4.14 g plant⁻¹ produced by genotype BRA 052023 g plant⁻¹, with an average value of 2.72 g plant⁻¹. Overall, root dry weight was 97% higher at the higher N rate compared to lower N rate. Fageria (2009) reported that N fertilization improved root dry weight in upland rice. The positive effect of N on root dry matter has been previously documented (Eghball et al., 1993; Costa et al., 2002). Similarly, Kuchenbuch and Barber (1987) and Stoffel et al. (1995) reported that root growth varies considerably with environmental conditions.

Root dry weight of 20 upland rice genotypes was also significantly influenced by P and genotype treatments (Table 1.10). Root dry weight of 20 upland rice genotypes at low P level varied from 2.00 to 5.68 g plant⁻¹, with an average value of 3.41 g plant⁻¹. At high P level, root dry weight varied from 2.43 to 8.55 g plant⁻¹, with an average value of 4.01 g plant⁻¹. The increase in dry weight at high P level was about 18% compared to low P level. Fageria and Baligar (1997) and Baligar et al. (1998) have reported increase in root dry weight of upland rice with the addition of P in Brazilian Oxisols. Similarly, these authors also reported that higher P level roots had more fine hairs compared to lower P level. Hence, roots at higher P level had higher capacity to uptake nutrients and water compared to lower P level.

1.8   SURFACE AREA

Besides length and weight, surface area is an important parameter of root system in crop plants. The form of root systems and their development conditions greatly affect the surface area of roots. The surface area of roots has a high positive correlation

TABLE 1.10

Root Dry Weight of 20 Upland Rice Genotypes as Influenced by Nitrogen and Phosphorus Fertilization

Genotype	Root Dry Weight (g Plant^{-1})			
	0 mg N kg^{-1}	300 mg N kg^{-1}	25 mg P kg^{-1}	200 mg P kg^{-1}
BRA 01506	0.92a	0.40f	3.92ab	3.22c
BRA 01596	0.87a	0.45f	2.78ab	2.73c
BRA 01600	1.14a	1.03ef	2.81ab	3.03c
BRA 025535	1.33a	3.25abcd	3.12ab	4.30c
BRA 02601	1.11a	3.73ab	4.42ab	3.20c
BRA 032033	1.12a	2.41cd	3.70ab	3.62c
BRA 032039	1.24a	3.77ab	2.91ab	4.36c
BRA 032048	1.05a	3.62abc	3.96ab	3.91c
BRA 032051	1.31a	2.82bcd	2.00b	2.58c
BRA 042094	1.78a	2.31de	2.82ab	3.92c
BRA 042156	1.19a	2.84bcd	2.50b	2.91c
BRA 042160	1.51a	3.33abcd	5.68a	8.32ab
BRA 052015	1.67a	2.83bcd	3.91ab	2.98c
BRA 052023	1.48a	4.14a	4.69ab	8.55a
BRA 052033	1.56a	3.38abcd	2.23b	2.43c
BRA 052034	1.78a	2.58bcd	3.18ab	3.99c
BRA 052045	1.75a	3.66abc	3.07ab	3.08c
BRA 052053	1.65a	2.72bcd	2.57ab	3.87c
BRS Primavera	1.68a	2.49bcd	3.56ab	5.21bc
BRS Sertaneja	1.45a	2.68bcd	4.36ab	3.92c
Average	1.38	2.72	3.41	4.01
F-Test				
N or P level	*		NS	
Genotype (G)	**		**	
N × G or P × G	**		*	

NS, nonsignificant.

*, **Significant at the 5% and 1% probability levels, respectively. Means in the same column followed by the same letter are not significantly different at the 5% probability level by Tukey's test. Low and high N levels and low and high P levels values were compared in the same lines.

with the amount of nutrient absorption (Takenaga, 1995). Various studies show that 90%–95% or more of the root length of an intact plant is made up of roots <0.6 mm in diameter (Zobel, 2003, 2005b). The surface area of a root system is significantly affected by the radii of the roots (Wang et al., 2011). Claassen and Steingrobe (1999) suggested that the specific soil volume, out of which a nutrient diffuses to plant roots, is negatively related to the root radius, indicating that a thinner root system has a larger surface area, which may account for the observation that thinner roots increased the availability of Zn due to more through exploration of the soil (Dong et al., 1995).

1.9   ROOT DEPTH AND DENSITY

Rooting depth and density or distribution pattern in soil determines the zone of water and nutrient availability to plants and also resistance to biotic and abiotic stresses (Perfect et al., 1987; Castronguay et al., 2006). Karcher et al. (2008) reported that enhanced water uptake through increased root size and depth is one of the most desirable drought tolerance mechanisms for crop plants, as this allows the plants to fully utilize available soil water resources and prolong the need for supplement irrigation. Nearly 80% of the variation in shoot P content of two canola (*Brassica napus* L.) genotypes was explained by root length under greenhouse conditions (Solaiman et al., 2007). Hammond et al. (2009) compared a wide range of *Brassica oleracea* L., entries and found that P use efficiency indices often were correlated with root traits, such as lateral root number and growth rate. Hence, knowledge of root growth and distribution in soil is very important. Plants with root systems concentrated in the topsoil are more likely to exhibit superior nutrient acquisition efficiency than plants with more deeply distributed roots (Ge et al., 2000; Wang et al., 2010), whereas deeply rooted plants tend to resist periods of water stress better (Ho et al., 2005). Differences in root lengths, dry weight of roots at different soil depths, and extent of rooting at the seedling stage among wheat cultivars were related to differences in yield and ability to escape drought (Hurd, 1974). Upland rice cultivars, which are more drought tolerant than lowland cultivars, have deeper and more prolific rooting systems (Steponkus et al., 1980; Fageria, 2010; Fageria et al., 2011). Deep rooting of bean cultivars was positively associated with seed yield, crop growth, cooler canopy temperature, and soil water extraction, provided soil types did not restrict rooting potential (Sponchiado et al., 1989). Large differences in root morphology and distribution exist between genotypes of plant species (O'Toole and Bland, 1987; Romer et al., 1988; Atkinson, 1991; Gahoonia and Nielsen, 2004).

Root growth and rooting depth of crop plants can be restricted because of physical and chemical impediments (Gregory, 2006a). Almost all the roots growing through soil experience some degree of mechanical impedance, and if continuous pores of appropriate size do not already exist then the root tip region must exert sufficient force to deform the soil (Gregory, 2006b). Roots are often larger than the water-filled pores at field capacity (i.e., pores with diameter $<60\,\mu m$), so that freely draining pores are the main spaces in which roots can grow. In addition to packing density of soil particles, soil strength is also affected by soil water content so that the depth of rain infiltration influences rooting depth (Gregory, 2006b).

Because about 90% of the total NH_4^+, P, and K uptake and root length of flooded rice cultivars occur within 20 cm of the soil surface, samples collected for routine soil tests should be taken within the level of 20 cm (Teo et al., 1995a). Lowland rice plant develops surface matting of roots in the oxygenated zone near soil surfaces soon after application of flood waters (Fageria et al., 2003). The presence of these roots in surface soil layers may contribute to high amounts of nutrients measured in the upper 20 cm of soil. Using the Claassen–Barber model to predict nutrient uptake by maize grown in silt loam soil, >90% of K and P uptake occurred in the top 20 cm soil depth (Schenk and Barber, 1980). Silberbush and Barber (1984) reported that about 80% of P and 54% of K uptake by soybean was from 0 to 15 cm depth. Duriex et al. (1994) reported that more than half of the root length of maize was located in the surface 0–20 cm depth at all sampling times during a season. Costa et al. (2009) reported that the major part of corn roots were concentrated in the topsoil layer (0–10 cm) where higher concentration of P was located. Roots of the peanut (*Arachis hypogaea* L.) cultivar Florunner penetrated to depths of up to 280 cm when grown in a sandy soil, and the most extensive root growth occurred in the top 30 cm (Boote et al., 1982). Sharratt and Cochran (1993) reported that 85% and 95% of the root mass of barley was located in the interrows of the top 20 and 40 cm of soil, respectively. Welbank and Williams (1968) also found that nearly 80% of barley roots occupied the uppermost 15 cm of soil.

Root density usually decreases with depth (Kalisz et al., 1987; Klepper and Rickman, 1990). The highest concentration of roots, especially fine roots, occurs near the soil surface where conditions are more favorable for root growth (Singh and Sainju, 1998). These layers are usually rich in organic matter, nutrients, CEC, and porosity and are low in bulk density (Sainju and Good, 1993). Sainju and Good (1993) and Sainju and Kalisz (1990) found an exponential decline in root density with soil depth, reflecting declining fertility, reduced organic matter concentration, and decreasing aeration.

Maximum root densities of peanut were within the top 30 cm soil depth, which was located in the region above a tillage pan (Robertson et al., 1979). Even though the fibrous peanut roots below the usual harvesting depth of 30 cm constituted a small fraction of root weight, these roots may be important for water absorption (Boote et al., 1982).

A study conducted by Stone and Pereira (1994a,b) of four common bean cultivars and three upland rice cultivars to evaluate rooting depth grown in an Oxisol of central Brazil showed that 70% of roots were concentrated in the top 20 cm layer, and about 90% were concentrated in the top 40 cm soil depth of both crops. According to Beyrouty et al. (1988), nearly 100% of the root length of lowland rice grown on Crowley soil of Arkansas was measured in the upper 40 cm of soil. The same authors reported that 80%–90% of the root length was measured between 0 and 20 cm soil depths at maximum tillering. As high as 80% of the total weight of the roots of barley was recovered from the top 15 cm of soil, ~12% was between 15 and 30 cm deep, and ~8% was between 30 and 60 cm (Welbank et al., 1974).

The distribution of roots in the soil profiles is important in determining uptake of water and nutrients. Roots are more abundant in the upper soil layer (Wilhelm et al., 1982; Qin et al., 2004). The global average distribution of root mass for all biomes and vegetation types was 30% in the upper 0.1 m, 50% in the upper 0.2 m, and 75% in the upper 0.4 m (Gregory, 2006b). Soil moisture and soil type determine rooting depth of crop plants. Adequate soil moisture content improves root depth of crop plants. Similarly, sandy soils have deeper rooting depth compared to clay or loam soils. Gregory (2006b) reported that root length and mass decreased exponentially with depth. Gerwitz and Page (1974) first proposed this model after reviewing literature on vegetable crops, cereals, and grasses and it has been widely adopted since (Greenwood et al., 1982; Robertson et al., 1993; Zhuang et al., 2001; Zuo et al., 2004; Wu et al., 2005).

1.10   ROOT DEVELOPMENT RELATIVE TO PLANT GROWTH STAGE

Root development varies with stages of plant growth and development. This information can guide researchers to the appropriate time for root growth observations during crop growth cycles. Dry matter accumulation by root systems typically follows the sigmoidal pattern with increasing plant age commonly observed with shoots (Gregory, 1994). Flowering initiation appears to be a particularly important developmental stage after which assimilates are required predominantly to fill the developing grain and little assimilate for roots (Gregory and Atwell, 1991; Jensen, 1994; Gregory et al., 1997). In cereals, the mass of the root system rarely increases after flowering and may decrease substantially depending on soil conditions (Mengel and Barber, 1974). Most legumes are much less determinate than cereals so that the demand for assimilate by the grain increases gradually. Root weight of many legumes continues to increase during flowering and early grain filling (Gregory et al., 1997). Gregory (1994) discussed the changes in root:total plant mass for crops of wheat, corn, soybean, and lupine at a range of sites. Typically, root weight was about 0.3–0.4 during early growth and decreased as the crops started to grow rapidly to values of about 0.1–0.2 at flowering in cereals and during pod filling in the legumes.

The most rapid development of maize roots occurs during the first 8 weeks after planting (Anderson, 1987). As maize plants age, growth of roots generally increases at slower rates than shoots (Baligar, 1986). After silking, maize root length declines (Mengel and Barber, 1974). This decline in root length after silking presumably is due to the high C demand of grain, resulting in enhanced translocation of C and N to grain, including some C and N that roots would normally obtain (Wiesler and Horst, 1993).

During the first week after the emergence of grain sorghum plants, the root system contains 20%–30% of the dry matter of the whole plant (Jones, 1985). During the next few days, the root system grows rapidly, and by the eight-leaf stage, about 40% of the total plant dry weight is in the root system (Maizlish et al., 1980). Subsequently, the shoot grows at a relatively more rapid rate than the root system, and the percentage of total plant dry weight in the root system decreases gradually to about 10% maturity (Jones, 1985). During the early stages of corn and sorghum growth, vertical root growth is more rapid than horizontal root growth. Rate of vertical root growth ranging from 1.3 to 6 cm day^{-1} has been reported (Allmaras et al., 1975; Myers, 1980).

Peanut RLD and root weight density increased at each soil depth increment from planting to 80 days after planting (Ketring and Reid, 1993). Roots had penetrated to depths of 120 cm 40–45 days after planting, and spread laterally to 46 cm in midfurrow. The 0–15 cm depth increment had the highest mean RLD, which increased to a maximum at 2.1 cm^{-3} 80 days after planting (Ketring and Reid, 1993). This meant that peanut roots were established both deeply and laterally in the soil profile early in the growing season. This would be advantageous in drought environments and helpful for water management.

Sunflower (*Helianthus annuus* L.) rooting depth reached 1.88 m at the beginning of disk flowering and 2.02 m at the completion of disk flowering (Jaffar et al., 1993). In a review of depth development of roots with time for 55 crop species (Borg and

Grimes, 1986), it was shown that maximum rooting depth for most crop species was generally achieved at physiological maturity. Kaspar et al. (1984) noted that the rate of soybean root depth penetration reached a maximum during early flowering and declined during seed fill. However, some root growth was observed throughout the reproductive stage until physiological maturity.

Slaton et al. (1990) studied root growth dynamics of lowland rice and found that maximum root growth rates were reached between active tillering and panicle initiation, and maximum root length was reached by early booting. Beyrouty et al. (1987) noted that the most rapid rate of root and shoot growth in flooded rice occurred before panicle initiation, which corresponds to the plant transition between vegetative and reproductive growth. Approximately 77% and 81% of total shoot and root biomass, respectively, was achieved before panicle initiation. Following panicle initiation, elongation of roots and shoots increased only slightly until harvest (physiological maturity). Beyrouty et al. (1988) also reported that lowland rice root growth was most rapid during vegetative growth, with maximum root length occurring at panicle initiation. Root length either plateaued or declined during reproductive growth.

The author studied root dry weight of five crop species during their growth cycle (Table 1.11). The root dry weight of the five crop species was significantly influenced by plant age. The root dry weight of rice increased linearly with the advancement of plant age. The remaining four crop species' root dry weight increased in a quadratic fashion with the increasing plant age. Root as well as shoot growth followed similar pattern in the five crop species (Table 1.12 and Figures 1.25 through 1.31). Higher root and shoot growth of corn compared to other crop species may be related to its C_4 photosynthetic pathway. Wang et al. (2011) reported that root system formation

TABLE 1.11

Root Dry Weight of Five Crop Species during Their Growth Cycle

Plant Age in Days	Upland Rice (g Plant⁻¹)	Corn (g Plant⁻¹)	Wheat (g Plant⁻¹)	Soybean (g Plant⁻¹)	Dry Bean (g Plant⁻¹)
21	0.05	0.22	0.06	0.08	0.27
40	0.53	3.43	0.10	0.40	2.45
61	1.25	7.81	0.17	1.08	3.01
78	2.51	8.04	0.27	1.26	1.49
97	3.62	5.64	0.17	1.40	1.27
118	4.40	4.64			

Regression Analysis

Upland rice vs. root dry wt. $(Y) = -1.2473 + 0.0478X$, $R^2 = 0.9254$**
Corn plant age vs. root dry wt. $(Y) = -6.4108 + 0.3553X - 0.0022X^2$, $R^2 = 0.8652$**
Wheat plant age vs. root dry wt. $(Y) = -0.0955 + 0.0077X - 0.000048X^2$, $R^2 = 0.7002$**
Soybean plant age vs. root dry wt. $(Y) = -0.6792 + 0.0368X - 0.00015X^2$, $R^2 = 0.9434$**
Dry bean age vs. root dry wt. $(Y) = -2.3708 + 0.1655X - 0.0013X^2$, $R^2 = 0.6829$**

Crops were harvested at physiological maturity. Upland rice and corn were matured at 118 days. Wheat, soybean, and dry bean were matured at 97 days.

**Significant at the 1% probability level.

TABLE 1.12

Shoot Dry Weight of Five Crop Species during Their Growth Cycle

Plant Age in Days	Upland Rice (g Plant⁻¹)	Corn (g Plant⁻¹)	Wheat (g Plant⁻¹)	Soybean (g Plant⁻¹)	Dry Bean (g Plant⁻¹)
21	1.34	0.86	0.11	0.41	2.29
40	1.72	12.73	0.35	3.21	6.35
61	4.56	39.07	1.62	14.18	18.91
78	11.00	86.33	3.52	17.23	26.88
97	17.39	50.59	3.39	17.13	23.69
118	27.53	59.82			

Regression Analysis

Upland rice vs. shoot dry wt. $(Y) = 0.5833 \exp.(0.0340X)$, $R^2 = 0.9662$**

Corn plant age vs. shoot dry wt. $(Y) = -6.4108 + 0.3553X - 0.0022X^2$, $R^2 = 0.8652$**

Wheat plant age vs. shoot dry wt. $(Y) = -0.0955 + 0.0077X - 0.000048X^2$, $R^2 = 0.7002$**

Soybean plant age vs. shoot dry wt. $(Y) = -0.6792 + 0.0368X - 0.00015X^2$, $R^2 = 0.9434$**

Dry bean age vs. shoot dry wt. $(Y) = -2.3708 + 0.1655X - 0.0013X^2$, $R^2 = 0.6829$**

Crops were harvested at physiological maturity. Upland rice and corn were matured at 118 days. Wheat, soybean, and dry bean were matured at 97 days.

**Significant at the 1% probability level.

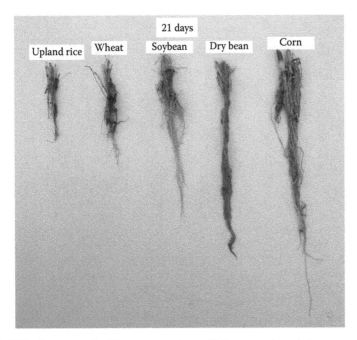

FIGURE 1.25   Root growth of five crop species at 21 days growth period.

FIGURE 1.26 Growth of tops of five crop species at 21 days growth period.

FIGURE 1.27 Root growth of five crop species at 61 days growth period.

FIGURE 1.28   Growth of tops of five crop species at 61 days growth period.

proceeds in close coordination with shoot growth. Accordingly, root growth and its functions are regulated tightly by the shoot through materials cycling between roots and shoots.

1.11   CROP YIELD VERSUS ROOT GROWTH

Crop yield is the economic part of the plant used for human or animal consumption and is measured as grain or dry matter quantity per unit land area (Fageria, 1992). In defining yield, the term potential yield is also often used. Potential yield is an estimate of the upper limit of yield increase that can be obtained from a crop plant under ideal environmental and management conditions (Fageria, 1992). Advances in environmental physics and plant physiology have allowed calculation to be made of maximum potential crop yields. Gales (1983) reported that a maximum potential

FIGURE 1.29   Root growth of five crop species at 97 days growth period.

yield of spring barley in Britain could be achieved at 10–11 Mg ha^{-1} and winter wheat at 11–13 Mg ha^{-1}. These calculations are based on modern cultivars in conditions of average sunshine and they use the best recorded values of photosynthetic efficiency (about 4.5%) and harvest index (about 0.5). The figure for photosynthetic efficiency represents the energy stored in the crop as a percentage of the energy in the photosynthetically active radiation (PAR) (Gales, 1983). The calorific value of the crop dry matter is taken as 17.5 kJ g^{-1} and the PAR is 0.5 of the total incident radiation. Maximum potential yield could be raised if some of the characteristics of the crop were improved. The harvest index, for example, might be increased to 0.65, or the photosynthetic efficiency might be increased by a decrease in respiration losses (Gales, 1983). Grain yields of lowland rice over 13 Mg ha^{-1} have been reported in farmer's field in the subtropical environment of Yunnan, China and the temperate environment of Yanco, Australia (Ying et al., 1998). The yield of a crop is influenced by several factors and their interactions. These factors are broadly grouped as climatic, soil, plant, and socioeconomic. They are highly variable from region to region, year to year, and season to season within the year, which accounts for the variability of crop yield from region to region and year to year.

Limitations to grain yield caused by inadequate rooting mainly occur when soil conditions such as compaction or root damage by soilborne pathogens prevent the plants from accessing the potentially available water and nutrients in the soil (Hoad et al., 2001). Fageria and Moreira (2011) studied the relationship between maximum

FIGURE 1.30   Growth of tops of five crop species at 97 days growth period.

FIGURE 1.31   Root growth of upland rice and corn at 118 days growth period.

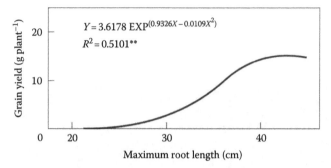

FIGURE 1.32 Relationship between maximum root length and grain yield of upland rice. **Significant at the 1% probability level. (From *Adv. Agron.*, 110, Fageria, N.K. and Moreira, A., The role of mineral nutrition on root growth of crop plants, 251–331, 2011, Elsevier.)

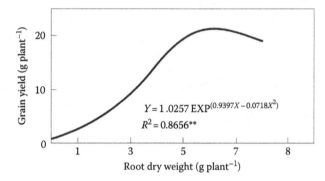

FIGURE 1.33 Relationship between root dry weight and grain yield of upland rice. **Significant at the 1% probability level. (From *Adv. Agron.*, 110, Fageria, N.K. and Moreira, A., The role of mineral nutrition on root growth of crop plants, 251–331, 2011, Elsevier.)

root length and root dry weight and grain yield of upland rice (Figures 1.32 and 1.33). Grain yield increased in a quadratic fashion with increasing root length or root dry weight, and root dry weight was a better predictor than root length of yield. Similarly, the author studied the relationships between root length and root dry weight and shoot dry weight of tropical legume cover crops (Figures 1.34 and 1.35). There was a significant increase in shoot dry weight of legume cover crops with increasing root length and dry weight, and as with upland rice, root dry weight was a better predictor than root length of shoot dry weight. Barber and Silberbush (1984) studied the relationship between root length and soybean yield and concluded that yield was significantly related to total root length at the R6 (full seed) stage. These authors concluded that root growth is important in determining the nutrient supply to the shoot which, in turn, affects crop yield. Similarly, Thangaraj et al. (1990) reported that RLD of lowland rice at flowering was directly proportional to grain yield.

Leon and Schwang (1992) found that yield stability of oats and barley cultivars was related to their total root length. Barraclough (1984) found that total root length of winter wheat was positively correlated to grain yield. This raises the possibility

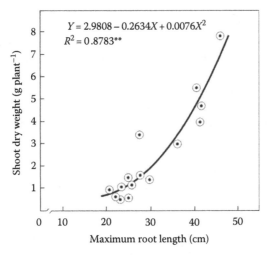

FIGURE 1.34 Relationship between maximum root length of tropical cover crops and shoot dry weight. **Significant at the 1% probability level. (From *Adv. Agron.*, 110, Fageria, N.K. and Moreira, A., The role of mineral nutrition on root growth of crop plants, 251–331, 2011, Elsevier.)

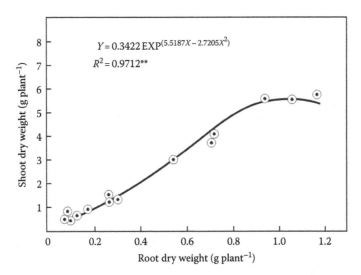

FIGURE 1.35 Relationship between root dry weight of tropical cover crops and shoot dry weight. **Significant at the 1% probability level. (From *Adv. Agron.*, 110, Fageria, N.K. and Moreira, A., The role of mineral nutrition on root growth of crop plants, 251–331, 2011, Elsevier.)

that the selection and breeding of crop genotypes with extensive root systems may contribute to more efficient use of nutrients, especially immobile nutrients like P and yield stability (Gahoonia and Nielsen, 2004). Beyrouty (2002) reported a positive linear relationship between total root length developed in the 40 cm of soil layer and grain yield of eight rice cultivars grown in the field.

1.12 CONCLUSIONS

The root system is an important organ of plants. It absorbs water and nutrients from soil and supplies to plant tops for their metabolic activities. Roots also give plants a mechanical support and supply hormones that help plants in many physiological and biochemical processes associated with growth and development. Roots are also responsible for mitigation of greenhouse gases by storing a large amount of C in the soils. A vigorous root system is responsible for the development of healthy plants and, consequently, higher yields. Roots that are left in the soil after crop harvest improve soil organic matter content, nitrogen cycle, and microbial activities. All these activities improve soil structure, soil water holding capacity, and water infiltration rate in the soil and reduce soil bulk density and soil erosion, which ultimately leads to higher soil productivity.

Root growth is controlled genetically and also influenced by environmental factors. Environmental factors that influence plant root growth are soil temperature, soil moisture content, solar radiation, and the physical, chemical, and biological properties of the soil. Most of the root biomass of annual crops is located in the 0–20 cm soil depth. This may be associated with a large amount of organic matter, nutrient accumulation, and water availability in the topsoil layer compared to lower soil depths.

The root system of cereals is fibrous, whereas the root system of legumes is taproot type. Cereals have a more vigorous root system compared with legumes. Overall, root contributes 10%–20% of total plant weight. Deeper and vigorous root system helps in the absorption of water from deeper soil layers during water shortage or drought period. When environmental conditions are favorable, smaller root system or less vigorous root system can produce maximum economic yield. Root length as well as root weight has positive association with grain yield in crops. However, based on R^2 values, root dry weight has higher correlation with grain yield compared with maximum root length. Root growth in crop plants is parallel to shoot growth during crop growth cycle. Crop plants having C_4 photosynthetic pathway produce more vigorous root system compared to crop plants with C_3 photosynthetic pathway.

REFERENCES

Allmaras, R. R., D. R. Linden, and C. E. Clapp. 2004. Corn-residue transformation into root and soil carbon as related to nitrogen, tillage, and stover management. *Soil Sci. Soc. Am. J.* 68:1366–1375.

Allmaras, R. R., W. W. Nelson, and W. B. Voorhees. 1975. Soybean and corn rooting in southwestern Minnesota: II. Root distributions and related water inflow. *Soil Sci. Soc. Am. Proc.* 39:771–777.

Anderson, E. L. 1987. Corn root growth and distribution as influenced by tillage and nitrogen fertilization. *Agron. J.* 79:544–549.

Andrews, R. E. and E. I. Newman. 1970. Root density and competition for nutrients. *Oecolog. Plantar.* 5:319–334.

Anil, L., J. Park, R. H. Philipps, and F. A. Miller. 1998. Temperature intercropping of cereals for forage: A review of the potential for growth and utilization with particular reference to the UK. *Grass Forage Sci.* 53:301–317.

Atkinson, D. 1991. Influence of root system morphology and development on the need for fertilizers and the efficiency of use. In: *Plant Roots: The Hidden Half*, eds., Y. Eshel and U. Kafkaki, pp. 411–451. New York: Marcel Dekker.

Balesdent, J. and J. M. Balabane. 1996. Major contribution of roots to soil carbon storage inferred from maize cultivated soils. *Soil Biol. Biochem.* 28:1261–1263.

Baligar, V. C. 1986. Interrelationships between growth and nutrient uptake in alfalfa and corn. *J. Plant Nutr.* 9:1391–1404.

Baligar, V. C., N. K. Fageria, and M. Elrashidi. 1998. Toxicity and nutrient constraints on root growth. *HortScience* 33:960–965.

Barber, S. A. 1979. Corn residue management and soil organic matter. *Agron. J.* 71:625–627.

Barber, S. A. 1995. *Soil Nutrient Bioavailability*, 2nd edn. New York: Wiley.

Barber, S. A. and Silberbush, M. 1984. Plant root morphology and nutrient uptake. In: *Roots, Nutrient and Water Influx, and Plant Growth*, eds., S. A. Barber and D. R. Bouldin, D. M. Kral, and S. L. Hawkins, pp. 65–88. Madison, WI: CSSA, ASA, and SSSA.

Barlow, P. W. 1986. Adventitious roots of whole plants: Their forms, functions and evolution. In: *New Root Formation in Plants and Cuttings*, ed., M. B. Jackson, pp. 67–110. Dordrecht, the Netherlands: Martinus Nijhoff.

Barraclough, P. B. 1984. The growth and activity of winter wheat roots in the field: Root growth of high yielding crops in relation to shoot growth. *J. Agric. Sci.* 103:439–442.

Belford, R. K. and F. K. G. Henderson. 1985. Measurement of growth of wheat roots using a TV camera system in the field. In: *Wheat Growth and Modeling*, ed., W. Day and R. K. Atrin, pp. 99–105. New York: Plenum Publishing.

Benjamin, J. G., A. D. Halvorson, D. C. Nielsen, and M. M. Mikha. 2010. Crop management effects on crop residue production and changes in soil organic carbon in the central great plains. *Agron. J.* 102:990–997.

Beyrouty, C. A. 2002. Ecophysiology of roots of aquatic plants. In: *Plant Roots: The Hidden Half*, 3rd edn., eds., Y. Waisel, A. Eshel, and U. Kafka, pp. 1007–1024. New York: Marcel Dekker.

Beyrouty, C. A., B. R. Wells, R. J. Norman, J. N. Marvel, and J. A. Pillow. 1987. Characterization of rice roots using a minirhizotron technique. In: *Minirhizotron Observation Tubes: Methods and Applications for Measuring Rhizosphere Dynamics*, Spec. publ. 50, ed., H. M. Taylor, pp. 99–108. Madison, WI: American Society of Agronomy.

Beyrouty, C. A., B. R. Wells, R. J. Norman, J. N. Marvel, and J. A. Pillow. 1988. Root growth dynamics of a rice cultivar grown at two locations. *Agron. J.* 80:1001–1004.

Bland, W. L. and W. A. Dugas. 1988. Root length density from minirhizotron observations. *Agron. J.* 82:1024–1026.

Blaser, B. C., L. R. Gibson, J. W. Singer, and J. L. Jannink. 2006. Optimizing seeding rates for winter cereal grains and frost-seeded red clover intercrops. *Agron. J.* 98:1041–1049

Bohm, W. 1979. *Methods of Studying Root Systems*. New York: Springer-Verlag.

Bolinder, M. A., D. A. Angers, M. Giroux, and M. R. Laverdiere. 1999. Estimating C inputs retained as soil organic matter from corn (*Zea mays* L.). *Plant Soil* 215:85–91.

Bonos, S. A., D. Rush, K. Highnight, and W. A. Meyer. 2004. Selection for deep root production in tall fescue and perennial ryegrass. *Crop Sci.* 44:1770–1775.

Boone, R. D. 1994. Light-fraction soil organic matter: Origin and contribution to net nitrogen mineralization. *Soil Biol. Biochem.* 26:1459–1468.

Boote, K. J., J. R. Stansell, A. M. Schubert, and J. F. Stone. 1982. Irrigation, water use, and water relationships. In: *Peanut Science and Technology*, eds., H. E. Pattee and C. T. Yong, pp. 164–205. Yoakum, TX: American Peanut Research and Education Society.

Borg, H. and D. W. Grimes. 1986. Depth development of roots with time: An empirical description. *Trans. Am. Soc. Agric. Eng.* 29:194–197.

Bottner, P., M. Pansu, and Z. Sallih. 1999. Modeling the effect of active roots on soil organic matter turnover. *Plant Soil* 216:15–25.

Bragg, P. L., G. Govi, and R. Q. Cannell. 1983. A comparison of methods, including angled and vertical minirhizotrons for studying root growth and distribution in a spring oat crop. *Plant Soil* 73:435–440.

Butnor, J. R., J. A. Doolittle, K. H. Johnsen, L. Samuelson, T. Stokes, and L. Kress. 2003. Utility of ground-penetrating radar as a root biomass survey tool in forest systems. *Soil Sci. Soc. Am. J.* 67:1607–1615.

Buyanovsky, G. A. and G. H. Wagner. 1997. Crop residue input to soil organic matter in the Sanborn field. In: *Soil Organic Matter in Temperate Ecosystems: Long-Term Experiments in North America*, eds., E. A. Pal et al., pp. 73–83. Boca Raton, FL: CRC Press.

Cakmak, I., C. Hengeler, and H. Marschner. 1994. Changes in phloem export of sucrose in leaves in response to phosphorus, potassium and magnesium deficiency in bean plants. *J. Exp. Bot.* 45:1251–1257.

Caldwell, M. M. 1987. Plant architecture and resource competition. In: *Ecological Studies*, Vol. 61, eds., E. D. Schulzer and H. Zwolfer, pp. 164–179. New York: Springer.

Castronguay, Y., S. Laberge, E. C. Brummer, and J. J. Volenec. 2006. Alfalfa winter hardiness: A research retrospective and integrated perspective. *Adv. Agron.* 90:203–265.

Christopher, S. F., R. Lal, and U. Mishra. 2009. Regional study of no-till effects on carbon sequestration in the Midwest United States. *Soil Sci. Soc. Am. J.* 73:207–216.

Claassen, N. and B. Steingrobe. 1999. Mechanistic simulation models for a better understanding of nutrient uptake from soil. In: *Mineral Nutrition of Crops: Fundamental Mechanisms and Implications*, ed., Z. Rangel, pp. 327–367. New York: The Haworth Press.

Collins, H. P., J. L. Smith, S. Fransen, A. K. Alva, C. E. Kruger, and D. M. Granatstein. 2010. Carbon sequestration under irrigated switgrass (*Panicum virgatum* L.) production. *Soil Sci. Soc. Am. J.* 74:2049–2058.

Cook, R. J. 1988. Biological control and holistic plant-health care in agriculture. *Am. J. Altern. Agric.* 3:51–62.

Costa, C., L. M. Dwyer, X. Zhou, P. Dutilleul, C. Hamel, L. M. Reid, and D. L. Smith. 2002. Root morphology of contrasting maize genotypes. *Agron. J.* 94:96–101.

Costa, S. E. V. G., E. D. Souza, I. Anghinoni, J. P. C. Flores, E. G. Cao, and M. J. Holzschuh. 2009. Phosphorus and root distribution and corn growth as related to long-term tillage systems and fertilizer placement. *R. Bras. Ci. Solo* 33:1237–1247.

Davidson, R. L. 1969. Effect of root/leaf temperature differentials on root/shoot ratios in some pasture grasses and clover. *Ann. Bot.* 33:561–569.

Dong, B., Z. Rengel, and R. D. Graham. 1995. Root morphology of wheat genotypes differing in zinc efficiency. *J. Plant Nutr.* 18:2761–2773.

Duriex, R. P., E. J. Kamprath, W. A. Jackson, and R. H. Moll. 1994. Root distribution of corn: The effect of nitrogen fertilization. *Agron. J.* 86:958–962.

Eghball, B. and J. W. Maranville. 1993. Root development and nitrogen influx of corn genotypes grown under combined drought and nitrogen stresses. *Agron. J.* 85:147–152.

Eghball, B., J. R. Settimi, J. W. Maranville, and A. M. Parkhurst. 1993. Fractal analysis for morphological description of corn roots under nitrogen stress. *Agron. J.* 85:287–289.

Eissenstat, D. M. 1992. Costs and benefits of constructing roots of small diameter. *J. Plant Nutr.* 15:763–782.

Eissenstat, D. M., C. E. Wells, R. D. Yanai, and J. L. Whitbck. 2000. Building roots in a changing environment: Implications for root longevity. *New Phytol.* 147:33–42.

Entry, J. A., R. E. Sojka, and G. E. Shewmaker. 2002. Management of irrigated agriculture to increase organic carbon storage in soils. *Soil Sci. Soc. Am. J.* 66:1957–1964.

Essau, K. 1977. *Anatomy of Seed Plants*, 2nd edn., New York: Wiley.

Evans, L. T. and I. F. Wardlaw. 1976. Aspects of the comparative physiology of grain yield in cereals. *Adv. Agron.* 28:301–359.

Exner, D. N. and R. M. Cruse. 2001. Profitability of crop rotations in Iowa in a stress environment. *J. Iowa Acad. Sci.* 108:84–89.

Fageria, N. K. 1989. *Tropical Soils and Physiological Aspects of Crops*, 425 pp. Brasilia, Brazil: EMBRAPA/CNPAF Document 18.

Fageria, N. K. 1992. *Maximizing Crop Yields*. New York: Marcel Dekker.

Fageria, N. K. 2007. Yield physiology of rice. *J. Plant Nutr.* 30:843–879.

Fageria, N. K. 2009. *The Use of Nutrients in Crop Plants*. Boca Raton, FL: CRC Press.

Fageria, N. K. 2010. Root growth of upland rice genotypes as influenced by nitrogen fertilization. Paper presented at *the 19th World Soil Science Congress*, August 1–6, 2010, Brisbane, Queensland, Australia: Australian Society of Soil Science.

Fageria, N. K. and V. C. Baligar. 1997. Upland rice genotypes evaluation for phosphorus use efficiency. *J. Plant Nutr.* 20:499–509.

Fageria, N. K. and V. C. Baligar. 2005. Enhancing nitrogen use efficiency in crop plants. *Adv. Agron.* 88:97–185.

Fageria, N. K., V. C. Baligar, and R. B. Clark. 2006. *Physiology of Crop Production*. New York: Haworth.

Fageria, N. K., V. C. Baligar, and C. A. Jones. 2011. *Growth and Mineral Nutrition of Field Crops*, 3rd edn. Boca Raton, FL: CRC Press.

Fageria, N. K. and A. Moreira. 2011. The role of mineral nutrition on root growth of crop plants. *Adv. Agron.* 110:251–331.

Fageria, N. K. and A. B. Santos. 2007. Yield physiology of dry bean. *J. Plant Nutr.* 31:983–1004.

Fageria, N. K., N. A. Slaton, and V. C. Balihra. 2003. Nutrient management for improving lowland rice productivity and sustainability. *Adv. Agron.* 80:63–152.

Fisher, N. M. and R. J. Dunham. 1984. Root morphology and nutrient uptake. In: *The Physiology of Tropical Field Crops*, ed., R. Goldsworthy, pp. 85–117. New York: John Wiley & Sons.

Fitter, A. H. 1985. Functional significance of root morphology and root system architecture. In: *Ecological Interactions in Soil*, ed., A. H. Fitter et al., pp. 87–106. Oxford, U.K.: Blackwell.

Fredeen, A. L., I. M. Rao, and N. Terry. 1989. Influence of phosphorus nutrition on growth and carbon partitioning in *Glycine max*. *Plant Physiol.* 89:225–230.

Fujita, K., K. G. Ofosubudu, and S. Ogata. 1992. Biological nitrogen fixation in mixed legume-cereal cropping systems. *Plant Soil* 141:155–175.

Gahoonia, T. S. and N. E. Nielsen. 2004. Root traits as tools for creating phosphorus efficient crop varieties. *Plant Soil* 260:47–57.

Gale, W. J., C. A. Cambardella, and T. B. Bailey. 2000. Surface residue and root-derived carbon in stable and unstable aggregates. *Soil Sci. Soc. Am. J.* 64:196–201.

Gales, K. 1983. Yield variation of wheat and barley in Britain in relation to crop growth and soil conditions-A review. *J. Sci. Food Agric.* 34:1085–1104.

Ge, Z. Y., G. Rubio, and J. P. Lynch. 2000. The importance of root gravitropism for inter-root competition and phosphorus acquisition efficiency: Results from a geometric simulation model. *Plant Soil* 218:159–171.

Gerwitz, A. and E. R. Page. 1974. An empirical mathematical model to describe plant root system. *J. Appl. Ecol.* 11:773–782.

Greenwood, D. J., A. Gerwitz, D. A. Stone, and A. Barnes. 1982. Root development of vegetable crops. *Plant Soil* 68:75–96.

Gregory, P. J. 1994. Root growth and activity. In: *Physiology and Determination of Crop Yield*, eds., K. J. Boote, J. M. Bennett, T. R. Sinclair, and G. M. Paulsen, pp. 65–93. Madison, WI: ASA, CSSA, and SSSA.

Gregory, P. J. 2006a. *Plant Roots: Growth, Activity and Interactions with Soil*. Oxford, U.K.: Blackwell Scientific.

Gregory, P. J. 2006b. Roots, rhizosphere and soil: The route to better understanding of soil science? *Europ. J. Soil Sci.* 57:2–12.

Gregory, P. J. and B. J. Atwell. 1991. The fate of carbon in pulse-labelled crops of barley and wheat. *Plant Soil* 136:205–213.

Gregory, P. J. and J. Eastham. 1996. Growth of shoots and roots, and interception of radiation by wheat and lupin crops on a shallow, duplex soil in response to time of sowing. *Aust. J. Agric. Res.* 47:427–447.

Gregory, P. J., J. A. Palta, and G. R. Batts. 1997. Root systems and root:mass ratio-carbon allocation under current and projected atmospheric conditions in arable crops. *Plant Soil* 187:221–228.

Groff, P. A. and D. R. Kaplan. 1988. The relation of root systems to shoot systems in vascular plants. *Bot. Rev.* 54:387–422.

Hammond, J. P., M. R. Broadley, P. J. White, G. J. King, H. C. Bowen, R. Hayden, M. C. Meacham, A. Mead, T. Overs, W. P. Spracklen, and D. J. Greenwood. 2009. Shoot yield drives phosphorus use efficiency in *Brassica oleracea* and correlates with root architecture traits. *J. Exp. Bot.* 60:1953–1968.

Hinsinger, P., A. G. Bengough, D. Vetterlein, and I. M. Young. 2009. Rhizosphere: Biophysics, biogeochemistry and ecological relevance. *Plant Soil* 321:117–152.

Ho, M. D., J. C. Rosas, K. M. Brown, and J. P. Lynch. 2005. Root architectural tradeoffs for water and phosphorus acquisition. *Funct. Plant Biol.* 32:737–748.

Hoad, S. P., G. Russell, M. E. Lucas, and I. J. Bingham, I. J. 2001. The management of wheat, barley, and oat root systems. *Adv. Agron.* 74:193–246.

Huggins, D. R. and D. J. Fuchas. 1997. Long-term N management effects on corn yield and soil C of an aquic Haplustoll in Minnesota. In: *Soil Organic Matter in Temperate Ecosystems: Long-Term Experiments in North America*, eds., E. A. Pal et al., pp. 121–128. Boca Raton, FL: CRC Press.

Hurd, E. A. 1974. Phenotype and drought tolerance in wheat. *Agric. Meteorol.* 14:39–55.

Jackson, R. B., J. Canadell, J. R. Ehleringer, H. A. Mooney, O. E. Sala, and E. D. Schulze. 1996. A global analysis of root distribution for terrestrial biomes. *Oecologia* 108:389–411.

Jaffar, M. N., L. R. Stone, and D. E. Goodwin. 1993. Rooting depth and dry matter development of sunflower. *Agron. J.* 85:281–286.

Jaleel, C. A., P. Manivannan, A. Wahid, M. Farooq, H. J. Al-Juburi, R. Somasundarama, and R. P. Vam. 2009. Drought stress in plants: A review on morphological characteristics and pigments composition. *Int. J. Agric. Biol.* 11:100–105.

Jastrow, J. D., R. M. Miller, and J. Lussenhop. 1998. Contributions of interacting biological mechanisms to soil aggregate stabilization in restored prairie. *Soil Biol. Biochem.* 30:905–916.

Jensen, B. 1994. Rhizodeposition by field-grown winter barley exposed to $^{14}CO_2$ pulse labelling. *Appl. Soil Ecol.* 1:1010.

Jobbagy, E. G. and R. B. Jackson. 2000. The vertical distribution of organic carbon and its relation to climate and vegetation. *Ecol. Appl.* 10:423–436.

Johnson, J. M. F., R. R. Allmaras, and D. C. Reicosky. 2006. Estimating source carbon from crop residues roots and rhizodeposits using the national grain-yield database. *Agron. J.* 98:622–636.

Jones, C. A. 1985. C_4 *Grasses and Cereals: Growth, Development, and Stress Response.* New York: John Wiley & Sons.

Kalisz, P. J., R. W. Zimmerman, and R. N. Muller. 1987. Root density, abundance, and distribution in the mixed mesophytic forest of eastern Kentucky. *Soil Sci. Soc. Am. J.* 51:220–225.

Karcher, D. E., M. D. Richardson, K. Hignight, and D. Rush. 2008. Drought tolerance of tall fescue populations selected for high root/shoot ratios and summer survival. *Crop Sci.* 48:771–777.

Kaspar, T. C., H. M. Taylor, and R. M. Shibles. 1984. Taproot elongation rates of soybean cultivars in the glasshouse and their relation to field rooting depth. *Crop Sci.* 24:916–920.

Ketring, D. L. 1984. Root diversity among peanut genotypes. *Crop Sci.* 24:229–232.

Ketring, D. L. and J. L. Reid. 1993. Growth and peanut roots under field conditions. *Agron. J.* 85:80–85.

Klepper, B. 1990. Root growth and water uptake. In: *Irrigation of Agricultural Crops*, eds., B. A. Stewart and D. R. Nielsen, pp. 281–322. Madison, WI: ASA, CSSA, and SSSA.

Klepper, B. 1992. Development and growth of crop root systems. *Adv. Soil Sci.* 19:1–25.

Klepper, B., R. K. Belford, and R. W. Rickman. 1984. Root and shoot development in winter wheat. *Agron. J.* 76:117–122.

Klepper, B. and T. C. Kaspar. 1994. Rhizotrons: Their development and use in agricultural research. *Agron. J.* 86:745–753.

Klepper, B. and R. W. Rickman. 1990. Modeling crop root growth and function. *Adv. Agron.* 44:113–132.

Kong, A. Y. Y. and J. Six. 2010. Tracing root vs. residue carbon into soils from conventional and alternative cropping systems. *Am. Soc. Soil Sci. J.* 74:1201–1210.

Kuchenbuch, R. O. and S. A. Barber. 1987. Yearly variation of root distribution with depth in relation to nutrient uptake and corn yield. *Commun. Soil Sci. Plant Anal.* 18:255–263.

Lemos, R. and R. Lal. 2005. Bioenergy crops and carbon sequestration. *Crit. Rev. Plant Sci.* 24:1021.

Leon, J. and K. U. Schwang. 1992. Description and application of a screening method to determine root morphology traits of cereals cultivars. *Z. Acker. Pflanzenbau* 169:128–134.

Leskovarant, D. I. and P. J. Stofella. 1995. Vegetable seedling root systems: Morphology, development, and importance. *HortScience* 30:1153–1159.

Liao, M., I. R. P. Filley, and J. A. Palta. 2004. Early vigorous growth is a major factor influencing nitrogen uptake in wheat. *Funct. Plant Biol.* 31:121–129.

Ma, Z., C. W. Wood, and D. I. Bransby. 2000. Impacts of soil management on root characterization of switgrass. *Biomass Bioenergy* 18:105–112.

Maizlish, N. A., D. D. Fritton, and W. A. Kendall. 1980. Root morphology and early development of maize at varying levels of nitrogen. *Agron. J.* 72:25–31.

Makela, A. A. and R. P. Sievanen. 1987. Comparison of two shoot-root partitioning models with respect to substrate utilization and functional balance. *Ann. Bot.* 59:129–140.

Manivannan, P., C. A. Jaleel, B. Sankar, A. Kishorekumar, R. Somasundaram, G. M. A. Lakshmanan, and R. Panneerselvam. 2007. Growth, biochemical modifications and praline metabolism in *Helianthus annuus* L. as induced by drought stress. *Colloids Surf. B: Biointerfaces* 59:141–149.

Marschner, H. 1998. Role of root growth, arbuscular mycorrhiza, and root exudates for the efficiency in nutrient acquisition. *Field Crops. Res.* 56:203–207.

McLaughlin, S. B. and L. A. Kszos. 2005. Development of switchgrass (*Panicum virgatum*) as a bioenergy feedback in the United States. *Biomass Bioenergy* 28:515–535.

McLaughlin, S. B. and M. E. Walsh. 1998. Evaluating environmental consequences of producing herbaceous crops for bioenergy. *Biomass Bioenergy* 14:317–324.

McMichael, B. L. and H. M. Taylor. 1987. Applications and limitations of rhizotrons and minirhizotrons. In: *Minirhizotron Observation Tubes: Methods and Applications for Measuring Rhizosphere Dynamics*, ed., H. M. Taylor, pp. 1–13. Madison, WI: ASA, CSSA, and SSSA.

McWilliam, J. R. 1986. The national and international importance of drought and salinity effects of agricultural production. *Aust. J. Plant Physiol.* 13:1–13.

Melhuish, F. M. and A. R. G. Lang. 1968. Quantitative studies of roots in soil: I. Length and diameters of cotton roots in a clay-loam soil by analysis of surface ground blocks of resin-impregnated soil. *Soil Sci.* 106:16–22.

Mengel, K. 1985. Dynamics and availability of major nutrients in soils. *Adv. Soil Sci.* 2:66–131.

Mengel, D. B. and S. A. Barber. 1974. Development and distribution of the corn root system under field conditions. *Agron. J.* 68:341–344.

Merrill, S. D. 1992. Pressurized-well minirhizotron for field observation of root growth. *Agron. J.* 84:755–758.

Meyer, W. S. and H. D. Baris. 1985. Nondestructive measurement of wheat roots in large undisturbed and replaced clay soil cores. *Plant Soil* 85:237–247.

Michulmas, D. G., W. K. Lauenroth, J. S. Singh, and C. V. Cole. 1985. Root turnover and production by ^{14}C dilution: Implications of carbon portioning in plants. *Plant Soil* 88:353–365.

Murphy, J. A., M. G. Hendricks, P. E. Rieke, A. J. M. Smucker, and B. E. Branham. 1994. Turfgrass root systems evaluated using the minirhizotron and video recording methods. *Agron. J.* 86:247–250.

Myers, R. J. K. 1980. The root system of a grain sorghum crop. *Field Crops Res.* 3:53–64.

Newman, E. I. 1966. A method of estimating the total length of root in a sample. *J. Appl. Ecol.* 3:139–145.

Norby, R. J. and M. F. Cotrufo. 1998. A question of litter quality. *Nature* 396:17–18.

Noulas, C., M. Liedgens, P. Stamp, I. Alexiou, and J. M. Herrera. 2010. Subsoil root growth of field grown spring wheat genotypes (*Triticum estivum* L.) differing in nitrogen use efficiency. *J. Plant Nutr.* 33:1887–1903.

O'Toole, J. C. and W. L. Bland. 1987. Genotypic variation in crop plant root systems. *Adv. Agron.* 41:91–145.

Otto, R., P. C. O. Trivelin, H. C. J. Franco, C. E. Faroni, and A. C. Vitti. 2009. Root system distribution of sugarcane as related to nitrogen fertilization, evaluated by two methods: Monolith and probes. *R. Bras. Ci. Solo* 33:601–611.

Perfect, E., R. D. Miller, and B. Burton. 1987. Root morphology and vigor effects on winter heaving of established alfalfa. *Agron. J.* 79:1061–1067.

Poorter, H. and O. Nagel. 2000. The role of biomass allocation in the growth response of plants to different levels of light, CO_2, nutrients and water: A quantitative review. *Aust. J. Plant Physiol.* 27:595–607.

Puget, P. and L. E. Drinkwater. 2001. Short-term dynamics of root and shoot derived carbon from a leguminous green manure. *Soil Sci. Soc. Am. J.* 65:771–779.

Qin, R., P. Stamp, and W. Richner. 2004. Impact of tillage on root systems of winter wheat. *Agron. J.* 96:1523–1530.

Reicosky, D. C. and F. Forcella. 1998. Cover crop and soil quality interactions in agroecosystems. *J. Soil Water Conserv.* 53:224–229.

Robertson, M. J., S. Fukai, G. L. Hammer, and M. M. Ludlow. 1993. Modeling root growth of grain sorghum using the CERES approach. *Field Crops Res.* 33:113–130.

Robertson, W. K., L. C. Hammond, J. T. Johnson, and G. M. Prine. 1979. Root distribution of corn, soybeans, peanuts, sorghum and tobacco in fine sands. *Soil Crop Sci. Soc. Fla. Proc.* 38:54–59.

Romer, W. J., J. Augustin, and G. Schilling. 1988. The relationship between phosphate absorption and root length in nine wheat cultivars. *Plant Soil* 111:199–201.

Rufty Jr., T. W., D. W. Israel, R. J. Volk, J. Qui, and T. Sa. 1993. Phosphate regulation of nitrate assimilation in soybean. *J. Exp. Bot.* 44:879–891.

Russell, R. S. 1977. *Plant Root Systems: Their Function and Interaction with the Soil.* London, U.K.: McGraw-Hill Book Co. Ltd.

Sainju, U. M. and R. E. Good. 1993. Vertical root distribution in relation to soil properties in New Jersey Pinelands. *Plant Soil* 150:87–97.

Sainju, U. M. and P. J. Kalisz. 1990. Characteristics of coal bloom horizons in undisturbed forest soils in eastern Kentucky. *Soil Sci. Soc. Am. J.* 54:879–882.

Schenk, M. K. and S. A. Barber. 1980. Potassium and phosphorus uptake by corn genotype grown in the field as influenced by root characteristics. *Plant Soil* 54:65–75.

Schlesinger, W. W. 1990. Evidence from chronosequence studies for a low carbon storage potential of soils. *Nature* 348:232–234.

Sharp, R. E. and M. E. LeNoble. 2002. ABA, Ethylene and the control of shoot and root growth under water stress. *J. Exp. Bot.* 53:33–37.

Sharratt, B. S. and V. L. Cochran. 1993. Skip-row and equidistant-row barley with nitrogen placement; Yield, nitrogen uptake and root density. *Agron. J.* 85:246–250.

Silberbush, M. and S. A. Barber. 1984. Phosphorus and potassium uptake of field grown soybeans predicted by a simulation model. *Soil Sci. Soc. Am. J.* 48:592–596.

Singer, J. W. and W. J. Cox. 1998. Agronomics of corn production under different crop rotation in New York. *J. Prod. Agric.* 11:462–468.

Singh, B. P. and U. M. Sainju. 1998. Soil physical and morphological properties and root growth. *HortScience* 33:966–971.

Slaton, N. A. and C. A. Beyrouty. 1992. Rice shoot response to root confinement within a membrane. *Agron. J.* 84:50–53.

Slaton, N. A., C. A. Beyrouty, B. R. Wells, R. J. Norman, and E. E. Gbur. 1990. Root growth and distribution of two short-season rice genotypes. *Plant Soil* 126:269–278.

Smucker, A. J. M. 1993. Soil environmental modifications of root dynamics and measurement. *Annu. Rev. Phytopathol.* 31:191–216.

Solaiman, Z., P. Marschner, D. Wang, and Z. Rengel. 2007. Growth, P uptake and rhizosphere properties of wheat and canola genotypes in an alkaline soil with low P availability. *Biol. Fertil. Soils* 44:143–153.

Sponchiado, B. N., J. W. White, J. A. Castillo, and P. G. Jones. 1989. Root growth of four common bean cultivars in relation to drought tolerance in environments with contrasting soil types. *Exp. Agric.* 25:249–257.

St. Aubin, G., M. J. Canny, and M. E. McCully. 1986. Living vessel elements in the late metaxylem of sheathed maize roots. *Ann. Bot.* 58:577–588.

Steponkus, P. J., J. M. Cutler, and J. C. O'Toole. 1980. Adaptation to water stress in rice. In: *Adaptation of Plants to Water and High Temperature Stress*, eds., N. C. Turner and P. J. Kramer, pp. 401–418. New York: John Wiley & Sons.

Stoffel, S., R. Gutser, and N. Claassen. 1995. Root growth in an agricultural landscape of the "Tertiar-Hugelland." *Agrobiol. Res.* 48:330–340.

Stoffella, P. J. and B. A. Kahn. 1986. Root system effects on lodging of vegetable crops. *HortScience* 21:960–963.

Stoffella, P. J., R. F. Sandsted, R. W. Zobel, and W. L. Hymes. 1979. Root characteristics of some black beans. I. Relationship of root size to lodging and seed yield. *Crop Sci.* 19:823–826.

Stone, L. F. and A. L. Pereira. 1994a. Rice-common bean rotation under sprinkler irrigation: Effects of row spacing, fertilization and cultivar on growth, root development and water consumption of common bean. *Pesq. Agropec. Bras. Brasilia* 29:939–954.

Stone, L. F. and A. L. Pereira. 1994b. Rice-common bean rotation under sprinkler irrigation: Effects of row spacing, fertilization and cultivar on growth, root development and water consumption of rice. *Pesq. Agropec. Bras. Brasilia* 29:1577–1592.

Szumigalski, A. R. and R. C. V. Acker. 2006. Nitrogen yield and land use efficiency in annual sole crops and intercrops. *Agron. J.* 98:1030–1040.

Tahir, M. N. H., M. Imran, and M. K. Hussain. 2002. Evaluation of sunflower (*Helianthus annuus* L.) inbred lines for drought tolerance. *Int. J. Agric. Biol.* 3:398–400.

Takenaga, H. 1995. Internal factors in relation to nutrient absorption. In: *Science of the Rice Plant*, Vol. 2, eds., T. Matsuo, K. Kumazawa, R. Ishii, K. Ishihara, and H. Hirata, pp. 294–309. Tokyo, Japan: Food and Agriculture Policy Research Center.

Tennant, D. 1975. A test of a modified line intersect method of estimating root length. *J. Ecol.* 63:955–1001.

Teo, Y. H., C. A. Beyrouty, and E. E. Gbur. 1995a. Evaluation of a model to predict nutrient uptake by field grown rice. *Agron. J.* 87:7–12.

Teo, Y. H., C. A. Beyrouty, R. J. Norman, and E. E. Gbur. 1995b. Nutrient uptake relationship to root characteristics of rice. *Plant Soil* 171:297–302.

Thangaraj, M., J. C. O'Tolle, and S. K. De Datta. 1990. Root response to water stress in rainfed lowland rice. *Exp. Agric.* 26:287–296.

Thomas, A. L. and T. C. Kaspar. 1997. Maize nodal root response to time of soil ridging. *Agron. J.* 89:195–200.

Tisdall, J. M. and J. M. Oades. 1979. Stabilizing of soil aggregates by the root systems of ryegrass. *Aust. J. Soil Res.* 29:729–743.

Tufekcioglu, A., A. W. Raich, T. M. Osenhart, and R. C. Schulz. 2003. Biomass, carbon and nitrogen dynamics of multi-species riparian buffers within an agricultural watershed in Iowa, USA. *Agrofor. Ecosyst.* 57:187–198.

Upchurch, D. R. 1985. Relationship between observation in minirhizotrons and true root length density. PhD dissertation. Lubbock, TX: Texas Tech University, (Dissertation Abstract 85-28594).

Upchurch, D. R. and J. T. Ritchie. 1983. Battery operated color video camera for root observations in minirhizotrons. *Agron. J.* 76:1015–1017.

Van Noordwijk, M. 1993. Roots: Length, biomass, production and mortality. In: *Tropical Soil Biology and Fertility: A Handbook of Methods*, eds., J. M. Anderson and J. S. I. Ingram, pp. 132–144. Wallingford, U.K.: CAB International.

Van Noordwijk, M. and G. Brouwer. 1991. Review of quantitative root length data in agriculture. In: *Plant Roots and Their Environment*, eds., B. L. McMichael and H. Persson, pp. 515–524. Amsterdam, the Netherlands: Elsevier Scientific Publishers BV.

Waisel, Y., A. Eshel, and U. Kafkafi. 2002. *Plant Roots: The Hidden Half*. Madison, WI: Marcel Dekker.

Wander, M. M. and X. Yang. 2000. Influence of tillage on the dynamics of loose and occluded particulate and humified organic matter fractions. *Soil Biol. Biochem.* 32:1151–1160.

Wang, Y., R. Amundson, and S. Trumbore. 1999. The impact of land use change on c turnover in soils. *Global Biogeochem. Cycles* 13:47–57.

Wang, H., Y. Inukai, and A. Yamauchi. 2011. Root development and nutrient uptake. *Crit. Rev. Plant Sci.* 25:279–301.

Wang, X., X. Yan, and H. Liao. 2010. Genetic improvement for phosphorus efficiency in soybean: A radical approach. *Ann. Bot.* 106:215–222.

Wardlaw, I. A. 1990. The control of carbon partitioning in plants. *New Phytol.* 116:341–381.

Watson, C. A., J. M. Ross, U. Bagnaresi, G. F. Minnota, F. Roffi, and D. Atkinson. 2000. Environmental induced modifications to root longevity in *Lolium perenne* and *Trifolium repens*. *Anna. Bot.* 85:397–401.

Weaver, J. E. 1926. *Root Development of Field Crops*. New York: McGraw-Hill.

Welbank, P. J., M. J. Gibb, P. J. Taylor, and E. D. Williams. 1974. Root growth of cereal crops. In: *Rothamsted Experimental Station Report for 1973, Part 2*. Rothamsted Experiment Station, Harpenden, U.K., pp. 26–66.

Welbank, P. J. and E. D. Williams. 1968. Root growth of a barley crop estimated by sampling with portable powered soil coring equipment. *J. Appl. Ecol.* 5:447.

Wiesler, F. and W. J. Horst. 1993. Differences among maize cultivars in the utilization of soil nitrate and the related losses of nitrate through leaching. *Plant Soil* 151:193–203.

Wilhelm, W. W., L. N. Mielke, and C. R. Fenster. 1982. Root development of winter wheat as related to tillage practice in western Nebraska. *Agron. J.* 74:85–88.

Willey, R. W. 1979. Intercropping: Its importance and research needs. Part 1. Competition and yield advantages. *Field Crop Abstr.* 32:1–10.

Wilson, J. B. 1988. A review of evidence on the control of shoot:root ratio, in relation to models. *Ann. Bot.* 61:433–449.

Wu, L., M. B. McGechan, C. A. Watson, and J. A. Baddeley. 2005. Developing existing plant root system architecture models to meet future agricultural challenges. *Adv. Agron.* 85:181–219.

Yamauchi, A., Y. Kono, and J. Tatsumi. 1987a. Qualitative analysis of root system structure of upland rice and maize. *Jpn. J. Crop Sci.* 56:608–617.

Yamauchi, A., Y. Kono, and J. Tatsumi. 1987b. Comparison of root system structures of 13 species of cereals. *Jpn. J. Crop Sci.* 56:618–631.

Yaseen, M. and S. S. Malhi. 2011. Exploitation of genetic variability among wheat genotypes for tolerance to phosphorus deficiency stress. *J. Plant Nutr.* 34:665–699.

Ying, J., S. Peng, Q. He, H. Yang, C. Yang, R. M. Visperas, and K. G. Cassman. 1998. Comparison of high yield rice in tropical and subtropical environments. I. Determinants of grain and dry matter yields. *Field Crops Res.* 57:71–84.

Yoshida, S. 1981. *Fundamentals of Rice Crop Science.* Los Banos, Philippines: IRRI.

Zan, C. S., J. W. Fyles, P. Girouard, and R. A. Samson. 2001. Carbon sequestration in perennial bioenergy, annual corn and uncultivated systems in southern Quebec. *Agric. Ecosyst. Environ.* 86:135–144.

Zhuang, J., G. R. Yu, and K. Nakayama. 2001. Scaling of root length density of maize in the field profile. *Plant Soil* 235:135–142.

Zobel, R. W. 1991. Root growth and development. In: *The Rhizosphere and Plant Growth*, eds., D. L. Keister and P. B. Cregan, pp. 61–71. Dordrecht, the Netherlands: Kluwer Academic Publishers.

Zobel, R. W. 2003. Sensitivity analysis of computer based diameter measurement from digital images. *Crop Sci.* 43:583–591.

Zobel, R. W. 2005a. Primary and secondary root systems. In: *Roots and Soil Management: Interactions between Roots and the Soil*, eds., R. W. Zobel and S. F. Wright, pp. 3–14. Madison, WI: ASA, CSSA, and SSSA.

Zobel, R. W. 2005b. Tertiary root systems. In: *Roots and Soil Management: Interactions between Roots and the Soil*, eds., R. W. Zobel and S. F. Wright, pp. 35–56. Madison, WI: ASA, CSSA, and SSSA,

Zonta, E., F. C. Brasil, S. R. Goi, M. M. T. Rosa. 2006. Root system and its interaction with edaphic factors. In: *Mineral Nutrition of Plants*, ed., M. S. Fernandes, pp. 7–52. Viçosa, Brazil: Brazilian Soil Science Society.

Zuo, Q., F. Jie, R. Zhang, and L. Meng. 2004. A generalized function of wheats root length density distribution. *Vadose Zone J.* 3:271–277.

2 Uptake of Nutrients by Roots

2.1 INTRODUCTION

Nutrient uptake by roots is related to mineral nutrition, which includes the supply, absorption, and utilization of nutrients essential for the growth and yield of crop plants (Fageria and Oliveira, in press). Mineral nutrition alone has contributed significantly to increased crop yields during the twentieth century. Borlaug and Dowswell (1994) reported that 50% of the increase in crop yields worldwide during the twentieth century was due to the application of chemical fertilizers. The authors also reported that during the twenty-first century, the essential plant nutrients would be the single most important factor limiting crop yields, especially in developing countries. Borlaug and Dowswell (1997) state that science-based commercial agriculture is more of a twentieth-century invention. Loneragan (1997) states that knowledge generated during the twentieth century in the field of mineral nutrition has had an impact on current food production and provided information needed for further advances for the twenty-first century. Nutrient uptake is regulated by growth rate and nutrient availability. Actively growing plants usually exhibit a high nutrient uptake rate in order to meet demands for optimum growth (Marin et al., 2011). Optimum growth is a result of high carbon fixation in the leaves, which in turn supplies the roots with carbohydrates and energy for nutrient uptake. Internal nutritional status may also regulate uptake rate by the synthesis of signal molecules that are transported from the shoot to the root (Marin et al., 2011).

During 1950–1990, grain yields of cereals such as wheat (*Triticum aestivium* L.), maize (*Zea mays* L.), and rice (*Oryza sativa* L.) tripled worldwide. Wheat yields in India, for example, increased by nearly 400% from 1960 to 1985, and yields of rice in Indonesia and China more than doubled. This vastly increased production resulted from high-yielding varieties, improved irrigation facilities, and the use of chemical fertilizers. The results were significant in Asia and Latin America, where the term *green revolution* was used to describe the process (Brady and Weil, 2002). The increase in the productivity of annual crops with the application of fertilizers and lime in the Brazilian Cerrado or savanna region during the 1970s and 1980s is another example of twentieth century expansion of the agricultural frontier in acid soils (Borlaug and Dowswell, 1997; Fageria et al., 2008).

Most plant nutrients have limited mobility in soils. Hence, roots must grow deep enough into soils to absorb the nutrients, and nutrient availability to plants may be limited by a soil zone that restricts root growth. In contrast to water deficiencies, nutrient deficiencies usually do not cause plant death, but may severely limit plant growth and yield (Unger and Kaspar, 1994). The supply of essential plant nutrients in adequate amount and proportion is a major factor responsible for improving the yield of food crops in all types of soils, under all agroecological conditions. If nutrients are not supplied in adequate amount, it may reduce crop yields. Similarly, nutrients supplied in excess of plants' needs may create environmental pollution. Nutrient pollution of surface and groundwaters from agricultural land use can pose human health risks and has been shown to contribute to aquatic system degradation (Carpenter et al., 1998; Brady and Weil, 2002; Starovoytov et al., 2010). Nitrogen pollution has been linked to the use of inorganic N fertilizers and animal manure application; however, N mineralization from legume crops is also prone to offsite movement when nutrient release is not synchronized with crop nutrient uptake (Ranells and Wagger, 1997; Rosecrance et al., 2000; Crews and Peoples, 2004). Ongoing improvement of nutrient recommendations for crops is necessary because of changes in yield levels, cultivars, and production practices and economic conditions (Wortmann et al., 2009).

The soil's ability to supply sufficient nutrient to plants in labile, exchangeable, and moderately available forms plays a pivotal role in achieving maximum economic yield of crops (Soil Science Society of America, 1997). Furthermore, the supply and absorption of chemical elements needed for growth and metabolism may be defined as nutrition, and the chemical elements required by an organism are termed nutrients (Fageria and Baligar, 2005a). Seventeen nutrients are essential for plant growth and development. Essential nutrients may be defined as those without which plants cannot complete their life cycle, that are irreplaceable by other elements, and that are directly involved in plant metabolism. The essential nutrients such as carbon (C), hydrogen (H), oxygen (O), nitrogen (N), phosphorus (P), potassium (K), calcium (Ca), magnesium (Mg), and sulfur (S) are known as macronutrients, whereas, iron (Fe), manganese (Mn), zinc (Zn), copper (Cu), boron (B), molybdenum (Mo), chlorine (Cl), and nickel are termed micronutrients (Fageria et al., 2002, 2010; Fageria and Baligar, 2005a; Fageria, 2009).

Macronutrients are required in large quantities by plants compared to micronutrients. Micronutrients have also been called minor or trace elements, indicating that their concentrations in plant tissues are minor or in trace amounts relative to the macronutrients (Mortvedt, 2000). The accumulation of micronutrients by plants generally follows the order of Mn > Fe > Zn > Cu > B > Mo > Ni. This order may change with plant species and growth conditions like flooded rice (Fageria et al., 2002). Chlorine has been referred to as a micronutrient even though its concentration in plant tissue is often equivalent to that of macronutrients. Other micronutrients commonly found beneficial to the growth of some plants are silicon (Si), sodium (Na), vanadium (V), and cobalt (Co). These nutrients stimulate the growth of certain plants, but are not considered essential according to the Arnon and Stout (1939) definition of essentiality. Possibly, other essential micronutrients will be discovered in the future because of recent advances in solution culture techniques and the availability of highly sensitive analytical instruments (Fageria et al., 2002).

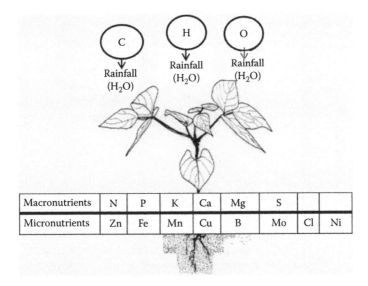

Macronutrients	N	P	K	Ca	Mg	S		
Micronutrients	Zn	Fe	Mn	Cu	B	Mo	Cl	Ni

FIGURE 2.1 Essential nutrients for plant growth. (Adapted from Wilcox, G.E. and Fageria, N.K., Nutrient deficiencies in dry bean and their corrections, Goiania, Brazil: EMBRAPA-CNPAF, Technical Bulletin: 5, p. 21, 1977.)

Based on physicochemical properties, the essential micronutrients are metals, except B and Cl. Even though micronutrients are required in small quantities by field crops, their influence is as large as that of macronutrients in crop production. Plants obtain C, H, and O from air and water, and the remaining nutrients as inorganic ions or oxides are absorbed from soil solution by growing plant roots (Figure 2.1). Micronutrients are normally constituents of prosthetic groups that catalyze redox processes by electron transfer such as with the transition elements Cu, Fe, Mn, and Mo, and form enzyme–substrate complexes by coupling enzyme with substrate (Fe and Zn) or enhance enzyme reactions by influencing molecular configurations between enzyme and substrate (Zn) (Römheld and Marschner, 1991; Fageria, 2009; Fageria et al., 2010).

Nutritional disorders (deficiency/toxicity) limit crop yields in all types of soils around the world. For example, in Brazilian Oxisols and Inceptisols, deficiency of N, P, K, and Zn limits the yield of almost all field crops (Fageria and Baligar, 2008; Fageria, 2009; Fageria et al., 2010). Figure 2.2 shows the importance of N application in increasing the yield of five lowland rice genotypes grown on a Brazilian Inceptisol. Similarly, Figure 2.3 shows the importance of N application in increasing yield of three upland rice genotypes grown on a Brazilian Oxisol. It is clear from Figure 2.2 that lowland rice genotypes responded differently to N fertilization. Two genotypes had quadratic responses and the other three had linear responses. Similarly, the results of Figure 2.3 show that the yields of upland rice genotypes increased with increasing levels of N until an adequate level in the soil was achieved. The objective of this chapter is to provide information on the latest advances on the forms of nutrients absorbed by roots and their functions, nutrient disorder diagnostic techniques, nutrient requirements, and use efficiency in crop plants.

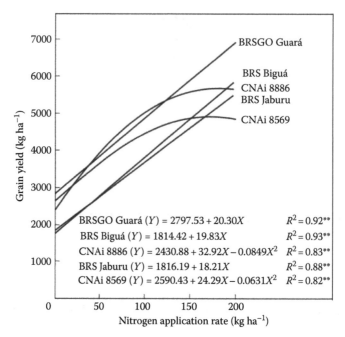

FIGURE 2.2 Response of five lowland rice genotypes to N fertilization grown on a Brazilian Inceptisol. **Significant at the 1% probability level. (From Fageria, N.K. et al., *J. Plant Nutr.*, 31, 1121, 2008.)

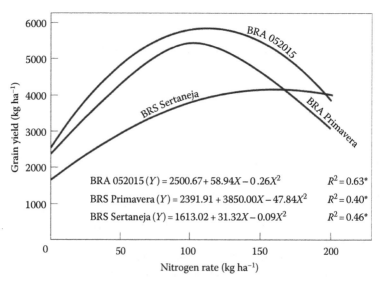

FIGURE 2.3 Response of three upland rice genotypes to P fertilization grown on a Brazilian Oxisol. *Significant at the 5% probability level.

2.2 FORMS OF ABSORPTION AND FUNCTIONS OF ESSENTIAL NUTRIENTS IN PLANTS

Nutrient absorption by plants is usually referred to as ion uptake or ion absorption because it is in the ionic form that the nutrients are absorbed by roots. Cations and anions may be absorbed independently and may not be absorbed in equal quantities; however, electroneutrality must be maintained in the plant and in the growth medium. Therefore, ionic relationship achieves a major importance in plant nutrition. The majority of cations in plant tissues are in the inorganic form, predominantly K^+, Ca^{2+}, and Mg^{2+}, and the majority of anions are in the organic form. The organic ions are synthesized within the tissue, while inorganic ions are absorbed from the growth medium. Generally, the monovalent cations are absorbed rapidly, whereas, the divalent cations, especially Ca^{2+}, are absorbed more slowly. Similarly, the monovalent anions are generally absorbed more rapidly than the polyvalent anions (Fageria et al., 2006). The so-called physiological acidity or alkalinity of a salt depends upon which ion of the salt, the cation or the anion, is most absorbed rapidly. Thus, a salt like K_2SO_4 would be physiologically acid, since the K would enter the roots more rapidly than the SO_4^{2-}. By the same token, $CaCl_2$ should be physiologically alkaline, since the Ca^{2+} enters slowly and the Cl^- rapidly (Fageria and Baligar, 2005a).

Each of these essential nutrients performs a specific biophysical or biochemical function within plant cells. Nutrients like N, P, and K increase leaf area in cereals as well as legumes and, consequently, the photosynthetic process. Figure 2.4 shows the response of a Brazilian upland rice cultivar BRS Sertaneja to N fertilization. The growth of rice plants increased with increasing N rates. Based on regression equation, the maximum grain yield was obtained with the application of 380 mg N kg^{-1} (Fageria et al., 2011b). Similarly, the growth of upland rice cultivar BRS Sertaneja was much better at 200 mg P kg^{-1} soil treatment compared to 25 mg P kg^{-1} soil treatment (Figure 2.5). The growth of upland rice cultivar BRS Sertaneja also improved with the addition of K fertilization (Figure 2.6). In addition, root growth of crop

FIGURE 2.4   Response of upland rice cultivar BRS Sertaneja to N fertilization.

FIGURE 2.5 Growth of upland rice cultivar BRS Sertaneja with low and high P fertilization.

FIGURE 2.6 Growth of upland rice cultivar BRS Sertaneja without and with K fertilization.

plants also improved with the addition of an adequate amount of macronutrients in the soil (Figures 2.7 through 2.9). To obtain higher yields, essential nutrients should be present in adequate concentrations and in available form as well as in appropriate balance. An imbalance results when the supply of nutrients are either excess or inadequate. Table 2.1 shows forms of absorption and functions of essential plant nutrients.

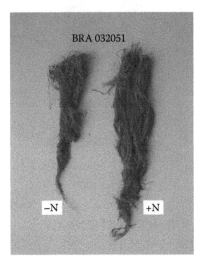

FIGURE 2.7 Root growth of upland rice genotype BRA 032051 without and with N fertilization.

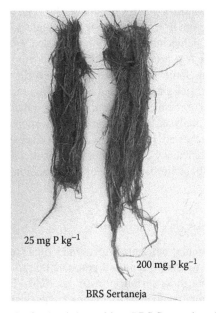

FIGURE 2.8 Root growth of upland rice cultivar BRS Sertaneja at low and high P levels.

FIGURE 2.9   Root growth of upland rice cultivar BRS Sertaneja without and with K fertilization.

2.3   NUTRIENT-DEFICIENCY/SUFFICIENCY DIAGNOSTIC TECHNIQUES

Diagnostic techniques for nutrient deficiency/sufficiency refer to the methods for identifying nutrient deficiencies, toxicities, or imbalances in the soil plant system (Fageria and Baligar, 2005a). Nutritional deficiency can occur when there is insufficient nutrient in the medium or when it cannot be absorbed and utilized by plants as the result of unfavorable environmental conditions. Table 2.2 shows soil conditions associated with nutrient deficiencies. Nutritional deficiencies are common in almost all field crops worldwide (Fageria et al., 2011a). The magnitude varies from crop to crop and region to region. Some cultivars are more susceptible to nutritional deficiencies than others even within a crop species. There are four methods to assess nutrient deficiency/sufficiency of mineral nutrients for plant growth: (1) visual symptoms, (2) soil testing, (3) plant analysis, and (4) crop growth response (Fageria and Baligar, 2005a). The four approaches are becoming widely used separately or collectively as nutrient availability or deficiency or sufficiency diagnostic aids. They are extremely helpful, yet are not without limitations. A brief discussion of these techniques is given in this section.

2.3.1   Visual Symptoms

When the supply of particular nutrients is at an inadequate level in the soil, or when plant roots are not able to absorb required amounts due to unfavorable conditions in the rhizosphere, plants show certain growth disorders. These disorders may be expressed as reduced height, reduced tillering in cereals, leaf discoloration, reduced

TABLE 2.1
Forms of Absorption and Functions of Essential Nutrients in Plants

Nutrient	Forms Taken Up by Plants	Functions
Carbon	CO_2	Basic molecular component of carbohydrates, proteins, lipids, and nucleic acids
Hydrogen	H_2O	Hydrogen plays a central role in plant metabolism. Important in ionic balance and as main reducing agent and plays a key role in energy relations of cells
Oxygen	H_2O, O_2	Oxygen is somewhat like carbon in that it occurs in virtually all organic compounds of living organisms
Nitrogen	NH_4^+, NO_3^-	Nitrogen is a component of many important organic compounds ranging from proteins to nucleic acids
Phosphorus	$H_2PO_4^-, HPO_4^{2-}$	Central role in plants is in energy transfer and protein metabolism
Potassium	K^+	Helps in osmotic and ionic regulation. Potassium functions as a cofactor or activator for many enzymes of carbohydrate and protein metabolism
Calcium	Ca^{2+}	Calcium is involved in cell division and plays a major role in the maintenance of membrane integrity
Magnesium	Mg^{2+}	Component of chlorophyll and a cofactor for many enzymatic reactions
Sulfur	SO_4^{2-}	Sulfur is somewhat like phosphorus in that it is involved in plant cell energetic process
Iron	Fe^{2+}, Fe^{3+}	An essential component of many heme and nonheme Fe enzymes and carriers, including the cytochromes (respiratory electron carriers) and the ferredoxins. The latter are involved in key metabolic functions such as N fixation, photosynthesis, and electron transfer
Zinc	Zn^{2+}	Essential component of several dehydrogenases, proteinases, and peptidases, including carbonic anhydrase, alcohol dehydrogenase, glutamic dehydrogenase, and malic dehydrogenase, among others
Manganese	Mn^{2+}	Involved in the O_2-evolving system of photosynthesis and is a component of the enzymes arginase and phosphotransferase
Copper	Cu^{2+}	Constituent of a number of important oxidase enzymes, including cytochrome oxidase, ascorbic acid oxidase, and lactase, and important in photosynthesis, and protein and carbohydrate metabolism
Boron	$B(OH)_3^0$	Activates certain dehydrogenase enzymes, involved in carbohydrate metabolism, synthesis of cell wall components, and essential for cell division and development
Molybdenum	Mo	An essential component of nitrate reductase and N_2 fixation enzymes and required for normal assimilation of N

(continued)

TABLE 2.1 (continued)
Forms of Absorption and Functions of Essential Nutrients in Plants

Nutrient	Forms Taken Up by Plants	Functions
Chlorine	Cl⁻	Essential for photosynthesis and as an activator of enzymes, involved in splitting water, and functions in osmoregulation of plants growing on saline soils
Nickel	Ni²⁺	Nickel is a constituent of plant enzyme urease, the enzyme that catalyzes the degradation of urea to carbon dioxide and ammonia

Sources: Fageria, N.K. and Gheyi, H.R., *Efficient Crop Production*, Federal University of Paraiba, Campina Grande, Brazil, 1999; Marschner, H., *Mineral Nutrition of Higher Plants*, 2nd edn., Academic Press, New York, 1995; Mengel, K. et al., *Principles of Plant Nutrition*, 5th edn., Kluwer Academic Publishers, Dordrecht, the Netherlands, 2001; Brady, N.C. and Weil, R.R., *The Nature and Properties of Soils*, 13th edn., Prentice Hall, Upper Saddle River, NJ, 2002; Fageria, N.K. and Baligar, V.C., Nutrient availability, in *Encyclopedia of Soils in the Environment*, ed., Hillel, D., pp. 63–71, Elsevier, San Diego, CA, 2005a; Fageria, N.K. et al., *Physiology of Crop Production*, The Haworth Press, New York, 2006; Fageria, N.K., *The Use of Nutrients in Crop Plants*, CRC Press, Boca Raton, FL, 2009; Fageria, N.K. et al., *Growth and Mineral Nutrition of Field Crops*, 3rd edn., CRC Press, Boca Raton, FL, 2011a.

root growth, and reduced growth of newly emerging parts of the plant. Table 2.3 summarized the symptoms of nutritional deficiency in crop plants. Figures 2.10 and 2.11 show N-deficiency symptoms in upland rice and dry bean grown on a Brazilian Oxisol. Similarly, Figures 2.12 and 2.13 show P-deficiency symptoms in upland rice and dry bean, respectively. Figures 2.14 and 2.15 show K- and Zn-deficiency symptoms in upland rice. Visual symptoms are the cheapest nutritional disorders diagnostic technique compared to other three methods. However, it needs a lot of experience on the part of the observer, because deficiency symptoms are confused with drought, insects and disease infestation, herbicide damage, soil salinity, and inadequate drainage problems. Sometimes, a plant may be on borderline with respect to deficiency and adequacy of a given nutrient. In this situation there are no visual symptoms, but the plant is not producing at its capacity. This condition is frequently called *hidden hunger.*

Deficiency symptoms normally occur over an area and not on an individual plant. If a symptom is found on a single plant, it may be due to disease or insect injury or a genetic variation. Also, the earlier symptoms are often more useful than late, mature symptoms. Some nutrients are relatively immobile in the plant, while others are more mobile. In general, deficiency symptoms caused by immobile nutrients first appear on the younger or upper leaves. The older leaves do not show any symptoms because immobile nutrients do not move or translocate from older to newer leaves. Immobile nutrients are calcium, zinc, boron, copper, iron, manganese, and molybdenum. In contrast, when there is a deficiency of a mobile nutrient, the symptoms first

TABLE 2.2
Soil Conditions Inducing Nutrient Deficiencies for Crop Plants

Nutrient	Conditions Inducing Deficiency
Nitrogen	Excess leaching with heavy rainfall, low organic matter content of soils, burning the crop residue, and low fertilizer application rate in the crop plants
Phosphorus	Acidic, organic, leached, and calcareous soils; high rate of liming and antagonistic interactions with other nutrients
Potassium	Sandy, organic, leached, and eroded soils; high liming application, intensive cropping system
Calcium	Acidic, alkali, or sodic soils
Magnesium	Similar to calcium
Sulfur	Low organic matter content of soils; use of N and P fertilizers containing no sulfur, burning the crop residue
Iron	Calcareous soils; soils high in P, Mn, Cu, or Zn; high rate of liming
Zinc	Highly leached acidic soils, calcareous soils; high levels of Ca, Mg, and P in the soils
Manganese	Calcareous silt and clay, high organic matter, calcareous soils and high soil pH as a result of liming acid soils
Copper	Sandy soils; high liming rate in acid soils
Boron	Sandy soils, naturally acidic leached soils, alkaline soils with free lime
Molybdenum	Highly podzolized soils; well-drained calcareous soils
Nickel	Similar to molybdenum

Sources: Fageria, N.K. and Gheyi, H.R., *Efficient Crop Production*, Federal University of Paraiba, Campina Grande, Brazil, 1999; Marschner, H., *Mineral Nutrition of Higher Plants*, 2nd edn., Academic Press, New York, 1995; Mengel, K. et al., *Principles of Plant Nutrition*, 5th edn., Kluwer Academic Publishers, Dordrecht, the Netherlands, 2001; Brady, N.C. and Weil, R.R., *The Nature and Properties of Soils*, 13th edn., Prentice Hall, Upper Saddle River, NJ, 2002; Fageria, N.K. and Baligar, V.C., Nutrient availability, in *Encyclopedia of Soils in the Environment*, ed., Hillel, D., pp. 63–71, Elsevier, San Diego, CA, 2005a; Fageria, N.K. et al., *Physiology of Crop Production*, The Haworth Press, New York, 2006; Fageria, N.K., *The Use of Nutrients in Crop Plants*, CRC Press, Boca Raton, FL, 2009; Fageria, N.K. et al., *Growth and Mineral Nutrition of Field Crops*, 3rd edn., CRC Press, Boca Raton, FL, 2011a.

appear on the older leaves of the plant. This is because the mobile nutrients move out of the older leaves to the younger part of the plant. Mobile nutrients are nitrogen, phosphorus, potassium, and magnesium. Sulfur may behave as mobile or immobile nutrient. However, in rice plants, sulfur deficiency first appears in younger leaves.

In conclusion, the use of visible symptoms has the advantage of direct field application without the need of costly equipment or laboratory support services, as is the case with soil and plant analysis. A disadvantage is that sometimes it is too late to correct the deficiency of a given nutrient because the disorder is identified when it is too severe to produce visible symptoms. For some disorders, considerable yield loss may have already occurred by the time visible symptoms appear. Further, several publications are available in which nutritional disorders have been described and illustrated with color photographs for important field crops (Wallace, 1961; Ishizuka, 1978; Yoshida, 1981; Grundon, 1987; Bennett, 1993; Fageria et al., 2011a).

TABLE 2.3

General Description of Nutrient-Deficiency Symptoms in Field Crops

Nutrient	Symptoms
N	Chlorosis starts in old leaves; in cereals tillering is reduced; under field conditions, if deficiency is severe, whole crop appears yellowish and growth is stunted
P	Growth is stunted; purple orange color of older leaves; new leaves dark green; in cereals tillering is drastically reduced
K	Older leaves may show spots or marginal burn starting from tips; increased susceptibility to diseases, drought, and cold injury
Ca	New leaves become white; growing points die and curl
Mg	Marginal or interveinal chlorosis with pinkish color of older leaves; sometimes leaf rolling like drought effect; plants susceptible to winter injury
S	Chlorosis of younger leaves; under severe deficiency whole plant becomes chlorotic and similar to appearance in N deficiency
Zn	Rusting in strip of older leaves with chlorosis in fully matured leaves; leaf size is reduced
Fe	Interveinal chlorosis of younger leaves; under severe deficiency whole leaf becomes first yellow and finally white
Mn	Similar to iron deficiency; at advanced stage necrosis develops instead of white color
Cu	Chlorosis of young leaves, rolling, and dieback
Mo	Mottled pale appearance in young leaves; bleaching and withering of leaves
B	Pale green tips of blades, bronze tints; death of growing points
Ni	Nickel-deficient plants show burning of leaf tips

Sources: Fageria, N.K. and Gheyi, H.R., *Efficient Crop Production*, Federal University of Paraiba, Campina Grande, Brazil, 1999; Marschner, H., *Mineral Nutrition of Higher Plants*, 2nd edn., Academic Press, New York, 1995; Mengel, K. et al., *Principles of Plant Nutrition*, 5th edn., Kluwer Academic Publishers, Dordrecht, the Netherlands, 2001; Brady, N.C. and Weil, R.R., *The Nature and Properties of Soils*, 13th edn., Prentice Hall, Upper Saddle River, NJ, 2002; Fageria, N.K. and Baligar, V.C., Nutrient availability, in *Encyclopedia of Soils in the Environment*, ed., Hillel, D., pp. 63–71, Elsevier, San Diego, CA, 2005a; Fageria, N.K. et al., *Physiology of Crop Production*, The Haworth Press, New York, 2006; Fageria, N.K., *The Use of Nutrients in Crop Plants*, CRC Press, Boca Raton, FL, 2009; Fageria, N.K. et al., *Growth and Mineral Nutrition of Field Crops*, 3rd edn., CRC Press, Boca Raton, FL, 2011a.

Readers may refer to these publications to get acquainted with nutrient deficiencies/ toxicities symptoms in important field crops.

2.3.2 SOIL TESTING

In a broad sense, soil testing is any chemical or physical measurement that is made on a soil. The main objective of soil testing is to measure soil nutrient status and lime requirements in order to make fertilizer and lime recommendations for profitable farming. Soil testing is an important tool in high-yield farming but produces best

FIGURE 2.10   Upland rice plants with N deficiency (−N) and without N deficiency (+N).

FIGURE 2.11   Dry bean plants with N deficiency (left) and without N deficiency (right).

results only when used in conjugation with other good farming practices. "There is good evidence that the competent use of soil tests can make a valuable contribution to the more intelligent management of the soil," according to the USA National Soil Test Workgroup in its 1951 report, and is still applicable today. Soil testing involves collecting soil samples, preparation for analysis, chemical or physical analysis, interpretation of analysis results, and finally making fertilizers and lime recommendations for the crops to be grown. A detailed description of these soil-testing components can be found in several books or articles.

FIGURE 2.12   Upland rice plants with P deficiency (left) and without P deficiency (right).

FIGURE 2.13   Dry bean plants with P deficiency (left) and without P deficiency (right).

The use of soil analysis as a fertilizer recommendation method is based on the existence of a functional relationship between the amount of nutrient extracted from the soil by chemical methods and crop yield. When a soil analysis test shows a low level of a particular nutrient in a given soil, the supply of that nutrient is expected to increase crop yield. Generally, nutrient analysis is arbitrarily classified as very low, low, adequate, high, and excess. Under very low nutrient levels, relative crop yield is expected to be less than 70% and a larger application of fertilizer for soil-building purposes is required. After the application of the nutrient, growth response is expected to be

FIGURE 2.14   Upland rice plants with K-deficiency symptoms.

FIGURE 2.15   Upland rice plants with Zn-deficiency symptoms.

dramatic and profitable. Under the low fertility level, relative yield is expected to be
70%–90%. Under this situation, annual application of fertilizer is necessary to produce maximum response and increase soil fertility. The increased yield justifies the
cost of fertilization. When soil analysis test shows adequate level, relative crop yield
under this situation is expected to be 90%–100%. Normal annual applications to produce maximum yields are recommended. In this case, more fertilizer may increase

yields slightly but the added yield would not pay back the expense of the additional fertilizer. Under high level of nutrient, there is no increase in yield. Under this situation, small application is recommended to maintain soil level. The amount suggested may be doubled and applied in alternate years. When soil test shows very high or excess of a nutrient, yield may be reduced due to toxicity or imbalances of nutrients. Under this situation, there is no need to apply nutrient until level drops back to low range. To get such nutrient level and yield relationship, it is necessary to conduct fertilizer yield trials in several locations in a given agroecological region for different crops. Some specific recommendations for soil analysis are summarized as follows:

1. Soil samples must be representative of the land area in question. It is recommended to take a minimum of one composite sample per 12–15 ha for lime and fertilizer recommendations. A representative soil sample must comprise 15–20 subsamples from a uniform field with no major variation in slope, drainage, or past fertilizer history. Any of these listed factors, if changed, will have an effect on the number of samples and unit area from which the sample is obtained.
2. Depth of sampling for mobile nutrients like nitrogen should be of 60 cm and for immobile nutrients like P, K, Ca, and Mg, 15–20 cm sampling depth can give satisfactory results. For pasture crops, sampling depth of 10 cm is normally sufficient to evaluate nutrient status and making liming and fertilizer recommendations.
3. Selecting appropriate extractor. Three extracting solutions, 0.05N HCl + 0.025N H_2SO_4 (Mehlich 1), 0.03N NH_4F + 0.025N HCl (Bray-P1), and 0.5N $NaHCO_3$ at pH 8.5 (Olsen), are the most commonly used extractants for P at the present time and are generally adequate to cover the broad range of soils. Commonly used extractants for K, Ca, and Mg are double acid (Mehlich 1), 1M NH_4Ac at pH 7, and NaOAc at pH 4.8. Multielement extracting reagents are replacing the more familiar single-element extractants. After mixing with an appropriate aliquot of soil, the obtained extract is assayed by an inductively coupled plasma emission spectrometer. A flow injection analyzer is another multielement analyzer capable of assaying these soil extracts.
4. Optimum soil test values for macro- and micronutrients vary from soil to soil, crop to crop, and extract to extract. But normally about >10 mg P kg^{-1}, >50 mg K kg^{-1}, >600 mg Ca kg^{-1}, >120 mg Mg kg^{-1}, and >12 mg S kg^{-1} can produce satisfactory results for most soils and crops. For micronutrients, the critical values reported are Fe 2.5–5 mg kg^{-1}, for Mn 4–8 mg kg^{-1}, for Zn 0.8–3 mg kg^{-1}, for B 0.1–2 mg kg^{-1}, for Cu 0.5–2 mg kg^{-1}, and for Mo 0.2–0.5 mg kg^{-1}, respectively (Fageria, 1992).
5. The pH of agricultural soils is in the range of 4–9. It is difficult to define optimum pH values of different plant species. Most food crops grow well in acid soils if pH is around 6. Lime is called the foundation of crop production or workhouse in acid soils.
6. Fertilizer field trials are the important part of the overall diagnostic process of nutrient disorder, because the only way to establish the fertilizer requirement accurately is by field experiments under which the full interplay of soil, plant, climate, and management factors can occur.

2.3.3 Plant Analysis

Plant analysis is based on the concept that the concentration of an essential element in a plant or part of the plant indicates the soil's ability to supply that nutrient. This means that it is directly related to the quantity in the soil that is available to the plant. For annual crops, the primary objective of plant analysis is to identify nutritional problems or to determine or monitor the nutrient status during the growing season. If deficiency is identified early in the growth stage of a crop, it is possible to correct it during the current season. Otherwise, steps should be taken to correct it in the succeeding crop. Plant analysis can be useful for the prediction of nutrient needs in perennial crops, usually for the year following the time of sampling and analysis. Like soil analysis, plant analysis also involves plant sampling, plant tissue preparation, analysis, and the interpretation of analytical results. All these steps are important for a meaningful plant analysis program.

Many factors such as soil, climate, plant, and their interaction affect the absorption of nutrients by growing plants. However, the concentrations of the essential nutrients are maintained within rather narrow limits in plant tissues. Such consistency is thought to arise from the operation of delicate feedback systems that enable plants to respond in a homeostatic fashion to environmental fluctuations. Fageria et al. (2011a) presented average adequate concentrations for essential nutrients in important field crops. No doubt these values vary with soil, climate, crops, and management practice, and it is very difficult or not logical to make generalizations. Generalized values (Fageria et al., 2011a) do give the reader some idea about what adequate levels of nutrients are in crop plants.

For the interpretation of plant analysis results, a critical nutrient concentration concept was developed. This concept is widely used now in the interpretation of plant analysis results for nutritional disorders diagnostic purposes. Critical nutrient concentration is usually designated as a single point within the bend of the curve when crop yield is plotted against nutrient concentration, where the plant nutrient status shifts from deficient to adequate. The critical nutrient concentration has been defined in several ways: (1) the concentration that is just deficient for maximum growth; (2) the point where growth is 10% less than the maximum; (3) the concentration where plant growth begins to decrease; and (4) the lowest amount of element in the plant accompanying the highest yield. The aforementioned definitions are similar, but not identical. It is well known that there is a good relationship between the uptake of a nutrient and the yield of a given crop. This is illustrated in Figures 2.16 and 2.17, which show a highly significant correlation ($R^2 = 0.49**$ and $R^2 = 0.95**$) between N uptake (concentration × dry weight) in shoot and grain of lowland rice plants (*O. sativa* L.) grown on an Inceptisol in the central part of Brazil. The N uptake in shoot had a quadratic relationship with grain yield in the range of 14–90 kg ha^{-1} (Figure 2.16), whereas, N uptake in grain had a linear association in the range of 20–75 kg N ha^{-1} (Figure 2.17). These results suggest that higher grain yield in rice is associated with higher N uptake in the grain. Fageria and Baligar (2001) reported significant increase in the grain yield of lowland rice with N accumulation in the grain. In addition to critical nutrient concentration approach, tissue nutrient status can also be determined by the diagnosis and recommendation integrated system and composition nutrient diagnosis.

Nutrient concentrations in plants are generally expressed in mg kg^{-1} or g kg^{-1}. Sometimes, there is confusion between concentration and accumulation.

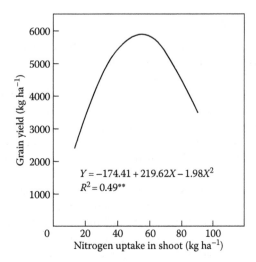

FIGURE 2.16   Relationship between N uptake in shoot and grain yield of lowland rice geno-types. **Significant at the 1% probability level. (From Fageria, N.K. et al., *J. Plant Nutr.*, 32, 1965, 2009.)

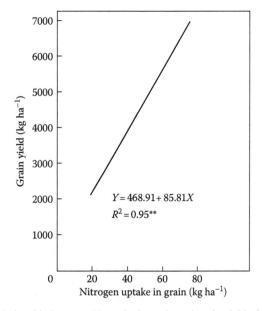

FIGURE 2.17   Relationship between N uptake in grain and grain yield of lowland rice geno-types. **Significant at the 1% probability level. (From Fageria, N.K. et al., *J. Plant Nutr.*, 32, 1965, 2009.)

Concentration is content per unit dry weight of plant tissue, whereas accumulation is concentration multiplied by dry matter. Nutrient accumulation is generally expressed in kg ha^{-1}. Data related to the quantity of nutrients accumulated during a cropping season would give an idea of fertility depletion and help in managing soil fertility for the succeeding crops. The best time to determine nutrient accumulation is at flowering or at harvest when accumulation is expected to be maximum. Grains as well as straw should be analyzed and their weights per unit area should be determined to calculate total accumulation (Fageria et al., 2008, 2011a).

2.3.4 CROP GROWTH RESPONSE

Soil and plant analyses are the common practices for identifying nutritional deficiencies in crop production. The best criterion for diagnosing nutritional deficiencies in annual crops, however, is through evaluating crop responses to applied nutrients. If a given crop responds to an applied nutrient in a given soil, this means that the nutrient is deficient for that crop. Figure 2.18 shows the response of lowland rice to N fertilization in a Brazilian Inceptisol. Similarly, Figure 2.19 shows the

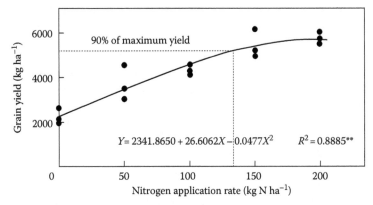

FIGURE 2.18 Grain yield of lowland rice as influenced by N application rate. Values are averages of 12 genotypes and 2 year field trial. **Significant at the 1% probability level. (From Fageria, N.K. et al., *J. Plant Nutr.*, 31, 1121, 2008.)

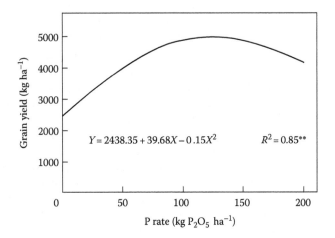

FIGURE 2.19 Grain yield of upland rice as influenced by P fertilization. **Significant at the 1% probability level.

response of upland rice to P fertilization in a Brazilian Oxisol. Based on these two figures, it can be concluded that Brazilian Inceptisol was deficient in N and Oxisol was deficient in P. The relative decrease in yield in the absence of a nutrient, as compared to an adequate soil fertility level, can give an idea of the magnitude of nutrient deficiency. For example, research conducted on an Oxisol of central Brazil provided evidence that P deficiency was the primary yield-limiting factor for annual production of crops such as upland rice, common bean, corn, soybean, and wheat (Fageria and Baligar, 1997a).

2.4   UPTAKE OF NUTRIENTS BY CROP PLANTS

Knowledge of the uptake of nutrients by crop plants is an important aspect in maintaining soil fertility for sustainable crop production. The amount of nutrients removed by a crop species or cultivar of the same species should be replenished to sustain crop productivity over a longer duration. The removal of nutrients in harvested plant biomass has been used to estimate nutrient requirements of annual crops (Fageria, 2009). Nutrient requirements of crop plants varied with soil type, crop species, and yield level. To determine the nutrient uptake of a crop species or cultivar, weight of straw and grain should be measured at harvest and chemical analysis of plant materials (grain and straw separately) should be done. Results of straw and grain yield of grain crops are generally expressed in kg ha^{-1} or Mg ha^{-1}. Nutrient uptake results are expressed in kg ha^{-1} for macronutrients and g ha ha^{-1} for micronutrients for field experiments. When nutrient uptake is determined for a particular crop species, its yield should be above average. Lower yield does not represent actual value of nutrient uptake or requirement.

As yields increase, so do nutrient exports from soil. In the United States, for example, the harvest of major crops annually removes about 7.8 Tg of N (Tg = Teragram = 10^{12} g) (excluding N_2-fixing crops like alfalfa, soybean, and peanut), 2.3 Tg of P, and 6.7 Tg of K, with removals increasing by roughly 1% per year (International Plant Nutrition Institute, 2010). Results related to straw and grain yield and nutrient uptake by important food crops are presented in Tables 2.4 through 2.6. In lowland rice, the uptake of macronutrients was in the order of K > N > Ca > P > Mg, and the uptake of micronutrients was in the order of Mn > Fe > Zn > Cu > B in the straw and grain. In upland rice, the nutrient uptake pattern was in the order of K > N > Ca > Mg > P > Mn > Fe > Zn > Cu > B. In dry bean, the nutrient uptake pattern was in the order of N > K > Ca > P > Mg > Fe > Zn > Mn > Cu. In corn, the nutrient uptake pattern was in the order of N > K > Ca > Mg > P > Fe > Mn > Zn > B > Cu. In soybean, the nutrient uptake pattern was in the order of N > K > Ca > Mg > P > Fe > Mn > Zn > Cu (Table 2.6). From these results it can be concluded that in cereals as well as in legumes, N and K uptakes were maximum and Cu and B uptakes were minimum. To produce 1 MT (metricton) of grain in lowland as well as upland rice, K requirement was maximum followed by N (Table 2.7). In legumes, maximum requirement to produce 1 MT of grain was N followed by K. Overall, phosphorus had the maximum use efficiency in the cereals as well as legume plants among macronutrients. In micronutrients, the maximum use efficiency for grain production was of B in rice and Cu in corn, dry bean, and soybean.

TABLE 2.4
Straw and Grain Yield and Nutrients Uptake by Lowland and Upland Rice Grown on Brazilian Inceptisol and Oxisol

Yield/Nutrient Uptake	Lowland Rice		Upland Rice	
	Straw	Grain	Straw	Grain
Yield (kg ha^{-1})	9423	6389	6343	4568
N (kg ha^{-1})	65	86	56	70
P (kg ha^{-1})	15	15	3	10
K (kg ha^{-1})	156	20	150	56
Ca (kg ha^{-1})	26	5	23	4
Mg (kg ha^{-1})	15	7	12.5	5
Zn (g ha^{-1})	546	224	161	138
Cu (g ha^{-1})	77	102	35	57
Mn (g ha^{-1})	4724	369	1319	284
Fe (g ha^{-1})	2553	505	654	117
B (g ha^{-1})	69	33	53	30

Source: Fageria, N.K. et al., *Growth and Mineral Nutrition of Field Crops*, 3rd edn., CRC Press, Boca Raton, FL, 2011a.

TABLE 2.5
Straw and Grain Yield and Nutrients Uptake by Dry Bean and Corn Grown on a Brazilian Oxisol

Yield/Nutrient Uptake	Dry Bean		Corn	
	Straw	Grains	Straw	Grain
Yield (kg ha^{-1})	1893	3.56	11873	8501
N (kg ha^{-1})	13.47	119.35	72	127
P (kg ha^{-1})	1.69	12.32	5	17
K (kg ha^{-1})	35.17	61.47	153	34
Ca (kg ha^{-1})	16.57	8.02	33	8
Mg (kg ha^{-1})	7.17	6.09	21	9
Zn (g ha^{-1})	49.09	122.49	184	192
Cu (g ha^{-1})	8.4	37.91	53	12
Mn (g ha^{-1})	25.7	45.55	452	82
Fe (g ha^{-1})	896.79	396.77	2048	206
B (g ha^{-1})			103	43

Source: Fageria, N.K. et al., *Growth and Mineral Nutrition of Field Crops*, 3rd edn., CRC Press, Boca Raton, FL, 2011a.

TABLE 2.6

Yield of Straw and Grain and Nutrients Uptake by Soybean Grown on a Brazilian Oxisol

Yield/Nutrient Uptake	Straw	Grain
Yield (kg ha^{-1})	3518	4003
N (kg ha^{-1})	38	280
P (kg ha^{-1})	2	12
K (kg ha^{-1})	58	78
Ca (kg ha^{-1})	31	13
Mg (kg ha^{-1})	20	10
Zn (g ha^{-1})	29	169
Cu (g ha^{-1})	33	60
Mn (g ha^{-1})	117	120
Fe (g ha^{-1})	187	373

Source: Fageria, N.K. et al., *Growth and Mineral Nutrition of Field Crops*, 3rd edn., CRC Press, Boca Raton, FL, 2011a.

TABLE 2.7

Macro- and Micronutrients Requirement to Produce 1 MT Grain of Important Food Crops

Nutrient	Lowland Rice	Upland Rice	Dry Bean	Corn	Soybean
N (kg)	24	28	37	23	79
P (kg)	5	3	5	3	4
K (kg)	28	45	27	22	34
Ca (kg)	5	6	7	5	11
Mg (kg)	3	4	4	4	8
Zn (g)	121	65	48	44	50
Cu (g)	28	20	13	8	23
Mn (g)	797	351	20	63	59
Fe (g)	479	169	365	265	140
B (g)	16	18	18	17	30

Source: Fageria, N.K. et al., *Growth and Mineral Nutrition of Field Crops*, 3rd edn., CRC Press, Boca Raton, FL, 2011a.

Determination of nutrient uptake in crop plants during the growth cycle is also important to know when the crop requires a maximum amount of nutrients. To achieve this objective, plant samples should be taken at different growth stages during the crop growth cycle. Results related to dry matter yield and nitrogen and phosphorus uptake during the growth cycle of corn, upland rice, soybean, and dry bean are presented in Figures 2.20 through 2.22. The dry weight of the shoot of rice, dry bean, corn, and soybean was significantly ($P < 0.01$) influenced with the advancement of the age of the crop plants. Dry matter production increased up to 102 days in rice and then decreased, up to 84 days in dry bean and then decreased, up to 84 days in corn and then decreased, and up to 120 days in soybean and then decreased. The decrease in dry weight at harvest of four crop species was related to the translocation of photosynthetic material to grain during the interval from flowering to harvest (Fageria and Baligar, 2005b). In dry bean and soybean, decrease in dry matter was also partially related to falling off the leaves. Such decreases in dry matter have been reported by Fageria (2009). The uptake of N and P in the shoot (without grain) of four crop species followed the dry matter production pattern as expected.

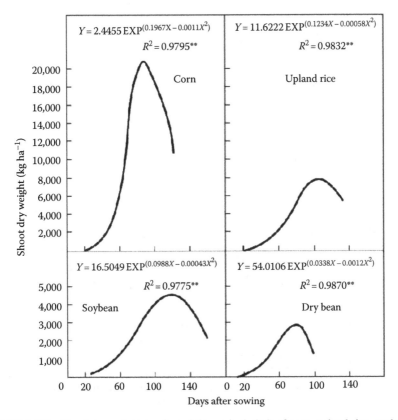

FIGURE 2.20 Shoot dry weight (grain weight not included) of corn, upland rice, soybean, and dry bean as a function of plant age. **Significant at the 1% probability level. (Adapted from Fageria, N.K., *Commun. Soil Sci. Plant Anal.*, 35, 961, 2004.)

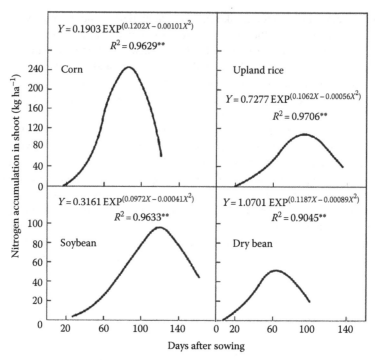

FIGURE 2.21 Nitrogen uptake in corn, upland rice, soybean, and dry bean as a function of plant age. **Significant at the 1% probability level. (Adapted from Fageria, N.K., *Commun. Soil Sci. Plant Anal.*, 35, 961, 2004.)

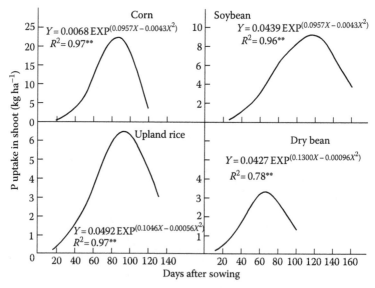

FIGURE 2.22 Phosphorus uptake in corn, upland rice, soybean, and dry bean as a function of plant age. **Significant at the 1% probability level. (Adapted from Fageria, N.K., *Commun. Soil Sci. Plant Anal.*, 35, 961, 2004.)

2.5   NUTRIENT INTERACTION

Nutrient interaction in crop plants is probably one of the most important factors affecting yields of annual crops. Knowledge of interactions among essential plant nutrients is important in formulating a balanced supply of fertilizers to crop plants. Nutrient interaction may be positive, negative, and no interaction. It can be evaluated by determining yield and yield components under different levels of nutrients as well as by the uptake of nutrients by crop plants under variable rates of fertilizers. Interactions occur when the supply of one nutrient affects the absorption and utilization of another nutrient (Robson and Pitman, 1983; Wilkinson et al., 2000). Nutrient interactions affect plant growth and development only when supply of a determined nutrient is too low compared to applied ones. In other words, yield decrease will only occur when supply of some nutrients falls below the critical level. If the soil or growth medium has sufficient supply of other essential nutrients compared to add one, plant growth will not be affected adversely, even though the uptake of some nutrients may decrease. Hence, plant growth or yield is considered a better criterion for evaluating nutrient interactions in crop plants.

Soil, plant, and climatic factors can influence interaction. In nutrient interaction studies, all other factors should be at an optimum level except the variation in the level of nutrient under investigation. Nutrient interaction can occur at the root surface or within the plant. Interactions at the root surface are due to the formation of chemical bonds by ions and precipitation or complexes. One example of this type of interaction is the liming of acid soils, which decreases the concentration of almost all the micronutrients except molybdenum (Fageria et al., 2006). The second type of interaction is between ions whose chemical properties are sufficiently similar such that they compete for site of absorption, transport, and function on plant root surface or within plant tissues. Such interactions are more common between nutrients of similar size, charge, and geometry of coordination and electronic configuration (Robson and Pitman, 1983). The interaction varies from nutrient to nutrient and crop species to species and sometimes among cultivars of same species.

The author studied the interaction among N, P, and K, using upland rice as a test crop. There was significant $N \times P \times K$ interaction for plant height, shoot dry weight, grain yield, and grain harvest index (GHI) (Table 2.8). Hence, the response of these plant characteristics is associated with an adequate rate of N, P, and K fertilization. Plant height varied from 24.33 cm in the $N_0P_0K_0$ treatment to 111.58 cm in the $N_1P_1K_2$ (N = 150 mg kg^{-1}, P = 100 mg kg^{-1}, and K = 200 mg kg^{-1}) treatment, with an average value of 74.92 cm. Shoot dry weight varied from 0.30 g plant^{-1} in the $N_0P_0K_0$ treatment to 28.66 g plant^{-1} in the $N_2P_2K_2$ (N = 300 mg kg^{-1}, P = 200 mg kg^{-1}, and K = 200 mg kg^{-1}) treatment, with an average value of 10.65 g plant^{-1}. Fageria and Baligar (1997a) reported significant increase in rice plant height and shoot dry weight with the addition of N, P, and K fertilization in a Brazilian Oxisol.

Grain yield varied from 0 to 23.01 g plant^{-1} in the $N_0P_0K_0$ and $N_2P_2K_2$ treatments, respectively, with an average value of 6.78 g plant^{-1}. Plants that did not receive P fertilization but received adequate rate of N and K did not produce panicle or grain. Hence, it can be concluded that P is the most yield-limiting nutrient in highly weathered Brazilian Oxisol. Fageria and Baligar (1997b) and Fageria (2009) have reported

TABLE 2.8

Influence of N, P, and K Treatments on Plant Height, Shoot Dry Weight, Grain Yield, and Grain Harvest Index of Upland Rice

N, P, and K Treatments	Plant Height (cm)	Shoot Dry Weight (g Plants⁻¹)	Grain Yield (g Plant⁻¹)	Grain Harvest Index
N0P0K0	24.33i	0.30i	0.00j	0.00g
N0P0K1	26.00i	0.40hi	0.00j	0.00g
N0P0K2	46.83f	0.86hi	0.00j	0.00g
N0P1K0	75.00e	3.82gh	2.32ij	0.38abcde
N0P1K1	75.17de	8.55ef	3.05ij	0.26f
N0P1K2	91.92bc	7.30ef	3.75i	0.34bcdef
N0P2K0	78.42de	8.56ef	3.20ij	0.27f
N0P2K1	79.92de	8.17ef	3.34ij	0.29ef
N0P2K2	83.33cde	8.86e	4.05hi	0.31def
N1P0K0	39.21fgh	0.52hi	0.00j	0.00g
N1P0K1	43.33fg	0.49hi	0.00j	0.00g
N1P0K2	35.08ghi	0.54hi	0.00j	0.00g
N1P1K0	85.25cde	12.75d	5.93ghi	0.32cdef
N1P1K1	106.83a	16.32c	10.71ef	0.39abcde
N1P1K2	111.58a	17.69c	12.64de	0.42abc
N1P2K0	93.67bc	15.35cd	10.21ef	0.40abcd
N1P2K1	111.08a	18.66c	12.35cd	0.43ab
N1P2K2	109.08a	22.19b	16.64bc	0.43ab
N2P0K0	27.75hi	0.35hi	0.00j	0.00g
N2P0K1	30.75hi	0.44hi	0.00j	0.00g
N2P0K2	35.83fghi	5.24fg	0.00j	0.00g
N2P1K0	86.67cd	15.29cd	7.56fgh	0.33bcdef
N2P1K1	110.67a	22.38b	16.20bcd	0.42abc
N2P1K2	110.58a	22.89b	18.63b	0.45a
N2P2K0	92.75bc	16.73c	8.82fg	0.35abcdef
N2P2K1	103.33ab	24.28b	18.75b	0.44ab
N2P2K2	108.42a	28.66a	23.01a	0.45a
Average	74.92	10.65	6.78	0.25
F-test				
N	**	**	**	**
P	**	**	**	**
K	**	**	**	**
N × P	**	**	**	**
N × K	**	**	**	**
P × K	**	**	**	**
N × P × K	**	**	**	**
CV (%)	4.69	9.98	16.31	9.66

Source: Fageria, N.K. and Oliveira, J.P., in press.

**Means within same column followed by the same letter do not differ significantly at the 5% probability level by Tukey's test.

similar results. Grain yield results also showed that there is strong positive interaction among N, P, and K fertilization in upland production. This type of interaction is widely reported in the literature (Wilkinson et al., 2000). These authors also reported that increasing N rate increases the demand for other nutrients, especially P and K, and higher yields were obtained at the highest rates of N, P, and K. Wilson (1993) also confirmed the generalization that the response to one nutrient depends on the sufficiency level of other nutrients. Yield reductions were found when high levels of one nutrient were combined with low levels of the other nutrients (Wilkinson et al., 2000). Alleviating the yield-depressing effect of excessive macronutrient supply involved removing the limitation of a low supply of other nutrients.

The GHI varied from 0 in the treatment that did not receive P to 0.45 in the treatment $N_2P_2K_2$, with an average value of 0.25. GHI is an important index in determining the partitioning of dry matter between shoot and grain. Fageria (2009) reported that rice GHI is influenced by environmental factors, including mineral nutrition. Fageria and Baligar (2005b) reported that variation in rice GHI is from 0.23 to 0.50. However, Kiniry et al. (2001) reported that rice GHI values varied greatly among cultivars, locations, seasons, and ecosystems, and ranged from 0.35 to 0.62. The GHI values of modern crop cultivars are commonly higher than old traditional cultivars for major field crops (Ludlow and Muchow, 1990). The limit to which GHI can be increased is considered to be about 0.60 (Austin et al., 1980). Hence, cultivars with low harvest indices would indicate that further improvement in the partitioning of biomass would be possible (Fageria and Baligar, 2005b). On the other hand, cultivars with harvest indices between 0.50 and 0.60 would probably not benefit by increasing harvest index (Sharma and Smith, 1986).

Zhu et al. (2011) studied variation and interrelations among essential mineral nutrients in forage wheat leaves at Zadoks et al.'s (1974) 30 growth stage (before stem elongation), when plants were at the cattle-grazing stage. Results of this study showed that 28 out of 55 pairs of comparison had correlations at a significant (P, 0.05) or highly significant (P, 0.01) level. Twenty six of 28 significant correlations were positive, and the remaining two significant correlations were negative. These results indicated that most of the minerals play a positive role in regulating other minerals and only P is antagonistic to Fe and Ni in the wheat foliage. Magnesium and Ca showed a coexisting relationship with very high $r = 0.93^{**}$ (**Significant at the 1% probability level) correlation. Among other major minerals, K was highly significantly correlated with P ($r = 0.89^{**}$), and S had a correlation coefficient higher than 0.60 with Ca, Mg, and Na. These authors reported that Cu had no significant correlation with any mineral except Zn, suggesting that Cu might play a relatively independent role from that of other minerals. In contrary, Zn had significant correlations with all of the tested minerals except Cu, Mn, and Ni, suggesting Zn might play a universal role in biological metabolisms in wheat.

2.6   NUTRIENT USE EFFICIENCY

Efficiency is defined as output divided by input for an activity or process. The higher the output of an activity, the higher is the efficiency of the system. In case of nutrient use efficiency (NUE), output is the economic yield of crop plants (grain yield in case of food crops) and input is the quantity of nutrient applied. NUE varies with crop species or genotypes within species, climatic conditions, soil type, biological activities in the growth medium, and yield level (Fageria et al., 2008, 2011a). According to Fageria and Baligar (2005b), NUE can be defined as the maximum economic yield produced per unit of nutrient applied, absorbed, or utilized by the plant to produce grain and straw. However, NUE has been defined in several ways in the literature, although most of them denote the ability of a system to convert inputs into outputs. Definitions of NUE have been grouped or classified as agronomic efficiency, physiological efficiency, agrophysiological efficiency, apparent recovery efficiency, and utilization efficiency (Fageria and Baligar, 2001, 2003, 2005b). The determination of NUE in crop plants is an important approach to evaluate the fate of applied chemical fertilizers and their role in improving crop yields. The N use efficiencies are calculated by using the following formulas (Fageria and Baligar, 2003; Fageria, 2009; Fageria et al., 2011a):

$$\text{Agronomic efficiency } (AE)(\text{kg kg}^{-1}) = \frac{GY_f - GY_{uf}}{N_a}$$

where
  GY_f is the grain yield of the fertilized plot (kg)
  GY_{uf} is the grain yield in the unfertilized plot (kg)
  N_a is the quantity of nutrient applied (kg)

$$\text{Physiological efficiency } (PE)(\text{kg kg}^{-1}) = \frac{BY_f - BY_{uf}}{N_{af} - N_{auf}}$$

where
  BY_f is the total biological yield (grain plus straw) of the fertilized plot (kg)
  BY_{uf} is the total biological yield in the unfertilized plot (kg)
  N_{af} is the nutrient accumulation in the fertilized plot in grain and straw (kg)
  N_{auf} is the nutrient accumulation in the unfertilized plot in grain and straw (kg)

$$\text{Agrophysiological efficiency } (APE)(\text{kg kg}^{-1}) = \frac{GY_f - GY_{uf}}{N_{af} - N_{auf}}$$

where
  GY_f is the grain yield in the fertilized plot (kg)
  G_{uf} is the grain yield in the unfertilized plot (kg)
  N_{af} is the nutrient accumulation by straw and grain in the fertilized plot (kg)
  N_{auf} is the nutrient accumulation by straw and grains in the unfertilized plot (kg)

$$\text{Apparent recovery efficiency } (ARE)(\%) = \frac{N_{af} - N_{auf}}{N_a} \times 100$$

where
  N_{af} is the nutrient accumulation by the total biological yield (straw plus grain) in the fertilized plot (kg)
  N_{auf} is the nutrient accumulation by the total biological yield (straw plus grain) in the unfertilized plot (kg)
  N_a is the quantity of nutrient applied (kg)

$$\text{Utilization efficiency } (UE)(\text{kg kg}^{-1}) = PE \times ARE$$

TABLE 2.9

Nitrogen Use Efficiencies in Lowland Rice as Affected by N Fertilizer

N Rate (kg ha^{-1})	AE (kg kg^{-1})	PE (kg kg^{-1})	APE (kg kg^{-1})	ARE (%)	UE (kg kg^{-1})
30	35	156	72	49	76
60	32	166	73	50	83
90	22	182	75	37	67
120	22	132	66	38	50
150	18	146	57	34	50
180	16	126	51	33	42
210	13	113	46	32	36
Average	23	146	63	39	58
R^2	0.93**	0.62**	0.87**	0.82**	0.90**

Source: Fageria, N.K. and Baligar, V.C., *Commun. Soil Sci. Plant Anal.*, 32, 1405, 2001.

AE, agronomic efficiency; PE, physiological efficiency; APE, agrophysiological efficiency; ARE, apparent recovery efficiency; and UE, utilization efficiency.

**Significant at the 1% probability level.

The aforementioned five N use efficiencies for lowland rice were calculated, and the values are presented in Table 2.9. On average, all N use efficiencies were higher at lower N rates and decreased at higher N rates. This indicated that rice plants were unable to absorb N when applied in excess because their absorption mechanisms might have been saturated (Fageria and Baligar, 2005b). Under these conditions, the possibility exists for more N being subject to loss by NH_3 volatilization, leaching, and denitrification. Nitrogen recovery efficiency in annual crops averages only about 42% and 29% in developed and developing countries, respectively (Raun and Johnson, 1999). Similarly, Fageria and Baligar (2005b) reported that N recovery efficiency in crop plants is usually less than 50% worldwide. This low nutrient recovery efficiency is associated with the loss of applied nutrients through leaching, volatilization, denitrification, and soil erosion (Fageria and Baligar, 2005b). In addition, the use of inadequate crop management practices and biotic and abiotic stresses are also responsible for low NUE.

Phosphorus use efficiencies of lowland rice using thermophosphate (insoluble in water) as a P source were calculated, and the results are presented in Table 2.10. Agronomic, physiological, agrophysiological, recovery, and utilization efficiencies decreased with increasing P rates. Across P rates, 10.3 kg rice grain yield was produced with the application of 1 kg P. Similarly, 509.2 kg dry matter (straw plus grain) was produced with the accumulation of 1 kg P in the grain plus straw. In the case of agrophysiological efficiency, across P rates, 324.4 kg grain yield was produced with the accumulation of 1 kg P in the grain plus straw. Average recovery

TABLE 2.10

Phosphorus Use Efficiency in Lowland Rice under Different P Rates

P Rate (kg ha^{-1})	AE (kg kg^{-1})	PE (kg kg^{-1})	APE (kg kg^{-1})	RE (%)	UE (kg kg^{-1})
131	15.5	604.4	300.9	6.3	39.6
262	12.7	536.8	269.5	5.1	27.1
393	10.5	521.8	477.4	3.8	19.7
524	6.8	443.6	277.8	3.4	14.8
655	6.2	439.3	296.6	2.7	10.9
Average	10.3	509.2	324.4	4.3	22.4
R^2	0.52**	0.23^{NS}	0.02^{NS}	0.43**	0.50**

Source: Fageria, N.K. and Barbosa Filho, M.P., *Commun. Soil Sci. Plant Anal.*, 38, 1289, 2007.

Values are averaged across 2 years and thermophosphate (yoorin) was P source.

NS, nonsignificant.

**Significant at the 1% probability level.

efficiency was 4.3% and utilization efficiency was 22.4 kg grain yield with the utilization of 1 kg P. The highest efficiency is usually obtained with the first increment of nutrient, with additional increments providing smaller increases (Fageria et al., 2011a). Singh et al. (2000) reported that agrophysiological efficiencies in lowland rice varied from 235 to 316 kg grain per kg P. Similarly, Witt et al. (1999) reported an agrophysiological efficiency value of 385 grain per kg P when all production factors were at normal levels. Sahrawat and Sika (2002) reported an apparent recovery of applied P by rice in an Ultisol in the range of 4.8%–11%. The low recovery efficiency may be associated with a high rate of P fixation in these soils by iron and aluminum oxides (Abekoe and Sahrawat, 2001). Agronomic efficiency of N and P in upland rice and P in lowland rice is given in Figures 2.23 through 2.25, respectively. Nitrogen and P use efficiency significantly varied from genotype to genotype. In upland rice, P use efficiency was much higher compared with N use efficiency.

Phosphorus use efficiency for water-soluble triple phosphorus was also determined for lowland rice genotypes (Table 2.11). Genotypes were significantly different in absorption and utilization of P. However, highly significant differences among genotypes were for agronomic efficiency (grain yield per unit P applied), physiological efficiency (grain + straw yield per unit P uptake in grain plus straw), and agrophysiological efficiency (grain yield per unit P uptake in the grain plus straw). Internal plant use efficiency of P (dry matter plus grain yield per unit uptake of P) is higher than N and K internal use efficiencies (Fageria et al., 2003). However, P recovery efficiency is much lower than N and K recovery efficiency (Fageria et al., 2003). The lower recovery efficiency of P (<20%) may be associated

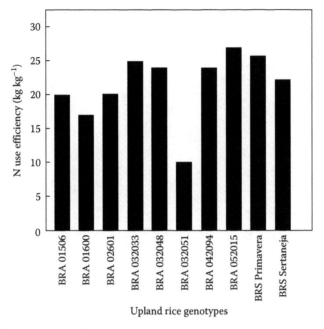

FIGURE 2.23   Agronomic efficiency of N use in upland rice genotypes.

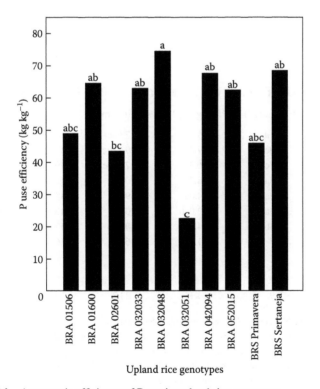

FIGURE 2.24   Agronomic efficiency of P use in upland rice genotypes.

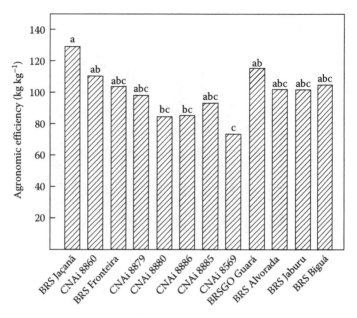

FIGURE 2.25 Agronomic efficiency of P use in lowland rice genotypes. (From Fageria, N.K. et al., *J. Plant Nutr.*, 34, 371, 2011c.)

TABLE 2.11
Phosphorus Use Efficiency of Seven Lowland Rice Genotypes

Genotype	AE (mg mg⁻¹)	PE (mg mg⁻¹)	APE (mg mg⁻¹)	ARE (%)	UE (mg mg⁻¹)
CNAi 8859	31.1d	905.3c	173.3c	19.5a	160.9c
CNAi 8860	53.9c	917.3c	291.7bc	18.6a	170.8bc
CNAi 8870	50.7cd	976.3c	276.3bc	19.1a	184.6ab
CNAi 8879	86.8ab	1488.7a	701.3a	12.8b	187.1ab
CNAi 8880	88.0ab	1474.0a	699.7a	12.9b	187.5ab
CNAi 8886	94.3a	1382.0ab	700.3a	13.5b	186.5ab
CNAi 8885	69.4bc	1124.0bc	401.3b	17.4ab	195.0a
Average	67.7	1181.0	463.4	16.3	181.8
F-test					
Genotype	**	**	**	*	*
CV (%)	17.9	14.1	23.2	16.7	5.5

Source: Fageria, N.K. et al., *J. Plant Nutr.*, 34, 1087, 2011d.

AE, agronomic efficiency; PE, physiological efficiency; APE, agrophysiological efficiency; ARE, apparent recovery efficiency; and UE, utilization efficiency.

*, **Significant at the 5% and 1% probability levels, respectively. Means followed by the same letter in the same column are not significantly different at the 5% probability level by Tukey's test.

TABLE 2.12
Relationship between Phosphorus Use Efficiencies (×) and Grain Yield of Lowland Rice Genotypes

Plant Variable	Regression Equation	R^2
AE vs. grain yield	$Y = 0.00087 + 1.1998X$	1.0
PE vs. grain yield	$Y = -75.1156 + 0.1861X - 0.000043X^2$	0.6595**
APE vs. grain yield	$Y = 1.1368 + 0.2566X - 0.00014X^2$	0.9537**
ARE vs. grain yield	$Y = 188.3038 - 7.7668X + 0.0692X^2$	0.4604**
UE vs. grain yield	$Y = -1615.8230 + 17.6541X - 0.0455X^2$	0.5235**

Source: Fageria, N.K. et al., *J. Plant Nutr.*, 34, 1087, 2011d.

AE, agronomic efficiency; PE, physiological efficiency; APE, agrophysiological efficiency; ARE, apparent recovery efficiency; and UE, utilization efficiency.

**Significant at the 1% probability level.

with its higher fixation by Fe, Al, and Mn oxides (Sanchez, 1976). All the five P use efficiencies had significant positive association with grain yield, except apparent recovery efficiency (Table 2.12). Hence, improving agronomic efficiency, physiological efficiency, agrophysiological efficiency, and utilization efficiency can improve grain yield in lowland rice.

Three P use efficiencies (agronomic efficiency, utilization efficiency, and apparent recovery efficiency) were determined for soybean grown on a Brazilian Oxisol (Table 2.13). Agronomic efficiency (kg grain per kg P applied) decreased with increasing P rate, except Arad source of P (Table 2.13). In Arad also, agronomic efficiency was maximum at lowest P rate. Higher agronomic efficiency at lower P rate indicates better P utilization by soybean at low rate. This type of results is common in nutrient-efficiency studies in crop plants (Fageria, 1992). Jarrell and Beverly (1981) reported that in any experiment with a nutritional variable, plants grown at the lowest nutrient concentrations would inevitably have the highest utilization quotient because of a dilution effect. Overall, P utilization efficiency (kg grain plus straw per kg P accumulated in grain plus straw) was highest in the Yoorin and lowest in Arad phosphate. Higher utilization efficiency in Yoorin may be associated with a higher uptake of P in shoot and grain under this treatment, compared to two other sources of P.

Overall, apparent recovery efficiency of P was similar in superphosphate and Yoorin and was lower in Arad. The lower P recover efficiency in Arad was due to low uptake in straw and grain compared to two other sources of P. Overall, phosphorus recovery efficiency was about 10% in three P sources of fertilization. This lower efficiency was associated with P immobilization by Fe and Al oxides in the Oxisol under investigation. Phosphorus recovery efficiency of less than 20% has been reported in Brazilian Oxisols by annual crops, including soybean. (Fageria et al., 2006)

TABLE 2.13

Phosphorus Use Efficiency in Soybean as Influenced by Phosphorus Source and Rate

P Rate (kg ha⁻¹)	AE (kg kg⁻¹)	UE (kg kg⁻¹)	ARE (%)
Simple Superphosphate			
17.5	25.85	186.3	13.8
35.0	20.27	106.5	11.0
52.5	17.96	78.2	10.5
Average	21.36	123.7	11.77
Yoorin			
17.5	25.21	207.3	15.8
35.0	22.89	106.7	10.5
52.5	11.45	74.0	8.3
Average	19.85	129.3	11.53
Arad			
17.5	13.13	170.4	7.9
35.0	2.5	79.4	2.8
52.5	10.92	62.4	7.1
Average	13.28	104.1	5.9

Source: Fageria, N.K. et al., *Commun. Soil Sci. Plant Anal.*, 42, 2716, 2011b.

Values are averages of first and second year crops.

AE, agronomic efficiency; UE, utilization efficiency; and ARE, apparent recovery efficiency.

Zinc recovery efficiency for upland rice genotypes grown on a Brazilian Oxisol was calculated (Table 2.14). It varied from 8.3% to 23.1%, with an average value of 13% across ten genotypes. Mortvedt (1994) reported that crop recovery of applied micronutrients is relatively low (5%–10%) compared to macronutrients (10%–50%). Such low recovery of applied micronutrients is due to their uneven distribution in a soil because of low application rates, reaction with soil to form unavailable products, and low mobility in soil (Fageria and Baligar, 2005c).

2.7   MECHANISMS INVOLVED IN NUTRIENT UPTAKE AND USE EFFICIENCY

Significant variation exists among crop species and genotypes of the same species in nutrient uptake and utilization (Gerloff and Gabelman, 1983; Baligar et al., 1990, 2001; Epstein and Bloom, 2005; Fageria and Baligar, 2005). For example, the siliceous and calcareous sandy soils of South Australia are considered severely deficient in micronutrients for the growth of wheat, oats (*Avena sativa* L.), or barley

TABLE 2.14

Zinc Recovery Efficiency in Upland Rice Genotypes Grown on a Brazilian Oxisol

Genotype	Zn Recovery Efficiency (%)
Bonança	15.0ab
Caipó	14.6ab
Canastra	12.4ab
Carajas	13.3ab
Carisma	11.8ab
CAN 8540	10.4b
CAA 8557	8.5b
Guarani	23.1a
Maravilla	13.1ab
IR 42	8.3b
Average	13.1

Source: Adapted from Fageria, N.K. and Baligar, V.C., *Pesq. Agropec. Bras.*, 40, 1211, 2005c.

Means followed by the same letter in the same column are not different at the 5% probability level by Tukey's test.

(*Hordeum vulgare* L.), but, the growth and yield of rye (*Secale cereale* L.) was optimal on these soils (Graham, 1984). The native vegetation in this area is fully adapted to these soils mainly due to their slow growth rate (Loneragan, 1978). The difference in nutrient uptake and utilization may be associated with better root geometry; the ability of plants to take up sufficient nutrients from lower or subsoil concentrations; plants' ability to solubilize nutrients in the rhizosphere; better transport, distribution, and utilization within plants; and balanced source–sink relationships (Graham, 1984; Baligar et al., 2001; Fageria and Barbosa Filho, 2001; Fageria and Baligar, 2003). The antagonistic (uptake of one nutrient is restricted by another nutrient) and synergistic (uptake of one nutrient is enhanced by other nutrients) effects of nutrients on NUE among various plant species and cultivars within species have not been well explored.

2.7.1 BETTER ROOT GEOMETRY

Plants having vigorous and extensive root systems can explore large soil volumes and absorb more water and nutrients under nutrient stress conditions and can increase crop yield and NUE (Merrill et al., 2002). The quantity of nutrient taken up by plants is largely influenced by root radius, mean root hair density, and length of root (Barber, 1995). The shape and extent of root systems influence the rate and pattern of nutrient uptake from soil. Vose (1990) states that rooting depth—lateral spreading, branching, and the number of root hairs—has major impact on plant nutrition. The configuration of the root system is influenced markedly by nutrient supply. Mineral excess and deficiency affects growth (dry mass, root:shoot ratio) and morphology (length, thickness, surface area, density) of roots and root hairs. Nutrient deficiency leads to much finer roots. When plants are N deficient, their roots branch more in regions where the soil is locally enriched with N (Scott-Russell, 1977). The configuration (root and root hair abundance and density, distribution, effective radius, and elongation) of root systems in relation to nutrient uptake is extensively covered in an earlier paper by Barley (1970). The amount of C and N supplied by roots can be significant for maintaining or improving soil organic matter and influencing NUE (Sainju et al., 2005). A well-developed root system may play a dominant role in soil C and N cycles (Gale et al., 2000; Puget and Drinkwater, 2001) and may have relatively greater influence on soil organic C and N levels than the aboveground plant biomass (Boone, 1994; Norby and Cotrufo, 1998). Roots can contribute from 400 to 1460 kg C ha^{-1} during a growing season (Qian and Doran, 1996; Kuo et al., 1997). Liang et al. (2002) found that maize roots contributed as much as 12% of soil organic C, 31% of water-soluble C, and 52% of microbial biomass C within a growing season. All of these chemical and biological changes in soils affected by root systems improve NUE in plants.

Cultivar differences in root size are quite common and have been related to differences in nutrient uptake (Caradus, 1990; Baligar et al., 1998; Fageria et al., 2006). Differences between white clover (*Trifolium repens* L.) populations and cultivars in P uptake per plant at low levels of P have been related to differences in root size and absolute growth rate (Caradus and Snaydon, 1986). Data in Table 2.15 show that root dry weight of common bean (*Phaseolus vulgaris* L.) genotypes varied from 1.54 to 3.14 g per three plants, a variation of twofold at 0 mg K kg^{-1} of soil.

TABLE 2.15

Root Dry Weight (RDW) and Maximum Root Length (MRL) of Six Common Bean Genotypes as Influenced by Potassium Levels Applied to a Brazilian Oxisol

	0 mg K kg^{-1}		200 mg K kg^{-1}	
Genotype	RDW (g/3 Plants)	MRL (cm)	RDW (g/3 Plants)	MRL (cm)
Apore	1.54	45	1.67	32
Perola	1.97	42	2.04	39
Ruda	1.94	44	2.30	35
IAC Carioca	3.14	45	1.70	38
Jalo Precoce	2.24	42	1.67	36
Safira	1.77	46	1.50	44
Average	2.10	44	1.81	37

Source: Fageria, N.K. et al., *J. Plant Nutr.*, 31, 1121, 2008.

At the 200 mg K kg^{-1} level, root dry weight varied from 1.50 to 2.30 g per three plants, a variation of 1.5-fold. Similarly, maximum root length varied from 42 to 46 cm at low K level and 32 to 44 cm at higher K level. At the higher K level, there was a slight decrease in the root length of all the genotypes and the root weight of three genotypes also decreased at the higher K level. However, at the higher K level, there were more root hairs than at the lower K level (visual observations). There is evidence for genotype diversity in root characteristics of many crops in response to environment and an increasing interest in using this diversity to improve agricultural production and, consequently, NUE (Barber, 1994; Gregory, 1994). Mineral deficiency and toxicity, mechanical impedance, moisture stress, oxygen stress, and temperature have tremendous effects on root growth, development, and function and, subsequently, the ability of roots to absorb and translocate nutrients (Barber, 1995; Marschner, 1995; Baligar et al., 1998; Mengel et al., 2001). Mineral deficiency induces considerable variations in the growth and morphology of roots, and such variations are strongly influenced by plant species and genotypes. Overall, the growth of the main axis is little affected by nutrient deficiency, but that of the lateral branches and their elongation rates appear to be substantially reduced. Baligar et al. (1998) summarized the effects of various essential elements as follows: nitrogen deficiency increases root hair length, increases or has no effect on root hair density, and reduces branching; P deficiency increases overall growth of roots and root hair length, increases the number of second-order laterals, and either increases or does not affect root hair density; K and Ca deficiencies reduce root growth; however, higher Mg levels reduce the dry mass of roots. These nutrient stress factors on nutrient efficiency in the plant have not been well explored. Baligar et al. (1998) states that low pH reduces root mass, root length, and root hair formation; in alkaline soils, ammonium toxicity causes severe root inhibition; and, in general, salinity leads to a reduction in mass and length of roots and dieback of laterals.

2.7.2 HIGHER RATE OF NUTRIENT ABSORPTION AT LOW RHIZOSPHERE CONCENTRATIONS

The capacity of some plant species or genotypes within species to absorb nutrients at a higher rate at low nutrient concentration of the growth medium is one of the mechanisms responsible for efficient nutrient use by plants. The V_{max} and K_m values according to Michaelis–Menten kinetics or enzyme kinetics are generally used to explain the rate of ion influx in plant roots (Barber, 1995). According to this hypothesis, when nutrient uptake rate is plotted against nutrient concentration, a quadratic response or increase is obtained and maximum rate of uptake is designated as V_{max} (Y-axis). Half of the maximum velocity line touching the uptake rate curve and corresponding concentration on the X-axis is designated by K_m. Lower K_m values (higher affinity) indicate a higher uptake rate of plants for a determined nutrient at low concentration. Figure 2.26 shows different uptake rates and K_m values of two genotypes designated A and B. Although, the two genotypes have similar V_{max} values, genotype A has a lower K_m value than genotype B, and, hence, genotype A will have higher

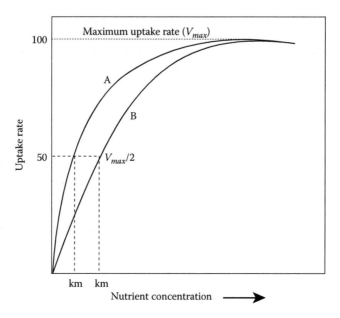

FIGURE 2.26 Hypothetical relationship between nutrient concentration and uptake rate in two genotypes and their K_m and V_{max} values. (From Fageria, N.K. et al., *J. Plant Nutr.*, 31, 1121, 2008.)

TABLE 2.16
Michaelis–Menten Constants for the Absorption of Phosphorus by Principal Crop Species in Nutrient Solution

Crop Species	Range of P Concentration (μM)	K_m in Mole (M) at Low Concentration	Reference
Barley	1–1000	5.4×10^{-6}	Andrew (1966)
Peanut	0.03–400	0.6×10^{-6}	Alagarswamy (1971)
Alfalfa	100–1000	2.0×10^{-6}	Baligar (1987)
Alfalfa	1–500	4.3×10^{-6}	Andrew (1966)
Wheat	0.1–1000	7.4×10^{-6}	Edwards (1970)
Rice	0.1–161	2.5×10^{-6}	Fageria (1973)
Rice	0.6–161	1.4×10^{-6}	Fageria (1974)
Corn	100–1000	2.2×10^{-6}	Baligar (1987)

uptake rates at low rhizosphere nutrient concentrations. In this case, genotype A is more efficient in nutrient uptake at lower rhizosphere nutrient concentration. Table 2.16 shows K_m values for P uptake by various plant species, and P uptake rate was in the order of peanut (*Arachis hypogaea* L.) > rice > alfalfa (*Medicago sativa* L.) > corn > Barley > wheat.

2.7.3   Ability of Plant to Solubilize Nutrients in Rhizosphere

Several chemical changes occur in the rhizosphere due to plant roots and soil environmental interactions. Among these changes, pH, oxidation potential, rhizodeposition, nutrient concentrations, and root exudates are prominent. These chemical changes in the rhizosphere significantly influence nutrient solubility and uptake by plants. Soil pH is one of the most important chemical properties that influence nutrient solubility and, hence, availability to plants. At lower pH (<5.5), the availability of most micronutrients is higher except Mo and decreases with increasing soil pH. This decrease is mostly associated with adsorption and precipitation processes. The availability of N as well as P is lower at lower pH and improves in a quadratic fashion with increasing pH to about 7.0. The increase in N availability is mainly associated with improved activity of N turnover by bacteria. The availability of P is associated with the neutralization of Al, Mn, and Fe compounds that immobilize this element at lower soil pH.

It is well known that the acidification of the rhizosphere can solubilize several low-soluble macronutrients (Riley and Barber, 1971; Barber, 1995) and micronutrients (Marschner, 1995; Hinsinger and Gilkes, 1996; Fageria et al., 2002). Bar-Yosef et al. (1980) reported that the root excretion of H^+ at the root surface is a mechanism for enhancing Zn uptake rather than the excretion of complexing agents. When more cations are absorbed, H^+ ions are released in the rhizosphere and pH decreases, and when more anions are absorbed, OH^- ions are released and pH increases (Barber, 1995; Mengel et al., 2001). The release of H^+ and OH^- ions in the rhizosphere is associated with maintaining cation and anion balance in plants during the ion uptake process. Enhanced reducing activity at root surfaces has been noted as root-induced responses to Fe deficiency in dicotyledonous and nongraminaceous monocotyledonous plants (Marschner, 1995).

Root activity alters the rhizosphere redox potential through respiratory oxygen consumption and ion uptake or exudation. In particular, root absorption and assimilation of NH_4^+ and NO_3^- consume 0.31 mol O_2 per mol of NH_4^+ and 1.5 mol O_2 per mol of NO_3^-, respectively (Bloom et al., 1992). Hence, when roots use NO_3^- as a nitrogen source, the rhizosphere redox potential declines more rapidly than when they use NH_4^+ (Bloom et al., 2003). The concentration of NH_4^+ and NO_3^- in the rhizosphere and the rhizosphere redox potential may be partially responsible for the observed large fluctuations in the relative availability of soil NH_4^+ and NO_3^- and in root growth (Jackson and Bloom, 1990).

Many other nutrient solubility or uptake processes occur in the rhizosphere and alter the redox potential. Redox reactions involve forms of Mn (Mn^{2+} and Mn^{4+}), Fe (Fe^{2+} and Fe^{3+}), and Cu (Cu^+ and Cu^{2+}) (Lindsay, 1979). However, the Fe and Mn redox reactions are considerably more important than those of Cu because of their higher concentrations in soil (Fageria et al., 2002). The primary source of electrons for biological redox reactions in soil is organic matter, but aeration, pH, and root and microbial activities also influence these reactions. Redox reactions in the rhizosphere can also be influenced by organic metabolites produced by roots and microorganisms.

Iron-efficient plants have the ability to respond to iron-deficiency stress by activating biochemical reactions that release compounds (phytosiderophores) to enhance

Fe uptake (Bienfait, 1988; Marschner, 1995). The reduction of rhizosphere pH due to the root excretion of H^+, the root exudation of organic acids (mainly phenolics), the enhanced root reduction of Fe^{3+} to Fe^{2+}, and activated root-reducing capacity at cell plasma membranes are responsible for overcoming Fe deficiency by dicotyledonous plants. In general, C_3 species are more productive under Fe-deficiency stress than C_4 species (Duncan and Carrow, 1999).

Root-induced rhizosphere chemical change has been reported to increase the availability of P to pigeon pea (*Cajanus cajan* L. Millsp.) (Ae et al., 1990). Roots of this plant release piscidic acid, which complexes Fe, and thereby free some of the tightly bound soil P. Hence, pigeon pea is successfully grown in P-deficient tropical soils (Radin and Lynch, 1994). Keerthisinghe et al. (2001) reported that white lupin (*Lupinus albus* L.) and pigeon pea have the ability to access fixed P and this is attributed to the exudation of organic acids into the rhizosphere. Under P-limiting conditions, white lupin exudes large quantities of citrate, and pigeon pea responds by increased exudation of malonic and piscidic acids. These organic acids increase the availability of P in acid soils, mainly by the chelation of Al and Fe bound to P and by suppressing the readsorption and precipitation of organic P. Extensive discussions of chemical changes in the rhizosphere and nutrient availability are given by Baligar et al. (1990), Darrah (1993), Barber (1995), Marschner (1995), Hinsinger (1998), Fageria and Stone (2006), and Fageria et al. (2002).

2.7.4 Better Distribution and Utilization of Nutrients within Plants

Better distribution of nutrients in parts of plant (root, shoot, and grain) reflects their use efficiency. In recent years, there have been major increases in the average yields of most crops. Most of this increase in yields has been accompanied by an increase in plant tissue that has high nutrient content such as grain compared to the lower nutrient content straw (Atkinson, 1990). Higher accumulation of N and P in grain improves yield and, consequently, leads to higher use efficiency of these nutrients (Fageria et al., 2006). The proportion of total plant N or P partitioned to grain is called N or P harvest index. Nutrient harvest index is defined as nutrient uptake in grain divided by nutrient uptake in grain plus straw. This index is very useful in measuring nutrient partitioning in crop plants, which provides an indication of how efficiently the plant utilizes acquired nutrients for grain production (Fageria and Baligar, 2005b). High nitrogen or P harvest index is associated with efficient utilization of N (Fawcett and Frey, 1983; Rattunde and Frey, 1986; Fageria et al., 2006).

Schmidt (1984) pointed out that new cultivar development may need to be directed toward the production of genotypes that exploit inputs most efficiently and not on genotypes that have superior yield only when high production inputs are used. Isfan (1993) reported that physiological efficiency of absorbed N (ratio of grain produced to the total N absorbed by the aboveground plant parts) may be used in a plant-breeding program to detect potentially high-yielding oat genotypes and to evaluate those capable of exploiting N input most efficiently. The physiological efficiency index of N is related to many physiological processes such as absorption nitrate reduction efficiency, nitrogen remobilization, translocation, assimilation, and storage (Novoa and Loomis, 1981). Figure 2.27 shows a relationship between N harvest index (NHI = N uptake in

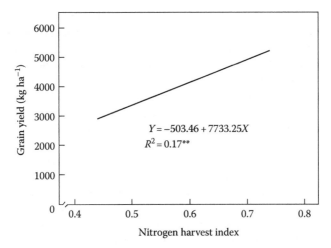

FIGURE 2.27 Relationship between NHI and grain yield of lowland rice. Values are averages of five genotypes. **Significant at the 1% probability level. (From Fageria, N.K. et al., *J. Plant Nutr.*, 31, 1121, 2008.)

the grain per N uptake in the grain plus straw) and grain yield of lowland rice. Grain yield increased linearly with increasing NHI within the range of 0.44–0.74.

Amounts of N or P remobilization from storage tissues influence grain N or P use efficiency and this varies among genotypes and appears to be under genetic control (Moll et al., 1982; Dhugga and Waines, 1989). Variation in nutrient harvest indices among crop species, or genotypes of the same species, is a useful trait in selecting crop genotypes for higher grain yield (Fageria and Baligar, 2005a). Dhugga and Waines (1989) reported that genotypes that accumulate little or no N after anthesis had low grain yields and low NHI.

Moll et al. (1982) reported that eight single-cross corn hybrids differed in N-efficiency traits and yield when grown in the field with low and high soil N. At low soil N, hybrid differences in N use efficiency were due largely to variation in the utilization of acquired N. At high soil N, hybrid differences were attributed to variation in N uptake efficiency. Differences in N translocation and remobilization to the grain were important only at low levels of soil N (Clark and Duncan, 1991). Inter- and intraspecific differences for NUE for macro- and micronutrients for sorghum (*Sorghum bicolor* L.), maize, alfalfa (*M. sativa* L.), and red clover (*Trifolium pratense* L.) have been reported (Baligar, 1987; Baligar and Fageria, 1997, 1999).

2.7.5 BETTER ALLOCATION OF DRY MATTER WITHIN PLANTS

Better distribution of dry matter in crop plants (shoot and grain) is generally associated with higher yields and, consequently, higher NUE. While the production and utilization of dry matter within a plant depend on each other, the regulation of the partitioning of dry matter into different plant parts is independent of the production of assimilate (Ho, 1988). This means, the partitioning of assimilate is genetically determined in crop plants. However, it is also influenced by environmental factors.

Dry matter distribution is measured by GHI. The GHI is the ratio of grain yield to total biological yield and is calculated with the help of the equation GHI = (grain yield per grain + straw yield). The term GHI was introduced by Donald (1962) and since has been considered an important trait for yield improvement in field crops. Values for GHI in cereals and legumes are normally less than 1. Although GHI is a ratio, it is sometimes expressed as a percentage.

Generally, dry matter is positively associated with grain yield (Fageria et al., 2004a). Figure 2.9 shows that grain yield of common bean increased significantly and quadratically with increasing shoot dry weight. Evans (1993, 1994) reported that yield increases in many cereals, legumes, and root crops during the twentieth century were due to increase in harvest indices of these crops. Austin (1994) reported that in rice, wheat, and barley, modern cultivars are short in stature and can have a GHI near 0.50. In contrast, old cultivars are taller and have harvest indices of 0.30 or lower. Hay (1995) reported that GHI of grain crops, particularly cereals, has increased with increasing crop yields during the latter half of the twentieth century. However, plant breeders have not sought to raise GHI, and probably any increase in this trait has been an unplanned secondary effort of breeding for grain yield (Araujo and Teixeira, 2003).

The GHI values of modern crop cultivars are commonly higher than those of old traditional cultivars for major field crops (Ludlow and Muchow, 1990). Genetic improvement in annual crops such as wheat, barley, corn, oat, rice, and soybean (*Glycine max* L. Merr.) has been reported due to increase in dry weight as well as GHI (Austin et al., 1980; Wych and Rasmusson, 1983; Wych and Stuthman, 1983; Cregan and Yaklich, 1986; Payne et al., 1986; Tollenaar, 1989; Feil, 1992; Peng et al., 2000). In potato (*Solanum tuberosum* L.), modern cultivars have plant dry weights 10 times that of the wild species (*Solanum demissum* L.). Harvest index (tuber dry weight as a proportion of plant weight) increased from 7% in wild species to 81% in modern cultivars (Inoue and Tanaka, 1978). Peng et al. (2000) reported that the genetic gain yield of rice cultivars released before 1980 were mainly due to improvement in GHI, while increases in total biomass were associated with yield trends for cultivars developed after 1980. Cultivars developed after 1980 had relatively high GHI values, but, further improvement in GHI was not achieved. These authors also reported that further increases in rice yield potential would be likely to occur through increasing biomass production rather than increasing GHI.

2.7.6 Balanced Source–Sink Relationship

Genetic and production physiological studies show that crop yield potentials are high and that they are not fully exploited (Fageria et al., 2006). Balanced source–sink relationships were vital for higher yields and, consequently, higher NUE in crop plants. However, neither source nor sink manipulation alone can improve crop yield indefinitely (Ho, 1988). Most plants have the ability to buffer any imbalance between source and sink activities by storing carbohydrates during periods of excess production and the mobilization of these reserves when the demands of growth exceed the supply of carbohydrates available through current photosynthesis (Evans and

Wardlaw, 1996). Both source activity and sink activity vary with plant development and are modified by environmental factors.

Biomass production in plants depends on photosynthesis. At the initial stage of plant growth, leaves function as sinks but with the advancement of age they serve as sources. Hence, leaves are the main site of photosynthesis and source of carbohydrate in plants; however, with advances in plant age, stems and inflorescence of some of cereals contribute substantially to photosynthetic activity (Evans and Wardlaw, 1976). Evans and Wardlaw (1996) reported that photosynthesis by glumes and young grains of wheat constitute an important source of assimilate as well as a means of recapturing respired CO_2. Ear photosynthesis throughout grain growth contributed 33% to grain growth requirements in one awned wheat cultivar and 20% in an unawned one (Evans and Rawson, 1970).

Panicles or heads in cereal, pods in legumes, and tubers in root crops are main sinks of photoassimilates. A small portion of photosynthetic product is also translocated to roots. The growing organs of plants are active sinks and these prevent the accumulation of photoassimilates in the sources, if source capacity is limited. Assimilated carbohydrates in the source as well as sink are lost through respiration and this loss is reportedly half of the total carbon assimilated in photosynthesis (Evans and Wardlaw, 1996). In modern cultivars, source capacity has been more limiting to yield than in the older ones (Evans and Wardlaw, 1996). During the twentieth century, both source and sink have been improved in important annual crops and this has made possible an improvement in yields (Ho, 1988). The capacity of dry matter production in leaves may either be higher or lower than the capacity of dry matter accumulation in other parts of the plant. Hence, at different times, either source- or sink-limiting situations may exist in crop production (Ho, 1988).

2.8 ROLE OF CROPS IN IMPROVING NUTRIENT USE EFFICIENCY UNDER BIOTIC AND ABIOTIC STRESSES

Abiotic (soil acidity, soil salinity/alkalinity, drought, water logging, high temperature, mineral deficiency/toxicities) and biotic (diseases, pests, weeds) stresses have tremendous effects on plant growth and development and the ability to take up and utilize nutrients more efficiently (Alam, 1999; Pessarakli, 1999; Baligar et al., 2001). NUE and yield of crops under stress could be enhanced by the selection/breeding of plants that have high NUE (acquisition, influx, transport, utilization, and remobilization), and the ability to interact effectively with environmental extremes (drought, solar radiation, temperature extremes). Plant species and genotypes/cultivars within species differ in optimal environmental requirements and their abilities to tolerate a particular stress.

In addition to low soil fertility, soil acidity is a serious problem worldwide. Soil acidity decreases crop productivity and NUE in crop plants. Reasons for poor crop productivity in acid soils are as follows: presence of elemental toxicities (Al, Mn, Fe, H) and deficiencies or unavailability of essential nutrients (N, P, Ca, Mg, K, Fe, Zn); physical constraints (compaction, hard pan); degraded and infertile soils (erosion, leaching and gaseous nutrient losses, low microbial activities); low recovery efficiency of applied nutrients during a single season (<50% for N, <20% for

P, 40%–70% for K, micronutrients 5%–10%) (Mortvedt, 1994; Raun and Johnson, 1999; Fageria, 2000a; Fageria and Baligar, 2001, 2005c); inadequate/nonuse of soil amendments (lime, fertilizers); adverse environmental conditions (high temperature, low rainfall, and high ET); existence of low plant species or cultivar tolerance to soil acidity; low potential for plant growth and NUE; planting of nonacid-tolerant plants with inefficient NUE; and high intensity of weeds, insect, and diseases (Baligar and Fageria, 1997, 1999).

Liming acid soils is a dominant and effective practice for improving crop yields and, consequently, NUE. Lime reduces toxic effects of hydrogen, aluminum, and manganese; improves soil biological activities, cation exchange capacity, P, Ca, and Mg availability and soil structure; promotes N_2 fixation; stimulates nitrification; and decreases availability of K, Mn, Zn, Fe, B, and Cu. Fageria et al. (2006) reported that in a Brazilian Oxisol, common bean yield increased significantly and quadratically with increasing soil pH. Low pH (excess Al) reduces root elongation and lateral root growth and greatly affects the absorption of nutrients. Hence, the use of acid-tolerant crop species and genotypes within species in combination with lime is an important strategy for reducing the cost of crop production on acid soils. Variation in acidity or Al tolerance among different crop species and genotypes of same species has been widely reported (Foy, 1984, 1992; Kochian, 1995; Baligar and Fageria, 1997, 1999; Fageria et al., 2004b; Yang et al., 2004). The grain yield of upland rice and soil chemical properties is significantly and negatively correlated with pH, Ca saturation, and bases saturation, and is significantly and positively correlated with Al, and H^+ Al (Table 2.17). Upland rice genotypes evaluated in this

---

TABLE 2.17

Correlation Coefficients (*r*) between Grain Yield and Soil Chemical Properties across 2 Soil Acidity Levels and 20 Upland Rice Genotypes

Soil Chemical Properties	Grain Yield (g Pot⁻¹)
pH in H_2O	−0.23*
Al (mmol$_c$ dm^{-3})	0.19*
H + Al (mmol$_c$ dm^{-3})	0.23*
Base saturation (%)	−0.21*
Ca saturation (%)	−0.21*

Source: Adapted from Fageria, N.K. et al., Response of upland rice genotypes to soil acidity, in *The Red Soils of China: Their Nature, Management, and Utilization*, eds., M.J. Wilson, Z. He, and X. Yang, pp. 219–237, Kluwer Academic Publishers, Dordrecht, the Netherlands, 2004b.
*Significant at the 5% probability level.

study were tolerant to soil acidity. Hence, developing soil-acidity-tolerant crop species or cultivars of the same species also deserves logical consideration during the twenty-first century.

Plant species and genotypes/cultivars within species differ widely in tolerance to soil acidity constraints (Foy, 1984, 1992; Baligar and Fageria, 1997, 1999). Acid soil tolerance of crop plants generally involves more than one mechanism. For example, plasma membrane/cell wall exclusionary responses via selective permeability/ polymerization, pH barrier (chelating ligands or mucilage) formation at root–soil interface, internal chelation by organic acids (carboxylic, citric, malic, and trans-aconitric) or metal-binding proteins and enzymes, and vacuole compartmentation of Al or Mn are involved (Duncan, 1994; Duncan and Carrow, 1999; Yang et al., 2004). Soil-acidity constraints also reduce the uptake of certain essential elements and increase that of others (Foy, 1984; Baligar and Fageria, 1997, 1999). Soil-acidity-tolerant plant species and genotypes are efficient in the absorption and utilization of nutrients (Baligar and Fageria, 1997, 1999). Foy (1984) lists the following mechanisms that plants have developed to maintain their nutrient requirements and to overcome soil-acidity constraints: raise the rhizosphere pH to reduce the toxic levels of Al, Mn, and H; increase microbial activities to enhance organic matter decomposition and thereby release inorganic nutrients for plant use and improve root rhizobial associations, thereby increasing nitrogen fixation; and efficient NUE for N, P, Ca, Mg, and K.

Saline soils contain an excess of neutral salts such as chloride and sulfates of Na^+, K^+, Ca^{2+}, and Mg^{2+} (Mengel et al., 2001). In saline soils, plants are affected by water deficit, ion toxicity (Cl, Na), and nutrient imbalances due to depression in uptake and transport (Grattan and Grieve, 1999a,b). In alkaline soils, Fe deficiency, B toxicity, and salinity are the most obvious problems for successful crop production. Salinity reduces root growth and permeability and, consequently, reduces water and nutrient uptake. Large differences in salt tolerance have been reported for plant species and cultivars within species (Maas, 1986; Marschner, 1995). Saline conditions greatly affect availability in soil, movement to root surface, the uptake, transport and partitioning of N, P, K, Ca, Mg, and micronutrients (Grattan and Grieve, 1999a,b). By the selection and breeding of soil-alkalinity- and soil-salinity-tolerant plant species and cultivars within species, along with improved best management practices (BMPs) to reduce salinity and alkalinity in soil, it is possible to achieve reasonable crop yields in high pH soils. Differences in tolerance to salinity have been reported among genotypes of maize, bean, and other species (Foy, 1992). Plant development and successful crop production in salt-affected soils depend on improved supply of adequate water and nutrients. Adapting soil management practices to reduce salt levels, improving water and nutrient status, and the selection or breeding of salt-tolerant species and cultivars are effective in improving NUE and crop yields in salt-affected soils. Epstein and Bloom (2005) have covered the physiology of salt stress. Further, they state that the mechanisms of salt tolerance and its genetic and molecular feature are currently under intense development and the knowledge developed could be useful in the future in the breeding of salt-tolerant crops.

Poor crop productivity in many soils of the world is due to deficiencies of essential elements and toxicities of metal elements. Inter- and intraspecific variation for plant growth and mineral composition has been well documented (Epstein and Jefferies, 1964; Gerloff and Gabelman, 1983; Vose, 1984; Clark, 1990). Genetic and physiological components of plants have profound effects on their abilities to absorb and utilize nutrients under various environmental and ecological conditions. Clark (1990) has covered extensively the mechanisms involved in genotypic variation in mineral nutrient uptake and utilization. Mineral deficiency and toxicity stress have a major effect on root growth and morphology (number, diameter, length, surface area, and distribution in soil) (Baligar et al., 1998). Such changes in root morphology and growth will have effects on plant ability to take up nutrients effectively from soil. Clark and Duncan (1993) suggested the use of the juvenile stage of plant growth in the selection of mineral stresses and the most common trait is yield (vegetative or grain). The best management strategy for overcoming mineral-stressed soil is to select and improve plants for production on abiotic-stressed soils with limited soil amendments (Clark and Duncan, 1993).

During the growth cycle, plants are subjected to drought (water deficit) of a very short period or longer in duration. Water stress during the growth cycles of plants adversely affects many physiological growth processes (photosynthesis, translocation of carbohydrates and growth regulators, ion uptake transport and assimilation, N_2 fixation, turgidity, respiration) and shoot and root morphology and growth (cell enlargement, leaf area, and root growth and extension) (Fageria et al., 2006). Water stress is primarily responsible for stomatal closure, thereby reducing assimilation and growth. Water stress reduces plant growth by reducing cell division and root enlargement and leads to a decline in ion transport to the root surface. In dry soil, nutrients are less mobile mainly because pores are filled with air and the pathways for nutrient flux from soil to root surface are less direct. Such conditions in soil limit ion flux to root surface by diffusion and mass flow (Barber, 1995; Pugnaire et al., 1999). The extent of drought injury to plants depends on the length of drought and the nature of the species and cultivars/genotypes within the species involved. Plants are known to have drought escape and drought avoidance or tolerance components to overcome drought injury. Species and varieties differences in drought avoidance are attributed to lower transpiration rate, rapid stomatal closure, ability to retain a high percentage of water, greater water uptake, greater root volume, and higher root–shoot ratio. Drought tolerance is attributed to avoidance of dehydration of guard cells, hydration tolerance of photosynthesizing cells, and decreased rate of protein loss. In plant selection process for drought tolerance, the morphological drought escape and avoidance and tolerance features in plants need to be harnessed (Baligar et al., 1990). Among various factors, temperature is the major uncontrollable factor that has a great impact on crop growth and production. The plant has adapted tolerance or avoidance mechanisms to overcome heat stress. High temperature stress leads to an insufficient supply of carbohydrates to root meristems whereas low temperature leads to poor or reduced shoot growth due to an insufficient supply of mineral nutrients and water (Marschner, 1995). In temperature extremes, nutrient availability and uptake are inhibited in addition to reduction in root growth. Epstein and Bloom (2005) state that rise in temperature alters integrity of biological membrane. At high temperature, nutrient influx cannot keep pace with nutrient efflux, which leads to increased nutrient leakage from roots. Low temperature reduces growth of shoots and roots and the mineral nutrition of plants (Cooper, 1973; Bowen, 1991). Inter- and intraspecific differences in plant dry matter yields' mineral composition and NUE at varying temperatures have been reported (Cooper, 1973; Bowen, 1991; Baligar et al., 1997).

Nutrient levels and their availability to plants in soil may affect plant susceptibility to insects and diseases. Plant diseases are greatly influenced by environmental factors, including deficiencies and/or toxicities of essential nutrients, and balanced nutrition has an important role in determining plant resistance or susceptibility to diseases (Fageria et al., 2011a). Global preharvest crop losses due to pathogens are 9%–15% of annual production. Mineral elements are directly involved in the defense mechanisms of plants as integral components of cells, substrates, enzymes, and electron carriers or as activators, inhibitors, and regulators of metabolism (Huber, 1980). Nutrient-stressed plants are often more susceptible to disease than those at a nutritional optimum, yet plants receiving a large excess of a required mineral may become predisposed to disease. Fageria et al. (2011a) summarized the nutrient element role in disease intensity in plants as follows: (1) high N increases plant susceptibility to obligate pathogens but decreases their susceptibility to facultative pathogens; (2) application of K, Ca, Mn, Fe, B, Cu, and Si to soil deficient in these elements usually increases the resistance; however, the effects of P and Zn are variable and there is not sufficient information available on Mg and S to reach definite conclusions; and (3) deficiency ranges of micronutrients are known to decrease disease resistance. Copper, B, and Mn are involved in the synthesis of lignin, and simple phenols and Si create physical barriers to pathogen invasion. The greatest benefits from nutrients are found with moderately susceptible or partially resistant cultivars. Salinity induces metabolic changes such as the accumulation of proline, glycine-betaine, Na^+ and Cl^- in plants; such changes are known to reduce aphid feeding on plants (Araya et al., 1991). Fageria and Scriber (2002) state that minerals and primary metabolites that are involved in basic plant processes are rarely considered responsible for plant resistance to insect attack, despite the major role they play in insect behavior. Overall, plants have lower average concentrations of N, S, P, Fe, Zn, and Cu and have equal or greater concentrations of Mg, K, Ca, and Mn than insects (Schoonhoven et al., 1998). Seasonal variations, inter- and intraspecific plant variations, and environmental factors (soil type, fertilizer regime) influence concentrations of elements in plants. Some information is available concerning plant N content and how it alters mechanisms of plant resistance to insect herbivores; however, information is lacking about the effects of other elements on insect herbivores (Fageria and Scriber, 2002). Overall, from the available data it appears that the influence of the levels of macro- and micronutrients in plants may have positive, negative, or no effects on insect damage in crop plants (Fageria and Scriber, 2002). Biotic and abiotic factors are known to alter growth and elemental concentrations and modify plant resistance, but how such relations affect insect and disease resistance in plant is not clearly understood. How insect attacks affect plants that have high NUE in the presence or absence of abiotic stress needs to be evaluated. Therefore, to overcome disease and insect pressure it is important to identify species and cultivars within species that are efficient in absorption and utilization of nutrients under abiotic stress. Such plant types will have greater ability to overcome abiotic stresses and achieve yield potentials.

There is need for establishing breeding programs to focus on developing cultivars with high NUE under specific stresses. BMPs such as the use of fertilizer and amendment (lime), proper crop rotations, increases in organic matter content, and control of erosion, insects, diseases, and weeds can significantly improve crop yields and optimize NUE. The development of new cultivars with high NUE coupled with BMPs with an integrated pest management strategy will contribute to economically viable and environmentally sustainable systems for the vast stress ecosystems of the world. Foy (1984, 1992) states that more emphasis should be given to plant–soil interactions and the breeding or selection of plants to fit the soils, and less emphasis should be placed on fitting all soils to meet the demands of all plants. Fitting all soils to meet the needs of all plants requires heavy input of amendments and fertilizers, which could increase the cost of cultivation and accelerate environmental degradation.

2.9 BREEDING FOR NUTRIENT USE EFFICIENCY

Selection and breeding nutrient-efficient species or genotypes within a species is justified in terms of reduction in the fertilizer input cost of crop production and also the reduced risk of contamination of soil and water. Through plant breeding, the genetic yield potential of wheat, soybean, corn, and peanuts has been improved by 40%–100% within the twentieth century (Gifford et al., 1984; Ho, 1988). Genetic variability among crop species and genotypes of the same species for macro- and micronutrients use or requirement is well documented (Clark and Duncan, 1991; Baligar and Fageria, 1999; Baligar et al., 2001; Fageria and Baligar, 2005; Hillel and Rosenzweig, 2005). Micronutrients are required in small amounts by crops and their requirements can often be easily met by planting efficient genotypes. Micronutrient efficiencies of existing plants should be weighed against the cost of breeding more efficient genotypes.

Considerable progress has been made in identifying crop species and genotypes within species for NUE, tolerance of elemental toxicity, and understanding the possible mechanisms involved (Foy, 1984, 1992; Graham, 1984; Maas, 1986; Clark and Duncan, 1991; Marschner, 1995; Baligar et al., 2001; Blamey, 2001; Okada and Fischer, 2001; Fageria et al., 2003, 2006; Yang et al., 2004; Epstein and Bloom, 2005; Fageria and Baligar, 2005). Plant traits and characteristics showing tolerance to essential nutrient deficiencies are numerous and have been reviewed (Baligar et al., 1990). Clark and Duncan (1993) suggested that the juvenile stage of plant growth is more desirable to evaluate plants for mineral stress tolerance. Further, they state that yield (vegetative or grain/seed/fruit) is probably the most common trait used to evaluate plants for tolerance to soil mineral stresses. However, progress has been limited in releasing crop cultivars having these traits. One good example of solving the nutrient-deficiency problem with breeding involves iron deficiency. This problem in calcareous soils has been overcome by selecting/breeding iron-efficient genotypes of corn, soybean, sorghum, and rice (Graham, 1984). When Brazilian Oxisols were limed at pH above 6.0 for growing legume crops such as common bean and soybean, iron precipitated and created Fe deficiency in a subsequent upland rice crop (Fageria et al., 2003). These soils are well supplied with iron; however, at higher pH (>6.0) iron is precipitated and its availability is low (Fageria et al., 2003). Under these conditions Fe availability is improved by decreased pH-reducing conditions and Fe chelators, root exudates by Fe-efficient genotypes (Graham, 1984).

Induced iron-deficiency chlorosis is widespread and is a major concern for plants growing on calcareous or alkaline soils due to their high pH and low availability of iron (Welch et al., 1991; Marschner, 1995). Planting iron-efficient genotypes is the best solution for correcting iron deficiency under these conditions (Fageria et al., 2003). Iron efficiency can range from monogenic to polygenic control, depending on species (Duncan, 1994). Both additive and dominant gene actions may be involved (Duncan and Carrow, 1999).

Breeding of more efficient plants for major nutrients such as N, P, and K, which are required in large amounts by crop plants for maximum economic yield, requires special attention. Authors and coworkers have conducted several field and greenhouse experiments using genotypes of rice, wheat, and common bean in Brazilian Inceptisols and Oxisols using different N and P rates (Fageria and Baligar, 1997b, 1999; Fageria, 1998, 2000b; Fageria et al., 2001; Fageria and Barbosa Filho, 2001). In these studies, inter- and intraspecific differences were observed for growth and N and P use efficiencies. When P level in the soil extracted by Mehlich 1 extracting was around 2 mg kg^{-1} of soil, either most of the genotypes did not produce or produced insignificant grain yield. Similarly, without the addition of N fertilizers, rice genotypes produced very low grain yield. Hence, the strategy should be to use efficient crop genotypes along with a judicious use of N, P, and K fertilizers.

In addition, although numerous studies have shown a wide range of genotypic differences among and within species for N-, P-, and K-efficiency traits, the genetics of these plant responses are not well understood and appear to be complex (Clark and Duncan, 1991). Most studies have indicated a genetic control. Heritabilities of some N-efficiency traits were relatively high, while others were low (Clark and Duncan, 1991). Clark and Duncan (1991) reported that P-efficiency traits are heritable, and could be used to improve germplasm for P nutrition. A prime example of success has been with white clover in New Zealand (Caradus, 1990). Root growth, morphology, ion uptake, and use efficiency should be considered when plants are to be improved for mineral nutrition traits involving K in breeding programs (Pettersson and Jensen, 1983; Clark and Duncan, 1991). Yield has long been classified as a character controlled by quantitative genetics, that is, one influenced by many genes with the effects of individual genes normally unidentified (Wallace et al., 1972). This means that yield improvement by the use of nutrient-efficient genotypes deserves special attention in relation to identifying physiological components causing varietal differences in economic yield and acquiring understanding of their genetic control. The high yields achieved in rice by incorporating short, erect, thick, dark-green leaves, and short stiff stems clearly demonstrate the merit of including physiological component traits in plant-breeding programs (Wallace et al., 1972; Fageria et al., 2006).

Richardson (2001) reported that soil P uptake can be increased by plant modification. The selection of plants for increased efficiency of P has been demonstrated with root morphology being particularly important (Lynch, 1995). Similarly, gene technologies offer opportunities for manipulating the structure and function of plant roots for improved acquisition of soil P (Richardson, 2001). Plant genes that regulate root branching have been isolated (Zhang and Forde, 1998) and the expression in plants of specific bacterial genes (i.e., encoding phytohormone activities) may offer new insights into the role of such genes in plant growth and development (Spena et al., 1992; Richardson, 2001). The cloning and characterization of plant and fungal phosphate transporter genes may also provide new possibilities for increasing plant P uptake (Smith et al., 2000; Richardson, 2001).

Plant selection for sustained production in water-deficit environments has received considerable attention for three or four decades (Blum, 1993), yet genotypes/cultivars with substantial drought tolerance still remain elusive. Blum (1993) puts the plant varieties into three categories: (1) those with uniform superiority over all environments, (2) those relatively better in poor environments, and (3) those relatively better in favored environments. However, so far no reliable genotypes or cultivars have evolved with considerable tolerance to drought. Progress in molecular biology to improve drought resistance/tolerance is constrained by ignorance in agronomy and crop physiology (Blum, 1993).

Molecular biology technology can be an important approach in isolation, identification, localization, and laboratory reproduction of gene(s) carrying desirable nutrient-efficiency traits (Clark and Duncan, 1991). In the twentieth century, genetic engineering techniques did not play a significant role in improving nutrient-efficient crop genotypes. However, its wide applicability or potential in the twenty-first century for improving nutrient efficiency in crop plants is highly predicted. In addition, recently, new possibilities have arisen to transfer desired traits (genes) not just between strains of the same species, but even from one species to another, thus greatly enlarging the range of potential genetic resources available to agricultural scientists (Hillel and Rosenzweig, 2005).

2.10 NUTRIENT REQUIREMENTS OF CONVENTIONAL AND TRANSGENIC PLANTS

Transgenic crops have been widely adopted over the past decade, and their cultivation continues to grow at a sustained pace (James, 2007). Its proponents claim that it has the potential to reconcile the needs of a growing population with the goal of conserving and promoting biodiversity. Hillel and Rosenzweig (2005) reported that the transgenic technique allows the development of transgenic plants whose purpose is to serve in the screening of new generations of agrochemicals that are intended to meet the requirements of higher specific activity (hence, decreased application amounts per unit area), greater selectivity (so that they may affect only the target organisms to be controlled), and higher biodegradability (so that they may not accumulate and persist in the environment). However, there are ecological and agricultural concerns associated with the cultivation of transgenic crops. Of particular importance with regard to risk assessment and property rights protection is the migration of transgenes to the same species (intraspecific gene flow), including neighboring nontransgenic crops and volunteer crop plants (Ellstrand, 2003; Willenborg and Van Acker, 2008;

Willenborg et al., 2009). Mechanisms of gene flow vary from plant species to plant species and range from the possibility of asexual propagation, short- or long-distance pollen dispersal mediated by insects or wind, and seed dispersal (Chandler and Dunwell, 2008). Chandler and Dunwell (2008) reported that there is ample evidence that gene flow from crops to related wild species occurred before the development of transgenic crops and this should be taken into account in the risk assessment process. There are many excellent articles that provide recent perspectives and literature reviews on gene flow in the context of genetically modified plants (Andow and Zwahlen, 2006; Chapman and Burke, 2006; Davis, 2006; Chandler and Dunwell, 2008). Genetically modified plants of many crops like corn, soybean, cotton, rice, sugarbeet, sunflower, sorghum, cucurbits, wheat, papaya, alfalfa, and *Brassica napus* have been released for cultivation in many parts of the world (Chandler and Dunwell, 2008).

Data are limited on the nutritional requirements of transgenic crop plants. Faria and Fageria (in press) compared the response of transgenic and nontransgenic dry bean genotypes to soil fertility (Tables 2.18 through 2.21). Genotypes evaluated

TABLE 2.18
Influence of Soil Fertility Genotype Treatments on Shoot Dry Weight, Grain Yield and Pods per Plant of Dry Bean Genotypes

Fertility/Genotype	Shoot Dry Weight (g Plant^{-1})	Grain Yield (g Plant^{-1})	Pods (Plant^{-1})
Low fertility	6.39a	10.06b	8.47b
High fertility	6.87a	15.58a	14.93a
Olathe Pinto	4.57c	12.07a	11.25a
Olathe Pinto 5.1[a]	5.23bc	11.22a	10.12a
Pérola	7.05abc	12.32a	13.08a
Pérola 5.1[a]	8.49a	14.47a	12.66a
BRS Pontal	6.56abc	11.34a	10.04a
BRS Pontal 5.1[a]	7.88ab	15.51a	13.04a
Mean	6.63	12.82	11.69
F-test			
Fertility level (FL)	NS	**	**
Genotype (G)	**	NS	NS
FL × G	NS	NS	NS
CV (%)	30.15	43.64	38.78

Source: Faria, J.C. and Fageria, N.K., *J. Plant Nutr.*, 34, in press.

Values of genotypes are across two fertility levels.

NS, nonsignificant.

[a] Genetically modified genotypes.

**Significant at the 1% probability level. Means followed by same letter in the same column (fertility and genotypes separate) are significantly not different by Tukey's test at the 5% probability level.

TABLE 2.19

Influence of Soil Fertility and Genotype Treatments on Seeds per Pod and 100 Grain Weight of Dry Bean Genotypes

Fertility/Genotype	Seeds per Pod	100 Grain Weight (g)	
		Low Fertility	High Fertility
Low fertility	4.00a	—	—
High fertility	3.60b	—	—
Olathe Pinto	3.30b	31.36ab	33.13a
Olathe Pinto 5.1[a]	3.12b	34.32a	35.20a
Pérola	3.30b	23.75c	32.66a
Pérola 5.1[a]	3.71b	32.85a	30.68a
BRS Pontal	4.46a	26.15bc	24.39b
BRS Pontal 5.1[a]	4.90a	26.11bc	23.06b
Mean	3.79	29.09	29.85
F-test			
Fertility level (FL)	**	NS	
Genotype (G)	**	**	
FL × G	NS	**	
CV (%)	10.72	9.05	

Source: Faria, J.C. and Fageria, N.K., *J. Plant Nutr.*, 34, in press.
NS, nonsignificant.
[a] Genetically modified genotypes.
**Significant at the 1% probability level. Means followed by same letter in the same column (fertility and genotypes separate) are significantly not different by Tukey's test at the 5% probability level.

were Olathe Pinto, Olathe 5.1 (genetically modified), BRS Pontal, BRS Pontal 5.1 (genetically modified), Perola, and Perola 5.1 (genetically modified). Fertility levels were 1 g fertilizer (5-30-15) per kg soil (low fertility level) and 2 g fertilizer (5-30-15) per kg soil (high fertility level). These fertility levels were designated as low and high, respectively. Grain yield, number of pods per plant, and seed per pod were significantly increased with the increase in soil fertility. Shoot dry weight, seed per pod, and 100 seed weight were also significantly influenced by genotype treatment. Fertility × genotypes interaction was significant for maximum root length and root dry weight, indicating that genotypes responded differently at two fertility levels in relations to these two traits. Shoot dry weight, number of pods per plant, and GHI had significant association with grain yield, indicating that by increasing these three traits grain yield can be increased. Grain yield efficiency index (GYEI) had significant linear association with grain yield. Hence, on the basis of GYEI, genotypes were classified as efficient (E), moderately efficient (ME), and inefficient in nutrient use. Three conventional genotypes (Olathe Pinto, BRS Pontal, and Pérola) and one genetically modified genotype (Olathe Pinto 5.1)

TABLE 2.20

Influence of Soil Fertility and Genotype Treatments on Grain Harvest Index and Maximum Root Length of Six Dry Bean Genotypes

Genotype	Grain Harvest Index		Maximum Root Length (cm)	
	Low Fertility	High Fertility	Low Fertility	High Fertility
Olathe Pinto	0.65a	0.75a	23.00b	31.25a
Olathe Pinto 5.1[a]	0.63a	0.68a	25.50ab	27.25a
Pérola	0.45a	0.73a	43.50a	27.25a
Pérola 5.1[a]	0.51a	0.68a	31.00ab	28.75a
BRS Pontal	0.67a	0.58a	35.75ab	27.25a
BRS Pontal 5.1[a]	0.63a	0.66a	34.00ab	28.00a
Mean	0.59b	0.68a	32.12a	28.29a
F-test				
Fertility level (FL)	**		NS	
Genotype (G)	NS		NS	
FL × G	*		*	
CV (%)	17.77		23.65	

Source: Faria, J.C. and Fageria, N.K., *J. Plant Nutr.*, 34, in press.

NS, nonsignificant.

[a] Genetically modified genotypes.

*, **Significant at the 5% and 1% probability levels, respectively. Means followed by same letter in the same column (fertility and genotypes separate) are significantly not different by Tukey's test at the 5% probability level.

were classified as moderately efficient and two genetically modified genotypes (Pérola 5.1 and BRS Pontal 5.1) were classified as efficient; none of the genotypes fell into the inefficient group.

Faria and Fageria (in press) concluded that out of three genetically modified dry bean genotypes, two produced higher grain yield compared with conventional genotypes. Genetic modification did not impose a negative response to nutrient addition. These two genotypes (BRS Pontal 5.1 and Pérola 5.1) were also classified as efficient in nutrient use based on grain efficiency index. These authors also compared the growth of tops (Figures 2.28 through 2.33) and roots (Figures 2.34 through 2.36) of transgenic and nontransgenic dry bean plants.

2.11 CONCLUSIONS

The root is the major organ for nutrient uptake in plants. Nutrients are absorbed in ionic forms by roots, and, hence, the process is known as ion uptake by plants. There are 17 nutrients essential for plant growth. These are C, H, O, N, P, K, Ca, Mg, and S, known as major nutrients. The minor nutrients are Zn, Cu, Fe, Mn, Fe, B, Mo, Cl, and Ni. All the essential nutrients are equally important for the growth

TABLE 2.21

Influence of Fertility and Genotype Treatments on Root Dry Weight and Specific Root Length of Six Dry Bean Genotypes

Fertility/Genotype	Specific Root Length (cm g⁻¹)	Root Dry Weight (g Plant⁻¹)	
		Low Fertility	High Fertility
Low fertility	77.22a		
High fertility	62.25b		
Olathe Pinto	83.13ab	0.25c	0.42ab
Olathe Pinto 5.1[a]	80.60abc	0.30c	0.39b
Pérola	49.48bc	0.86a	0.57ab
Pérola 5.1[a]	45.63c	0.68b	0.68a
BRS Pontal	73.71abc	0.40c	0.52ab
BRS Pontal 5.1[a]	85.86a	0.39c	0.38b
Mean	69.73	0.48a	0.49a
F-test			
Fertility level (FL)	*	NS	
Genotype (G)	**	**	
FL × G	NS	**	
CV (%)	34.04	20.81	

Source: Faria, J.C. and Fageria, N.K., *J. Plant Nutr.*, 34, in press.
NS, nonsignificant.
[a] Genetically modified genotypes.
*, **Significant at the 5% and 1% probability levels, respectively. Means followed by same letter in the same column (fertility and genotypes separate) are significantly not different by Tukey's test at the 5% probability level.

FIGURE 2.28 Growth of cultivar Olathe Pinto (conventional) at two fertility levels. (Left) low fertility and (right) high fertility.

FIGURE 2.29 Growth of cultivar Olathe 5.1 (genetically modified) under low (left) and high (right) fertility levels.

FIGURE 2.30 Growth of cultivar Pontal (conventional) at two fertility levels. (Left) low fertility and (right) high fertility.

FIGURE 2.31   Growth of cultivar BRS Pontal 5.1 (genetically modified) at two fertility levels. (Left) low fertility and (right) high fertility.

FIGURE 2.32   Growth of cultivar Pérola at two fertility levels. (Left) low fertility and (right) high fertility.

FIGURE 2.33 Growth of cultivar Pérola 5.1 (genetically modified) at two soil fertility levels. (Left) low fertility and (right) high fertility.

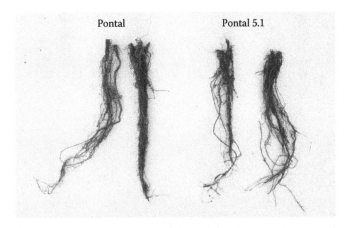

FIGURE 2.34 Root growth of cultivar Pontal (conventional) and Pontal 5.1 (genetically modified) at two fertility levels. Pontal, low fertility (left) and high fertility (right). Similarly, Pontal 5.1, low fertility (left) and high fertility (right).

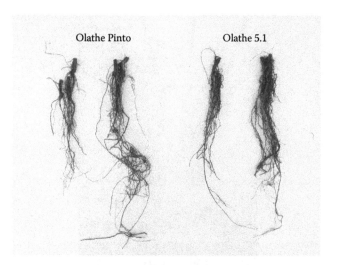

FIGURE 2.35 Root growth of cultivar Olathe Pinto (conventional) and Olathe 5.1 (genetically modified) at two fertility levels. Olathe Pinto, low fertility (left) and high fertility (right). Similarly, Olathe Pinto 5.1, low fertility (left) and high fertility (right).

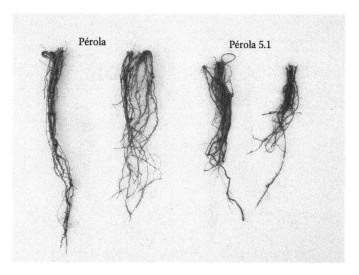

FIGURE 2.36 Root growth of cultivar perola (conventional) and genetically modified Pérola 5.1 at low fertility levels. Pérola, low fertility (left) and high fertility (right). Similarly, Pérola 5.1, low fertility (left) and high fertility (right).

and development of crop plants. The classification of nutrients as major and minor is made based on the quantity required. Major nutrients are required in higher amounts compared to minor nutrients. Visual symptoms, soil testing, plant tissue testing, and crop growth responses are nutrient-deficiency/sufficiency diagnostic techniques. Among these diagnostic techniques, soil testing is mainly used for making fertilizer and lime recommendations.

NUE is defined in several ways in the literature. We have provided five definitions of NUE in this chapter: agronomic efficiency (grain yield per unit of nutrient applied), physiological efficiency (biological yield per unit of nutrient uptake in the plant), agrophysiological efficiency (grain yield per unit of nutrient uptake in the plants), recovery efficiency (nutrient uptake in the plant per unit of nutrient applied), and utilization efficiency (physiological efficiency × recovery efficiency). Nutrient recovery efficiency in crop plants is less than 50% under most agroecological conditions. The lower nutrient efficiency is associated with loss of nutrients due to leaching, denitrification, immobilization, erosion, nutrient interaction, and uneven distribution in the soils.

Nutrient inputs in crop production systems have received special attention in recent years because of increasing fertilizer costs and concern about environmental pollution. Nutrients supplied by inorganic fertilizers make up the majority of plant nutrition requirements to sustain higher crop yields. The use of nutrient-efficient crop species or genotypes within species in combination with other improved crop production practices offers the best option for meeting the future food requirements of expanding world populations. Research efforts are also needed to generate more information on correct assessment of nutrient-deficiency/toxicity diagnosis in crop plants. Limited data are available on adequate rates, form and methods of nutrient application, and plant utilization efficiency, especially under field conditions.

Conventional and population breeding approaches were successful in the twentieth century and should continue to be important avenues of crop improvement programs in the twenty-first century. However, molecular genetic approaches, along with conventional plant-breeding methods should be applied more vigorously in developing nutrient-efficient crop species or genotypes/cultivars within species. Improved mineral nutrition traits in plants will help in reducing crop production costs and environmental pollution and should also benefit animal and human nutrition.

NUEs have improved over time. This improvement was associated with increasing yield per unit area with better crop management practices and developing crop genotypes of higher yield potentials. In conclusion, the efficient use of inorganic fertilizers is essential in today's agriculture and will be even more important in the years to come. Hence, nutrient-efficient plants will play a vital role in increasing crop yields per unit area and improve health and quality of life of humans in the twenty-first century.

REFERENCES

Abekoe, M. K. and K. L. Sahrawat. 2001. Phosphate retention and extractability in soils of the humid zone in West Africa. *Geoderma* 102:175–187.

Ae, N., J. Arihara, K. Okada, T. Yoshihara, and C. Johansen. 1990. Phosphorus uptake by pigeon pea and its role in cropping systems of the Indian subcontinent. *Science* (Washington, DC) 248:477–480.

Alagarswamy, G. 1971. Modelling of phosphate uptake by the groundnut plant (*Arachis hypogaea* L.). Doctoral thesis. Louvain, Belgium: Catholic University of Louvain.

Alam, S. M. 1999. Nutrient uptake by plants under stress conditions. In: *Handbook of Plant and Crops Stress*, ed., M. Pessarakli, pp. 285–313. New York: Marcel Dekker.

Andow, D. A. and C. Zwahlen. 2006. Assessing environmental risks of transgenic plants. *Ecol. Lett.* 9:196–214.

Andrew, C. S. 1966. A kinetic study of phosphate absorption by excised roots of *Stylosanthes humiles, Phaseolus lathyroides, Desmodium uncinatum, Medicago sativa* and *Hordeum vulgare. Aust. J. Agric. Res.* 17:611–624.

Araujo, A. P. and M. G. Teixeira. 2003. Nitrogen and phosphorus harvest indices of common bean cultivars: Implications for yield quantity and quality. *Plant Soil* 257:425–433.

Araya, F., O. Abarca, G. E. Zuniga, and L. J. Corcuera. 1991. Effects of NaCl on glycine-betaine and on aphids in cereal seedlings. *Phytochemistry* 30:1793–1795.

Arnon, D. I. and P. R. Stout. 1939. The essentiality of certain elements in minute quantity for plants with special reference to copper. *Plant Physiol.* 14:371–385.

Atkinson, D. 1990. Influence of root system morphology and development on the need for fertilizers and the efficiency of use. In: *Crops as Enhancers of Nutrient Use*, eds., V. C. Baligar and R. R. Duncan, pp. 411–451. San Diego, CA: Academic Press.

Austin, R. B. 1994. Plant breeding opportunities. In: *Physiology and Determination of Crop Yield*, eds., K. J. Boote, J. M. Bennett, T. R. Sinclair, and G. M. Paulsen, pp. 567–589. Madison, WI: ASA, CSSA, SSSA.

Austin, R. B., J. Bingham, R. D. Blackwell, L. T. Evans, M. A. Ford, C. L. Morgan, and M. Taylor. 1980. Genetic improvements in winter wheat yields since 1900 and associ-ated physiological changes. *J. Agric. Sci.* 94:675–689.

Baligar, V. C. 1987. Phosphorus uptake parameters of alfalfa and corn as influenced by P and pH. *J. Plant Nutr.* 10:33–46.

Baligar, V. C., R. R. Duncan, and N. K. Fageria. 1990. Soil-plant interactions on nutrient use efficiency in plants. An overview. In: *Crops as Enhancers of Nutrient Use*, eds., V. C. Baligar and R. R. Duncan, pp. 351–373. San Diego, CA: Academic Press.

Baligar, V. C. and N. K. Fageria. 1997. Nutrient use efficiency in acid soils: Nutrient manage-ment and plant use efficiency. In: *Plant Soil Interaction at Low pH*, eds., A. C. Monez, A. M. C. Furlani, R. E. Schaffert, N. K. Fageria, C. A. Roselem, and H. Cantarella, pp. 75–95. Campinas, Brazil: Brazilian Soil Science Society.

Baligar, V. C. and N. K. Fageria. 1999. Plant nutrient use efficiency: Towards the second paradigm. In: *Soil Fertility, Soil Biology and Plant Nutrition Interrelationships*, eds., J. O. Siqueira, F. M. S. Moreira, A. S. Lopes, L. R. G. Guilherme, V. Farquin, A. E. Furtini Neto, and J. G. Carvalho, pp. 183–204. Lavras, Brazil: Brazilian Soil Science Society/Federal University of Lavras.

Baligar, V. C., N. K. Fageria, and M. A. Elrashidi. 1998. Toxicity and nutrient constraints on root growth. *HortScience* 3:960–965.

Baligar, V. C., N. K. Fageria, and Z. L. He. 2001. Nutrient use efficiency in plants. *Commun. Soil Sci. Plant Anal.* 32:921–950.

Baligar, V. C., R. J. Wright, and N. K. Fageria. 1997. Differences in growth and nutrition of legumes in varying soil temperature and phosphorus. In. *Plant Nutrition for Sustainable Food Production and Environment*, eds., T. Ando, K. Fujita, T. Mae, H. Matsumoto, S. Mori, and J. Sekiy, pp. 133–134. Dordrecht, the Netherlands: Kluwer Academic Publishers.

Barber, S. A. 1994. Root growth and nutrient uptake. In: *Physiology and Determination of Crop Yield*, eds., K. J. Boote, J. M. Bennett, T. R. Sinclair, and G. M. Paulsen, pp. 95–99. Madison, WI: ASA, CSSA, SSSA.

Barber, S. A. 1995. *Soil Nutrient Bioavailability: A Mechanistic Approach*. New York: John Wiley & Sons Inc.

Barley, K. P. 1970. The configuration of the root system in relation to nutrient uptake. *Adv. Agron.* 22:159–201.

Bar-Yosef, B., S. Fishman, and H. Talpaz. 1980. A model of zinc movement to single roots in soils. *Soil Sci. Soc. Am. J.* 44:1272–1279.

Bennett, W. F. 1993. *Nutrient Deficiencies and Toxicities in Crop Plants*. St. Paul, MN: American Phytopathological Society.

Bienfait, H. F. 1988. Mechanisms in Fe-efficiency reactions of higher plants. *J. Plant Nutr.* 11:605–629.

Blamey, F. P. C. 2001. The role of root cell wall in aluminum toxicity. In: *Plant Nutrient Acquisition: New Perspectives*, eds., N. Ae, J. Arihara, K. Okada, and A. Srinivasan, pp. 201–226. Tokyo, Japan: Springer Verlag.

Bloom, A. J., P. A. Meyerhoff, A. R. Taylor, and T. L. Rost. 2003. Root development and absorption of ammonium and nitrate from the rhizosphere. *J. Plant Growth Regul.* 21:416–431.

Bloom, A. J., S. S. Sukrapanna, and R. L. Warner. 1992. Root respiration associated with ammonium and nitrate absorption and assimilation by barley. *Plant Physiol.* 99:1294–1301.

Blum, A. 1993. Selection for sustainable production in water-deficit environment. In. *International Crop Science I*, eds., D. R. Buxton, R. Shibles, R. A. Frosberg, B. L. Blad, K. H. Asay, G. M. Paulsen, and R. F. Wilson, pp. 343–347. Madison, WI: Crop Science Society of America.

Boone, R. D. 1994. Light-fraction soil organic matter: Origin and contribution to net nitrogen mineralization. *Soil Biol. Biochem.* 26:1459–1468.

Borlaug, N. E. and C. R. Dowswell. 1994. Feeding a human population that increasingly crowds a fragile planet. Paper presented at the *15th World Congress of Soil Science*, July 10–16, 1994, Acapulco, Mexico.

Borlaug, N. E. and C. R. Dowswell. 1997. The acid lands: One of the agricultures last frontiers. In: *Plant Soil Interactions at Low pH: Sustainable Agriculture and Forestry Production*, eds., A. C. Moniz, A. M. C. Furlani, R. E. Schaffert, N. K. Fageria, C. A. Rosolem, and H. Cantarella, pp. 5–15. Campinas, Brazil: Brazilian Soil Science Society.

Bowen, G. D. 1991. Soil temperature, root growth and plant function. In. *Plant Roots-the Hidden Half*, eds., Y. Waisel, A. Eshel, and U. Kafkafi, pp. 309–330. New York: Marcel Dekker.

Brady, N. C. and R. R. Weil. 2002. *The Nature and Properties of Soils*, 13th edn. Upper Saddle River, NJ: Prentice Hall.

Brown, L. R. 1997. Facing the challenge of food scarcity: Can we raise grain yields fast enough? In: *Plant Nutrition for Sustainable Food Production and Environment*, eds., T. Ando, K. Fujita, T. Mae, H. Matsumoto, S. Mori, and J. Sekiya, pp. 15–24. Dordrecht, the Netherlands: Kluwer Academic Publishers.

Caradus, J. R. 1990. Mechanisms improving nutrient use by crop and herbage legumes. In: *Crops as Enhancers of Nutrient Use*, eds., V. C. Baligar and R. R. Duncan, pp. 253–311. San Diego, CA: Academic Press.

Caradus, J. R. and R. W. Snaydon. 1986. Plant factors influencing phosphorus uptake by white clover from solution culture. I. Populations differences. *Plant Soil* 93:153–163.

Carpenter, S., N. F. Caraco, D. L. Correll, R. W. Howarth, A. N. Sharply, and V. H. Smith. 1998. Nonpoint pollution of surface waters with phosphorus and nitrogen. *Issues Ecol.* 3:1–12.

Chandler, S. and J. M. Dunwell. 2008. Gene flow, risk assessment and the environmental release of transgenic plants. *Crit. Rev. Plant Sci.* 27:25–49.

Chapman, M. A. and J. M. Burke. 2006. Letting the gene out of the bottle: The population genetics of genetically modified crops. *New Phytologist* 170:429–443.

Clark, R. B. 1990. Physiology of cereals for mineral nutrient uptake, use, and efficiency. In: *Crops as Enhancers of Nutrient Use*, eds., V. C. Baligar and R. R. Duncan, pp. 131–209. San Diego, CA: Academic Press.

Clark, R. B. and R. R. Duncan. 1991. Improvement of plant mineral nutrition through breeding. *Field Crop Res.* 27:219–240.

Clark, R. B. and R. R. Duncan. 1993. Selection of plants to tolerate soil salinity acidity, and mineral deficiencies. In: *International Crop Science I*, eds., D. R. Buxton, R. Shibles, R. A. Frosberg, B. L. Blad, K. H. Asay, G. M. Paulsen, and R. F. Wilson, pp. 371–379. Madison, WI: Crop Science Society of America.

Cooper, A. J. 1973. Root temperature and plant growth-A review. In. *Research Review No 4 Commonwealth Bureau of Horticulture and Plantation Crops*. Wallingford, U.K.: Commonwealth Agricultural Bureaux.

Cregan, P. B. and R. W. Yaklich. 1986. Dry matter and nitrogen accumulation and partitioning in selected soybean genotypes of different derivation. *Theor. Appl. Genet.* 72:782–786.

Crews, T. E. and M. B. Peoples. 2004. Legume versus fertilizer sources of nitrogen: Ecological tradeoff and human needs. *Agric. Ecosyst. Environ.* 102:279–297.

Crop Science Society of America. 1992. *Glossary of Crop Science Terms*. Madison, WI: Crop Science Society of America.

Darrah, P. R. 1993. The rhizosphere and plant nutrition: A quantitative approach. *Plant Soil* 156:1–20.

Davis, L. C. 2006. Genetic engineering, ecosystem change, and agriculture: An update. *Biotech. Mol. Biol. Rev.* 1:87–102.

Dhugga, K. S. and J. G. Waines. 1989. Analysis of nitrogen accumulation and use in bread and durum wheat. *Crop Sci.* 29:132–139.

Donald, C. M. 1962. In search of yield. *J. Aust. Inst. Agric. Sci.* 28:171–178.

Dudal, R. 1982. Land degradation in a world perspective. *J. Soil Water Conserv.* 37:345–249.

Duncan, R. R. 1994. Genetic manipulation. In: *Plant Environment Interactions*, ed., R. E. Wilkinson, pp. 1–38. New York: Marcel Dekker.

Duncan, R. R. and R. N. Carrow. 1999. Turfgrass molecular genetic improvement for abiotic/edaphi stress resistance. *Adv. Agron.* 67:233–306.

Edwards, D. G. 1970. Phosphate absorption and long-distance transport in wheat seedlings. *Aust. J. Biol. Sci.* 23:255–264.

Ellstrand, N. C. 2003. *Dangerous Liaisons? When Cultivated Plants Mate with Their Wild Relatives*. Baltimore, MD: John Hopkins University Press.

Epstein, E. and A. J. Bloom. 2005. *Mineral Nutrition of Plants: Principles and Perspectives*, 2nd edn. Sunderland, MA: Sinauer Associations, Inc.

Epstein, E. and R. L. Jefferies. 1964. The genetic basis of selective ion transport in plants. *Annu. Rev. Plant Physiol.* 15:169–184.

Evans, L. T. 1993. *Crop Evolution, Adaptation and Yield*. Cambridge, England: Cambridge University Press.

Evans, L. T. 1994. Crop physiology: Prospects for the retrospective science. In: *Physiology and Determination of Crop Yield*, eds., K. J. Boote, J. M. Bennett, T. R. Sinclair, and G. M. Paulsen, pp. 19–35. Madison, WI: ASA, CSSA, SSSA.

Evans, L. T. and H. M. Rawson. 1970. Photosynthesis and respiration by the flag leaf and components of the ear during grain development in wheat. *Aust. J. Biol. Sci.* 23:245–254.

Evans, L. T. and I. F. Wardlaw. 1976. Aspects of the comparative physiology of grain yield in cereals. *Adv. Agron.* 28:301–359.

Evans, L. T. and I. F. Wardlaw. 1996. Wheat. In: *Photoassimilate Distribution in Plants and Crops: Source-Sink Relations*, eds., E. Zamski and A. A. Schaffer, pp. 501–518. New York: Marcel Dekker.

Fageria, N. K. 1973. Uptake of nutrients by the rice plant from dilute solutions. Doctoral thesis. Louvain, Belgium: Catholic University of Louvain.

Fageria, N. K. 1974. Kinetics of phosphate absorption in intact rice plants. *Aust. J. Agric. Res.* 25:395–400.

Fageria, N. K. 1992. *Maximizing Crop Yields*. New York: Marcel Dekker.

Fageria, N. K. 1998. Phosphorus use efficiency by bean genotypes. *Revista Brasileira Eng. Agri. Amb.* 2:128–131.

Fageria, N. K. 2000a. Adequate and toxic levels of zinc for rice, common bean, corn, soybean and wheat production in cerrado soil. *Revista Brasileira Eng. Agri. Amb.* 4:390–395.

Fageria, N. K. 2000b. Potassium use efficiency of upland rice genotypes. *Pesq. Agropec. Bras.* 35:2115–2120.

Fageria, N. K. 2004. Dry matter yield and shoot nutrient concentrations of upland rice, common bean, corn, and soybean grown in rotation on an Oxisol. *Commun. Soil Sci. Plant Anal.* 35:961–974.

Fageria, N. K. 2009. *The Use of Nutrients in Crop Plants.* Boca Raton, FL: CRC Press.

Fageria, N. K. and V. C. Baligar. 1997a. Response of common bean, upland rice, corn, wheat, and soybean to soil fertility of an Oxisol. *J. Plant Nutr.* 20:1279–1289.

Fageria, N. K. and V. C. Baligar. 1997b. Upland rice genotypes evaluation for phosphorus use efficiency. *J. Plant Nutr.* 20:499–509.

Fageria, N. K. and V. C. Baligar. 1999. Phosphorus use efficiency in wheat genotypes. *J. Plant Nutr.* 22:331–340.

Fageria, N. K. and V. C. Baligar. 2001. Lowland rice response to nitrogen fertilization. *Commun. Soil Sci. Plant Anal.* 32:1405–1429.

Fageria, N. K. and V. C. Baligar. 2003. Methodology for evaluation of lowland rice genotypes for nitrogen use efficiency. *J. Plant Nutr.* 26:1315–1333.

Fageria, N. K. and V. C. Baligar. 2005a. Nutrient availability. In: *Encyclopedia of Soils in the Environment*, ed., D. Hillel, pp. 63–71. San Diego, CA: Elsevier.

Fageria, N. K. and V. C. Baligar. 2005b. Enhancing nitrogen use efficiency in crop plants. *Adv. Agron.* 88:97–185.

Fageria, N. K. and V. C. Baligar. 2005c. Growth components and zinc recovery efficiency of upland rice genotypes. *Pesq. Agropec. Bras.* 40:1211–1215.

Fageria, N. K. and V. C. Baligar. 2006. Nutrient efficient plants in improving crop yields in the twenty first century. Paper presented at the *18th World Soil Science Congress*, July 9–15, 2006, Philadelphia, PA.

Fageria, N. K. and V. C. Baligar. 2008. Ameliorating soil acidity of tropical Oxisols by liming for sustainable crop production. *Adv. Agron.* 99:345–399.

Fageria, N. K., V. C. Baligar, and R. B. Clark. 2002. Micronutrients in crop production. *Adv. Agron.* 77:185–268.

Fageria, N. K., V. C. Baligar, and R. B. Clark. 2006. *Physiology of Crop Production.* New York: The Haworth Press.

Fageria, N. K., V. C. Baligar, and C. A. Jones. 2011a. *Growth and Mineral Nutrition of Field Crops*, 3rd edn. Boca Raton, FL: CRC Press.

Fageria, N. K., V. C. Baligar, and Y. C. Li. 2008. The role of nutrient efficient plants in improving crop yields in the twenty first century. *J. Plant Nutr.* 31:1121–1157.

Fageria, N. K. and M. P. Barbosa Filho. 2001. Nitrogen use efficiency in lowland rice genotypes. *Commun. Soil Sci. Plant Anal.* 32:2079–2089.

Fageria, N. K. and M. P. Barbosa Filho. 2007. Dry matter and grain yield, nutrient uptake, and phosphorus use efficiency of lowland rice as influenced by phosphorus fertilization. *Commun. Soil Sci. Plant Anal.* 38:1289–1297.

Fageria, N. K., M. P. Barbosa Filho, and J. G. C. da Costa. 2001. Potassium use efficiency in common bean genotypes. *J. Plant Nutr.* 24:1937–1945.

Fageria, N. K., M. P. Barbosa Filho, and L. F. Stone. 2004a. Phosphorus nutrition for bean production. In: *Phosphorus in Brazilian Agriculture*, eds., T. Yamada and S. R. S. Abdalla, pp. 435–455. Piracicaba, Brazil: Brazilian Potassium and Phosphate Research Association.

Fageria, N. K., E. M. Castro, and V. C. Baligar. 2004b. Response of upland rice genotypes to soil acidity. In: *The Red Soils of China: Their Nature, Management and Utilization*, eds., M. J. Wilson, Z. He, and X. Yang, pp. 219–237. Dordrecht, the Netherlands: Kluwer Academic Publishers.

Fageria, N. K. and H. R. Gheyi. 1999. *Efficient Crop Production.* Campina Grande, Brazil: Federal University of Paraiba.

Fageria, N. K., A. Moreira, and C. Castro. 2011b. Response of soybean to phosphorus fertilization in Brazilian Oxiso. *Commun. Soil Sci. Plant Anal.* 42:2716–2723.

Fageria, N. K., A. Moreira, and A. M. Coelho. 2011c. Yield and yield components of upland rice as influenced by nitrogen sources. *J. Plant Nutr.* 34:371–386.

Fageria, N. K. and J. P. Oliveira. (in press). Nitrogen, phosphorus, and potassium interactions in upland rice. *J. Plant Nutr.* 34.

Fageria, N. K., A. B. Santos, and V. A. Cutrim. 2009. Nitrogen uptake and its association with grain yield in lowland rice. *J. Plant Nutr.* 32:1965–1974.

Fageria, N. K., A. B. Santos, and A. B. Heinemann. 2011d. Lowland rice genotypes evaluation for phosphorus use efficiency in tropical lowland. *J. Plant Nutr.* 34:1087–1095.

Fageria, N. K., A. B. Santos, and A. Moreira. 2010. Yield, nutrient uptake, and changes in soil chemical properties as influenced by liming and iron application in common bean in a no-tillage system. *Commun. Soil Sci. Plant Anal.* 41:1740–1749.

Fageria, N. K. and J. M. Scriber. 2002. The role of essential nutrients and minerals in insect resistance in crop plants. In: *Insect and Plant Defense Dynamics*, ed., T. N. Ananthakrishnan, pp. 23–54. Enfield, NH: Science Publisher.

Fageria, N. K., N. A. Slaton, and V. C. Baligar. 2003. Nutrient management for improving lowland rice productivity and sustainability. *Adv. Agron.* 80:63–152.

Fageria, N. K. and L. F. Stone. 2006. Physical, chemical and biological changes in rhizosphere and nutrient availability. *J. Plant Nutr.* 29:1–30.

Faria, J. C. and N. K. Fageria. (in press). Genetically modified and conventional dry bean genotype responses to soil fertility. *J. Plant Nutr.* 34.

Fawcett, J. A. and K. J. Frey. 1983. Association among nitrogen harvest index and other traits within two *Avena* species. *Proc. Iowa Acad. Sci.* 90:150–153.

Feil, B. 1992. Breeding progress in small grain cereals: A comparison of old and modern cultivars. *Plant Breeding* 108:1–11.

Foy, C. D. 1984. Physiological effects of hydrogen, aluminum and manganese toxicities in acid soils. In: *Soil Acidity and Liming*, 2nd edn., ed., F. Adams, pp. 57–97, Madison WI: SSSA, ASA, and CSSA.

Foy, C. D. 1992. Soil chemical factors limiting plant root growth. *Adv. Soil Sci.* 19:97–149.

Gale, W. J., C. A. Cambardell, and T. B. Bailey. 2000. Root derived carbon and the formation and stabilization of aggregates. *Soil Sci. Soc. Am. J.* 64:201–207.

Gerloff, G. C. and W. H. Gabelman. 1983. Genetic basis of inorganic plant nutrition. In: *Inorganic Plant Nutrition*. Encyclopaedia and Plant Physiology New Series, Vol. 15B, eds., A. Lauchli and R. L. Bieleski, pp. 453–480. New York: Springer Verlag.

Gifford, R. M., J. H. Thorne, W. Hitz, and R. T. Giaquinta. 1984. Crop productivity and photoassimilate partitioning. *Science* 225:801–808.

Gourley, C. J. P., D. L. Allan, and M. P. Russelle. 1994. Plant nutrient efficiency: A comparison of definitions and suggested improvement. *Plant Soil* 158:29–37.

Graham, R. D. 1984. Breeding for nutritional characteristics in cereals. In: *Advances in Plant Nutrition*, Vol. 1, eds., P. B. Tinker and A. Lauchi, pp. 57–102. New York: Praeger Publisher.

Grattan, S. R. and C. M. Grieve. 1999a. Salinity-mineral nutrient relations in horticultural crops. *Sci. Hort.* 78:127–157.

Grattan, S. R. and C. M. Grieve. 1999b. Mineral nutrient acquisition and response by plants grown in saline environments. In: *Handbook of Plant and Crops Stress*, ed., M. Pessarakli, pp. 203–229. New York: Marcel Dekker.

Gregory, P. J. 1994. Root growth and activity. In: *Physiology and Determination of Crop Yield*, eds., K. J. Boote, J. M. Bennett, T. R. Sinclair, and G. M. Paulsen, pp. 65–93. Madison, WI: ASA, CSSA, SSSA.

Grundon, N. J. 1987. *Hungry Crops: A Guide to Nutrient Deficiencies in Field Crops*. Brisbane, Queensland, Australia: Department of Primary Industries.

Gupta, R. J. and I. P. Abrol. 1990. Salt-affected soils: Their reclamation and management for crop production. *Adv. Soil Sci.* 11:223–288

Hay, R. K. M. 1995. Harvest index: A review of its use in plant breeding and crop physiology. *Ann. Appl. Biol.* 126:197–216.

Hillel, D. and C. Rosenzweig. 2005. The role of biodiversity in agronomy. *Adv. Agron.* 88:1–34.

Hinsinger, P. 1998. How do plant roots acquire mineral nutrients? Chemical processes involved in the rhizosphere. *Adv. Agron.* 64:225–265.

Hinsinger, P. and R. J. Gilkes. 1996. Mobilization of phosphate from phosphate rock and alumina-sorbed phosphate by the roots of ryegrass and clover as related to rhizosphere pH. *Eur. J. Soil Sci.* 47:533–544.

Ho, L. C. 1988. Metabolism and compartmentation of imported sugars in sink organs in relation to sink strength. *Annu. Rev. Plant Physiol. Plant Mol. Biol.* 39:355–378.

Huber, D. M. 1980. The role of mineral nutrition in defense In: *Plant Pathology: An Advanced Treatise*, eds., G. Horsfall and E. B. Cowling, pp. 381–406. New York: Academic Press.

Inoue, H. and A. Tanaka. 1978. Comparison of source and sink potentials between wild and cultivated potatoes. *J. Soil Sci. Manure Jpn.* 49:321–327.

International Plant Nutrition Institute. 2010. A preliminary nutrient use geographic information system for the U.S. IPNI Publication no. 30-3270. Norcross, GA: IPNI.

Isfan, D. 1993. Genotypic variability for physiological efficiency index of nitrogen in oats. *Plant Soil* 154:53–59.

Ishizuka, Y. 1978. *Nutrient Deficiencies of Crops.* Taipei, Taiwan: Food and Fertilizer Technology Center.

Jackson, L. E. and A. J. Bloom. 1990. Root distribution in relation to soil nitrogen availability in field grown tomatoes. *Plant Soil* 128:115–126.

James, C. 2007. Executive summary: Global status of commercialized biotech/GM crops: 2007. ISAAA Briefs No. 37-2007. Ithaca, NY: ISAAA.

Jarrell, W. M. and R. B. Beverly. 1981. The dilution effect in plant nutrition studies. *Adv. Agron.* 34:197–224.

Keerthisinghe, G., F. Zapta, P. Chalk, and P. Hocking. 2001. Integrated approach for improved P nutrition of plants in tropical acid soils. In: *Plant Nutrition: Food Security and Sustainability of Agroecosystems*, eds., W. J. Horst, M. K. Schenk, A. Burkert, N. Claassen, H. Flessa, W. B. Frommer et al., pp. 974–975. Dordrecht, the Netherlands: Kluwer Academic Publishers.

Kiniry, J. R., G. McCauley, Y. Xie, and J. G. Arnold. 2001. Rice parameters describing crop performance of four U.S. cultivars. *Agron. J.* 93:1354–1361.

Kochian, L. V. 1995. Cellular mechanisms of aluminum toxicity and resistance in plants. *Annu. Rev. Plant Physiol. Plant Mol. Biol.* 46:237–260.

Kuo, S., U. M. Sainju, and E. J. Jellum. 1997. Winter cover crop effects on soil organic carbon and carbohydrate. *Soil Sci. Soc. Am. J.* 61:145–152.

Lal, R. 1990. Soil erosion and land degradation: The global risks. *Adv. Soil Sci.* 11:129–172.

Liang, B. C., X. L. Wang, and B. L. Ma. 2002. Maize root-induced change in soil organic carbon pools. *Soil Sci. Soc. Am. J.* 66:845–847.

Lindsay, W. L. 1979. *Chemical Equilibrium in Soils.* New York: John Wiley & Sons.

Loneragan, J. F. 1978. The physiology of plant tolerance to low phosphorus availability. In: *Crop Tolerance to Suboptimal Land Conditions*, ed., G. A. Jung, pp. 329–343. Madison, WI: ASA.

Loneragan, J. F. 1997. Plant nutrition in the 20th and perspectives for the 21st century. In: *Plant Nutrition for Sustainable Food Production and Environment*, eds., T. Ando, K. Fujita, T. Mae, H. Matsumoto, S. Mori, and J. Sekiya, pp. 3–14. Dordrecht, the Netherlands: Kluwer Academic Publishers.

Ludlow, M. M. and R. C. Muchow. 1990. A critical evaluation of traits for improving crop yields in water-limited environments. *Adv. Agron.* 43:107–153.

Lynch, J. 1995. Root architecture and plant productivity. *Plant Physiol.* 109:7–13.

Maas, E. V. 1986. Salt tolerance of Plants. *Appl. Agric. Res.* 1:12–26.

Marin, M., L. A. Valdez-Aguilar, A. M. Castillo-Gonzalez, J. Pineda-Pineda, and J. J. G. Luna. 2011. Modeling growth and ion concentration of lilium in response to nitrogen:potassium:calcium mixture solution. *J. Plant Nutr.* 34:12–26.

Marschner, H. 1995. *Mineral Nutrition of Higher Plants*, 2nd edn. New York: Academic Press.

Mengel, K., E. A. Kirkbay, H. Kosegarten, and T. Appel. 2001. *Principles of Plant Nutrition*, 5th edn. Dordrecht, the Netherlands: Kluwer Academic Publishers.

Merrill, S. D., D. L. Tanaka, and J. D. Hanson. 2002. Root length growth of eight crop species in Haplustoll soils. *Soil Sci. Soc. Am. J.* 66:913–923.

Moll, R. H., E. J. Kamprath, and W. A. Jackson. 1982. Analysis and interpretation of factors which contribute to efficiency of nitrogen utilization. *Agron. J.* 74:562–564.

Mortvedt, J. J. 1994. Needs for controlled availability micronutrient fertilizers. *Fert. Res.* 38:213–221.

Mortvedt, J. J. 2000. Bioavailability of micronutrients. In: *Handbook of Soil Science*, ed., M. E. Sumner, pp. 71–88. Boca Raton, FL: CRC Press.

Norby, R. J. and M. F. Cotrufo. 1998. A question of litter quality. *Nature* (London) 396:17–18.

Novoa, R. and R. S. Loomis. 1981. Nitrogen and plant production. In: *Soil Water and Nitrogen in Mediterranean Type Environments*, eds., J. Monteith and C. Webb, pp. 177–204. The Hague, the Netherlands: Martinus Nijhoff/Dr. W. Junk Publishers.

Okada, K. and A. J. Fischer. 2001. Adaptation mechanisms of upland rice genotypes to highly weathered acid soils of South American Savannah's. In: *Plant Nutrient Acquisition: New Perspectives*, eds., N. Ae, J. Arihara, K. Okada, and A. Srinivasan, pp. 185–200. Tokyo, Japan: Springer Verlag.

Payne, T. S., D. D. Stuthman, R. L. McGraw, and P. P. Bregitzer. 1986. Physiological changes associated with three cycles of recurrent selection for grain yield improvement in oats. *Crop Sci.* 26:734–736.

Peng, S., R. C. Laza, R. M. Visperas, A. L. Sanico, K. G. Cassman, and G. S. Khush. 2000. Grain yield of rice cultivars and lines developed in the Philippines since 1966. *Crop Sci.* 40:307–314.

Pessarakli, M. ed. 1999. *Handbook of Plant and Crops Stress*. New York: Marcel Dekker.

Pettersson, S. and P. Jensen. 1983. Variation among species and varieties in uptake and utilization of potassium. *Plant Soil* 72:231–237.

Puget, P. and L. E. Drinkwater. 2001. Short-term dynamics of root and shoot derived carbon from a leguminous green manure. *Soil Sci. Soc. Am. J.* 65:771–779.

Pugnaire, F., L. Serrano, and J. Pardos. 1999. Constraints by water stress on plant growth. In: *Handbook of Plant and Crops Stress*, ed., M. Pessarakli, pp. 271–283. New York: Marcel Dekker.

Qian, J. H. and J. W. Doran. 1996. Available carbon released from crop roots during growth as determined by carbon-13 natural abundance. *Soil Sci. Soc. Am. J.* 60:828–831.

Radin, J. W. and J. Lynch. 1994. Nutritional limitations to yield: Alternatives to fertilization. In: *Physiology and Determination of Crop Yield*, eds., K. J. Boote, J. M. Bennett, T. R. Sinclair, and G. M. Paulsen, pp. 277–283. Madison, WI: ASA, CSSA, SSSA.

Ranells, N. N. and M. G. Wagger. 1997. Winter annual grass-legume bicultures for efficient nitrogen management in no-till corn. *Agric. Ecosyst. Environ.* 65:23–32.

Rattunde, H. F. and K. J. Frey. 1986. Nitrogen harvest index in oats: Its repeatability and association with adaptation. *Crop Sci.* 26:606–610.

Raun, W. R. and G. V. Johnson. 1999. Improving nitrogen use efficiency for cereal production. *Agron J.* 91:357–363.

Richardson, A. E. 2001. Prospects for using soil microorganisms to improve the acquisition of phosphorus by plants. *Aust. J. Plant Physiol.* 28:897–906

Riley, D. and S. A. Barber. 1971. Effect of ammonium and nitrate fertilization on phosphorus uptake as related to root-induced pH changes at the root soil-interface. *Soil Sci. Soc. Am. Proc.* 35:301–306.

Robson, A. D. and M. G. Pitman. 1983. Interaction between nutrients in higher plants. In: *Inorganic Plant Nutrition. Encyclopedia of Plant Physiology*, Vol. 15A, eds., A. Lauchli and R. L. Bieleski, pp. 147–180. New York: Springer Verlag.

Romheld, V. and H. Marschner. 1991. Functions of micronutrients n plants. In: *Micronutrients in Agriculture*, 2nd edn., eds., J. J. Mortvedt, F. R. Cox, L. M. Shuman, and R. M. Welch, pp. 297–328. Madison, WI: Soil Science Society of America.

Rosecrance, R. C., G. W. McCarty, D. R. Shelton, and J. R. Teasdale. 2000. Denitrification and N mineralization from hairy vetch (*Vicia villosa* Roth) and rye (*Secale cereal* L.) cover crop monocultures and bicultures. *Plant Soil* 227:283–290.

Sahrawat, K. L. and M. Sika. 2002. Direct and residual phosphorus effects on soil test values and their relationships with grain yield and phosphorus uptake of upland rice on an Ultisol. *Commun. Soil Sci. Plant Anal.* 33:321–332.

Sainju, U. M., B. P. Singh, and W. F. Whitehead. 2005. Tillage, cover crops, and nitrogen fertilization effects on cotton and sorghum root biomass, carbon, and nitrogen. *Agron J.* 97:1279–1290.

Sanchez, P. A. 1976. *Properties and Management of Soils in the Tropics*. New York: John Wiley & Sons.

Schmidt, J. W. 1984. Genetic contributions to yield grains in wheat. In: *Genetic Contributions to Yield Grains of Five Major Crop Plants*, Crop Science Society of America special publication 7, ed., W. R. Fehr, pp. 89–101. Madison, WI: ASA.

Schoonhoven, L. M., T. Jermy, and J. J. A. van Loon. 1998. *Insect-Plant Biology: Physiology to Evolution*. New York: Chapman and Hall.

Scott-Russell, R. 1977. *Plant Root Systems: Their Function and Interaction with the Soil.* London, U.K.: McGraw-Hill.

Sharma, R. C. and E. L. Smith. 1986. Selection for high and low harvest index in three winter wheat populations. *Crop Sci.* 26:1147–1150.

Singh, Y., A. Dobermann, B. Singh, K. F. Bronson, and C. S. Khind. 2000. Optimal phosphorus management strategies for wheat-rice cropping on a loamy sand. *Soil Sci. Soc. Am. J.* 64:1413–1422.

Smith, F. W., A. L. Rae, and M. J. Hawkesford. 2000. Molecular mechanisms of phosphate and sulphate transport in plants. *Biochim. Biophys. Acta* 1465:236–245.

Soil Science Society of America. 1997. *Glossary of Soil Science Terms*. Madison, WI: Soil Science Society of America.

Spena, A., J. J. Estruch, and J. Schell. 1992. On microbes and plants: New insights into phytohormonal research. *Curr. Opin. Biotechnol.* 3:159–163.

Starovoytov, A., R. S. Gallagher, K. L. Jacobsen, J. P. Kaye, and B. Bradley. 2010. Management of small grain residues to retain legume-derived nitrogen in corn cropping systems. *Agron. J.* 102:895–903.

Tollenaar, M. 1989. Genetic improvement in grain yield of commercial maize hybrids grown in Ontario from 1959 to 1988. *Crop Sci.* 29:1365–1371.

Unger, P. W. and T. C. Kaspar. 1994. Soil compaction and root growth: A review. *Agron. J.* 86:759–766.

Vose, P. B. 1984. Effects of genetic factors on nutritional requirement of plants. In: *Crop Breeding: A Contemporary Basis*, eds., P. B. Vose and S. G. Blixt, pp. 67–114. Oxford, U.K.: Pergamon Press.

Vose, P. B. 1990. Plant nutrition relationships at the whole-plant level. In: *Crops as Enhancers of Nutrient Use*, eds., V. C. Baligar and R. R. Duncan, pp. 65–80. San Diego, CA: Academic Press.

Wallace, T. 1961. *The Diagnosis of Mineral Deficiencies in Plants by Visual Symptom*, 2nd edn. New York: Chemical Publishing.

Wallace, D. H., J. L. Ozbun, and H. M. Munger. 1972. Physiological genetics of crop yield. *Adv. Agron.* 24:97–146.

Welch, R. M., W. H. Allaway, W. A. House, and J. Kubota. 1991. Geographic distribution of trace element problems. In: *Micronutrients in Agriculture*, 2nd edn., eds., J. J. Mortvedt, F. R. Cox, L. M. Shuman, and R. M. Welch, pp. 31–57. Madison, WI: SSSA.

Wilcox, G. E. and N. K. Fageria. 1977. Nutrient deficiencies in dry bean and their corrections. EMBRAPA-CNPAF, Technical Bulletin 5, 21 pp, Goiania, Brazil.

Wilkinson, S. R., D. L. Grunes, and M. E. Sumner. 2000. Nutrient interactions in soil and plant nutrition. In: *Handbook of Soil Science*, ed., M. E. Sumner, pp. 89–111. Boca Raton, FL: CRC Press.

Willenborg, C. J., E. C. Luschei, A. L. Brule-Babel, and R. C. Van Acker. 2009. Crop genotype and plant population density impact flowering phenology and synchrony between cropped and volunteer spring wheat. *Agron. J.* 101:1311–1321.

Willenborg, C. J. and R. C. Van Acker. 2008. The biology and ecology of hexaploid wheat (*Triticum aestivum* L.) and its implications for trait confinement. *Can. J. Plant Sci.* 88:997–1013.

Wilson, J. B. 1993. Macronutrient (NPK) toxicity and interactions in the grass *Festuca ovina*. *J. Plant Nutr.* 16:1151–1159.

Witt, C., A. Dobermann, S. Abulrachman, H. C. Gines, G. H. Wang, R. Nagarajan et al. 1999. Internal nutrient efficiencies of irrigated lowland rice in tropical and subtropical Asia. *Field Crop Res.* 63:113–138.

Wortmann, C. S., A. R. Dobermann, R. B. Ferguson, G. W. Hergert, C. A. Shapiro, D. D. Tarkalson, and D. T. Walters. 2009. High yielding corn response to applied phosphorus, potassium, and sulfur I Nebraska. *Agron. J.* 101:546–555.

Wych, R. D. and D. C. Rasmusson. 1983. Genetic improvement in malting barley cultivars since 1920. *Crop Sci.* 23:1037–1040.

Yang, X., W. Wang, Z. Ye, Z. He, and V. C. Baligar. 2004. Physiological and genetic aspects of crop plant adaptation to elemental stresses in acid soils. In: *The Red Soils of China: Their Nature, Management and Utilization*, eds., M. J. Wilson, Z. He, and X. Yang, pp. 171–218. Dordrecht, the Netherlands: Kluwer Academic Publishers.

Yoshida, S. 1981. *Fundamentals of Rice Crop Science*. Los Bãnos, Philippines: The International Rice Research Institute.

Zadoks, J. C., T. T. Chang, and C. F. Konzak. 1974. A decimal code for the growth stages of cereals. *Weed Sci.* 14:415–421.

Zhang, H. and B. G. Forde. 1998. An Arabidopsis MADS box gene that controls nutrient induced changes in root architecture. *Science* 279:407–409.

Zhu, X., H. Zhang, and L. Yan. 2011. Variation and interrelations among nutrient elements in wheat leaves used for forage. *J. Plant Nutr.* 34:1321–1329.

3 Absorption of Water by Roots

3.1 INTRODUCTION

Water is a precious natural resource for mankind. Shortage of freshwater resources is one of the main environmental problems in many regions of the world, and its importance will increase in the near future because of the growing pressure on water resources and the effects of climate change (Shannon et al., 2008; Lado and Ben-Hur, 2009). Agriculture worldwide is heavily dependent on water availability, making water management one of the most important components of modern agriculture. Water is the solvent and transport medium in natural systems. In general, the cultivation of plants is impossible without water. For normal growth and maturation of plants, the availability and preservation of sufficient soil moisture during the whole plant maturation period is required (Hamzei, 2011). Water is a major constituent of plants and microbes and a reactant or substrate in many important processes crucial for metabolic activity (Hinsinger et al., 2009). Another role of water is the maintenance of turgor, which is essential for cell enlargement and growth (Kramer and Boyer, 1995). The absorption of water in adequate amount by roots is the main factor for the optimal growth and development of plants.

The distribution of vegetation over the surface of the earth is controlled by the availability of water more than any other factor. Where there is adequate and well-distributed rainfall during the growing season, there is lush vegetation (Kramer, 1969). Examples are the rain forests of the tropics, especially the Amazon region of South America (Figures 3.1 through 3.4). Water is used in industries, in agriculture, and for domestic purpose; also for animals, water is one of the basic needs for survival. Globally, a major part of water is used in agricultural production ($\approx$70%). In Asia (India and China), irrigated agriculture uses about 80% of the region's freshwater. It is estimated that India will have a water deficit of 50% by 2030, while China would have a shortage of 25%. Climate change, rapid industrialization in India and China, water pollution, dietary shifts, and the drive to grow biofuels are also expected to escalate the water crisis in Asia and elsewhere. There clearly is mounting concern throughout the world regarding diminishing water supplies and the need for water conservation to overcome impending deficiencies of food and fiber at a time when the population is increasing, especially in developing countries.

Rajan et al. (2010) reported that in the semiarid Texas High Plains of the United States, the growth and yield of agricultural crops are inexorably linked to the amount of water available from precipitation and/or irrigation. Grain sorghum is a major crop grown under semiarid conditions in the United States and other parts of the world (Bandaru et al., 2006). In the U.S. southern Great Plains, dryland grain yields

FIGURE 3.1   Forest vegetation of Amazon forest in the state of Acre, Brazil.

FIGURE 3.2   Type of vegetation in the Amazon forest, state of Acre, Brazil.

are generally low and highly variable; the average yields from 1972 to 2004 were 2530 kg ha^{-1} for southwest Kansas, 2280 kg ha^{-1} for the north Texas High Plains, and 1860 kg ha^{-1} for the south Texas High Plains (National Agricultural Statistics Database, 2005).

Stone and Schlegel (2010) reported that inadequate precipitation is the primary factor that limits dryland crop production in the west-central Great Plains. Soil water deficits during grain fill in sorghum accounted for >300,000 Mg year^{-1} grain yield

FIGURE 3.3   Vegetation diversity Amazon forest in the state of Acre, Brazil.

FIGURE 3.4   Amazon forest clearing by bulldozer in the state of Acre, Brazil.

loss and was among the major causes of grain yield loss in Ethiopia (Mesfin et al., 2010). Increasing the productivity and profitability of dryland agriculture depends on achieving more efficient use of precipitation (i.e., getting more economic yield per unit of precipitation) (Peterson et al., 1996). Many studies have shown a linear relationship between crop dry mass or yield and the water used by the crop over the growing season (Sammis, 1981; Hanks and Rasmussen, 1982; Howell et al., 1984; Hay and Walker, 1989).

Root water uptake is regulated by several anatomical and physiological traits, such as the diameter and number of the xylem vessels, and by the solute content associated with osmotic adjustment (Cruz et al., 1992; Kramer and Boyer, 1995; Sonobe et al., 2011). Water is transported through the soil into the roots and plant xylem toward the plant canopy where it eventually is transpired into atmosphere to assimilate CO_2 (Jackson et al., 2000; Hopmans and Bristow, 2002). Roots provide the hydraulic continuity between soil and atmosphere and thereby play a key role in the global water cycle (Hinsinger et al., 2009). Roots are in contact with the surrounding soil by film on its surfaces or mucigel, which can also play a controlling role on water and nutrient absorption by plants (Hopmans and Bristow, 2002). A major part of water is absorbed by the root zone where the root hairs are located (generally a few millimeters or centimeters from the tip). As it is moved away from the root tips or root hair zone, the root surface becomes less permeable to water and its interior is more involved with conducting water upward to the stem (Nobel, 1974). For barley (*Hordeum vulgare* L.) and pumpkin (*Curcurbita pepo* L.), the maximum water uptake was reported to occur 3–8 mm behind the root tip (Clarkson et al., 1971; Sanderson, 1983; Kramer and Boyer, 1995). For tree roots maximum uptake was observed close to root tips or where lateral roots are emerging (Haussling et al., 1988). It is proved by direct measurement that water can move in either direction within the root system, depending on the direction of water potential gradients (Burgess et al., 2000), and there is no indication of a general rectifier like behavior of roots, that is, a higher resistance to water efflux compared with influx, from anatomical or physiological features (Hinsinger et al., 2009).

The role of roots in determining plant accessibility to water and nutrient received little attention throughout the early 1900s when the availability of nutrients to plants was defined by the use of chemical extractants and that of water was similarly defined by the equilibrium concepts of field capacity and permanent wilting point (Gregory, 2006). It was not until the mid-1950s that the idea of water and nutrient mobility superseded those of equilibrium and thermodynamics. For nutrients, the change of thinking came in a groundbreaking paper by Bray (1954) in which he introduced the concept that nutrient mobility was central to soil–plant relations and demonstrated that mobile nutrients such as nitrate moved to roots from large distances whereas adsorbed nutrients such as phosphorus moved only short distances (Gregory, 2006). The corollary of this was that the zones of competition for nutrients by roots differed depending upon the mobility of the nutrient. This change of thinking about the availability of nutrients to plants was paralleled by similar developments regarding the movement of water toward roots (Gardener, 1960; Gregory, 2006).

Currently, world food production depends heavily on irrigated agriculture (Bhattarai et al., 2005). Irrigation is essential to feeding the increasing world population. Evett and Tolk (2009) reported that it is underappreciated that a Blue Revolution—rapid intensification and expansion of irrigated areas and improvements in irrigation methods and management—was a key factor in the success of the Green Revolution of improved crop cultivars and fertilization. Crop yield of irrigated land is much higher compared to dryland or rain-fed cultivation (Table 3.1). This is due to adequate water supply and the use of other cultural practices at an adequate level by the farmers because of assured yield increase and economic return. Only 20% of the world's farmland is irrigated, but that farmland produces 40% of

TABLE 3.1
Average Dry Lands and Irrigated Yields
of Important Annual Crops

Crop Species	Dry Land Yield (kg ha^{-1})	Irrigated Yield (kg ha^{-1})	Ratio of Irrigated to Dry Land Yield
Corn	6654	8600	1.29
Wheat	2152	4641	2.16
Sorghum	3390	5838	1.72
Barley	2583	4358	1.69
Cotton	504	953	1.89
Soybean	2085	2421	1.16

Source: National Research Council, *Alternate Agriculture*, National Academic Press, Washington, DC, 1989.

the world's food supply, including 60% of cereal production (Howell, 2001; Bhattarai et al., 2005; Evett and Tolk, 2009). In fact, irrigated land is increasing worldwide, and for some very good reasons (Evett and Tolk, 2009). In arid areas, production depends almost entirely on irrigation, and irrigation is critical to improving water use efficiency (WUE) in semiarid regions (Musick et al., 1994). Even in subhumid regions, irrigation is increasingly adopted to prevent decline in yield or harvest quality due to short-term droughts (Evett and Tolk, 2009).

Our planet is bathed in water. But of all the water on earth ($\approx$1.4 billion km^3), only $\approx$3% is fresh and most of that is locked up in polar ice caps, glaciers, or underground reservoirs, leaving only a fraction available for humans and terrestrial ecosystems (Oki and Kanae, 2006; Jury and Vaux, 2007). South America has a vast water potential; the irrigated area in South America is less than 4% of the world total (Howell, 2001). It has a high potential of increasing irrigated areas due to availability of water. For example, Brazil has 12% of the world's potable water. Figure 3.5 shows the distribution of water

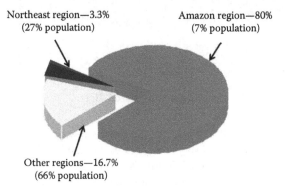

Northeast region—3.3%
(27% population)

Amazon region—80%
(7% population)

Other regions—16.7%
(66% population)

FIGURE 3.5 Water availability in Brazil. (Adapted from Paz, V. P. S., Teodoro, R. E. F., and Mendonça, F. C., *Rev. Bras. Eng. Agri. Amb.*, 4, 465, 2000.)

FIGURE 3.6   Acre river in the Amazon forest of Brazil in the state of Acre.

FIGURE 3.7   Amazon River in the state of Amazon, Brazil.

in the different regions of Brazil. Most of the Brazilian water is in the Amazon region (80%) where the population is very low. Figures 3.6 and 3.7 show river water in the state of Acre and Amazon in Brazil. Similarly, Figure 3.8 shows a waterfall in a mountain of the state of Espirito Santos in the Municipality of Marshall Rondon, Brazil.

In the future, irrigated land area will increase worldwide to meet the demand for better quality and quantity food and fiber. Most of the water used in irrigation is lost

FIGURE 3.8   Waterfall in the state of Espirito Santos, Municipality of Marshall Rondon, Brazil.

in the process of evapotranspiration (ET) and leaching. The leaching of irrigation water in the soil profile can lead to the contamination of water aquifers and water reservoirs and, consequently, to environmental pollution. Under these conditions, a greater knowledge of water absorption by roots, judicious use of water application to crop plants, and improving WUE in crop plants is required to keep pace with high demands of food and fiber in the future and to avoid environmental pollution. In addition, with increasing concern about the availability of water resources in both irrigated and rain-fed agriculture, there is renewed interest in trying to develop effective agricultural practices or farming systems to improve WUE in crop plants.

3.2   WATER REQUIREMENT OF ANNUAL CROPS

The importance of water to plant growth has been recognized for over 300 years. Despite this length of time, relatively little is known about specific water requirements for most annual field and horticultural crops grown under different agroecological conditions (Fageria et al., 2006). Another difficulty with plant water requirements is that much of the research deals with short-term responses to changing water status that may be transient and are eventually overridden by other biological changes that develop more slowly but have greater influences on yield. A sudden decrease in plant water status might induce stomatal closure, but adjustment by leaves to stresses may eventually result in stomata opening again (Passioura, 1994).

Water requirement is defined as the minimum amount of water required to provide optimal yield. The amount of water required is determined by yield levels, critical limits of deficiencies relative to yield, limits of tolerable yield reduction, size and permeability of plant evaporative surfaces, plant growth stage, and environmental factors affecting growth and transpiration (Spomer, 1985). Crop water demand and supply are influenced by meteorological and soil and plant physiological parameters. It is now understood that shortages of water at any stage in the crop life cycle will have consequences on yield, but more serious effects are likely when shortages occur after head/ear initiation (Fischer, 1973; Day et al., 1978; Fageria, 2007).

Plant water status and balances are important parameters for understanding the drought physiology of crop plants. Plant water status is the quantification of plant water conditions relative to requirements. One way to quantify plant water status is to measure water potential. Water potential is the physicochemical availability of water to participate in plant functions and determines tendencies for net water movement within plant systems. Plant water balances are the differences between plant or tissue water absorption and loss. When other conditions are favorable and water balances are positive, expansion growth occurs. However, positive water balances can also cause stresses resulting from water excesses. Whenever balances are disturbed (less water is absorbed than used), water deficits occur and the potential for expansion growth is immediately reduced (Hsio, 1973). The nature and extent of water balance effects depend on plant species or cultivar, plant part, plant growth status, yield, degree, duration, and timing of deficits, and interactions with other environmental factors (Spomer, 1985).

Progress in developing quantitative response functions to soil water deficits has been slow. Part of the problem may be that many studies have attempted to characterize water deficits with thermodynamic variables (Sinclair and Ludlow, 1985). Thermodynamic variables have not been related directly to leaf gas exchange or leaf expansion (Bennett et al., 1987; Joly and Hahn, 1989), and Ritchie (1981) proposed that responses of physiological processes to water deficits could be evaluated as functions of available soil water. This concept was refined by Sinclair and

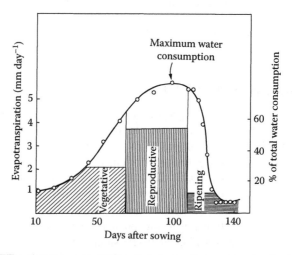

FIGURE 3.9 ET and water consumption during growing season of upland rice in central Brazil. (From Fageria, N.K., *Pesq. Agropec. Bras.*, 15, 259, 1980.)

Ludlow (1986) so as to express physiological processes as functions of transpirable soil water remaining in soil. These authors defined total transpirable soil water as differences between soil water contents at field capacity and soil water contents when transpiration of drought-stressed plants decreased to 10% or less than that of well-watered plants. Figure 3.9 shows ET and water consumption during the growing season of upland rice in central Brazil.

3.3 ESTIMATING CROP WATER USE

Water consumption is influenced by many factors such as the duration of crop growth, climatic conditions (temperature, precipitation, relative humidity, ET, and wind velocity), level of soil humidity, and soil structure (Karaaslan et al., 2007; Hamzei, 2011). Hence, the determination of accurate water requirement of a crop is very complex and difficult. However, numerous methods have been developed for estimating the water use (WU) of a growing crop (Rajan et al., 2010). Perhaps the most popular is that described by the Penman–Monteith equation (Van Bavel, 1966; Allen et al., 1989). This equation can be expressed as follows (Allen et al., 1989):

$$\lambda ET = \frac{\Delta(R_n - G) + \rho_a c_p (e_s - e_a)/r_a}{\Delta + \gamma(1 + r_s/r_a)}$$

where
   λ is the latent heat of vaporization in MJ kg^{-1}
   ET is the evapotranspiration rate in kg m^{-2} s^{-1}
   R_n is the net radiation in MJ m^{-2} s^{-1}
   G is the soil heat flux in MJ m^{-2} s^{-1}
   ρ_a is the air density in kg m^{-3}
   c_p is the specific of air at constant pressure in MJ kg^{-1}°C^{-1}
   $(e_s - e_a)$ is the vapor pressure deficit of the air in kPa
   Δ is the slope of the saturation vapor pressure curve in kPa°C^{-1}
   γ is the psychometric constant in kPa°C^{-1}
   r_s and r_a are the surface and aerodynamic resistances, respectively, in s m^{-1}

In this approach, proper assignment of the value for r_s allows the calculation of crop transpiration for conditions ranging from optimum to limited water supply (Van Bavel, 1967). When actual measurements of ET are available, as through studies involving lysimeters, it is possible to accurately calculate the surface resistance associated with a given set of growing conditions (Van Bavel, 1967; Hatfield, 1985; Howell et al., 1997). In many situations, however, the value of r_s may not be known. This difficulty has limited the practical application of the Pennman–Monteith method (Rajan et al., 2010).

It has been observed that the amount of water used by crops per unit ground area is related to the amount of plant canopy present, measured either by leaf area index (LAI) or ground cover (GC) (Rajan et al., 2010). As plants develop, there is an increase in mutual shading and interference among leaves within a plant canopy that causes plant transpiration to increase at a diminishing rate with increasing LAI and asymptotically leveling at LAI $> 4\,m^2\,m^{-2}$, progressively uncoupling transpiration from changes in LAI (Ritchie, 1972; Villalobos and Fereres, 1990; Sau et al., 2004; Hatfield et al., 2011). In general, crop WU tends to increase with increasing LAI up to a value of approximately $3\,m^2\,m^{-2}$, which represents full ground cover (GC = 1) for many crops (Chang, 1968; Ritchie, 1972; Bunting and Kassam, 1988).

3.4   WATER USE EFFICIENCY IN CROP PLANTS

Plant WUE is generally defined as the amount of biomass accumulated per unit of water used. This definition of WUE was first applied to irrigation research, where costs of applying water to plants are major economic considerations. In physiological terms, WUE has been defined as the ratio of C assimilated to water transpired. In more recent years, WUE can be estimated by C isotope discrimination techniques (Farquhar et al., 1989; Dingkuhn et al., 1991). The role of WUE in determining yield per unit land area is shown by the following equation (Passioura, 1983):

$$\text{Yield (g DM m}^{-2}) = \text{WUE (g DM g}^{-1}\,H_2O) \times \text{transpiration (g } H_2O\ m^{-2}) \times HI$$

where
   DM is dry matter
   HI is harvest index

WUE is a physiological trait associated with drought tolerance of plants (Mian et al., 1998). Wright et al. (1994) stated that WUE can contribute to crop productivity for plants grown in droughts. These authors also reported positive associations between WUE and total biomass yields in drought environments and suggested that WUE improvement in crop plants should result in superior yield performance if high HIs (harvest indexes) can be maintained. Water requirement, which is one index of WUE, varies among plant species and cultivars as well as among sites and years examined, even for the same plant species or cultivar. This means that WUE is determined by interactions between environmental and genetic factors. Hattendorf et al. (1988) reported that mean daily use rates for sunflower were 22% greater than mean rates for maize, grain sorghum, pearl millet, pinto bean, and soybean. WUE values of C_3 crops (pinto bean, soybean, and sunflower) were 90% less than those of C_4 crops (maize, sorghum, and pearl millet). Bauder and Ennen (1979) reported that established alfalfa and soybean used significantly greater amounts of water than spring wheat, pinto bean, or sunflower in North Dakota. In addition, Badaruddin and Meyer (1989) reported that legume crops used 10%–25% more seasonal water than wheat across environments, but WUE (kg DM ha^{-1} mm^{-1} water) of legumes was 0%–25% greater than that of wheat. Green manure and forage legumes generally had greater WU and WUE values than grain legumes, which was associated with their longer growing season and higher dry matter production. French and Schultz (1984a,b) studied the WUE of grain production in South Australia and concluded that potential yields of 20 kg of grain per mm of transpiration was possible in the Mediterranean climatic production zones of southeastern Australia.

Total water use (TWU) includes both transpiration and soil evaporation water (Ehdaie, 1995). Two models for WUE have been suggested: one with two components and another with three components, so that contributions from each component relative to WUE could be evaluated (Ehdaie, 1995). The two primary components of WUE were defined as evapotranspiration efficiency (ETE, ratio of total DM to TWU) and HI (ratio of grain yield to total DM). Thus, the two-component model for WUE can be expressed as follows (Ehdaie, 1995):

$$WUE = ETE \times HI$$

This model can be extended to a three-model equation when ET can be partitioned into transpired water and soil water evaporated. This extended model contains the following components: uptake efficiency (UE, ratio of total water transpired to TWU), transpiration efficiency (TE, ratio of TDM to total water transpired), and HI as previously defined (Ehdaie and Waines, 1993). This extended model for WUE can be expressed as follows (Ehdaie, 1995):

$$WUE = EU \times TE \times HI$$

Many methods exist for improving the efficiency of WU by plants, which have been summarized by Boyer (1996). WUE can be improved by increasing the efficiency of water delivery and timing of water application. This approach is widely adopted in many developed and developing countries and in other countries where water is becoming increasingly scarce. The per capita availability of water resources declined by 40%–60% in many Asian countries between 1955 and 1990 (Gleick, 1993). In 2025, per capita available water resources in these countries are expected to decline by 15%–54% compared to that available in 1990 (Guerra et al., 1998). The share of water to agriculture will decline at an even faster rate because of increasing competition for available water from urban and industrial sectors (Tuong and Bhuiyan, 1994). Under these situations, the approach of adapting plants to less available water may be an important strategy for crop production. Using drought-tolerant crop species or cultivars within species depends on understanding the specific biology of those plants to be used and whether these plants can be manipulated to achieve similar productivity with less water. This approach is complicated and will command considerably more research data for the different crops and agroclimatic regions than now available if successful crop productivity is to be achieved.

Genetic variations of WUE in many field crop plants such as soybean (Mian et al., 1996, 1998), wheat (Farquhar and Richards, 1984), barley (Hubick and Farquhar, 1989), peanut (Hubick et al., 1986), and sunflower (Virgona et al., 1990) have been reported in recent years. Nevertheless, the improvement of WUE in field crops through conventional breeding methods may not be practical because of the complexity and difficulty of measuring the WUE of large numbers of breeding lines grown in field conditions (Ismail and Hall, 1992). Indirect selection for improved WUE through molecular markers linked to quantitative trait loci (QTL) conditioning WUE in crop plants may prove to be useful (Mian et al., 1998). Differences among crop species for WUE are related to carboxylation pathways (e.g., C_4 species commonly

TABLE 3.2
WUE of Principal Food, Vegetable, and Fruit Crops in Brazil

Crop	Scientific Name	WUE (kg ha^{-1} mm^{-1})[a]	References
Rice	*Oryza sativa* L.	4.6	Coelho et al. (2009)
Dry bean	*Phaseolus vulgaris* L.	3.9	Coelho et al. (2009)
Corn	*Zea mays* L.	17.2	Coelho et al. (2009)
Wheat	*Triticum aestivum* L.	9.6	Coelho et al. (2009)
Soybean	*Glycine max* L. Merr.	5.7	Coelho et al. (2009)
Cotton	*Gossypium hirsutum* L.	6.8	Lima et al. (1999)
Garlic	*Allium sativum* L.	12.3	Lima et al. (1999)
Onion	*Allium cepa* L.	40.9	Lima et al. (1999)
Potato	*Solanum tuberosum* L.	38.2	Coelho et al. (2009)
Tomato	*Lycopersicon esculentum* Mill.	167.6	Coelho et al. (2009)
Banana	*Musa paradisiaca* L.	11.6	Coelho et al. (2009)
Watermelon	*Citrullus lanatus* (Thunb.) Matsum. and Nakai	29.8	Lima et al. (1999)
Cantaloupe	*Cucumis melo* L.	23.1	Coelho et al. (2009)
Grape	*Vitis vinifera* L.	37.7	Lima et al. (1999)

[a] Calculated on the basis of data of water consumed by the crops given by the authors and average yield of the irrigated crops.

have values twice that of C$_3$ species) and to energy requirements for the production of biomass containing different proportions of proteins, lipids, and carbohydrates (Ludlow and Muchow, 1990). Similarly, apparent differences in WUE between cultivars of the same species and among several food legumes can be related to differences in soil evaporation and the chemical composition of DM. Hatfield et al. (2001) reported a range of variations in measured WUE across a range of climates, crops, and soil management practices. WUE in principal food, vegetable, and fruit crops under Brazilian conditions is presented in Table 3.2. Similarly, data related to dry matter production and water used by pearl millet, corn, and peanut under Nigerian conditions are presented in Table 3.3. Bednarz et al. (2002) reported that Georgia cotton requires about 460 mm of appropriately timed water to maximize yields.

3.4.1 MANAGEMENT PRACTICES TO IMPROVE WATER USE EFFICIENCY IN CROP PLANTS

The demand for water will likely increase manyfold in the future due to an increase in irrigated area to improve the food and fiber production for a large and growing world population. It is estimated that the world population will be around 9 billion by the year 2050. Hence, improving WUE in crop production will be an important strategy to conserve this precious natural resource. Soil and crop management practices can improve WUE in crop plants. At least half of the increase in precipitation use efficiency (PUE) over the last century can be attributed to improved agronomic management (Turner, 2004), although Lopez-Bellido et al. (2007) reported that some

TABLE 3.3
Water Used, Dry Matter Produced, and WUE
of Three Field Crops at Samau, Nigeria

Crop Parameter	Pearl Millet	Corn	Peanut
Crop growth duration (days)	85	117	125
Total water used (mm)	330	486	438
Average water used day^{-1} (mm)	3.9	4.2	3.5
Dry matter produced (Mg ha^{-1})	22.5	19.1	8.4
Average dry matter day^{-1} (kg ha^{-1})	264	163	67
WUE (g dry matter kg^{-1} water)	6.6	3.9	1.7

Sources: Kassam, A.H. and Kowal, J.M., *Agric. Meteor.*, 15, 333, 1975;
Kassam, A.H. et al., *Trop. Agric.*, 52, 105, 1975; Kowal, J.M.
and Kassam, A.H., *Agric. Meteor.*, 12, 391, 1973.

authors consider one-third attributable to new cultivars and two-thirds to improvements due to increasing soil water storage, varying crop transpiration, or rising root density. Strategies to increase crop growth in the vegetative phase might be expected to increase WUE when evaporation and vapor pressure deficits are low, since soil evaporation would be reduced by additional canopy cover (Angus and Herwaarden, 2001). In many dryland environments, crops do not use all the water available in the soil profile because of restrictions to root growth. These restrictions may be physical, chemical, or biological (Lopez-Bellido et al., 2007). Agronomic practices that reduce the physical impedance to root growth can benefit dryland crop yields in water-limited environments (Hatfield, 2001; Turner, 2004).

The soil environment remaining after certain crops synergistically improves the WUE of subsequent crops (Anderson, 2005; Tanaka et al., 2005). The root system of some crops penetrates deeper, providing more biopores for a subsequent crop (Turner, 2004). The WUE of some crops can be improved by more intensive cropping systems (Nielsen et al., 2002). Important management practices that can be adopted to attain high WUE in crop plants are maintenance of organic matter (OM) content of the soil, conservation tillage, crop rotation, planting crop species or genotypes within species requiring less water, intercropping, maintaining adequate soil fertility, and the adoption of appropriate irrigation methods (Angus and Herwaarden, 2001; Turner, 2004; Nielsen et al., 2005).

3.4.1.1 Maintenance of Soil Organic Matter

OM refers to the solid, nonmineral portions of the soil, originating from plant and animal residues (Aust and Lea, 1991). According to the Soil Science Society of America (2008), soil organic matter (SOM) can be defined as the organic fraction of the soil exclusive of undecayed plant and animal residues. Hayes and Swift (1983) defined "organic matter" as the term used to refer more specifically to the nonliving components, which are a heterogeneous mixture, composed largely of products resulting from microbial and chemical transformations of organic debris. This transformation, known collectively as the humification process, gives rise to humus, a mixture of substances that have a degree of resistance to further microbial attack. An adequate amount of OM in the soil plays an important role in improving the physical, chemical, and biological properties of soil and, consequently, maintaining sustainability of cropping systems. In agricultural systems, maintenance of SOM has long been recognized as a strategy to reduce soil degradation (Mikha and Rice, 2004).

One of the most important effects of OM addition to the soil is that it changes the soil's water retention characteristics, which is generally related positively to crop production. A reduction in available water capacity is considered the foremost contributing factor in the loss of soil productivity caused by erosion. This reduction in available water capacity is attributable to changes induced in the soil water–holding characteristics of the root zone or by reduction in the depth (thickness) of the rooting zone (Bauer and Black, 1992). Most of the OM is generally concentrated in the plow layer of the soils (Fageria et al., 1991). Frye et al. (1982), in Kentucky, reported that the available water-holding capacity of an eroded Maury soil (fine-silty, mixed, misic typic paleudalf) was 4–6.1 lower in the upper 15 cm on a volume basis than its noneroded counterpart. Biswas and Khosla (1971) also observed a similar increase in soil's water retention characteristics and hydraulic conductivity from applying farmyard manure to soils over a 20 year period. Mays et al. (1973) found an increase from 11.1% to 15.3% in water content corresponding to the 0.33 bar (in modern terminology the Pascal [Pa] is the unit of pressure, 1 bar = 10^5 Pa = 10^2 kPa = 0.1 MPa) suction of a silt loam soil after an application of 327 MT ha^{-1} of municipal compost for 2 years.

Gupta et al. (1977) also reported that the amount of water retained at 15 bars (1.5 MPa) increased linearly with the increase in sludge addition OM in a coarse sandy soil. Scoot and Wood (1989) reported a linear increase in water retention of Crowley silt loam soil with increasing OM content. OM content alone could account for 84.4% of the variability of water retained at 10 kPa. The slope of the line indicates that each 10 g kg^{-1} of OM could account for an increase of 5.6% by volume of water retained (Scoot and Wood, 1989). Martens and Frankenberger (1992) studied the effects of different organic amendments on soil physical parameters and water infiltration rates on irrigated soil. Incorporation of three loadings (75 Mg ha^{-1} each) of poultry manure, sewage sludge, barley straw, and alfalfa to an Arlington soil for 2 years increased soil respiration rates (139%–290%), soil aggregate stability (22%–59%), organic C content (13%–84%), soil saccharide content (25%–41%), soil moisture content (3%–25%), and decreased soil bulk density (7%–11%). The changes in the physical properties of soil resulted in significantly increased cumulative water infiltration rates (18%–25%) in the organic-amended plots as compared with unamended plots. The increase in water retention of soil due to the addition of OM may be related to the following factors: (1) decreased bulk density and increased total porosity, (2) change in the aggregate size distribution (which may change the pore-size distribution), and (3) increased absorptive capacity of the soil (increase in total surface area) (Fageria and Gheyi, 1999).

Lado et al. (2004a) reported that the saturated hydraulic conductivity of soil with high OM content (3.5%) was higher than that of low OM content (2.3%). Similarly, Lado et al. (2004b) reported that in sandy loam soil, an increase of OM content from

2.3% to 3.5% reduced the aggregate breakdown, soil dispersivity, and the seal formation at the soil surface under raindrop impact conditions. These authors suggested
that the final infiltration values were lower in the low OM soil than in the high OM
soil because (1) there was more extensive breakdown and dispersion of the aggregate
at the surface of the low OM soil than at that of the high OM soil, so that a more
continuous crust was formed on the former soil, and (2) the rearrangement of the
detached and dispersed particles in the crust differed between the two soils, so that a
thicker, higher-density crust was formed on the low than on the high OM soil.

3.4.1.2 Conservation Tillage

Adopting conservation tillage or minimum tillage is also an important strategy
for improving WUE and the reduction of drought in crop plants. No-till (NT) and
minimum-till (MT) systems are more efficient than conventional-till (CT) systems
for conserving precipitation in crop production (Aase and Schaefer, 1996; Peterson
et al., 1996; McGee et al., 1997; Tanaka and Anderson, 1997; Halvorson et al., 1999).
The maintenance of previous crop residues on soil surfaces is associated with reduced
soil erosion, soil temperature, and increased soil water content (Wilhelm et al., 1986;
Hatfield et al., 2001). In addition, crop residue can reduce soil water evaporation,
seed zone soil moisture loss during the fallow period (Aase and Pikul, 1995; Riar
et al., 2010).

Conservation tillage is defined as any tillage sequence whose objective is to minimize or reduce the loss of soil and water; operationally, a tillage or tillage and
planting combination that leaves a 30% or greater cover of crop residues on the
surface (Soil Science Society of America, 2008). The terms "Minimum tillage,"
"no-tillage," and "zero tillage" are also used in the literature. According to the Soil
Science Society of America (2008), minimum tillage is defined as the minimum use
of primary and/or secondary tillage necessary for meeting crop production requirements under the existing soil and climatic conditions, usually resulting in fewer tillage operations for conventional tillage. Similarly, minimum tillage or zero tillage
is defined as a procedure whereby a crop is planted directly into the soil with no
primary or secondary tillage since harvest of the previous crop. In this process,
usually a special planter is necessary to prepare a narrow shallow seedbed immediately surrounding the seed being planted. No-tillage is sometimes practiced in
combination with subsoiling to facilitate seeding and early root growth, whereby the
surface residue is left virtually undisturbed except for a small slot in the path of the
subsoil shank. Conservation tillage, minimum tillage, or no-tillage is widely adopted
in developed as well as developing countries in recent years for crop production. It
is projected that conservation tillage will be practiced on 75% of cropland in the
United States by 2020 (Lal, 1997). Kern and Johnson (1993) reported that increasing
conservation tillage to 76% of planted cropland would change agricultural systems
from C sources to C sinks. Unger and Baumhardt (1999) reported soil water storage
data from 1939 to 1997 for Bushland, TX, and found that conservation tillage compared to conventional tillage during an 11 month fallow period increased the average
plant-available soil water at planting from 100 to 170 mm. Stewart and Steiner (1990)
further showed that sorghum grain yields were increased on an average of 15 kg ha^{-1}
for each additional millimeter of seasonal evaporation at Bushland; so the amount of

stored soil water is extremely important in this region for dryland crop production (Bandaru et al., 2006).

There is a general concept that tillage decreases aggregate stability by increasing mineralization of OM and exposing aggregates to additional raindrop impact energies (Tisdall and Oades, 1982; Elliott, 1986; Angers et al., 1992; Amezketa, 1999; Balesdent et al., 2000; Park and Smucker, 2005). All these processes lead to the lowering of the water retention capacity of the soil and, consequently, a decrease of WUE in crop plants. Tillage promotes SOM loss through crop residue incorporation into soil, physical breakdown of residues, and disruption of macroaggregates (Beare et al., 1994; Paustian et al., 2000; Six et al., 2000; Wright and Hons, 2004). In contrast, conservation or no-tillage reduces soil mixing and soil disturbance, which allows SOM accumulation (Blevins and Frye, 1993). Many studies have shown that conservation tillage improves soil aggregation and aggregate stability (Beare et al., 1994; Six et al., 1999). Conservation or minimum tillage promotes soil aggregation through enhanced binding of soil particles as a result of greater SOM content (Jastrow, 1996; Paustian et al., 2000; Six et al., 2002). Microaggregates often form around the particles of undecomposed SOM, providing protection from decomposition (Gupta and Germidia, 1988; Gregorich et al., 1989; Six et al., 2002; Wright and Hons, 2004). Microaggregates are more stable than macroaggregates, and thus, tillage is more disruptive of large aggregates than smaller aggregates, making SOM from large aggregates more susceptible to mineralization (Cambardella and Elloitt, 1993; Six et al., 2002; Wright and Hons, 2004). Since tillage often increases the proportion of microaggregates to macroaggregates, there may be less crop-derived SOM in conventional tillage than conservation or no-tillage (Six et al., 2000; Wright and Hons, 2004). Fungal growth and mycorrhizal fungi, which are promoted by no-tillage, contribute to the formation and stabilization of macroaggregates (Tisdall and Oades, 1982; Beare and Bruce, 1993). WUE efficiency in plants improves with improving aggregates.

Conservation tillage improves SOM content of the soil, which is responsible for higher water retention in the soil profile, and consequently improves WUE. Larger SOM accumulation in conservation tillage has been observed in intensive cropping systems where multiple crops are grown yearly (Ortega et al., 2002; Wright and Hons, 2004). The use of conservation tillage, including NT, is being considered as part of a strategy to reduce C loss from agricultural soils (Kern and Johnson, 1993; Paustian et al., 1997; Denef et al., 2004). Crop species also influence SOM accumulation in the soil. Residue quality often plays an important role in regulating long-term SOM storage (Lynch and Bragg, 1985). Crop residues that have low N concentration, such as wheat, generally decompose at slower rates than residues with higher N, such as sorghum and soybean (Franzluebbers et al., 1995; Wright and Hons, 2004). Since wheat residues often persist longer, they increase SOM more than sorghum or soybean (Wright and Hons, 2004).

Bauer et al. (2010) reported that conservation tillage resulted in an average 25% yield increase in cotton lint yield over conventional tillage during a 5 year drought. Conservation tillage can increase soil water, especially before the soil is covered with the crop canopy, and can increase crop yields compared to conventional tillage (Phillips et al., 1980). Higher yields with conservation tillage have been attributed to reduced evaporation from the soil surface (Lascano et al., 1994) and reduced runoff (Truman et al., 2003). Karl et al. (2009) reported that periods of drought are predicted to occur more often in the southeastern United States as a result of global climate change as greenhouse gases continue to accumulate in the atmosphere. Conservation tillage for crop production could be an important method to help mitigate the effects of climate change in the region if changes occur as predicted (Bauer et al., 2010). Hatfield et al. (2001) reported that it is possible to increase WUE by 25%–40% through soil management practices that involve tillage.

3.4.1.3 Crop Rotation

Adopting appropriate crop rotations is another water-efficient crop management practice. Crops should be managed in rotation sequence so that root systems fully exploit available water and nutrients (Karlen et al., 1994). Sadler and Turner (1994) suggested opportunistic cropping as a means for increasing agricultural sustainability through water conservation or by increasing productivity from applied water. Norwood (1995) also reported that irrigation WUEs were highest for rotated crops, and lowest for continuous crops. Peterson et al. (1996) suggested that a more efficient and profitable way for using summer precipitation was to use spring-planted crops in rotation with winter wheat instead of summer fallow periods. Crop rotations employing diversity in plant WU, rooting patterns, and crop types (broadleaf versus grass plants) generally had increased crop yields compared with monoculture systems. Rotation effects may arise from beneficial effects on soil moisture, microbes, nutrients, and structure, and from decreases in diseases, insects, weeds, and phytotoxic compounds (Bezdicek and Granatstein, 1989; Crookston et al., 1991). In the drought-prone areas of some regions, the conventional approach has been to employ fallow rotation systems, especially for wheat (Amir and Sinclair, 1996). A major benefit of fallow years is the increase of water availability through storage without crop growth (Unger, 1994; Thomas et al., 1995). Wheat–fallow rotation systems increase water storage and WUE compared to continuous wheat systems (Bolton, 1981; Bonfil et al., 1999). Sunflower (deep-rooted and intermediate water user) can extract soil water from root zones below that normal for small-grain crops (Alessi et al., 1977; Unger, 1984). Therefore, sunflower has the potential to improve WUE in rotation with small-grain crops. Sunflower is also a desirable warm-season crop for inclusion in more intensive dryland crop rotations because it provides diversity to rotations and is considered to be drought tolerant (Unger, 1984; Halvorson et al., 1999).

Conventional monoculture agriculture systems can reduce the quality of soils by the loss of OM and structure because of the low level of organic inputs and regular disturbance from tillage practices (Acosta-Martinez et al., 2004). Crop rotation may have many positive effects on soil quality and, consequently, on WUE and crop production. Crop rotation is defined as a planned sequence of crops growing in a regularly recurring succession on the same area of land, as compared to continuous culture of one crop or growing a variable sequence of crops (Soil Science Society of America, 2008). Bullock (1992) defined crop rotation as a system of growing different types of crops in a recurrent succession and in an advantageous sequence on the same land. Crop rotations are a key component of successful organic arable systems (Robson et al., 2002). Rotations can be optimized to conserve and recycle

TABLE 3.4

Fundamental Prerequisite of an Appropriate Crop Rotation

Prerequisite	Advantage
Including legume crops in rotation	Supply N to succeeding crops
Including crops of different root architecture	Improve soil physical and chemical properties, like nutrient uptake from deeper soil layers, soil porosity, aeration and drainage, and improve SOM content
Including appropriate cover/green manure crops in the rotation	Protect soil from erosion, control weeds, supply OM, conserve moisture, improve soil hydraulic conductivity, control diseases and insects
Include crops with different resistances to diseases and insects	Break cycle of diseases and insects, reduce host plant presence in rotation
Include crops suited to a given agroecological region	Better economic return and ecological viability

nutrients and minimize pest, disease, and weed problems (Lampkin, 1990; Robson et al., 2002). Appropriate crop rotation has significant influence on SOM content of soils. The results of long-term field trials in Illinois (United States) showed that crop rotation influenced the content of SOM (Odell et al., 1984). The level of soil C and N was highest in the maize–oats–clover, and lowest in the permanent corn rotation (Mengel et al., 2001). The fundamental prerequisites of an appropriate crop rotation are listed in Table 3.4.

Crop rotations under CT that provide residues with low C/N ratios stimulate the decomposition of native SOM to a greater extent than rotations providing residues with high C/N ratios (Ghidey and Alberts, 1993; Sisti et al., 2004; Wright and Hons, 2004). Under NT, crop rotations have been shown to have minimal effect on native SOM decomposition (Sisti et al., 2004). Wright and Hons (2004) also reported that greater differences in SOM between crop species occurred under CT rather than NT, especially in subsurface soil. Wani et al. (1994) reported that green manures and organic amendments in crop rotations provided a measurable increase in SOM quality and other soil quality attributes compared with continuous cereal systems.

Crop rotations have positive effects on soil properties related to the higher C inputs and diversity of plant residues to soils in comparison with continuous systems (Miller and Dick, 1995; Moore et al., 2000; Acosta-Martinez et al., 2004). Conservation tillage increases soil organic C (Franzluebbers et al., 1995; Acosta-Martinez et al., 2004), microbial biomass (Franzluebbers et al., 1995), and modifies the soil microbial community (Acosta-Martinez et al., 2004). All these changes in the soil due to crop rotations improve WUE in crop plants.

3.4.1.4 Intercropping

Intercropping is defined as the growing of two or more crops simultaneously on the same area of land (Fageria, 1992). The crops are not necessarily sown at exactly the same time, but they are usually simultaneous for a sustainable part of their growing periods. Yields of two or more intercrops are added to give a combined intercrop yield. Intercropping has several names in the literature. The most common names are mixed intercropping, row intercropping, strip cropping, relay cropping, and alley cropping (Fageria, 1992). Intercropping includes crops of different cycles as well as different growth habits, especially root growth patterns. All these crop characteristics lead to higher WUE.

3.4.1.5 Crop Species/Genotypes

Crop species or genotypes within species differ in their water requirements. Some crop species can produce good yield with less water compared to other crop species. Also there is difference in water requirements among genotypes of the same crop species. For example, sorghum is a C_4 (plants that have four-carbon malic or aspartic acid photosynthesis pathway) plant and exhibits excellent resistance to high temperatures and drought and low input levels (Doggett, 1988). Sorghum is often planted in marginal environments with little input of water or fertilizers (Burke et al., 2010).

There are significant differences among genotypes of same crop species in water requirement or drought tolerance. For example, upland rice is commonly planted in the central part of Brazil and 1–2 weeks of drought is a common phenomenon during the crop growth cycle (Fageria, 1980). Heinemann and Stone (2009) studied the effects of water deficit on grain yield and panicle number of four upland rice cultivars (Table 3.5). There was large variation in the grain yield of cultivars at irrigated as well as water-deficit conditions. However, at water deficit, grain yield was much reduced compared to non-water-deficit conditions. Similarly, panicle numbers were also reduced at water-deficit conditions compared to non-water-deficit conditions. Similarly, Guimarães et al. (2008, 2010) also reported significant differences among upland rice genotypes' grain yields under water stress and non-water stress conditions.

New crops with high WUE and increased drought tolerance are being sought for production in arid regions (Hamzei, 2011). One plant species with excellent potential as an alternative to more traditional crops is canola (*Brassica napus* L.). Canola is one of the most important edible oil sources for developed as well as developing

TABLE 3.5
Grain Yield and Panicle Number of Upland Rice Cultivars under Irrigated and Water Stress Conditions

Cultivar	Grain Yield (kg ha⁻¹)		Panicle Number (m⁻²)	
	Irrigated	Water Deficit	Irrigated	Water Deficit
Curinga	3463	2358	260	223
Guarani	3142	314	225	104
Primavera	3284	504	183	176
Soberana	2260	53	218	158

Source: Adapted from Heinemann, A.B. and Stone, L.F., *Pesq. Agropec. Trop.*, 39, 134, 2009.

countries (Al-Barrak, 2006). The WUE of canola is reported to be 8.3–11.4 kg ha^{-1} mm^{-1} in the semiarid regions of Canada (Al-Barrak, 2006).

3.4.1.6 Optimal Soil Fertility

Optimal soil fertility is defined as the soil having essential plant nutrients in adequate amounts and proportion for the growth of crop plants. If the soil has optimal fertility and crop production factors are at an adequate level, WUE in crop plant is higher. Hatefield et al. (2001) reported that the addition of N and P has an indirect effect on WU through the physiological efficiency of the plant. These authors further reported that WUE can be increased by 15%–25% through the modification of nutrient management practices. Other crop production factors are soil water availability, cultivars, control of diseases, insects, and weeds. If crop plants are infested with diseases, insects, and weeds, the photosynthetic process of plants will not be normal and WUE will decrease. Data presented in Table 3.6 show that grain yield, PUE, WU, and WUE were improved by a wheat crop with the addition of N fertilization at an adequate rate. Nitrogen management is linked to WU rates in cropping systems (Lopes-Bellido et al., 2007). Hatfield et al. (2001) reported that modifying nutrient management practices, such as N rate, can increase WUE by 15%–25%. Amir et al. (1991) reported that WUE is affected by N supply only when N becomes a deficient factor in the system, that is, when water is not a limiting factor. The adoption of N has an indirect effect on WU through the physiological efficiency of the plant (Hatfield et al., 2001). The largest increase in WUE in Mediterranean rain-fed crops is obtained by altering the balance between evaporation and transpiration. Rapid crop development, achieved through N fertilizer application through a higher planting density, can result in a substantial reduction in soil evaporation and a corresponding increase in transpiration and grain yield (Debaeke and Aboudrare, 2004).

TABLE 3.6
Grain Yield PUE, WU, and WUE in a Rain-Fed Wheat Crop under Different N Rates

N Rate (kg ha^{-1})	Grain Yield (kg ha^{-1})	PUE (kg ha^{-1} mm^{-1})	WU (mm)	WUE (kg ha^{-1} mm^{-1})
0	3018d	7.8c	483b	6.8c
50	3571c	9.1b	503a	7.6b
100	3939b	9.9a	503a	8.3a
150	4018a	10.1a	506a	8.5a

Source: Adapted from Lopez-Bellido, R.J. et al., *Agron. J.*, 99, 66, 2007.
Means followed by the same letter in the same column are not significantly different at the 5% probability level by LSD test. PUE = Grain yield divided by harvest to harvest precipitation, WU = R + SW$_{harvest}$ − SW$_{planting}$, where R is rainfall over a defined period, and SW is soil water content (0–90 cm) at planting and harvest, and WUE = grain yield divided by WU.

3.4.1.7   Use of Cover Crops

Cover crops can be defined as close-growing crops that provide soil protection and soil improvement between periods of normal crop production, or between trees in orchards and vines in vineyards (Fageria et al., 2005). Cover crops are grown not for market purposes, but when plowed under and incorporated into the soil, cover crops may be referred to as green manure crops. Cover crops are sometimes called catch crops. Planting cover crops before or between main crops as well as between trees or shrubs of plantation crops can improve the physical, chemical, and biological properties of soil and, consequently, lead to improved soil health and yield of principal crops (Fageria et al., 2005).

ET is one of the most important foundations of global biological and food production (Falkenmark and Rockstrom, 2004). It also plays a key role in global climate change, as soil moisture, vegetative productivity, the C cycle, and water budgets are all directly affected by ET. Of the total ET, evaporation from the soil surface constitutes a considerable proportion. Therefore, evaporation is regarded as a major path of water loss from soil and a major constraint to rain-fed agricultural production (Jalota and Prihar, 1990; Eberbach and Pala, 2005).

Conserving soil moisture with cover crop residues is widely reported (Smith et al., 1987; Sustainable Agriculture Network, 1998; Fageria et al., 2005). Cover crop residues left on soil improve the infiltration of rain water and also reduce evaporative losses, resulting in less moisture stress during drought periods. Grass-type cover crops such as rye, barley, wheat, and sorghum (Sudan grass) have been reported to be very effective in soil moisture conservation (Sustainable Agriculture Network, 1998). Gallaher (1977) showed that soil remained wetter and crop yields were higher when rye was left as surface mulch than when aboveground parts of the rye were removed in a conservation-tillage system. Daniel et al. (1999) reported that rye had the highest biomass of several cover crop species tested and soil had higher water content under rye. The greatest differences in water content between mulched and bare soils can be expected during short dry periods (7–14 days), not longer ones (Smith et al., 1987).

3.4.1.8   Improved Irrigation Efficiency

In modern agriculture, irrigation is one of the most important practices that sustain civilization. On average, irrigated crop yields are double those from unirrigated lands. To meet the needs of the world's estimated 8 billion population by 2025, the irrigated area must expand more than 20% and irrigated crop yields must improve by 40% above current yields. The Food and Agricultural Organization (FAO) has anticipated an expansion of irrigated land of about 242 million ha in 2030 in 93 developing countries and that agricultural water withdrawals will increase by approximately 14% during 2000–2030 to meet the food demand (Hamzei, 2011). Furthermore, irrigation is fundamental to maintaining the sustainability of cropping systems. Irrigation is defined as the intentional application of water to the soil, usually for the purpose of crop production (Soil Science Society of America, 2008). Efficiency is a measure of the output obtained from a given input. Hillel and Rawitz (1972) defined agronomic or economic irrigation efficiency as the marketable yield or financial return per amount of water applied or money invested in the water supply. The quality of irrigation application at field scale can be measured as the irrigation efficiency that is defined as the crop ET divided by the total water applied as irrigation plus precipitation. The Soil Science Society of America (2008) defined irrigation efficiency in several ways: (1) the ratio of the water actually consumed by crops on an irrigated area to the amount of water applied to the area, (2) the ratio of water infiltration to total water applied, and (3) the ratio of water profile storage increase to total water applied. Although it is difficult to arrive at reliable statistics, it has been estimated that the average efficiency of irrigation WU in the western United States is around 50% (Hillel and Rawitz, 1972). Hence, improving irrigation efficiency is important not only to water economy but also efficient management of irrigation can minimize leaching losses of highly soluble nutrients (i.e., nitrate) through the soil profile below the rooting zone. Data in Table 3.7 show irrigated area and irrigation efficiency in different continents or countries.

Optimum irrigation scheduling based on WU patterns and crop response to water deficit can potentially improve WUE (Amer et al., 2009). Traditionally, there are three basic approaches to determine in situ crop irrigation needs. The micrometeorological approach uses weather data (i.e., daily sunshine, temperature, rainfall) and specific crop (root depth, leaf area, drought tolerance) and soil factors (water retention, hydraulic conductivity) to determine a theoretical soil water balance; irrigation is then applied whenever the soil water deficit falls below a predefined threshold (Allen et al., 1998; Nadler et al., 2003). Similarly, a soil hydraulic approach has been developed that uses a direct measure of changes in either the soils volumetric water content (θ) (neutron scattering, TDR, gypsum blocks) or the soils matric potential

---

TABLE 3.7

Irrigated Area (Million Ha) and Irrigation Efficiency in Different Continents or Countries

Continent/Country	Irrigated Area	Irrigation Efficiency (%)
China	48.0	39
India	45.1	40
Rest of Asia	61.3	32
North America	21.6	53
Latin America	16.2	45
Europe	16.7	56
Middle East/North Africa	22.6	60
Rest of Africa	6.1	48
World	234.0	43

Sources: Adapted from Seckler, D. et al., World water demand and supply 1990 to 2025: Scenarios and issues, International Water Management Institute Research Report 19, International Water Management Institute, Colombo, Sri Lanka, 1998; Wood, S. et al., *Pilot Analysis of Global Ecosystems: Agroecosystems*, World Resource Institute, Washington, DC, 2001; Jury, W.A. and Vaux, H.J., *Adv. Agron.*, 95, 1, 2007.

(tensiometers) to schedule irrigation. In both cases, understanding the impact and dynamics of the root-zone soil moisture status is a prerequisite to determine the threshold value of soil water deficit that each crop can tolerate before water stress begins to affect crop growth and productivity (Nadler et al., 2003).

Causape et al. (2006) reported that depending on the soil properties and irrigation management, irrigation efficiency at the field level can be low to moderate under surface irrigation (53%–79%) but can reach high values in well-managed sprinkler irrigation systems (94%). Asia is where the largest irrigated land in the world is located and considerably more food could be produced with less water. Guerra et al. (1998) reported that only 30%–65% of water released at the headworks reaches intended field inlets in various countries of Asia. This means much improvement in irrigation efficiency on this continent could be made.

Irrigation methods can affect the root-zone soil environment (Zhang et al., 2008; Kong et al., 2010), which has a considerable effect on root extension and subsequent nutrient or water uptake along with the resulting crop growth and production (Cornish et al., 1984; Martinez et al., 2008). Crops can be irrigated by several methods. Different irrigation methods can result in different water and nutrient use efficiencies and, ultimately, plant growth (Kong et al., 2010). The most common methods for irrigating crops are furrow irrigation, central pivot irrigation, check basin irrigation, drip irrigation, spray irrigation, border strip irrigation, and subirrigation. The Soil Science Society of America (2008) defined these irrigation methods as follows:

1. Furrow irrigation: irrigation in which water is applied between crop rows in furrows made by tillage implements. Furrow irrigation is commonly used in arid, semiarid, and subhumid regions to apply supplemental water to row crops (Benjamin et al., 1997). Water is usually applied to each furrow in the field, but some researchers have proposed irrigating alternate furrows instead of every furrow in a field to increase WUE (Fischbach and Mulliner, 1974; Musick and Dusek, 1974; Crabtree et al., 1985). Small yield losses were recorded for sugar beet, sorghum, potato, and soybean (Musick and Dusek, 1974; Crabtree et al., 1985) for alternate-furrow irrigation systems compared with every-furrow irrigation, and irrigation WU decreased by 30%–50%. Presently, most of the world's irrigation is by furrow, where the irrigation efficiency (IE) (expressed as the ratio of crop WU to applied irrigation water) stands at only 50%–60% and is associated with large losses of water that at times lead to significant waterlogging and salinization (Jensen et al., 1990). Furrow-irrigated crops frequently suffer temporal hypoxia after irrigation, most especially on heavy clay soils. In addition, furrow irrigation, in general, supplies copious but infrequent quantities of water (Bhattarai et al., 2005).

   Recently, a raised-bed furrow-irrigation system has been adopted by many farmers in Mexico, India, and other countries (Tripathi et al., 2005; Aggarwal et al., 2006; Limon-Ortega et al., 2006; Kong et al., 2010). This system, developed in Mexico, uses a set number of rows planted on beds and has been shown to be an effective alternative to conventional planting

and irrigation. Compared with conventional systems, raised-bed planting can enhance crop WUE by 30% of savings in water consumption in
some years, promote nitrogen uptake, and increase nitrogen use efficiency
(NUE) in wheat (Zhang et al., 2007; Wang et al., 2009; Kong et al., 2010).
The bed planting can also improve the spatial light distribution within the
wheat field and eliminate soil surface crusting problems (Wang et al., 2004,
2009; Zhang et al., 2007). The bed planting can also improve the spatial
light distribution within the wheat field and eliminate the soil surface crusting problems (Wang et al., 2004, 2009; Zhang et al., 2007). Additionally,
this practice has led to reduced lodging and decreased the incidence of
some stem or leaf diseases, such as the sharp eye spot and powdery mildew
(Wang et al., 2004). Kong et al. (2010) reported that furrow irrigation in a
raised-bed planting system improved root-zone soil moisture and nutrient
regimes, decreased soil bulk density by 8.5%–10.4%, increased soil respiration by 3.2%–10.4%, and root dry weight by 2.8%–3.7% in wheat crop.

2. Central pivot irrigation: automated sprinkler irrigation achieved by auto-
matically rotating the sprinkler pope or boom, supplying water to the sprinkler heads or nozzles, as a radius from the center of the field to be irrigated.
Water is delivered to the center or pivot point of the system. Center-pivot
sprinkler irrigation is well suited to the region where water is a far more
restricted resource for irrigated agriculture than land. Widespread growth
in the use of center-pivot sprinkler systems in this area has made knowledge about crop WU for management and system design even more critical,
since the area is dependent on declining groundwater resources and on
low, highly variable precipitation (Howell et al., 1998). The central pivot
irrigation has dominated cotton irrigation methods in Georgia (USA) and
are used on close to half of the hectarage in Georgia (Hook et al., 2004;
Ritchie et al., 2009).

3. Check basin irrigation is defined as the irrigation method in which water is
applied rapidly to relatively level plots surrounded by levees. The basin is a
small check.

4. Drip irrigation is defined as the method of irrigation in which water is
slowly applied to the soil surface through small emitters having low discharge orifices. With the adoption of drip irrigation technology, the area
under irrigation could be almost doubled, or the current production level
could be achieved, with as little as half of today's global irrigation water
allocation (Bhattarai et al., 2005).

5. Subsurface drip irrigation is defined as the application of water below the
soil surface through emitters, with discharge rates generally in the same
range as drip irrigation. This method of water application is different from
and not to be confused with subirrigation where the root zone is irrigated by
water table control. Bhattarai et al. (2005) reported that theoretically, global
adoption of subsurface drip irrigation in irrigated production systems would
mean a 40% increase in food production, if the supply of water remained
the same for agriculture. Ritchie et al. (2009) reported that subsurface
drip irrigation system is used as a water-efficient alternative to overhead

irrigation in many crops. Sorensen et al. (2000) reported that subsurface drip-irrigated cotton provided high yields in Georgia (USA) compared to overhead irrigation. Other subsurface drip irrigation studies conducted in the Coastal Plain have reported subsurface drip irrigation cotton yields equal to or better than yields from overhead sprinkler irrigation (Sorensen et al., 2000; Whitaker et al., 2008).

6. Spray irrigation is the application of water by a small spray or mist to the soil surface, where travel through the air becomes instrumental in the distribution of water. The use of sprinkler irrigation systems can reduce water loss substantially at the point of delivery to field plots and, consequently, improve WUE. For example, the shift from graded furrow to sprinkler irrigation (an important regional transition) of predominantly center-pivot sprinkler (Musick et al., 1988) has reduced water applications and has contributed to sustained irrigated crop productivity on the Texas High Plains in the United States (Musick et al., 1990; Howell et al., 1998).

7. Border strip irrigation is the method of irrigation in which water is applied at the upper end of a strip with earth borders to confine water to the strip.

8. Subirrigation is defined as the method of irrigation where the root zone is irrigated by water table control.

The suitability of the methods of irrigation depends on the available quantity of water, source of water, size of the area to be irrigated, soil type, crop species, climatic conditions, and economic conditions of the farmers. However, irrigation efficiency can be improved with the adoption of adequate measures to avoid the loss of water during storage and conveyance and the use of optimal quantity and frequency of irrigation for a determined crop species. Optimal irrigation frequency and quantity depend upon soil, crop, and climatic conditions. Water percolation through the soil profile is one of the major losses of irrigation water. Water infiltration loss can be minimized through improvement and stabilization of soil aggregation. One of the measures to achieve this objective is addition or improvement in the SOM content of the soil (Fageria, 2002). In addition, evaporation from the open reservoir is a major loss of water from storage facilities in arid regions. Not only does evaporation cause water loss, but since surface water in dry regions generally contains appreciable amounts of soluble salts, evaporation also causes deterioration of water quality (Hillel and Rawitz, 1972).

One important factor that can be taken into consideration in the planning stage is the cross-sectional shape of the reservoir. The most favorable shape is that which exposes the minimum amount of surface per unit volume of water stored. Hence, a deep reservoir is obviously preferable to a shallow one (Hillel and Rawitz, 1972). Surface runoff of irrigation water can be minimized with the use of sprinkler irrigation and drip irrigation methods and the quantity of water applied should not exceed field capacity. Furrow irrigation does not allow precise control on the amount of water that infiltrates the field, although it can be modified somewhat by duration of irrigation and other irrigation factors. Adequate land leveling is also important in the uniform distribution of irrigation water in the field plots and improvement in the irrigation efficiency. Nonuniform irrigation creates a major management trade-off between maximizing crop production and minimizing potential groundwater degradation.

3.4.1.9 Adopting Controlled Deficit Irrigation

Controlled deficit irrigation is an approach that supplies water at a rate below the full crop water requirement and usually results in lower biomass yields than full irrigation (Carter and Sheaffer, 1983; Undersander, 1987; Grimes et al., 1992). Li et al. (2009) also reported that deficit irrigation (less than full irrigation) could improve agricultural WU, and the subsequent use of that water is possible for more efficient crop production. Adopting deficit irrigation in alfalfa in the United States is suggested as an example of water transfer from irrigation to urban population and improved WUE (Lindenmayer et al., 2011). Water saving potential from alfalfa is high because it is a high WU crop produced on 12% of the irrigated land in the United States (Lindenmayer et al., 2011). Established alfalfa may be adapted to deficit irrigation with drought avoidance mechanisms such as deep rooting (Peterson, 1972) and drought-induced dormancy (Robinson and Massengale, 1968; Peterson, 1972; Lindenmayer et al., 2011).

3.5 IMPORTANCE OF WATER TO MINERAL NUTRITION OF CROP PLANTS

Although the uptake of nutrients and the absorption of water are independent processes in the roots, the necessity for available water in both the plant and the soil for growth and nutrient transport makes them intimately related. Soil water availability in the potential tension ranges of 0.01–1.0 MPa (MPa = megapascal × 0.1 = bar) is essential for nearly all processes promoting nutrient acquisition (Viets, 1972). For example, reactions affecting nutrient concentrations in soil solutions, nutrient transport by diffusion and mass flow to root surfaces, and nutrient absorption by roots are affected by water potential. Nutrients are absorbed by roots as ions and water is responsible for their transport from soil to the root system by diffusion and/or mass flow. In dry soils, nutrient transport to the roots is reduced and, consequently, nutrient uptake rate decreases. In addition, in dry soils root growth is also reduced, which may limit nutrient uptake to plants.

Nutrients are essential to plant growth and maximum plant growth is achieved when the nutrient availability coincides with water availability (Amer et al., 2009). Nutrient availability in soils is at maximum when water content is near field capacity. Field capacity is defined as water that soil will hold against percolation after several days of drainage, and usually ranges from 0.01 to 0.03 MPa water potential (Viets, 1972). The Soil Science Society of America (2008) defined field water capacity as the content of water, on a mass or volume basis, remaining in a soil 2 or 3 days after having been wetted with water and after free drainage is negligible. Wilting point is another index used in indicating the water content of the soil for plant growth. Wilting point is defined as the water content of a soil when indicator plants growing in that soil wilt and fail to recover when placed in a humid chamber

(Soil Science Society of America, 2008). Soil water contents near field capacity allow for the best combinations of sufficient air space for oxygen diffusion, greatest amounts of nutrients in soluble forms, greatest cross-sectional areas for the diffusion of ions and mass flow of water, and most favorable conditions for root extension (Fageria et al., 2006).

Water content in the soil varies with soil texture. It can fluctuate through an eightfold range for sandy soils and nearly a twofold range for clay soils between the wilting percentage (1.5 MPa) (Viets, 1972) and saturation. Little experimental data are available for changes in the concentration of plant-available nutrients as soils dry from field capacity to wilting range with evaporation or water extraction by roots. However, nutrients that are released in soil solution with the mineralization of OM like N, S, and P can significantly decrease with the drying of the soil from field capacity to wilting point. This may happen due to decrease in the activities of soil microorganisms that are responsible for the ammonification of OM and the nitrification of ammonium (Viets, 1972). In water-stressed plants, N and P levels are reduced, the younger leaves showing the greatest reduction, and P showing earlier and more pronounced reduction than N. In both nutrients, there is a tendency for migration from leaf to stem when water stress is imposed (Slatyer, 1969).

Water stress also reduces grain yield, LAI, and NUE in wheat (Table 3.8). Over all, the grain yield of wheat was 42% higher in the irrigated treatment compared with dry treatment. Similarly, overall LAI was 29% lower at dry treatment compared to irrigated treatment. Overall, the NUE was 64% higher at the irrigated treatment compared with dry treatment. Similarly, higher response of perennial grasses to N fertilizer in the presence of adequate water was reported by Power (1990). According to this author, with relatively high water availability, a very large growth response to increasing fertilizer N rate was observed. As water deficit increased, the magnitude of this response decreased. Finally, when water was very limited, dry matter production by the grass was quite low, and little or no response to N fertilizer was observed.

TABLE 3.8
Grain Yield, LAI, and NUE in Wheat as Influenced by Water Level

N Rate	Irrigated			Dry		
(kg ha⁻¹)	Grain Yield (kg ha⁻¹)	LAI	NUE	Grain Yield (kg ha⁻¹)	LAI	NUE
0	2600	1.2	—	1900	1.0	—
75	5100	2.5	33	3200	2.4	17
300	9000	5.5	21	6700	3.9	16
Average	5567	3.1	27	3933	2.4	16.5

Source: Adapted from Lawlor, D.W., *J. Exp. Bot.*, 46, 1449, 1995.

$$\text{NUE (kg kg}^{-1}) = \frac{\text{Grain yield at N level in kg} - \text{Grain yield at zero N level in kg}}{\text{N rate in kg}}$$

3.6   WATER DEFICITS VERSUS SALINITY

Water deficits affect soil chemistry primarily through their effect on the solubility and precipitation of salts and minerals in the soil. As the soil dries, the concentration of soluble salts in the soil solution increases (Power, 1990). If this concentration exceeds about 5 dS m^{-1} (2–3 dS m^{-1} on dry soil basis), the osmotic potential of the soil solution may begin to affect plant–water relations and the subsequent growth of salt-sensitive crops adversely (Tayler, 1983). This level of salinity is not unusual in dryland soils (Power, 1990). If sodium is present in the soil in significant quantity (in excess of about 6% of the cation exchange capacity), soil structure and aggregation may begin to disintegrate, resulting in dispersed soil. Under such conditions, hydraulic conductivity is greatly reduced, soil resistance increases, surface crusting may become severe, and water infiltration is restricted (Power, 1990). Consequently, plant growth in such sodic soils is restricted by greater water deficits, greater soil resistance, and reduced hydraulic conductivity.

3.7   WATER DEFICITS VERSUS SOIL MICROBIOLOGY

The influence of water on soil microbes is significant, which affects the solubilization of nutrients and their availability to plants. This is especially true of N and P, because most of N and appreciable part of P in a soil are derived from the biological decomposition of OM (Doran and Smith, 1987). Linn and Doran (1984) have shown that such biological processes as N mineralization, nitrification, and CO_2 production increase as the percentage of the soil pores filled with water. In dry soils, these processes decrease and the availability of nutrients that are released through mineralization decreases. Okon and Kapulnik (1986) have reported that the moisture status of corn and sorghum was favorably affected by *Azospirillum* inoculation. Similarly, Sarig et al. (1988) demonstrated that inoculation resulted in higher leaf water potentials, lower canopy temperatures, and greater stomatal conductance and transpiration. Inoculated plants extracted more soil water, particularly from deeper layers, indicating that yield increases resulting from inoculation are primarily from improved soil moisture utilization (Summer, 1990).

3.8   DROUGHT

Drought stress is a major constraint to crop production and yield stability in many regions of the world. Various studies have demonstrated that drought can adversely affect many aspects of plant growth and physiological metabolism, including height, dry matter production, leaf area, grain number, grain size, grain yield, and photosynthesis (Goodman and Newton, 2005; Adejare and Umebese, 2007; Jeroni et al., 2007; Singh et al., 2008; Wu et al., 2008; Ge et al., 2010; Pahlavan-Rad et al., 2010, 2011). Decreased CO_2 assimilation in drought-affected plants is due to stomata closure, which is the main cause of decreasing transpiration (Molnar et al., 2004). Decreased cell turgor pressure as a result of drought is also important because cell growth depends on cell wall extensibility and protection. Plants under stress conditions selectively uptake some inorganic ions and produce some organic compounds for combating reduced cell turgor pressure (Zhu et al., 2005). Soil water deficits are estimated to depress agricultural crop yields in the United States by about 70%, compared with maximum achievable yields (Boyer, 1982). Similar problems are encountered worldwide. For example, the major constraint to bean production in many developing countries is drought, which affects 73% of land areas planted with beans in Latin America (Ramirez-Vallejo and Kelly, 1998). Approximately, 32% of wheat-growing regions in developing countries experience some type of drought stress during the growing season (Ginkel et al., 1998).

Drought, whether intermittent or terminal, can be confounded with high temperatures in certain locations and aggravated by shallow soils and/or root-rotting pathogens. Droughts are inevitable and recur frequently throughout the world, despite improved abilities to predict their onset and modify their impact. Drought remains the single-most important factor affecting world security and stability of land resources from which food is derived (McWilliam, 1986). Drought is a meteorological term that means lack of precipitation over prolonged periods of time (Hale and Orcutt, 1987). It is a meteorological and hydrological event involving precipitation, evaporation, and soil water storage (McWilliam, 1986). Kobata (1995) defined droughts as environmental situations where decreases in soil moisture or soil moisture potentials occur in the rooting zones of crops. Droughts denote periods of time without appreciable precipitation, during which the water content of soils is reduced to such an extent that plants suffer from lack of water (Larcher, 1995). Frequently, but not invariably, dryness of soil is coupled with strong evaporation caused by dryness of air and high levels of radiation. However, shortages of precipitation alone are not sufficient to cause aridity. Dryness results from the combinations of low precipitation and high evaporation. In dry regions, drought is of such regular and prolonged occurrence that annual evaporation may exceed total annual precipitation. Such climates are called arid as opposed to humid climates in regions with surplus precipitation (Thornthwaite, 1948). About one-third of the earth's continental area has rain deficits, and half of this (about 12% of land area) is so dry that annual precipitation is less than 250 mm, which is not even a quarter of potential evaporation. Reduction in soil water potential decreases the water potential of whole plants or parts of plant organs. Drought influences biological and economic yields by inhibiting various physiological functions because of water deficiency. Drought is the most prevalent environmental stress, and limits crop production on about 28% of the world's land area (Dent, 1980). Most higher plants (annual crops) are exposed to varying degrees of water stress at some stage of their ontogeny. The type of water stress may vary from small fluctuations in atmospheric humidity and net radiation in more mesic habitats to extreme soil water deficits and low humidity in arid environments (Morgan, 1984). The impact of water stress is a function of duration, crop growth stage, type of crop species or cultivar, soil type, and management practices. Drought is sporadic in nature, resulting in drastic crop losses even in humid climates (Dunphy, 1985).

The term "water deficit" will be used here to indicate that water pressures in the rhizosphere are sufficiently negative to reduce water availability for transpiration and plant expansion growth to suboptimal levels (Neumann, 1995). Water deficits have major effects on plant photosynthetic capacity. Reduction in photosynthetic activity and leaf senescence are well documented and adversely affect crop yield

(Gerik et al., 1996). Other adverse effects of water deficit on plant photosynthetic capacity include reduction in cell growth and enlargement, leaf expansion, assimilate translocation, and transpiration (Hsiao, 1973). Protein metabolism and the synthesis of amino acids are both soon impaired. One enzyme most strongly inhibited from water deficiencies is nitrate reductase. Even brief periods of negative water balance cause nitrate reductase activities to decrease by 20% or even by 50% after longer periods of water deficit (Hsiao, 1973). This is why drought leads to increases in nitrate contents of plants that have been fertilized with N fertilizers (Larcher, 1995). Although the consequences of water stress are well known, our ability to genetically manipulate these processes and improve drought tolerance is limited.

3.8.1 CROP YIELD RELATIVE TO WATER STRESS

Yield performances of genotypes grown under water stresses are reflective of both plant responses to stress and potential yield levels. Soil water deficits are common in the productivity of most crops and can have substantial negative impacts on growth and development. Four main aspects of plant behavior relative to drought that can readily be linked with yield are the modification of leaf area, root growth, efficiency by which leaves exchange water for CO_2, and processes involved in setting and filling of seeds (Passioura, 1994).

Since plant growth is primarily based on cell expansion, steady-state equations that empirically model relative contributions of hydraulic and wall mechanical factors to rates of single-cell expansion growth have been studied by Lockhart (1965). These equations provide useful starting points for considering the biophysical control of growth. The equation showing effects of hydraulic parameters on cell expansion can be expressed as

$$\text{Growth rate} = L_p(\Psi_0 - \Psi_i)$$

where

L_p is the hydraulic conductivity of plasma membranes
Ψ_0 is the water pressure of external water sources
Ψ_i is the water pressure of cell solutions

According to this equation, growth rates (i.e., rates of irreversible cell volume increases) are proportional to rates of water uptake by expanding cells. Water uptake occurs when Ψ_i values are more negative than Ψ_0, and the rates of water uptake are further affected by L_p.

A second equation of Lockhart (1965) indicates that the rates of cell expansion are also limited by the yielding properties of expanding cell walls:

$$\text{Growth rate} = m(P - Y)$$

where

m is the extensibility coefficient of cell walls
P is cell turgor pressure
Y is threshold pressure

In this case, growth rate is proportional to differences between P and Y. The Y values are minimum turgor pressures required to initiate irreversible expansion of cell walls. The term $(P - Y)$ represents growth effective turgor pressure acting on expanding cell walls. Growth rate is also modulated by m values. The aforementioned equations can also be combined (Neumann, 1995) as given:

$$\text{Growth rate} = (mL_p/m + L_p) \times (\Psi_0 - \Psi_\pi - Y)$$

where Ψ_π is cell osmotic pressure and the other terms the same as defined earlier. This latter equation does not directly include turgor pressure. However, cell turgor pressure is related to Ψ_π by the relationship $P = \Psi_i = \Psi_\pi$. Thus, the cellular adjustment of Ψ_π will interactively affect Ψ_i and turgor. It is also important to understand that turgor pressure in protoplasts is the Newtonian counterforce of tensile forces exerted on protoplasts by expanding cell walls (wall stress). Thus, changes in cell wall characteristics (e.g., m or Y) affect growth in addition to affecting turgor pressure and Ψ_i (Cosgrove, 1993; Neumann, 1995). More detailed information on growth inhibitory effects by water deficits has been discussed elsewhere (Taiz, 1984; Boyer, 1985; Cosgrove, 1993; Pritchard, 1994; Neumann, 1995).

It is well understood that various processes of photosynthesis are affected by internal water deficits in rice. This was observed specifically by the inhibition of leaf blade elongation (Turner et al., 1986), the reduction of photosynthetic rates per unit leaf area (Ishihara and Saito, 1983), and the acceleration of wilting (O'Toole and Moya, 1976). Relative to morphogenesis of rice panicles, which strongly affects dry matter distribution into grain, reductions in the water potentials of leaf blades beyond certain limits lead to delays in young panicle development (Tsuda, 1986), pollen development was inhibited by the reduction of water potentials in flag leaves (Namuco and O'Toole, 1986), the exertion rate of panicles during heading decreased, and the number of sterile grains increased (Curz and O'Toole, 1984). In plants other than rice, the inhibition of various physiological processes by drought conditions was affected indirectly by the dehydration of organs (Morgan, 1980; Schulze, 1986). Sensitivity of rice to water deficits was greatest during the reproductive phases of growth (O'Toole, 1982). In developing practical field screening systems for reproductive phase drought resistance in rice, Garrity and O'Toole (1995) assessed canopy temperature responses among wide ranges of rice germoplasm, and related their results to other plant characters related to drought resistance. Negative relationships were observed between grain yield and midday canopy temperatures on the date at which each entry flowered. Cultivars that exhibited midday temperature values >34°C at anthesis had essentially no yield. Fertility of spikelets was similarly related to midday canopy temperatures and decreased below 30% as anthesis midday canopy temperatures exceeded 34°C (Garrity and O'Toole, 1995).

Establishment and activity of legume rhizobium symbiosis are known to be sensitive to drought stresses, which can have important negative effects on yield (Zablotowicz et al., 1981; Kirda et al., 1989). Most research on mechanisms of drought effects on N_2 fixation has focused on nitrogenase activity rather than on nodule formation and growth. Several studies have reported that N_2 fixation by soybean is more sensitive to soil dehydration than leaf gas exchange for plants grown under controlled growth conditions (Durand et al., 1987; Serraj and Sinclair, 1996; Purcell et al., 1997) and field conditions (Sinclair et al., 1987; Serraj et al., 1997). Sall and Sinclair (1991) and Serraj and Sinclair (1998) reported genetic variability in N_2 fixation sensitivity to drought stress among soybean cultivars, and that plant responses of N_2 fixation rates to drought stress were related to nodule formation and growth.

3.8.2 CROP GROWTH STAGES SENSITIVE TO DROUGHT

Drought can occur anytime during the crop growth cycle. However, its negative effect on plant growth and, consequently, yield depends on crop growth stage. Grain yield losses in field crops are higher if the stress coincides with the more drought-sensitive stages of crop growth, such as flowering and grain filling (Badu-Apraku et al., 2011). Water stress during seed filling may shorten the seed-filling period and reduce yield as shown for soybean (Brevedan and Egli, 2003), wheat (Nicolas et al., 1984), corn (NeSmith and Ritchie, 1992), rice (Yang et al., 2002), and pearl millet (Bieler et al., 1993). Burke et al. (2010) identified the growth stages in sorghum that are critical in understanding drought tolerance: GS1, seedling establishment (early vegetative stage); GS2, preflowering (panicle differentiation to flowering); and GS3, postflowering (grain fill to physiological maturity of grain). Water stress at GS2 and GS3 affects panicle size, grain number, and grain yield. Retention of green leaf area at maturity, known as stay-green, is used as an indicator of postanthesis drought resistance in sorghum programs in the United States and Australia (Borrell et al., 2000). Craufurd et al. (1993) reported that water stress during the booting and flowering stages resulted in grain yield reduction of up to 85%.

Growth stages about 10 days before and after flowering are the most critical growth stages for water stress effects on upland rice. Different growth stages of an upland rice cultivar having 130 days growth cycle from sowing to physiological maturity grown on a Brazilian Oxisol are shown in Figure 3.10. Flowering stage in this cultivar occurs about 100 days after sowing. Hence, the most critical growth stage for drought in this cultivar is 90 days before flowering and 110 days after flowering. Drought stresses during flowering caused larger decreases in yield than similar stresses during vegetative or grain ripening stages (Fageria, 2007). Fukai and Cooper (1996) reported that rice grain yield was decreased at rates of 2% per day when 15 day stress periods (when dawn leaf water potentials were less than −1.0 MPa) occurred during panicle development. Unfilled grain numbers increase sharply with stress during late panicle development. Water stresses reduce assimilate production per plant, spikelets per panicle, filled grain percentages, and individual grain weights. However, unfilled grain numbers appear most susceptible to reduced assimilate availability. Assuming reductions of 2% grain yield per day with delays in termination of 15 day stress periods, 20 day differences of about 40% were noted if plants had the same yield potential as plants grown under non-limiting conditions.

It is likely that rice cultivars with differences in phenology will react differently to drought stresses depending on the timing of stress development (Maurya and O'Toole, 1986). Phenology is important in determining grain yield responses because early maturing cultivars often escape severe drought stresses, while late

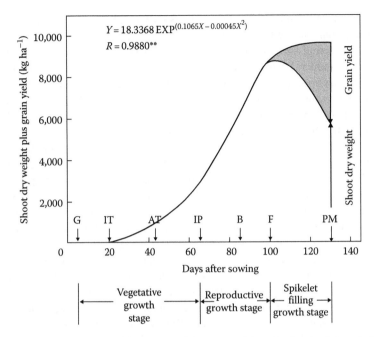

FIGURE 3.10   Shoot dry weight accumulation and grain yield of upland rice during the growth cycle of the crop in central Brazil. G, germination; IT, initiation of tillering; AT, active tillering; IP, initiation of panicle primordia; B, booting; F, flowering; and PM, physiological maturity. (From Fageria, N.K., *J. Plant Nutr.*, 30, 843, 2007.)

maturing cultivars may not escape drought stresses. These results indicate that genotypes should be compared for drought resistance/susceptibility within similar phenology groups, or that genotypic variation in phenology should be corrected statistically before genotypic differences in drought tolerance. Alternatively, it may be possible to implement strategies for staggered plantings of different genotypes so that plants would flower about the same time (Fukai and Cooper, 1996).

Yields of modern crop cultivars have generally increased with little change in aboveground biomass production, and these increases are attributed to increases in HI (Gifford, 1986). Yield increases have occurred without much change in the amount of water used, and this has resulted in the natural improvement of WUE (Richards et al., 1993). Among many factors, HI depends on the relative proportions of pre- and postanthesis biomass and the mobilization of preanthesis assimilate to grain. Severe water deficits at critical growth stages like flowering can greatly decrease seed numbers and HI.

Water stress during any stage of soybean development can reduce yield, but extents of yield reductions depend on the stage of development. Negative effects of stress are particularly important during flowering, seed set, and seed filling (Souza et al., 1997). Water stress during seed filling reduces yield by reducing seed size (Vieira et al., 1992). These reductions in seed size have been associated with early maturity (Ashley and Ethridge, 1978) and shorter seed-filling periods (Meckel et al., 1984; Smiciklas et al., 1989). It is generally assumed that shorter seed-filling periods result from accelerated leaf senescence (Souza et al., 1997). Leaf senescence in soybean, characterized by declines in leaf N, chlorophyll, and photosynthesis, begins early during seed filling and usually ends by physiological maturity (Egli and Crafts-Brandner, 1996). During leaf senescence, C and N are redistributed from leaves and other vegetative plant parts to developing seeds. This redistributed N can be a significant source of N to seeds by contributing 20%–100% of mature seed N in cultivars with different maturities (Zeiher et al., 1982). Increasing source–sink ratios may reduce the effects of moisture stress on seed size.

Effects of drought stress on grain yield may be assessed in cereal crops in terms of yield components, some of which can be more important than others depending on the intensity of stress and growth stage at which stress develops (Johnson and Kanemasu, 1982; Giunta et al., 1993). While intense drought mainly affects the number of kernels per unit area through decreases in fertility, mild droughts may cause only decreases in grain weight (Giunta et al., 1993). The reduction in the number of spikes per unit area and the number of kernels per spike has been associated with preanthesis water stresses (Day and Intalap, 1970). Water deficits around anthesis may lead to reduced yields by decreasing spike numbers, spikelet numbers, and the fertility of surviving spikelets (Aspinall, 1984). Water stress during grain filling, especially if accompanied by high temperature, is common for plants grown in Mediterranean environments (Royo and Blanco, 1998). These stresses hasten leaf senescence, reduce the duration of grain filling, and reduce grain weights (Davidson and Birch, 1978; Austin, 1989), presumably by reducing assimilate supply to developing kernels. Reduced assimilate supply has been related to reduced photosynthesis (Denmead and Miller, 1976), and early senescence. When cereal crops are subjected to severe postanthesis drought stress, kernel growth becomes increasingly dependent on contributions from vegetative reserves (Austin et al., 1980; Blum et al., 1983a,b). Reductions in grain yield and yield components due to droughts can be calculated with the formula (Blum et al., 1983a)

$$\text{Reduction in grain yield } (\%) = [(C - S)/C] \times 100$$

where C and S are the values of variables in control and stress treatments, respectively. Examples of variables are grain yield and yield components (such as panicles per unit area, 1000 grain weight or panicle length in case of rice).

The effects of water deficits on crop growth and yield depend on the degrees of stress and developmental stages at which stress occurs (Hsiao and Acevedo, 1974; Sullivan and Eastin, 1974). Low water potentials during anthesis and early grain fill can decrease yields in grain crops. Photosynthesis is inhibited under these conditions and carbohydrate reserves may become limited for grain growth (Westgate and Boyer, 1985). Wien et al. (1979) reported that 2 week drought stresses during vegetative or flowering growth stages had no significant effect on seed yield of cowpeas. Conversely, Shouse et al. (1981) reported that the most sensitive growth stages to drought were flowering and pod filling, with yield reduction of 35%–69% depending on timing and length of drought treatments in cowpeas. Soil water deficits during vegetative growth had fewer effects on crop and seed yields of cowpeas than during flowering or pod filling. Turk and Hall (1980), studying moisture stress effects on field-grown cowpeas during vegetative, flowering, and pod-filling growth stages relative to plant water potential, dry matter production, seed yield, and WUE, noted that moisture stress during flowering and pod-filling growth stages had the most serious influences on reduced seed production. However, moisture stress during vegetative stages had the least effect on seed yields. Values of WUE were reduced when irrigation was withheld during flowering and pod-filling stages, and no irrigation during vegetative stages increased the efficiency of WU by cowpea plants as seed yields were not reduced. Labanauskas et al. (1981) reported that the most severe seed yield reductions in cowpeas occurred when treatment plots were subjected to water stress during both flowering and pod-filling growth stages. Seed yield reductions of 67% were noted when water stress occurred during flowering and pod filling. No significant reductions in seed yield were noted when water stress occurred during vegetative stages.

In contrast to many crop plants including cowpea, potato tuber quality and yield can be reduced when water stresses occur during any part of the growing season (Adams and Stevenson, 1990). Thus, continuous water supplies are recommended from tuber initiation to maturity (Miller and Martin, 1983; Porter et al., 1999). Potato plants are very sensitive to moisture stress during periods from tuber initiation to shortly before foliage senescence (Singh, 1969). Water stress during tuber initiation has been reported to reduce the number of tubers produced per plant (Miller and Martin, 1985). Lynch and Tai (1989) reported that differential tolerances among potato cultivars to moisture stress are likely associated with differences in sensitivity during the ontogeny of yield development. These authors also reported that potato cultivars were also very sensitive to moisture stress during both tuber initiation and tuber-sizing growth stages, with greatest effects being during tuber sizing.

Water stress during vegetative development of maize decreased dry matter yields at harvest by reducing leaf area development, stalk dry matter accumulation, and the potential grain-filling capacity of plants (Wilson and Allison, 1978; Eck, 1984; Lorens et al., 1987). NeSmith and Ritchie (1992) reported that severe water stresses during the R3 and R5 (late grain filling) growth stages decreased total dry matter accumulation and grain yield equally. These authors also reported that corn yield can be reduced by as much as 90% if drought stress occurs between a few days before tassel emergence and the beginning of grain filling. Under induced moisture stress from about the tassel-emergence stage to the end of the crop cycle of corn, Badu-Apraku et al. (2005) observed yield reduction of 62% relative to the well-watered treatment. Wilson and Allison (1978) also noted that mild water stress during grain filling reduced total dry matter accumulation, but not grain yield because of remobilization of assimilates from stover to grain. These authors also reported that maize yield reductions were greatest when water potentials were low around anthesis, since embryo sacs may abort (Moss and Downey, 1971). Westgate (1994) reported that large yield reductions for maize resulted when drought occurred during flowering and early seed development. These yield reductions were primarily due to decreases in seed numbers per plant (Boyle et al., 1991). Inhibition of flower development, failure of embryo fertilization, and abortion of zygotes all contributed to decreases in seed numbers (Westgate and Boyer, 1986). Water deficits at flowering may also decrease maize grain yields even if pollination occurs (Schussler and

Westgate, 1994). Westgate and Boyer (1985) concluded that early kernel development was highly dependent on current supplies of assimilate in maize plants because reserves were not sufficient to maintain kernel growth during low water potentials.

Singh (1995) reported that water stress during the flowering and grain-filling stages of growth reduced seed yields and weights, and accelerated maturity of dry bean. Miller and Burke (1983) noted linear relationships between water applied during flowering and grain filling and the yield of dry bean grown on sandy soil. The most sensitive growth stages to drought of important annual crops are listed in Table 3.9.

Frederick et al. (1990) reported that yield reduction in corn due to drought stress depends on numerous factors such as the stage of plant development, the severity and duration of water deficiency, the susceptibility of the hybrids, as well as vulnerability to soil drought. Recep (2004) reported that grain yield losses of 66%–93% should be expected as a result of prolonged water stress during tasseling and ear formation stages. Canola (*B. napus* L.) seed yield and seed oil content generally increase with the amount of water received (Al-Jaloud et al., 1996). Although canola is commonly grown in most of the world under rain-fed farming systems, the production of canola under irrigation has been a common practice in some parts of the world, such

---

TABLE 3.9
Moisture-Sensitive Growth Stages for Selected Crop Species

Crop Species	Moisture-Sensitive Growth Stage
Alfalfa	Just after cutting for fodder and for seed production at flowering
Dry bean	Flowering and pod filling
Cotton	Flowering and boll formation
Peanut	Flowering and yield formation, particularly during pod setting
Corn	Flowering and grain filling
Lentils	Flowering
Millet	Flowering and heading
Onion	Bulb enlargement, particularly during rapid bulb growth and for seed production at flowering
Pepper	Flowering
Safflower	Flowering and seed filling
Sorghum	Flowering
Soybean	Flowering and pod filling
Sunflower	Flowering and bud development
Tobacco	Period of rapid growth, yield formation, and ripening
Tomato	Flowering and yield formation
Watermelon	Flowering, fruit filling, vegetative period, particularly during vine development
Wheat	Flowering and grain filling
Rice	Around flowering particularly 1 week before and 1 week after flowering

Source: Compiled from various sources by Fageria, N.K. and Gheyi, H.R., *Efficient Crop Production*, Federal University of Paraiba, Campina Grande, Brazil, 1999.

---

as Australia since 1970 (McCaffery, 2004) and the cooler, drier parts of Montana, United States (Bauder, 2006). Tesfamariam et al. (2010) reported that canola seed and oil yield are most sensitive to water stress at flowering and less sensitive during the vegetative and seed-filling stages.

3.8.3 ROOT CHARACTERISTICS RELATED TO DROUGHT STRESS

Healthy and functional roots are critical to survival and productivity of crop plants as they serve the combined functions of anchorage, support, and water and mineral nutrient uptake. Root growth, development, and function are largely determined by genetic information but respond substantially to available carbohydrate supply and local environmental conditions encountered during growth. Soil attributes that affect the extent and morphology of growing roots include water content, strength or mechanical impedance, thermal regime, and structural development (Baligar et al., 1998; Wraith and Wright, 1998).

Roots that extend beyond normal rooting zones for more extensive extraction of available soil water provide plants greater potential to increase yield under droughty conditions (Mambani and Lal, 1983). According to Boyer (1996), deep rooting probably accounts for major differences in drought tolerances between plant species. Maize and sorghum roots grow to soil depths of 2–3 m, and these plants grow and remain green when surrounding short grasses with shallow roots often become brown because of soil dehydration. The depth and extent of the rooting of various plant species have been listed by Taylor and Terrell (1982). Jordan and Miller (1980) claimed that sorghum root length densities in the lower rooting zones were insufficient to meet evaporative demands for plants grown in the Great Plains region of Texas. These authors predicted that increasing root length densities to 2 cm cm^{-3} below 50 cm soil depths would increase annual water uptake by 6 cm and reduce water stress during the moisture-sensitive grain-filling stage of growth. Root length densities of 0.5–1.2 m were noted for drought-tolerant soybean cultivars grown under water-limited conditions, and no root proliferation occurred below 0.8 m in drought-sensitive lines (Sponchiado et al., 1989). Greater total allocation to root biomass and greater allocation over the 60–120 cm depth were associated with drought tolerance among contrasting wheat cultivars (Hurd, 1974).

Genotypic differences have been reported among soybeans with more root efficiency for extracting available soil water reserves (Hudak and Patterson, 1996). It was noted that in the field screening of several exotic soybean genotypes, one genotype (PI 416937) maintained leaf turgidy during drought imposed at pod fill and yield was depressed less than that of other genotypes (Carter and Rufty, 1993). Subsequent investigations confirmed that PI 416937 root growth was tolerant to high concentrations of Al and that midday transpiration rates were higher than that of another contrasting genotype (Forrest) grown under soil water deficits (Sloane et al., 1990). Hudak and Patterson (1996) also reported that the PI 416937 advantage for drought tolerance may have resided in its ability to exploit upper soil horizons (above 68 cm) for water from networks of fibrous roots. Localized measurements of soil moisture tension revealed that the PI 416937 rate of soil desiccation was often slower than that of Forrest, indicating that larger total soil volumes were being explored by PI 416937 compared to Forrest. In addition, PI 416937 appeared to have greater lateral spread of roots. PI 416937 may provide soybean breeders with the ability to add diversity to root morphology of existing soybean cultivars. The incorporation of highly branched, fibrous roots like those of PI 416937 into other soybean cultivars might enhance the ability to utilize water resources where normal root accessibility is limited. Limited water accessibility often occurs within interrow areas and/or below chemical or physical soil barriers (Hudak and Patterson, 1996).

Comparisons between monocotyledonous and dicotyledonous plant species have also pointed out discrepancies between root density and water uptake. Mason et al. (1983) reported that sunflower extracted as much water from irrigated clay loams as maize and sorghum plants despite having only half the total root density. Hodgson and Chan (1984) reported that safflower extracted water at greater depths than wheat when both species were grown on poorly structured Vertisols. These authors inferred that safflower roots had greater capacity to penetrate and extract water from subsoils, again at lower densities. Bremner et al. (1986) noted that sunflower extracted substantially more water from subsoils of spodosols than sorghum, although maximum depths of water extraction were similar for both species. Hamblin and Tennant (1987) compared WU of spring wheat, barley, lupin, and field pea grown on four different soil types under drought-stressed conditions, and concluded that total root lengths per unit ground area for cereals were consistently 5–10 times larger than those for grain legumes.

Greater concentrations of starch in roots have been reported to be associated with drought tolerance in cotton (De Souza and Da Silva, 1987). Greater starch contents of perennial cotton appeared to be associated with greater root/shoot dry weight ratios and drought tolerance, which imparted survival mechanisms well suited to the cyclic availabilities of water in northeastern Brazil (Wells, 2002).

3.9 MANAGEMENT STRATEGIES FOR REDUCING DROUGHT

Drought is the most important environmental factor that limits crop production throughout the world. Crop management strategies for reducing water deficits can be grouped into two major categories: (1) relate cultural practices to improved WUE of plants to reduce impacts of drought on crop production, and (2) plant drought-tolerant crop species or cultivars that produce satisfactorily in drought-prone regions and soils. Under strategy No. 1, developing and adopting practices that will use water efficiently by plants grown with irrigation would be of concern, as the traditional solution to agricultural water shortages is irrigation. Wherever irrigation water is available, crop yields can be increased significantly by using higher technologies in crop production with less risk to farmers. Agriculture is the largest consumer of water and accounts for 72% of total WU in the world and 87% in developing countries (Barker et al., 1999). In recent years, the growing scarcity and competition for water has been witnessed worldwide, and opportunities for developing new water resources for irrigation are limited. Irrigated agriculture has traditionally consumed more than two-thirds of available water supply. As demand for industrial, municipal, and other uses rises, less water will be available for irrigation. Thus, ways must be found to increase productivity with limited amounts of water if food security is to be maintained (Barker et al., 1999). With increasing demands on water resources, greater efficiency is needed in irrigated agriculture.

Internal drainage losses from root zones can be reduced or managed to improve irrigation efficiency. Keys to reducing internal drainage through irrigation water management are improved uniformity of applied water and decreased average depths of water applied (Hanson, 1987). Tanji and Hanson (1990) reported that proper irrigation management can reduce overirrigation, and proper maintenance can control the uniformity of application. Subsurface drip irrigation systems can be adopted to reduce root zone drainage and improve WUE. Darusman et al. (1997) reported that adopting subsurface drip irrigation can attain nearly maximum grain yields of maize in western Kansas with significant decreases in the amount of irrigation water used (75% ET) compared to full irrigation (100% ET). Decreases in seasonal irrigations are accompanied by significant reductions in internal drainage losses from root zones.

Changes in water management practices in Japan from traditional continuous flooding to intermittent irrigation in flooded rice have been an important strategy for improving WUE. Some forms of intermittent irrigation similar to the Japanese systems are now practiced in many Asian rice production schemes where water supply is limited. These systems maintain yields close to maximum if weeds are satisfactorily controlled. As much as 30%–50% of the water used when standing water is maintained can be saved (Greenland, 1997).

The foremost concern in arid and semiarid areas is water availability and its efficient use. In dryland areas in North America, the dominant constraint to wheat production is limited water, especially where high evaporative demand coincides with low rainfall (Campbell et al., 1993; Musick et al., 1994; Weinhold et al., 1995). Although crop yields under dryland conditions are related to seasonal rainfall, WUE can be substantially improved by crop management practices (Keatinge et al., 1986; Cooper and Gregory, 1987; Harris et al., 1991). The introduction of supplemental irrigation to winter-grown cereals can potentially stabilize and increase yields, as well as increase WUE received from both rainfall and irrigation (Oweis et al., 1992). Supplemental irrigation is practiced in 40% of rain-fed wheat areas in Syria and has contributed substantially to cereal productivity (Oweis et al., 1998).

The adoption of deficit irrigation in crop production is another important strategy in improving crop yields under drought conditions. Controlled deficit irrigation is an approach that supplies water at a rate below the full crop water requirement and usually results in lower biomass yields than full irrigation (Carter and Sheaffer, 1983; Grimes et al., 1992; Lindenmayer et al., 2011). Established alfalfa may be adapted to deficit irrigation with drought avoidance mechanisms such as deep rooting (Peterson, 1972) and drought-induced dormancy (Robinson and Massengale, 1968; Peterson, 1972). Alfalfa is a high water use crop (Schneekloth and Andales, 2009) grown on 12% of the irrigated land in the United States (National Agricultural Statistics Service, 2007); controlled deficit irrigation of alfalfa has been proposed as a source of water transfer from agriculture (Putnam et al., 2005).

Another significant factor to avoid water stress is planting date. Research in the Mediterranean climate of Australia indicated that delaying sowing after the optimum time (coinciding with the onset of seasonal rains) consistently reduced crop yields (French and Schultz, 1984a). Using a simulation model, estimated potential wheat yields would decline by 4.2% per week in Syria with sowing delayed after November in normal years (Stapper and Harris, 1989). Early maturing soybean cropping systems in southeastern United States have generally called for early planting as a management practice for success in productivity. The primary rationale for producer adoption of early planting systems in this area has been to avoid late-summer droughts (Kane and Grabau, 1992; Savoy et al., 1992; Mayhew and Caviness, 1994). In water balance studies, pearl millet WUE (defined as yield/ET) was noted to be very low compared with WUE of other cereals (Klaij and Vachaud, 1992). Low WUE values can be addressed by at least three management options: (1) moderate fertilizer addition (20 kg N ha^{-1} and 9 kg P ha^{-1}), (2) moderate increases in plant population (10,000 hills ha^{-1}), and (3) the use of high water-efficient genotypes (Payne, 1997).

More efficient WU has been reported with more intensive cropping systems (Black et al., 1981). When precipitation storage efficiency improves, as is possible with MT and NT, producers can increase successes of crop yields more intensively than with crop–fallows (Halvorson and Reule, 1994; Peterson et al., 1996; McGee et al., 1997). Aase and Schaefer (1996) reported that NT annual-cropped spring wheat was more profitable and productive than spring wheat–fallow with 356 mm precipitation in northeast Montana. Introducing optimum combinations of improved technologies or management practices such as pest control and nutrient management can enhance crop yields and outputs per unit water transpired (Barker et al., 1999).

Matching phenology to water supply is another important strategy to reduce drought stress in crop plants. Genotypic variation in growth duration is an obvious means for matching seasonal transpiration with water supply, and thus maximizing water transpired (Ludlow and Muchow, 1990). Early flowering plants tend to provide higher yields and greater yield stability than later flowering plants if rains do not occur during the latter half of the growing season. Moreover, if cultivars can escape drought during the critical reproductive growth stages, HI is generally improved. The development of short-duration cultivars provides benefits where rainfall is reasonably predictable. In unpredictable environments, potentially transpirable water may be left in the soil at plant maturity when sufficient moisture comes, and yield is sacrificed (Ludlow and Muchow, 1990).

Foliar applications of aqueous methanol have been reported to increase yield, accelerate maturity, and reduce drought stress and irrigation requirements in crops grown in arid environments under elevated temperatures and in direct sunlight (Nonomura and Benson, 1992). These authors reported that spraying C_3 plants with 100–500 mL L^{-1} methanol solutions doubled plant growth and crop yields in several species.

It has been estimated that 30%–60% of total water supplied to soils in semi-arid regions is lost to evaporation (Cooper et al., 1983; French and Schultz, 1984b). Water uptake efficiency is a measure of the overall ability of plants to absorb water from soil and to reduce evaporation from soil. These traits are associated with root characteristics, early growth habits, and canopy closure. These characteristics are considered to possess adaptive values for droughty environments (Ludlow and

Muchow, 1990). As water evaporation decreases, larger fractions of water become available for plant transpiration and total dry matter production is increased.

Genetic variation is a key determinant in successful adaptation to environmental stresses (Stanca et al., 1992). Maqbool et al. (2001) reported that over the past six decades, efforts in plant breeding have enhanced the genetic potential of cereals and many sorghum hybrids tolerant to drought have been released. Studies that compared old and new corn hybrids reported little change in corn yield potential under irrigated conditions, but dryland corn yields have increased because of increased drought tolerance (Duvick and Cassman, 1999; Duvick, 2005). Assefa and Staggenborg (2010) reported that the yield focus of sorghum hybrid development was effective in dryland sorghum production, probably because of the intentional or inadvertent selection of hybrids with better drought tolerance.

Sall and Sinclair (1991) reported the presence of genetic variability in N_2-fixation sensitivity to drought stress among soybean cultivars. Indication that N_2-fixation tolerance may be associated with ureide levels in plants has been noted (Serraj and Sinclair, 1997). Sinclair and Serraj (1995) reported major differences in N_2-fixation response to drought among grain legume species. Legume species producing ureides were more sensitive to drought stress than amide-producing plants, which may mean that ureides are involved in N_2-fixation sensitivity to water deficits. Selection in stress environments may not necessarily target specific individual genes governing single components of the stress response mechanism, but may act on multiple loci (Allard et al., 1993; Duncan and Carrow, 1999). The genetic improvement of plant drought resistance by selection for yield under water stress is possible, but prolonged and problematic procedures often occur. Recent developments in understanding the physiological responses of plants to water stress and their associations with plant productivity allow scientists to embark on experimental selection programs that employ physiological selection criteria for drought resistance. While appreciable numbers of plant physiological attributes have been identified as components of drought resistance, their interrelationships with plant productivity have not been clarified and should be investigated further. Even if expected progress is not attained by such selection experiments, the resulting information should be extremely valuable as guidance for future research and development (Blum, 1983). Ceccarelli and Grando (1996) also reported that progress in cereal breeding for dry areas has been less successful than breeding for favorable environments. Breeding for drought resistance is complex because stress environments are intrinsically erratic in nature (Blum et al., 1983a,b), and successes of cultivars are not predictable (Ceccarelli and Grando, 1996).

Planting deep-rooted crops is also an important strategy for improving crop yields of plants grown under water deficits (Sharpley et al., 1998; Halvorson et al., 2001). Sunflower can extract water at soil depths of 3 m, which indicates that this plant can be included in crop sequences where water has amassed below normal rooting depths of most crops (Jones and Johnson, 1983). Stone et al. (2002) reported that water depletion fronts advanced downward at greater rates and to deeper depths with sunflower (3.1 m) than with sorghum (2.5 m). As noted before, sunflower's ability to produce under low-rainfall conditions is aided by its relatively deep root system (Connor and Hall, 1997) and relatively low water use requirement for initial seed yield (Nielsen, 1998).

Early vigorous growth has been a characteristic for increasing the yields of wheat grown in rain-fed environments (Asseng et al., 2003). Early vigorous growth resulting in enhanced leaf areas has been suggested as a way to increase growth when temperatures and vapor pressure deficits are low, thereby increasing the transpiration efficiency of crops (Fischer, 1979). It has been reported that WUE (includes both water evaporation from soil and crop transpiration) for grain yields increased by as much as 25% because of early vigor (Siddique et al., 1990; López-Castañeda and Richards, 1994).

The use of silicon (Si) in adequate rate is also reported to improve drought tolerance in crop plants (Liang et al., 2003, 2005; Sonobe et al., 2011). Gao et al. (2006) reported that Si application to corn decreased leaf transpiration under water stress and thereby improved the plant water status. On the other hand, Si application to sorghum plants increased leaf transpiration rates (stomatal conductance) and, consequently, alleviated the reduction of photosynthetic rates by water stress (Hattori et al., 2005). Hattori et al. (2008) and Sonobe et al. (2011) reported that Si application increases water uptake in sorghum. The use of Si activates antioxidant processes in plants under water stress (Kaya et al., 2006; Gong et al., 2008), which enhances drought tolerance.

3.10 CONCLUSIONS

Water is one of the most important limiting factors in crop production. Adequate moisture in the soil is important for the normal growth and development of plant roots and water and nutrient uptake. Excess water as well as water deficits reduce root growth and most physiological and biochemical processes in the plants. Adequate soil moisture in the soil for plant growth is near field capacity. Available water near field capacity favors microbial and soil solution reactions, transport of ions by diffusion and mass flow to the root, and root development into unexploited soil. All these processes are inhibited as the soil dries toward the wilting range. Water scarcity is expected worldwide in the future due to change in climate, and more water will be used in food and fiber production to meet the demands of an increasing world population. Under these conditions, conserving water and improving WUE are essential. Some management practices that can improve WUE are as follows: maintaining SOM content, conservation tillage, planting water-efficient crop species or genotypes within species, using cover crops in the cropping systems, and improving irrigation efficiency.

Drought is very common in all agroecological regions, which reduces crop yields. The effect of drought is more pronounced during the reproductive and grain-filling stages of crop plants. For example, in cereals, yield is severely reduced when drought occurs around flowering. Drought around flowering can increase grain sterility as well as grain weight. The most important strategy to reduce the effects of drought is investment in agricultural infrastructures, including irrigation, and improved technology, management, and more tolerant crop cultivars will be required to help improve and stabilize agricultural production. To optimize crop yields in irrigated environments, irrigation should be timed in a way that nonproductive soil water evaporation and drainage losses are minimized and that the possible inevitable water deficits coincide with least sensitive growth periods. In rain-fed environments, management practices should be tailored in ways that water availability matches crop water needs.

REFERENCES

Aase, J. K. and J. L. Pikul, Jr. 1995. Crop and soil response to long-term tillage practices in the northern Great Plains. *Agron. J.* 87:652–656.

Aase, J. K. and G. M. Schaefer. 1996. Economics of tillage practices and spring wheat and barley crop sequence in the northern Great Plains. *J. Soil Water Conserv.* 51:167–170.

Acosta-Martinez, V., T. M. Zobeck, and V. Allen. 2004. Soil microbial, chemical and physical properties in continuous cotton and integrated crop-livestock systems. *Soil Sci. Soc. Am. J.* 68:1875–1884.

Adams, S. S. and W. R. Stevenson. 1990. Water management, disease development and potato production. *Am. Potato J.* 67:3–11.

Adejare, F. B. and C. E. Umebese. 2007. Effect of water stress at different stages of growth on stomata complex, tolerance and yield of two cultivars of *Glycine max*. Merril. *South African J Bot.* 73:323–323.

Aggarwal, P., K. K. Choudhry, A. K. Singh, and D. Chakraborty. 2006. Variation in soil strength and rooting characteristics of wheat in relation to soil management. *Geoderma* 136:353–363.

Al-Barrak, K. M. 2006. Irrigation interval and nitrogen level effects on growth and yield of canola (*Brassica napus* L.). *Sci. J. King Faisal Univ.* 7:87–103.

Alessi, J., J. F. Power, and D. C. Zimmerman. 1977. Sunflower yield and water use as influenced by planting date, population, and row spacing. *Agron. J.* 69:465–469.

Al-Jaloud, A. A., G. Hussian, S. Karimula, and A. H. Ali-Hamidi. 1996. Effect of irrigation and nitrogen on yield components of two rapeseed cultivars. *Agric. Water Manage.* 30:57–68.

Allard, R. W., P. Garcia, L. E. Saenz de Miera, and V. M. Perez. 1993. Evolution of multilocus structure in *Avena hirtula* and *Avena barbata*. *Genetics* 135:1125–1139.

Allen, R. G., M. E. Jensen, J. L. Wright, and R. D. Burman. 1989. Operational estimates of reference evapotranspiration. *Agron. J.* 81:650–662.

Allen, R. G., L. S. Pereira, D. Raes, and M. Smith. 1998. Crop evapo-transpiration. Guidelines for computing crop water requirements. FAO Irrigation and Drainage Paper 56. FAO, Rome, Italy.

Amer, K. H., S. A. Midan, and J. L. Hatfield. 2009. Effect of deficit irrigation and fertilizer on cucumber. *Agron. J.* 101:1556–1564.

Amezketa, E. 1999. Soil aggregate stability: A review. *J. Sustainable Agric.* 14:83–151.

Amir, J., J. Krikun, D. Orion, J. Putter, and S. Klitman. 1991. Wheat production in an arid environment: I. Water use efficiency, as affected by management practices. *Field Crops Res.* 27:351–364.

Amir, J. and T. R. Sinclair. 1996. A straw mulch system to allow continuous wheat production in an arid climate. *Field Crops Res.* 27:365–376.

Anderson, R. L. 2005. Are some crops synergistic to following crops? *Agron. J.* 97:7–10.

Angers, D. A., A. Pesant, and J. Vigneux. 1992. Early cropping induced changes in soil aggregation, organic matter, and microbial biomass. *Soil Sci. Soc. Am. J.* 56:115–119.

Angus, J. F. and A. F. V. Herwaarden. 2001. Increasing water use and water use efficiency in dryland wheat. *Agron. J.* 93:290–298.

Ashley, D. A. and W. J. Ethridge. 1978. Irrigation effects on vegetative and reproductive development of three soybean cultivars. *Agron. J.* 70:467–471.

Aspinall, D. 1984. Water deficit and wheat. In: *Control of Crop Productivity*, ed., C. J. Pearson, pp. 91–110. Sydney, New South Wales, Australia: Academic Press.

Assefa, Y. and S. A. Staggenborg. 2010. Grain sorghum yield with hybrid advancement and changes in agronomic practices from 1957 through 2008. *Agron. J.* 102:703–706.

Asseng, S., N. C. Turner, T. Botwright, and A. G. Condon. 2003. Evaluating the impact of a trait for increased specific leaf area on wheat yields using a crop simulation model. *Agron. J.* 95:10–19.

Aust, W. M. and R. Lea. 1991. Soil temperature and organic matter in a disturbed-forested wetland. *Soil Sci. Soc. Am. J.* 55:1741–1746.

Austin, R. B. 1989. Maximizing crop production in water-limited environments. In: *Drought Resistance in Cereals*, ed., F. W. G. Baker, pp. 13–25. London, U.K.: CAB International.

Austin, R. B., C. L. Morgan, M. A. Ford, and R. D. Blackwell. 1980. Contributions to the grain yield from pre-anthesis assimilation in tall dwarf barley phenotypes in two contrasting seasons. *Ann. Bot.* 45:309–319.

Badaruddin, M. and D. W. Meyer. 1989. Water use by legumes and its effect on soil water status. *Crop Sci.* 29:1212–1216.

Badu-Apraku, B., R. O. Akinwale, S. O. Ajala, A. Menkir, M. A. B. Fakorede, and M. Oyekunle. 2011. Relationship among traits of tropical early maize cultivars in contrasting environments. *Agron. J.* 103:717–729.

Badu-Apraku, B., M. A. B. Fakorede, A. Menkir, A. Y. Kamara, and S. Dapaah. 2005. Screening maize drought tolerance in the Guinea savanna of West and Central Africa. *Cereal Res. Commun.* 33:533–540.

Balesdent, J., C. Chenu, and M. Balabane. 2000. Relationship of soil organic matter dynamics to physical protection and tillage. *Soil Tillage Res.* 53:215–230.

Baligar, V. C., N. K. Fageria, and M. A. Elrashidi. 1998. Toxicity and nutrient constraints on root growth. *Hort. Sci.* 33:960–965.

Bandaru, V., B. A. Stewart, R. L. Baumhardt, S. Ambati, C. A. Robinson, and A. Schlegel. 2006. Growing dryland grain sorghum in clumps to reduce vegetative growth and increase yield. *Agron. J.* 98:1109–1120.

Barker, R., D. Dawe, T. P. Tuong, S. I. Bhuiyan, and L. C. Guerra. 1999. The outlook for water resources in the year 2020: Challenges for research on water management in rice production. In: *Assessment and Orientation Towards the 21st Century. Proceedings of the 19th Session of the International Rice Commission*, ed., D. V. Tran, pp. 96–109. September 7–9, 1998, Cairo, Egypt: IRRI.

Bauder, J. W. 2006. The right strategy for irrigating your canola crop. Available at http://waterquality.montana.edu/docs/irrigation/canolastrategy. Shtmi (verified January 15, 2010). The Department of land Resources and Environmental Sciences, Montana State University, Bozeman, MT.

Bauder, J. W. and M. J. Ennen. 1979. Crop water use-how does sunflower rate? *The Sunflower* 5:10–11.

Bauer, A. and A. L. Black. 1992. Organic carbon effects on available water capacity of three soil textural groups. *Soil Sci. Soc. Am. J.* 56:248–254.

Bauer, P. J., B. A. Fortnum, and J. R. Frederick. 2010. Cotton responses to tillage and rotation during the turn of the century drought. *Agron. J.* 102:1145–1148.

Beare, M. H. and R. R. Bruce. 1993. A comparison of methods for measuring water-stable aggregates: Implications for determining environmental effects on soil structure. *Geoderma* 56:87–104.

Beare, M. H., M. L. Cabrera, P. F. Hendrix, and D. C. Coleman. 1994. Aggregate protected and unprotected organic matter pools in conventional and no-tillage soils. *Soil Sci. Soc. Am. J.* 58:787–795.

Bednarz, C. W., J. E. Hook, R. Yager, S. Cromer, D. Cook, and I. Griner. 2002. Cotton crop water use and irrigation scheduling. In: *Georgia Cotton Research Extension Report*, ed., A. S. Culpepper, pp. 61–64. Athens, GA: University of Georgia.

Benjamin, J. G., L. K. Porter, H. R. Duke, and L. R. Ahuja. 1997. Corn growth and nitrogen uptake with furrow irrigation and fertilizer bands. *Agron. J.* 89:609–612.

Bennett, J. M., T. R. Sinclair, R. E. Muchow, and S. R. Costello. 1987. Dependence of stomatal conductance on leaf water potential, turgor potential, and relative water content in field-grown soybean and maize. *Crop Sci.* 27:984–990.

Bezdicek, D. F. and D. Granatstein. 1989. Crop rotation efficiencies and biological diversity in farming systems. *Am. J. Altern. Agric.* 4:111–118.

Bhattarai, S. P., N. Su, and D. J. Midmore. 2005. Oxygation unlocks yield potentials of crops in oxygen limited soil environments. *Adv. Agron.* 88:313–377.

Bieler, P., L. K. Fussel, and F. R. Bidinger. 1993. Grain growth of *Pennisetum glauscm* (L.). *Field Crops Res.* 31:41–45.

Biswas, T. D. and B. K. Khosla. 1971. Building up of organic matter status of soil and its relation to the soil physical properties. *International Symposium on Soil Fertility Evaluation Proceedings*, Vol. 1, pp. 831–842.

Black, A. L., P. L. Brown, A. D. Halvorson, and F. H. Siddoway. 1981. Dryland cropping strategies for efficient water use to control saline seeps in the northern Great Plains, U. S. A. *Agric. Water Manage.* 4:295–311.

Blevins, R. L. and W. W. Frye. 1993. Conservation tillage: An ecological approach to soil management. *Adv. Agron.* 51:33–78.

Blum, A. 1983. Genetic and physiological relationships in plant breeding for drought resistance. *Agric. Water Manage.* 7:195–205.

Blum, A., J. Mayer, and G. Golan. 1983a. Chemical desiccation of wheat plants as a simulator of post-anthesis stress. II. Relations to drought stress. *Field Crops Res.* 6:149–155.

Blum, A., H. Poiarkova, G. Golan, and J. Mayer. 1983b. Chemical desiccation of wheat plants as a simulator of post-anthesis stress. I. Effects on translocation and kernel growth. *Field Crops Res.* 6:51–58.

Bolton, F. 1981. Optimizing the use of water and nitrogen through soil and crop management. *Plant Soil.* 58:231–247.

Bonfil, D., I. Mufradi, S. Klitman, and S. Asido. 1999. Wheat grain yield and soil profile water distribution in a no-till arid environment. *Agron. J.* 91:368–373.

Borrell, A. K., G. L. Hammer, and R. G. Henzell. 2000. Does maintaining green leaf area in sorghum improve yield under drought? II. Dry matter production and yield. *Crop Sci.* 40:1037–1048.

Boyer, J. S. 1982. Plant productivity and environment. *Science* 218:443–448.

Boyer, J. S. 1985. Water transport. *Annu. Rev Plant Physiol.* 36:473–516.

Boyer, J. S. 1996. Advances in drought tolerance in plants. *Adv. Agron.* 56:187–218.

Boyle, M. G., J. S. Boyer, and P. W. Morgan. 1991. Stem infusion of liquid culture medium prevents reproductive failure of maize at low water potential. *Crop Sci.* 31:1246–1252.

Bray, R. H. 1954. A nutrient mobility concept of soil–plant relationships. *Soil Sci.* 78:9–22.

Bremner, P. M., G. K. Preston, and G. C. Fazekas. 1986. A field comparison of sunflower and sorghum in a long drying cycle. *Aust. J. Agric. Res.* 37:483–493.

Brevedan, R. E. and D. B. Egli. 2003. Short period of water stress during seed filling, leaf senescence and yield in soybean. *Crop Sci.* 43:2083–2088.

Bullock, D. G. 1992. Crop rotation. *Crit. Rev. Plant Sci.* 11:309–326.

Bunting, A. H. and A. H. Kassam. 1988. Principles of crop water use, dry matter production, and dry matter partitioning that govern choices of crops and systems. In: *Drought Research Priorities for Dryland Tropics*, eds., F. R. Bidinger and C. Johnsen, pp. 43–61. Patancheru, India; ICRISAT.

Burgess, S. S. O., J. S. Pate, M. A. Adams, and T. E. Dawson. 2000. Seasonal water acquisition and redistribution in the Australian woody phreatophyte, *Banksia prionotes*. *Ann. Bot.* 85:215–224.

Burke, J. J., C. D. Franks, G. Burow, and Z. Xin. 2010. Selection system for the stay-green drought tolerance trait in sorghum germplasm. *Agron. J.* 102:1118–1122.

Cambardella, C. A. and E. T. Elliott. 1993. Carbon and nitrogen distribution in aggregates from cultivated and native grassland soils. *Soil Sci. Soc. Am. J.* 57:1071–1076.

Campbell, C. A., F. Selles, R. P. Zentner, and B. G. McConkey. 1993. Available water and nitrogen effects on yield components and grain nitrogen of zero-till spring wheat. *Agron. J.* 85:114–120.

Carter, T. E. and T. W. Rufty. 1993. Soybean plant introductions exhibiting drought and aluminum tolerance. In: *Adaptation of Food Crops to Temperature and Water Stress: Proceedings of an International Symposium*, August 13–18, 1992, ed., C. G. Kuo, pp. 335–346. Taipei, Taiwan: Asian Vegetables Research and Development Center.

Carter, P. R. and C. C. Sheaffer. 1983. Alfalfa response to soil water deficits. I. Growth, forage quality, yield, and water-use efficiency. *Crop Sci.* 23:669–675.

Causape, J., D. Quilez, and R. Aragues. 2006. Groundwater quality in CR-V irrigation district (Bardenas I, Spain): Alternative scenarios to reduce offsite salt and nitrate contamination. *Agric. Water Manage.* 84:281–289.

Ceccarelli, S. and S. Grando. 1996. Drought as a challenge for the plant breeder. *Plant Growth Regul.* 20:149–155.

Chang, J. 1968. *Climate and Agriculture: An Ecological Survey*. Chicago, IL: Aldine Publisher.

Clarkson, D. T., A. W. Robards, and J. Sanderson. 1971. The tertiary endodermis in barley roots: Fine structure in relation to radial transport of ions and water. *Planta* 96:292–305.

Coelho, E. F., M. A. Coelho Filho, and A. J. P. Silva. 2009. Agricultura irrigada: otimização da eficiência de irrigação e do uso da água. In: *Simpósio nacional sobre o uso da água na agricultura*, 3., 2009, Passo Fundo. Disponível em: http://www.upf.br/coaju/download/Eugenio.pdf. Acesso em: 11 set. 2009.

Connor, D. J. and A. J. Hall. 1997. Sunflower physiology. In: *Sunflower Technology and Production*, ed., A. A. Schneiter, pp. 113–182. Agron. Monograph. 35, Madison, WI: ASA, CSSA, and SSSA.

Cooper, P. J. M. and P. J. Gregory. 1987. Soil water management in the rainfed farming systems of the Mediterranean region. *Soil Use Manage.* 3:57–62.

Cooper, P. J. M., J. D. H. Keating, and H. Hughes. 1983. Crop evapotranspiration: A technique for calculation of its components by field measurements. *Field Crops Res.* 7:299–312.

Cornish, P. S., H. B. So, and J. R. McWilliam. 1984. Effects of soil bulk density and water regime on root growth and uptake of phosphorus by ryegrass. *Aust. J. Agric. Res.* 35:631–644.

Cosgrove. D. J. 1993. Wall extensibility: Its nature, measurement and relationship to plant cell growth. Transley Review 46. *New Phytol.* 124:1–23.

Crabtree, R. J., A. A. Yassin, I. Kargougou, and R. W. McNew. 1985. Effects of alternate-furrow irrigation: Water conservation on the yields of two soybean cultivars. *Agric. Water Manage.* 10:253–264.

Craufurd, P. Q., D. J. Flower, and J. M. Peacock. 1993. Effect of heat and drought stress on sorghum (*sorghum bicolor* L.). I. Panicle development and leaf appearance. *Exp. Agric.* 29:61–76.

Crookston, R. K., J. E. Kurle, P. J. Copeland, J. H. Ford, and W. E. Lueschen. 1991. Rotational cropping sequence affects yield of corn and soybean. *Agron. J.* 83:108–113.

Cruz, R. T., W. R. Jordan, and M. C. Drew. 1992. Structural changes and associated reduction of hydraulic conductance in roots of *Sorghum bicolor* L. following exposure to water deficit. *Plant Physiol.* 99:203–212.

Curz, R. T. and J. C. O'Toole. 1984. Dryland rice response to an irrigation gradient at flowering stage. *Agron. J.* 76:178–183.

Daniel, J. B., A. O. Abaye, M. M. Alley, C. W. Adcock, and J. C. Maitland. 1999. Winter annual cover crops in a Virginia no-till cotton production system: II. Cover crop and tillage effects on soil moisture, cotton yield, and cotton quality. *J. Cotton Sci.* 3:84–91.

Darusman, A. H. K., A. H. Khan, L. R. Stone, W. E. Spurgeon, and F. R. Lamm. 1997. Water flux below the root zone vs. irrigation in drip-irrigated corn. *Agron. J.* 89:375–379.

Davidson, J. L. and J. W. Birch. 1978. Response of a standard Australian and Mexican wheat to temperature and water stress. *Aust. J. Agric. Res.* 29:1091–1106.

Day, A. D. and S. Intalap. 1970. Some effects of soil moisture on the growth of wheat. *Agron. J.* 62:27–29.

Day, W., B. J. Legg, B. K. French, A. E. Johnston, D. W. Lawlor, and W. C. Jeffers. 1978. A drought experiment using mobile shelters: The effect of drought on barley yield, water use and nutrient uptake. *J. Agric. Sci.* 91:599–623.

De Souza, J. G. and J. V. Da Silva. 1987. Partitioning of carbohydrates in annual and perennial cotton (*Gossypium hirsutum* L.). *J. Exp. Bot.* 38:1211–1218.

Debaeke, P. and A. Aboudrare. 2004. Adaptation of crop management to water limited environments. *Eur. J. Agron.* 21:433–446.

Denef, K., J. Six, R. Merckx, and K. Paustian. 2004. Carbon sequestration in microaggregates of no-tillage soils with different clay mineralogy. *Soil Sci. Soc. Am. J.* 68:1935–1944.

Denmead, O. T. and B. D. Miller. 1976. Field studies of the conductance of wheat leaves and transpiration. *Agron. J.* 68:307–311.

Dent, F. J. 1980. Major production systems and soil-related constraints in Southeast Asia. In: *Properties for Alleviating Food Production in the Tropics*, ed., IRRI, pp. 79–106. Los Banos, Philippines: IRRI.

Dingkuhn, M., G. D. Farquhar, S. K. De Datta, and J. C. O'Toole. 1991. Discrimination of C-13 among upland rice having different water use efficiency. *Aust. J. Agric. Res.* 42:1123–1131.

Doggett, H. 1988. *Sorghum*, 2nd edn. New York: John Willey & Sons.

Doran, J. W. and M. S. Smith. 1987. Organic matter management and utilization of soil and fertilizer nutrients. In: *Soil Fertility and Organic Matter as Critical Components of Production Systems*, eds., R. F. Follet, J. W. B. Stewart, and C. V. Cole, pp. 53–72. Madison, WI: SSSA.

Duncan, R. R. and R. N. Carrow. 1999. Turfgrass molecular genetic improvement for abiotic/ edaphic stress resistance. *Adv. Agron.* 67:233–306.

Dunphy, E. J. 1985. Soybean on farm test report. North Carolina agric. Ext. Serv., Raleigh, NC.

Durand, J. L., J. E. Sheehy, and F. R. Minchin. 1987. Nitrogenase activity, photosynthesis and water potential in soybean plants experiencing water deprivation. *J. Exp. Bot.* 38:311–321.

Duvick, D. N. 2005. The contribution of breeding to yield advances in maize. *Adv. Agron.* 86:83–145.

Duvick, D. N. and K. G. Cassman. 1999. Post-green revolution trends in yield potential of temperate maize in the north-central United States. *Crop Sci.* 39:1622–1630.

Eberbach, P. and M. Pala. 2005. Crop row spacing and its influence on the partitioning of evaporation by winter grown wheat in northern Syria. *Plant Soil* 268:195–268.

Eck, H. V. 1984. Irrigated corn yield response to nitrogen and water. *Agron. J.* 76:421–428.

Egli, D. B. and S. J. Crafts-Brandner. 1996. Soybean. In: *Photoasimilate Distribution in Plants and Crops: Source-Sink Relationships*, eds., E. Zamski and A. A. Schaffer, pp. 595–623. New York: Marcel Dekker.

Ehdaie, B. 1995. Variation in water-use efficiency and its components in wheat: II. Pot and field experiments. *Crop Sci.* 35:1617–1626.

Ehdaie, B. and J. G. Waines. 1993. Variation in water use efficiency and its components in wheat. I. Well watered pot experiment. *Crop Sci.* 33:294–299.

Elliott, E. T. 1986. Aggregate structure and carbon, nitrogen, and phosphorus in native and cultivated soils. *Soil Sci. Soc. Am. J.* 50:627–633.

Evett, S. R. and J. A. Tolk. 2009. Introduction: Can water use efficiency be modeled well enough to impact crop management? *Agron. J.* 101:423–425.

Fageria, N. K. 1980. Upland rice response to phosphate fertilization as affected by water deficiency in cerrado soils. *Pesq. Agropec. Bras.* 15:259–265.

Fageria, N. K. 1992. *Maximizing Crop Yields.* New York: Marcel Dekker.

Fageria, N. K. 2002. Soil quality vs. environmentally based agricultural management practices. *Commun. Soil Sci. Plant Anal.* 33:2301–2329.

Fageria, N. K. 2007. Yield physiology of rice. *J. Plant Nutr.* 30:843–879.

Fageria, N. K., V. C. Baligar, and B. A. Bailey. 2005. Role of cover crops in improving soil and row crop productivity. *Commun. Soil Sci. Plant Anal.* 36:2733–2757.

Fageria, N. K., V. C. Baligar, and R. B. Clark. 2006. *Physiology of Crop Production.* New York: The Haworth Press.

Fageria, N. K. and H. R. Gheyi. 1999. *Efficient Crop Production.* Campina Grande, Brazil: Federal University of Paraiba.

Fageria, N. K., R. J. Wright, V. C. Baligar, and C. M. R. Sousa. 1991. Characterization of physical and chemical properties of varzea soils of Goias State of Brazil. *Commun. Soil Sci. Plant Anal.* 22:1631–1646.

Falkenmark, M. and J. Rockstrom. 2004. *Balancing Water for Human and Nature: The New Approach in Ecohydrology.* London, U.K.: Earthscan Press.

Farquhar, G. D., J. R. Ehleringer, and K. Hubic. 1989. Carbon isotope discrimination and photosynthesis. *Annu. Rev. Plant Physiol. Plant Mol. Biol.* 40:503–537.

Farquhar, G. D. and R. A. Richards. 1984. Isotopic composition of plant carbon correlates with water-use efficiency of wheat genotypes. *Aust. J. Plant Physiol.* 11:539–552.

Fischbach, P. E. and H. R. Mulliner. 1974. Every-other furrow irrigation of corn. *Trans. ASAE* 17:426–428.

Fischer, R. A. 1973. The effect of water stress at various stages of development on yield processes in wheat. In: *Plant Response to Climatic Factors,* ed., R. O. Slatyer, pp. 233–241. Paris, France: UNESCO.

Fischer, R. A. 1979. Growth and water limitation to dryland wheat yield in Australia: A physiological framework. *J. Aust. Inst. Agric. Sci.* 45:83–94.

Franzluebbers, A. J., F. M. Hons, and D. A. Zuberer. 1995. Soil organic carbon, microbial biomass, and mineralizable carbon and nitrogen in sorghum. *Soil Sci. Soc. Am. J.* 59:460–466.

Frederick J. R., E. F. Frederick, and J. D. Hesketh. 1990. Carbohydrate, nitrogen and dry matter accumulation and partitioning of maize hybrids under drought stress. *Ann. Bot.* 66:947–954.

French, R. J. and J. E. Schultz. 1984a. Water use efficiency of wheat in a Mediterranean type environment: I. The relation between yield, water use and climate. *Aust. J. Agric. Res.* 35:743–764.

French, R. J. and J. E. Schultz. 1984b. Water use efficiency of wheat in a Mediterranean type environment: II. Some limitations to efficiency. *Aust. J. Agric. Res.* 35:765–775.

Frye, W. W., S. A. Ebelhar, L. W. Murdock, and R. L. Blevins. 1982. Soil erosion effects on properties and productivity of two Kentucky soils. *Soil Sci. Soc. Am. J.* 46:1051–1055.

Fukai, S. and M. Cooper. 1996. Stress physiology in relation to breeding for drought resistance: A case study of rice. In: *Physiology of Stress Tolerance in Rice. Proceedings of the International Conference on Stress Physiology of Rice,* eds., V. P. Singh, R. K. Singh, B. B. Singh, and R. S. Ziegler, pp. 123–149. Los Banos, Philippines: IRRI.

Gallaher, R. N. 1977. Soil moisture conservation and yield of crops no-till planted in rye. *Soil Sci. Soc. Am. J.* 41:145–147.

Gao, X., C. Zou, L. Wang, and F. Zhang. 2006. Silicon decreases transpiration rate and conductance from stomata of maize plants. *J. Plant Nutr.* 29:1637–1647.

Gardner, W. R. 1960. Dynamic aspects of water availability to plants. *Soil Sci.* 89:63–73.

Garrity, D. P. and J. C. O'Toole. 1995. Selection for reproductive stage drought avoidance in rice, using infrared thermometry. *Agron.* J. 87:773–779.

Ge, T. D., F. G. Sui, S. Nie, N. B. Sun, H. Xiao, and C. L. Tong. 2010. Differential responses of yield and selected nutritional compositions to drought stress in summer maize grains. *J. Plant Nutr.* 33:1811–1818.

Gerik, T. J., K. L. Faver, P. M. Thaxton, and K. M. El-Zik. 1996. Late season water stress in cotton: Plant growth, water us, and yield. *Crop Sci.* 36:914–921.

Ghidey, F. and E. E. Alberts. 1993. Residue type and placement effects on decomposition: Field study and model evaluation. *Trans. ASAE* 36:1611–1617.

Gifford, R. M. 1986. Partitioning of photoassimilate in the development of crop yield. In: *Phloem Transport*, eds., J. Cronshaw, W. J. Lucas, and R. T. Giaquinta, pp. 535–549. New York: A. R. Liss.

Ginkel, M. V., D. S. Calhoun, G. Gebeyehu, A. Miranda, C. Tian-you, R. P. Lara, R. M. Trethowan, K. Sayre, J. Crossa, and S. Rajaram. 1998. Plant traits related to yield of wheat in early, late, or continuous drought conditions. *Euphytica* 100:109–121.

Giunta, F., R. Motzo, and M. Deidda. 1993. Effect of drought on yield and yield components of durum wheat and triticale in a Mediterranean environment. *Field Crops Res.* 33:399–409.

Gleick, P. H. 1993. *Water in Crisis: A Guide to the Worlds Fresh Water Resources.* New York: Oxford University Press.

Gong, H. J., K. M. Chen, Z. G. Zhao, G. C. Chen, and W. J. Zhou. 2008. Effects of silicon on defense of wheat against oxidative stress under drought at different development stages. *Biologia Plantarum* 52:592–596.

Goodman, B. A. and A. C. Newton. 2005. Effects of drought stress and its sudden relief on free radical processes in barley. *J. Sci. Food Agric.* 85:47–53.

Greenland, D. J. 1997. *The Sustainability of Rice Farming.* Los Banos, Philippines: IRRI.

Gregorich, E. G., R. G. Kachanoski, and R. P. Voroney. 1989. Carbon mineralization in soil size fractions after various amounts of aggregate disruption. *J. Soil Sci.* 40:649–659.

Gregory, P. J. 2006. Roots, rhizosphere and soil: The route to a better understanding of soil science? *Eur. J. Soil Sci.* 57:2–12.

Grimes, D. W., P. L. Wiley, and W. R. Shesley. 1992. Alfalfa yield and plant water relations with variable irrigation. *Crop Sci.* 32:1381–1387.

Guerra, L. C., S. I. Bhuiyan, T. P. Tuong, and R. Barker. 1998. Producing more rice with less water from irrigated systems. Discussion paper series no 29, Los Banos, Philippines: IRRI.

Guimarães, C. M., L. F. Stone, M. Lorieux, J. P. Oliveira, G. C. O. Alencar, and R. A. A. Dias. 2010. Infrared thermometry for drought phenotyping of inter and intra specific upland rice lines. *Rev. Bras. Eng. Agric. Amb.* 14:148–154.

Guimarães, C. M., L. F. Stone, and P. C. F. Neves. 2008. Production efficiency of rice cultivars with phenotypic diversity. *Rev. Bras. Eng. Agric. Amb.* 12:465–470.

Gupta, S. C., R. H. Dowdy, and W. E. Larson. 1977. Hydraulic and thermal properties of a sandy soil as influenced by incorporation of sewage sludge. *Soil Sci. Soc. Am. J.* 41:601–605.

Gupta, V. V. S. R. and J. J. Germida. 1988. Distribution of microbial biomass and its activity in different soil aggregate size classes as affected by cultivation. *Soil Biol. Biochem.* 20:777–786.

Hale, M. G. and D. M. Orcutt. 1987. *The Physiology of Plants under Stress.* New York: John Wiley & Sons.

Halvorson, A. D., A. L. Black, J. M. Krupinsky, S. D. Merrill, and D. L. Tanaka. 1999. Sunflower response to tillage and nitrogen fertilization under intensive cropping in a wheat rotation. *Agron. J.* 91:637–642.

Halvorson, A. D. and C. A. Reule. 1994. Nitrogen fertilizer requirements in an annual dryland cropping systems. *Agron. J.* 86:315–318.

Halvorson, A. D., B. J. Wienhold, and A. L. Black. 2001. Tillage and nitrogen fertilization influence grain and soil nitrogen in an annual cropping system. *Agron. J.* 93:836–841.

Hamblin, A. and D. Tennant. 1987. Root length density and water uptake in cereals and grain legumes: How well are they correlated? *Aust. J. Agric. Res.* 38:513–527.

Hamzei, J. 2011. Seed, oil, and protein yields of canola under combinations of irrigation and nitrogen application. *Agron. J.* 103:1152–1158.

Hanks, R. J. and V. P. Rasmussen. 1982. Predicting crop production as related to plant water stress. *Adv. Agron.* 35:193–215.

Hanson, B. R. 1987. A systems approach to drainage reduction. *Calif. Agric.* 41:19–24.

Harris, H. C., P. J. M. Cooper, and M. Pala. 1991. Soil and crop management for improved water use efficiency in rainfed areas. *Proceedings of an International Workshop*, Ankara, Turkey. May 15–19, 351, 1989. Aleppo, Syria: ICARDA.

Hatfield, J. L. 1985. Wheat canopy resistance determined by energy balance techniques. *Agron. J.* 77:279–283.

Hatfield, J. L., K. J. Boote, B. A. Kimball, L. H. Zisks, R. C. Izaurralde, D. Ort, A. M. Thomson, and D. Wolfe. 2011. Climate impacts on agriculture: Implications for crop production. *Agron. J.* 103:351–370.

Hatfield, J. L., T. J. Sauer, and J. H. Prueger. 2001. Managing soils to achieve greater water use efficiency: A review. *Agron. J.* 93:271–280.

Hattendorf, M. J., M. S. Redelfs, B. Amos, L. R. Stone, and R. E. Groin, Jr. 1988. Comparative water use characteristics of six row crops. *Agron. J.* 80:80–85.

Hattori, T., S. Inanaga, H. Araki, S. Morita, M. Luxova, and A. Lux. 2005. Application of silicone enhanced drought tolerance in *Sorghum biocolor. Physiol. Plant.* 123:459–466.

Hattori, T., K. Sonobe, H. Araki, S. Inanaga, P. An, and S. Morita. 2008. Silicon application improves water uptake by sorghum through the alleviation of stress-induced increase in hydraulic resistance. *J. Plant Nutr.* 31:1482–1495.

Haussling, M., C. A. Jorns, G. Lehmbecker, C. Hecht-Buchholz, and H. Marschner. 1988. Ion and water uptake in relation to root development in Norway spruce (*Picea abies* L.) *J. Plant Physiol.* 133:486–491.

Hay, K. M. R. and A. J. Walker. 1989. *An Introduction to the Physiology of Crop Yield*. Essex, U.K.: Longman Scientific & Technical.

Hayes, M. H. B. and R. S. Swift. 1983. The chemistry of soil organic colloids. In: *The Chemistry of Soil Constituents*, eds., D. J. Greenland and M. H. B. Hayes, pp. 179–320. New York: John Wiley & Sons.

Heinemann, A. B. and L. F. Stone. 2009. Effect of water deficit on the development and grain yield of four upland rice cultivars. *Pesq. Agropec. Trop.* 39:134–139.

Hillel, D. and E. Rawitz. 1972. Soil water conservation. In: *Water Deficits and Plant Growth: Plant Responses and Control of Water Balance*, Vol. 3, ed., T. T. Kozlowski, pp. 307–338. New York: Academic Press.

Hinsinger, P., A. G. Bengough, D. Vetterlein, and I. M. Young. 2009. Rhizosphere: Biophysics, biogeochemistry and ecological relevance. *Plant Soil* 321:117–152.

Hodgson, A. S. and K. Y. Chan. 1984. Deep moisture extraction and crack formation by wheat and safflower in a vertisol following irrigated cotton rotations. *Rev Rural Sci.* 5:299–304.

Hook, J. E., K. A. Harrison, and G. Hoogenboom. 2004. Ag water pumping: Statewide irrigation monitoring. Final report to Georgia Environmental Protection Division. Project Report 51. Atlanta, GA.

Hopmans, J. W. and K. L. Bristow. 2002. Current capabilities and future needs of root water and nutrient uptake modeling. *Adv. Agron.* 77:103–183.

Howell, T. A. 2001. Enhancing water use efficiency in irrigated agriculture. *Agron. J.* 93:281–289.

Howell, T. A., K. R. Davis, R. L. McCormick, H. Yamada, V. T. Walhood, and D. W. Meek. 1984. Water use efficiency of narrow row cotton. *Irrig. Sci.* 5:195–214.

Howell, T. A., J. L. Steiner, A. D. Schneider, S. R. Evett, and J. A. Tolk. 1997. Seasonal and maximum daily evapotranspiration of irrigated winter wheat, sorghum, and corn-Southern High Plains. *Trans. ASAE* 40:623–634.

Howell, T. A., J. A. Tolk, A. D. Schneider, and S. R. Evett. 1998. Evapotranspiration, yield, and water use efficiency of corn hybrids differing in maturity. *Agron. J.* 90:3–9.

Hsiao, T. C. 1973. Plant responses to water stress. *Annu. Rev. Plant Physiol.* 24:519–570.

Hsiao, T. C. and E. Acevedo. 1974. Plant responses to water deficits, water-use efficiency, and drought resistance. *Agric. Meteorol.* 14:59–84.

Huang, M., M. Shao, L. Zhang, and Y. Li. 2003. Water use efficiency and sustainability of different long-term crop rotation systems in the Loess Plateau of China. *Soil Tillage Res.* 72:95–104.

Hubick, K. T. and G. D. Farquhar. 1989. Carbon isotope discrimination and the ratio of carbon gained to water lost in barley cultivars. *Plant Cell Environ.* 12:795–804.

Hubick, K. T., G. D. Farquhar, and R. Shorter. 1986. Correlation between water-use efficiency and carbon isotope discrimination in diverse peanut germplasms. *Aust. J. Plant Physiol.* 13:803–816.

Hudak, C. M. and R. P. Patterson. 1996. Root distribution and soil moisture depletion pattern of a drought-resistant soybean plant introduction. *Agron. J.* 88:478–485.

Hurd, E. A. 1974. Phenotype and drought tolerance in wheat. *Agric. Meteorol.* 14:39–55.

Ishihara, K. and H. Saito. 1983. Relationships between leaf water potential and photosynthesis in rice plants. *JARQ* 17:81–86.

Ismail, M. A. and A. E. Hall. 1992. Correlation between water-use efficiency and carbon isotope discrimination in diverse cowpea genotypes and isogenic lines. *Crop Sci.* 32:7–12.

Jackson, R. B., J. S. Sperry, and T. E. Dawson. 2000. Root water uptake and transport: Using physiological processes in global predictions. *Trend Plant Sci.* 5:482–488.

Jalota, S. K. and S. S. Parihar. 1990. Bare soil evaporation in relation to tillage. *Adv. Soil Sci.* 12:187–216.

Jastrow, J. D. 1996. Soil aggregate formation and the accrual of particulate and mineral-associated organic matter. *Soil Biol. Biochem.* 28:656–676.

Jensen, M. E., W. R. Rangeley, and P. J. Dieleman. 1990. Irrigation trends in world agriculture. In: *Irrigation of Agricultural Crops*, eds., B. A. Stewart and D. R. Nielsen, pp. 32–63. Madison, WI: ASA.

Jeroni, G., A. Anunciacion, M. Hipólito, and F. Jaume. 2007. Photosynthesis and photoprotection responses to water stress in the wild-extinct plant *Lysimachia minoricensis*. *Environ. Exp. Bot.* 60:308–317.

Johnson, R. C. and E. T. Kanemasu. 1982. The influence of water availability on winter wheat yields. *Can. J. Plant Sci.* 62:831–838.

Joly, R. J. and D. T. Hahn. 1989. An empirical model for leaf expansion in cacao in relation to plant water deficit. *Ann. Bot.* 64:1–8.

Jones, O. R. and W. C. Johnson. 1983. Cropping practices: Southern Great Plains. In: *Dryland Agriculture*, eds., H. E. Dregne and W. O. Wills, pp. 365–385. Agron. Monogr. 23. Madison, WI: ASA, CSSA, and SSSA.

Jordan, W. R. and F. R. Miller. 1980. Genetic variability in sorghum root systems: Implications for drought tolerance. In: *Adaptation of Plants to Water and High Temperature Stress*, eds., N. C. Turner and P. J. Kramer, pp. 383–399. New York: Wiley.

Jury, W. A. and H. J. Vaux. 2007. The emerging global water crisis: Managing scarcity and conflict between water users. *Adv. Agron.* 95:1–76.

Kane, M. V. and L. J. Grabau. 1992. Early planted, early maturing soybean cropping system: Growth, development, and yield. *Agron. J.* 84:769–773.

Karaaslan, D., E. Boydak, S. Gercek, and M. Simsek. 2007. Influence of irrigation intervals and row spacing on some yield components of sesame growth in Harran region. *Asian J. Plant Sci.* 6:623–627.

Karl, T. R., J. M. Melillo, and T. C. Peterson. 2009. *Global Climate Change Impacts in the United States*. New York: Cambridge University Press.

Karlen, D. L., G. E. Varvel, D. G. Bullock, and R. M. Cruse. 1994. Crop rotations for the 21st century. *Adv. Agron.* 53:1–45.

Kassam, A. H. and J. M. Kowal. 1975. Water use, energy balance and growth of gero millet at Samaru, Northern Nigeria. *Agric. Meteor.* 15:333–342.

Kassam, A. H., J. M. Kowal, and C. Harkness. 1975. Water use and growth of groundnut at Samaru, Northern Nigeria. *Trop. Agric.* 52:105–112.

Kaya, C., L. Tuna, and D. Higgs. 2006. Effects of silicon on plant growth and mineral nutrition of maize grown under water stress conditions. *J. Plant Nutr.* 29:1469–1480.

Keatinge, J. D. H., M. D. Dennett, and J. Rodgers. 1986. The influence of precipitation regime on the management of dry areas in northern Syria. *Field Crops Res.* 13:239–249.

Kern, J. S. and M. G. Johnson. 1993. Conservation tillage impacts on national soil and atmospheric carbon levels. *Soil Sci. Soc. Am. J.* 57:200–210.

Kirda, C., S. K. A. Danso, and F. Zapata. 1989. Temporal water stress effects on nodulation, nitrogen accumulation and growth of soybean. *Plant Soil* 120:49–55.

Klaji, M. C. and G. Vachaud. 1992. Seasonal water balance of a sandy soil in Niger cropped with pearl millet, based on profile moisture measurements. *Agric. Water Manage.* 21:313–330.

Kobata, T. 1995. Drought resistance. In: *Science of the Rice Plant: Physiology*, Vol. 2, eds., T. Matsuo, K. Kumazawa, R. Ishii, K. Ishihara, and H. Hirata, pp. 474–483. Tokyo, Japan: Food and Agriculture Policy Research Center.

Kong, L., F. Wang, B. Feng, S. Li, J. Si, and B. Zhang. 2010. A root-zone soil regime of wheat: Physiological and growth responses to furrow irrigation in raised bed planting in northern China. *Agron. J.* 102:154–162.

Kowal, J. M. and A. H. Kassam. 1973. Water use, energy balance and growth of maize at Samaru, Northern Nigeria. *Agric. Meteor.* 12:391–406.

Kramer, P. J. 1969. *Plant and Soil Water Relationships: A Modern Synthesis*. New York: McGraw-Hill Book Company.

Kramer, P. J. and J. S. Boyer. 1995. *Water Relations of Plants and Soils*. San Diego, CA: Academic Press.

Labanauskas, C. K., P. Shouse, and L. H. Stolzy. 1981. Effects of water stress at various growth stages on seed yield and nutrient concentration of field grown cowpeas. *Soil Sci.* 131:249–256.

Lado, M. and M. Ben-Hur. 2009. Effects of irrigation with different effluents on saturated hydraulic conductivity of arid and semiarid soils. *Soil Sci. Soc. Am. J.* 74:23–32.

Lado, M., A. Paz, and M. Ben-Hur. 2004a. Organic matter and aggregate size interactions in saturated hydraulic conductivity. *Soil Sci. Soc. Am. J.* 68:234–242.

Lado, M., A. Paz, and M. Ben-Hur. 2004b. Organic matter and aggregate size interactions in infiltration, seal formation and soil loss. *Soil Sci. Soc. Am. J.* 68:935–942.

Lal, R. 1997. Residue management, conservation tillage and soil restoration for mitigating greenhouse effects by CO_2-enrichment. *Soil Tillage Res.* 43:81–107.

Lampkin, N. 1990. *Organic Farming*. Ipswich, U.K.: Farming Press Books.

Larcher, W. 1995. *Physiological Plant Ecology*, 3rd edn. New York: Springer.

Lascano, R. L., R. L. Baumhardt, S. K. Hicks, and J. L. Heilman. 1994. Soil and plant evaporation from strip-tilled cotton: Measurement and simulation. *Agron. J.* 86:987–994.

Lawlor, D. W. 1995. Photosynthesis, productivity and environment. *J. Exp. Bot.* 46:1449–1461.

Li, Q., M. Liu, J. Zhang, B. Dong, and Q. Bai. 2009. Biomass accumulation and radiation use efficiency of winter wheat under deficit irrigation regimes. *Plant Soil Environ.* 55:85–91.

Liang, Y. C., Q. Chen, Q. Liu, W. H. Zhang, and R. X. Ding. 2003. Exogenous silicon (Si) increases antioxidant enzyme activity and reduces lipid peroxidation in roots of salt-stressed barley (*Hordeum vulgare* L.). *J. Plant Physiol.* 160:1157–1164.

Liang, Y. C., J. Si, and V. Romheld. 2005. Silicon uptake and transport is an active process in *Cucumis sativus*. *New Phytologist* 167:797–804.

Lima, J. E. F. W., R. S. A. Ferreira, and D. Christofidis. 1999. O uso da irrigação no Brasil. In: *O estado das águas no Brasil*, ed., M. A. V. Freitas, pp. 73–101. Brasília: MME, MMA/ SRH, OMM.

Limon-Ortega, A., B. Govaerts, J. Deckers, and K. D. Sayre. 2006. Soil aggregate and microbial biomass in a permanent bed wheat-maize planting system after 12 years. *Field Crops Res.* 97:302–309.

Lindenmayer, R. B., N. C. Hansen, J. Brummer, and J. G. Pritchett. 2011. Deficit irrigation of alfalfa for water savings in the Great Plains and intermountain west: A review and analysis of the literature. *Agron. J.* 103:45–50.

Linn, D. M. and J. M. Doran. 1984. Aerobic and anaerobic microbial populations in no-till and plowed soils. *Soil Sci. Soc. Am. J.* 48:794–799.

Lockhart, J. A. 1965. An analysis of irreversible plant cell elongation. *J. Theor. Biol.* 8:264–275.

Lopez-Bellido, R. J., L. Lopez-Bellido, J. Benitez-Veja, and F. J. Lopez-Bellido. 2007. Tillage system, preceding crop, and nitrogen fertilization in wheat crop: II. Water utilization. *Agron. J.* 99:66–72.

López-Castañeda, C. and R. A. Richards. 1994. Variation in temperate cereals in rainfed environments: III. Water use and water use efficiency. *Field Crops Res.* 39:85–98.

Lorens, G. F., J. M. Bennett, and L. B. Loggale. 1987. Differences in drought resistance between two corn hybrids. II. Component analysis and growth rates. *Agron. J.* 79:808–813.

Ludlow, M. M. and R. C. Muchow. 1990. A critical evaluation of traits for improving crop yields in water-limited environments. *Adv. Agron.* 43:107–153.

Lynch, J. M. and E. Bragg. 1985. Microorganisms and soil aggregate stability. *Adv. Soil Sci.* 2:133–171.

Lynch, D. R. and G. C. C. Tai. 1989. Yield and yield component response of eight potato genotypes to water stress. *Crop Sci.* 29:1207–1211.

Mambani, B. and R. Lal. 1983. Response of upland rice cultivars to drought stress. III. Screening rice varieties by means of variable moisture along a toposequence. *Plant Soil* 73:73–94.

Maqbool, A. B., D. Prathibha, and B. S. Mariam. 2001. Biotechnology: Genetic improvement of sorghum (*Sorghum biocolor* L. Moench). *In Vitro Cell. Dev. Biol. Plant* 37:504–515.

Martens, D. A. and W. T. Jr. Frankenberger. 1992. Modification of infiltration rates in an organic-amended irrigated soil. *Agron. J.* 84:707–717.

Martinez, E., J. P. Fuentes, P. Silva, S. Valle, and E. Acevedo. 2008. Soil physiological properties and wheat root growth as affected by no-tillage and conventional tillage systems in a Mediterranean environment of Chile. *Soil Tillage Res.* 99:232–244.

Mason, W. K., W. S. Meyer, R. C. G. Smith, and H. d. Bars. 1983. Water balance of three irrigated crops on fine-textured soils of the Riverine plain. *Aust. J. Agric. Res.* 34:183–191.

Maurya, D. M. and J. C. O'Toole. 1986. Screening upland rice for drought tolerance. In: *Progress in Upland Rice Research, Proceedings of the 1985 Jakarta Conference*, ed., IRRI, pp. 245–261. Los Banos, Philippines: IRRI.

Mayhew, W. L. and C. E. Caviness. 1994. Seed quality and yield of early-planted, short-season soybean genotypes. *Agron. J.* 89:459–464.

Mays, D. A., G. L. Terman, and J. C. Duggan. 1973. Municipal compost: Effect on crop yield and soil properties. *J. Environ. Qual.* 2:89–91.

McCaffery, D. 2004. *Irrigated Canola: Management for High Yields*. Griffith, Australia: Grain Research and Development Corporation.

McGee, E. A., G. A. Peterson, and D. G. Westfall. 1997. Water storage efficiency in no-till dryland cropping systems. *J. Soil Water Conserv.* 52:131–136.

McWilliam, J. R. 1986. The national and international importance of drought and salinity effects on agricultural production. *Aust. J. Plant Physiol.* 13:1–13.

Meckel, L., D. B. Egli, R. E. Phillips, D. Radcliffe, and J. E. Leggett. 1984. Effect of moisture stress on seed growth in soybean. *Agron. J.* 76:647–650.

Mengel, K., E. A. Kirkby, Kosegarten, and T. Appel. 2001. *Principles of Plant Nutrition*, 5th edn. Dordrecht, the Netherlands: Kluwer Academic Publishers.

Mesfin, T., G. B. Tesfahunegn, C. S. Wortmann, M. Mamo, and O. Nikus. 2010. Skip-row planting and tie-ridging for sorghum production in semiarid areas of Ethiopia. *Agron. J.* 102:745–750.

Mian, M. A. R., D. A. Ashley, and H. R. Boerma. 1998. An additional QTL for water use efficiency in soybean. *Crop Sci.* 38:390–393.

Mian, M. A. R., M. A. Bailey, D. A. Ashley, R. Wells, T. E. Carter, Jr., W. A. Parrott, and H. R. Boerma. 1996. Molecular markers associated with water use efficiency and leaf ash in soybean. *Crop Sci.* 36:1252–1257.

Mikha, M. M. and C. W. Rice. 2004. Tillage and manure effects on soil and aggregate associated carbon and nitrogen. *Soil Sci. Soc. Am. J.* 68:809–816.

Miller, D. E. and D. W. Burke. 1983. Response of dry beans to daily deficit sprinkler irrigation. *Agron. J.* 75:775–778.

Miller, M. and R. P. Dick. 1995. Thermal stability and activities of soil enzymes influenced by crop rotations. *Soil Biol. Biochem.* 27:1161–1166.

Miller, D. E. and W. T. Martin. 1983. Effect of daily irrigation rate and soil texture on yield and quality of Russet Burbank potatoes. *Am. Potato J.* 60:745–757.

Miller, D. E. and M. W. Martin. 1985. Effect of water stress during tuber formation on subsequent growth and internal defects in Russet Burbank potatoes. *Am. Potto J.* 62:83–89.

Molnar, I., L. Gaspar, E. Sarvari, S. Dulari, B. Hoffmann, M. Molnar-Long, and G. Galiba. 2004. Physiological and morphological responses to water stresses in *Aegilus biuncialis* and *Triticum aestivum* genotypes with differing tolerance to drought. *Funct. Plant Biol.* 31:1149–1159.

Moore, J. M., S. Klose, and M. A. Tabatabai. 2000. Soil microbial biomass carbon and nitrogen as affected by cropping systems. *Biol. Fertil. Soil* 31:200–210.

Morgan, J. M. 1980. Osmotic adjustment in the spikelets and leaves of wheat. *J. Exp. Bot.* 31:655–665.

Morgan, J. M. 1984. Osmoregulation and water stress in higher plants. *Annu. Rev Plant Physiol.* 35:299–319.

Moss, G. I. and L. A. Downey. 1971. Influence of drought stress on female gametophyte development in corn and subsequent grain yield. *Crop Sci.* 11:368–372.

Musick, J. T. and D. A. Dusek. 1974. Alternate-furrow irrigating of fine textured soils. *Trans. ASAE* 17:289–294.

Musick, J. T., O. R. Jones, B. A. Stewart, and D. A. Dusek. 1994. Water yield relationship for irrigated and dryland wheat in the U. S. Southern plains. *Agron. J.* 86:980–986.

Musick, J. T., F. B. Pringle, W. L. Harman, and B. A. Stewart. 1990. Long-term irrigation trends: Texas High Plains. *Appl. Eng. Agric.* 6:717–724.

Musick, J. T., F. B. Pringle, and J. D. Walker. 1988. Sprinkler and furrow irrigation trends: Texas High Plains. *Appl. Eng. Agric.* 4:46–52.

Nadler, A., E. Raveh, U. Yermiyahu, and S. R. Green. 2003. Evaluation of TDR used to monitor water content in stem of lemon trees and soil and their response to water stress. *Soil Sci. Soc. Am. J.* 67:437–448.

Namuco, O. S. and J. C. O'Toole 1986. Reproductive stage water stress and sterility. I. Effects of stress during meiosis. *Crop Sci.* 26:317–321.

National Agricultural Statistics Database. 2005. Quick Stats: Agricultural Statistics Data Base. Available at www.nass.usda.gov/QuickStatus/ (accessed September 9, 2005; verified April 19, 2006). USDA, Washington, DC.

National Agricultural Statistics Service. 2007. Hay alfalfa (dry): National statistics. Available at http://www.nass.ysda.gov (modified April 3, 2008; accessed April 28, 2008; verified September 1, 2010). Washington, DC: U.S. Department of Agriculture.

National Research Council. 1989. *Alternate Agriculture*. Washington, DC: National Academic Press.

NeSmith, D. S. and J. T. Ritchie. 1992. Maize (*Zea mays L.*) response to a severe soil water-deficit during grain filling. *Field Crops Res.* 29:23–35.

Neumann, P. M. 1995. The role of cell wall adjustment in plant resistance to water deficits. *Crop Sci.* 35:1258–1266.

Nicolas, M. E., R. M. Gleadow, and M. J. Dalling. 1984. Effect of drought and high temperature on grain growth in wheat. *Aust. J. Plant Physiol.* 11:553–566.

Nielsen, D. C. 1998. Comparison of three alternative oilseed crops for the central Great Plains. *J. Prod. Agric.* 11:336–341.

Nielsen, D. C., P. W. Unger, and P. R. Millar. 2005. Efficient water use in dryland cropping systems in the Great Plains. *Agron. J.* 97:364–372.

Nielsen, D. C., M. E. Vigil, R. L. Anderson, R. A. Bowman, J. G. Benjamin, and A. D. Halvorson. 2002. Cropping system influence on planting water content and yield of winter wheat. *Agron. J.* 94:962–967.

Nobel, P. S. 1974. *Introduction to Biophysical Plant Physiology*. San Francisco, CA: W. H. Freeman and Company.

Nonomura, A. M. and A. A. Benson. 1992. The path of carbon in photosynthesis: Improved crop yields with methanol. *Proc. Natl Acad. Sci. U. S. A.* 89:9794–9798.

Norwood, C. A. 1995. Comparison of limited irrigated vs. Dryland cropping systems in the U. S. Great Plains. *Agron. J.* 87:737–743.

Odell, R. T., S. W. Melsted, and W. M. Walker. 1984. Changes in organic carbon and nitrogen of Morrow plot soils under different treatments, 1904–1973. *Soil Sci.* 137:160–171.

Oki, T. and S. Kanae. 2006. Global hydrological cycles and world water resources. *Science* 313:1068–1072.

Okon, Y. and Y. Kapulnik. 1986. Development and function of *Azospirillum*-inoculated roots. *Plant Soil* 90:3–16.

Ortega, R. A., G. A. Peterson, and D. G. Westfall. 2002. Residue accumulation and changes in soil organic matter as affected by cropping intensity in no-till dryland agroecosystems. *Agron. J.* 94:944–954.

O'Toole, J. C. 1982. Adaptation of rice to drought-prone environments. In: *Drought Resistance in Crops with Emphasis on Rice*, ed., IRRI, pp. 195–213. Los Banos, Philippines: IRRI.

O'Toole, J. C. and T. B. Moya. 1976. Genotypic variations in maintenance of leaf water potential rice. *Crop Sci.* 18:873–876.

Oweis, T., M. Pala, and J. Ryan. 1998. Stabilizing rainfed wheat yields with supplemental irrigation and nitrogen in a Mediterranean climate. *Agron. J.* 90:672–681.

Oweis, T., H. Zeidan, and A. Taimeh. 1992. Modeling approach for optimizing supplemental irrigation management. *Proceedings of the International Conference on Supplemental Irrigation and Drought Water Management*, Bari, Italy. Ist Agron. Mediterranean type environment. *Exp. Agric.* 32:339–349.

Pahlavan-Rad, M. R., S. A. R. Movahedi-Naeini, and M. Pessarakali. 2010. Response of wheat plants in terms of soil water content, bulk density, salinity, and root growth to different planting systems under various irrigation frequency. *J. Plant Nutr.* 33:874–888.

Pahlavan-Rad, M. R., S. A. R. Movahedi-Naeini, and M. Pessarakali. 2011. Nutrient uptake, soil and plant nutrient contents, and yield components of wheat plants under different planting systems and various irrigation frequencies. *J. Plant Nutr.* 34:1133–1143.

Park, E. J. and A. J. Smucker. 2005. Saturated hydraulic conductivity and porosity within macroaggregates modified by tillage. *Soil Sci. Soc. Am. J.* 69:38–45.

Passioura, J. B. 1983. Roots and drought resistance. In: *Plant Production and Management under Drought Conditions*, eds., J. F. Stone and W. Wills, pp. 265–280. New York: Elsevier Science Publishers.

Passioura, J. B. 1994. The yield of crops in relation to drought. In: *Physiology and Determination of Crop Yield*, eds., K. J. Boote, J. M. Bennett, T. R. Sinclair, and G. M. Paulsen, pp. 343–359. Madison, WI: ASA, CSSA, and SSSA.

Paustian, K., O. Andren, H. Janzen, R. Lal, P. Smith, G. Tian, H. Tiessen, M. V. Noordwijk, and P. Woomer. 1997. Agricultural soil as a C sink to offset CO_2 emissions. *Soil Use Manage.* 13:230–244.

Paustian, K., J. Six, E. T. Elliott, and H. W. Hunt. 2000. Management options for reducing CO_2 emissions from agricultural soils. *Biochemistry* 48:147–163.

Payne, W. A. 1997. Managing yield and water use of pearl millet in the Sahel. *Agron. J.* 89:481–490.

Paz, V. P. S., R. E. F. Teodoro, and F. C. Mendonça. 2000. Water resources, irrigated agriculture and the environment. *Rev. Bras. Eng. Agri. Amb.* 4:465–473.

Peterson, H. B. 1972. Water relationships and irrigation. In: *Alfalfa Science and Technology*, ed., C. H. Hanson, pp. 469–480. Madison, WI: ASA, CSSA, and SSSA.

Peterson, G. A., A. J. Schlegel, D. L. Tanaka, and O. R. Jones. 1996. Precipitation use efficiency as affected by cropping and tillage systems. *J. Prod. Agric.* 9:180–186.

Phillips, R. E., R. L. Blevins, G. W. Thomas, W. W. Frye, and S. H. Phillips. 1980. No-tillage agriculture. *Science* 208:1108–1113.

Porter. G. A., G. B. Opena, W. B. Bradbury, J. C. Mcburine, and J. A. Sisson. 1999. Soil management and supplemental irrigation effects on potato: I. Soil properties, tuber yield, and quality. *Agron. J.* 91:416–425.

Power, J. F. 1990. Fertility management and nutrient cycling. *Adv. Soil Sci.* 13:131–149.

Pritchard, J. 1994. Tansley review no. 68. The control of cell expansion in roots. *New Phytol.* 127:3–26.

Purcell, L. C., M. Silva, C. A. King, and W. H. Kim. 1997. Biomass accumulation and allocation in soybean associated with genotypic differences in tolerance of nitrogen fixation to water deficits. *Plant Soil* 196:101–113.

Putnam, D., S. Orlaff, B. Hanson, and H. Carlson. 2005. Controlled deficit irrigation of alfalfa in differing environments. In: *Abstracts 2005 International Annual Meeting*, Salt Lake City, UT, November 6–10, 2005. Madison, WI: ASA, CSSA, and SSSA.

Rajan, N., S. J. Maas, and J. C. Kathilankal. 2010. Estimating crop water use of cotton in the Texas High Plains. *Agron. J.* 102:1641–1651.

Ramirez-Vallejo, P. and J. D. Kelly. 1998. Traits related to drought resistance in common bean. *Euphytica* 99:127–136.

Recep, C. 2004. Effect of water stress at different development stages on vegetative and reproductive growth of corn. *Field Crops Res.* 89:1–16.

Riar, D. S., D. A. Ball, J. P. Yenish, S. B. Wuest, and M. K. Corp. 2010. Comparison of fallow tillage methods in the intermediate rainfall inland pacific northwest. *Agron. J.* 102:1664–1673.

Richards, R. A., C. Lopez-Castaneda, H. Gomez-Macpherson, and A. G. Condon. 1993. Improving the efficiency of water use by plant breeding and molecular biology. *Irrig. Sci.* 14:93–104.

Ritchie, J. T. 1972. Model for predicting evaporation from a row crop with incomplete cover. *Water Resour. Res.* 8:1204–1213.

Ritchie, J. T. 1981. Water dynamics in the soil–plant atmosphere system. *Plant Soil* 58:81–96.

Ritchie, G. L., J. R. Whitaker, C. W. Bednarz, and J. E. Hook. 2009. Subsurface drip and overhead irrigation: A comparison of plant boll distribution in upland cotton. *Agron. J.* 101:1336–1344.

Robinson, G. D. and M. A. Massengale. 1968. Effect of harvest management and temperature on forage yield, root carbohydrates, plant density and leaf area relationships in alfalfa. *Crop Sci.* 8:147–151.

Robson, M. C., S. M. Fowler, N. H. Lampkin, C. Leifert, M. Leitch, D. Robinson, C. A. Watson, and A. M. Litterick. 2002. The agronomic and economic potential of break crops for ley/arable rotations in temperate organic agriculture. *Adv. Agron.* 77:369–427.

Royo, C. and R. Blanco. 1998. Use of potassium iodide to mimic drought stress in triticale. *Field Crops Res.* 59:201–212.

Sadler, E. J. and N. C. Turner. 1994. Water relationships in a sustainable agricultural system. In: *Sustainable Agricultural System*, eds., J. L. Hatfield and D. L. Karlen, pp. 21–46. Boca Raton, FL: CRC Press.

Sall, K. and T. R. Sinclair. 1991. Soybean genotypic differences in sensitivity of symbiotic nitrogen fixation to soil dehydration. *Plant Soil* 133:31–37.

Sammis, T. W. 1981. Yield of alfalfa and cotton as influenced by irrigation. *Agron. J.* 73:323–329.

Sanderson, J. 1983. Water uptake by different regions of the barley root: Pathway of radial flow in relation to development of the endodermis. *J. Exp. Bot.* 34:240–253.

Sarig, S., A. Blum, and Y. Okon. 1988. Improvement of the water status and yield of field grown grain sorghum (*Sorghum bicolor*) by inoculation with *Azospirillum brasilense*. *J. Agric. Sci. Camb.* 110:271–277.

Sau, F., K. J. Boote, W. M. Bostick, J. W. Jones, and M. I. Minguez. 2004. Testing and improving evapotranspiration and soil water balance of the DSSAT crop models. *Agron. J.* 96:1243–1257.

Savoy, B. R., J. T. Cothren, and C. R. Shumway. 1992. Early-season production systems utilizing indeterminate soybean. *Agron. J.* 84:394–398.

Schneekloth, J. and A. Andales. 2009. Seasonal water needs and opportunities for limited irrigation for Colorado crops. Available at http://www.ext.colostate.edu/pubs/crops/04718. html (accessed May 6, 2008; verified September 1, 2010) Fort Collins, CO: Colorado State University Cooperative Extension.

Schulze, E. D. 1986. Carbon dioxide and water vapor exchange in response to drought in the atmosphere and in the soil. *Annu. Rev. Plant Physiol.* 37:247–274.

Schussler, J. R. and M. E. Westgate. 1994. Increasing assimilate reserves does not prevent kernel abortion at low water potential in maize. *Crop Sci.* 34:1569–1576.

Scoot, H. D. and L. S. Wood. 1989. Impact of crop production on the physical status of a typic Albaqualf. *Soil Sci. Soc. Am. J.* 53:1819–1825.

Seckler, D., U. Amarasinghe, D. Molden, R. Silva, and R. Barker. 1998. World water demand and supply 1990 to 2025: Scenarios and issues. International Water Management Institute Research Report 19, International Water Management Institute, Colombo, Sri Lanka.

Serraj, R., S. Bona, L. C. Purcell, and T. R. Sinclair. 1997. Nitrogen fixation response to water deficit in field grown Jackson soybean. *Field Crops Res.* 52:109–116.

Serraj, R. and T. R. Sinclair. 1996. Processes contributing to N_2-fixation insensitivity to drought in the soybean cultivar Jackson. *Crop Sci.* 36:961–968.

Serraj, R. and T. R. Sinclair. 1997. Variation among soybean cultivars in dinitrogen fixation response to drought. *Agron. J.* 89:963–969.

Serraj, R. and T. R. Sinclair. 1998. Soybean cultivar variability for nodule formation and growth under drought. *Plant Soil* 202:159–166.

Shannon, M. A., P. W. Bohn, M. Elimelech, J. G. Georgiadis, B. J. Marinas, and A. M. Mayes. 2008. Science and technology for water purification in the coming decades. *Nature* 452:301–310.

Sharpley, A., J. J. Meisinger, A. Breeuwsma, J. T. Sims, T. C. Daniel, and J. S. Schepers. 1998. Impacts of animal manure management on ground and surface water quality. In: *Animal Waste Utilization: Effective Use of Manure as a Soil Resource*, eds., J. L. Hatfield and B. A. Stewart, pp. 173–242. Chelsea, MI: Ann Arbor Press.

Shouse, P., S. Dasberg, W. A. Jury, and L. H. Stolzy. 1981. Water deficit effects on water poten-
    tial, yield and water use of cowpeas. *Crop Sci.* 73:333–336.
Siddique, K. H. M., D. Tennant, M. W. Perry, and R. K. Belford. 1990. Water use and water
    use efficiency of old and modern wheat cultivars in a Mediterranean type environment.
    Aust. J. Agric. Res. 41:431–447.
Sinclair, T. R. and M. M. Ludlow. 1985. Who taught plants thermodynamics? The unfulfilled
    potential of plant water potential. *Aust. J. Plant Physiol.* 12:213–217.
Sinclair, T. R. and M. M. Ludlow. 1986. Influence of soil water supply on the plant water
    balance of four tropical grain legumes. *Aust. J. Plant Physiol.* 13:329–341.
Sinclair, T. R., R. C. Muchow, J. M. Bennett, and L. C. Hammond. 1987. Relative sensitivity of
    nitrogen and biomass accumulation to drought in field grown soybean. *Agron. J.* 79:986–991.
Sinclair, T. R. and R. Serraj. 1995. Dinitrogen fixation sensitivity to drought among grain
    legume species. *Nature* 378:344.
Singh, G. 1969. A review of soil-moisture relationships in potatoes. *Am. Potato J.* 46:398–403.
Singh, S. P. 1995. Selection for water stress tolerance in interracial populations of common
    bean. *Crop Sci.* 35:118–124.
Singh, S., G. Singh, P. Singh, and N. Singh. 2008. Effect of water stress at different stages of
    grain development on the characteristics of starch and protein of different wheat variet-
    ies. *Food Chem.* 108:130–139.
Sisti, C. P. J., H. P. Santos, R. Kohhann, B. J. R. Alves, S. Urquiaga, and R. M. Boddey. 2004.
    Change in carbon and nitrogen stocks in soil under 13 years of conventional and zero
    tillage in southern Brazil. *Soil Tillage Res.* 76:39–58.
Six, J., E. T. Elliott, and K. Paustain. 1999. Aggregate and soil organic matter dynamics under
    conventional and no-tillage systems. *Soil Sci. Soc. Am. J.* 63:1350–1358.
Six, J., E. T. Elliott, and K. Paustain. 2000. Soil microaggregate turnover and microaggre-
    gate formation: A mechanism for c sequestration under no-tillage agriculture. *Soil Biol.
    Biochem.* 32:2099–2013.
Six, J., C. Feller, K. Denef, S. M. Ogle, J. C. Moraes, and A. Albrecht. 2002. Soil organic mat-
    ter, biota and aggregation in temperate and tropical soils-effects of no-tillage. *Agronomie*
    22:755–775.
Slatyer, R. O. 1969. Physiological significance of internal water relations in crop yield. In:
    Physiological Aspects of Crop Yield, ed., R. C. Dinauer, pp. 53–88. Madison, WI:
    American Society of Agronomy.
Sloane, R. J., R. P. Patterson, and T. E. Carter. 1990. Field drought tolerance of a soybean plant
    introduction. *Crop Sci.* 30:118–123.
Smiciklas, K. D., R. E. Mullen, R. E. Carlson, and A. D. Knapp. 1989. Drought-induced stress
    effect on soybean seed calcium and quality. *Crop Sci.* 29:1519–1523.
Smith, M. S., W. W. Frye, and J. J. Varco. 1987. Legume winter cover crops. *Adv. Soil Sci.*
    7:95–139.
Soil Science Society of America. 2008. *Glossary of Soil Science Terms.* Madison, WI: SSSA.
Sonobe, K., T. Hattori, P. An, W. Tsuji, A. E. Eneji, S. Kobayashi, Y. Kawamura, K. Tanaka,
    and S. Inanaga. 2011. Effect of silicon application on sorghum root response to water
    stress. *J. Plant Nutr.* 34:71–82.
Sorensen, R. B., F. S. Wright, and C. L. Butts. 2000. Subsurface drip irrigation system designed
    for research in row crop rotation. *Appl. Eng. Agric.* 17:171–176.
Souza, P. I., D. B. Egli, and W. P. Bruening. 1997. Water stress seed filling and leaf senescence
    in soybean. *Agron. J.* 89:807–812.
Spomer, L. A. 1985. Techniques for measuring plant water. *Hort. Sci.* 20:1021–1028.
Sponchiado, B. N., J. W. White, J. A. Castillo, and P. G. Jones. 1989. Root growth of four com-
    mon bean cultivars in relation to drought tolerance in environments with contrasting soil
    types. *Exp. Agric.* 25:249–257.

Stanca, A. M., V. Terzi, and L. Cattivelli. 1992. Biochemical and molecular studies of stress tolerance in barley. In: *Barley: Genetics, Biochemistry, Molecular Biology and Biotechnology*, ed., P. R. Shewry, pp. 277–288. Wallingford, U.K.: CAB International.

Stapper, M. and H. C. Harris. 1989. Assessing the productivity of wheat genotypes in a Mediterranean climate, using a crop simulation model. *Field Crops Res.* 20:129–152.

Stewart, B. A. and J. L. Steiner. 1990. Water-use efficiency. *Adv. Soil Sci.* 13:151–173.

Stone, L. R., D. E. Goodrum, A. J. Schlegel, M. N. Jaafar, and A. H. Khan. 2002. Water depletion depth of grain sorghum and sunflower in the central high plains. *Agron. J.* 94:936–943.

Stone, L. R. and A. J. Schlegel. 2010. Tillage and crop rotation phase effects on soil physical properties in the west-central Great Plains. *Agron. J.* 102:483–491.

Sullivan, C. Y. and J. D. Eastin. 1974. Plant physiological responses to water stress. *Agric. Meteorol.* 14:113–127.

Summer, M. E. 1990. Crop responses to *Azospirillum* inoculation. *Adv. Soil Sci.* 13:52–123.

Sustainable Agriculture Network. 1998. *Managing Cover Crops Profitably*, 2nd edn., 212 pp. Handbook Series Book 3, Beltsville, MD: Sustainable Agriculture Network.

Taiz, L. 1984. Plant cell expansion: Regulation of cell wall mechanical properties. *Annu. Rev. Plant Physiol.* 35:585–657.

Tanaka, D. L. and R. L. Anderson. 1997. Soil water storage and precipitation storage efficiency of conservation tillage systems. *J. Soil Water Conserv.* 52:363–367.

Tanaka, D. L., R. L. Anderson, and S. C. Rao. 2005. Crop sequencing to improve use of precipitation and synergize crop growth. *Agron. J.* 97:385–390.

Tanji, K. K. and B. R. Hanson. 1990. Drainage and return flows in relation to irrigation management. In: *Irrigation of Agricultural Crops*, eds., B. A. Stewart and D. R. Nielsen, pp. 1057–1087. Agron. Monogr. 30. Madison, WI : ASA, CSSA, and SSSA.

Taylor, H. M. 1983. Managing root systems for efficient water use: An overview. In: *Limitations to Efficient Water Use in Crop Production*, eds., H. M. Taylor, W. R. Jordan, and T. R. Sinclair, pp. 87–114. Madison, WI: ASA.

Taylor, H. M. and E. E. Terrell. 1982. Rooting pattern and plant productivity. In: *Handbook of Agricultural Productivity*, ed., M. Rechcigl, pp. 185–200. Boca Raton, FL: CRC Press.

Tesfamariam, E. H., J. G. Annandate, and J. M. Steyn. 2010. Water stress on winter canola growth and yield. *Agron. J.* 102:658–666.

Thomas, G. A., G. Gibson, R. G. H. Nielsen, W. D. Martin, and B. J. Radford. 1995. Effects of tillage, stubble, gypsum, and nitrogen fertilizer on cereal cropping on a red-brown earth in south-west Queensland. *Aust. J. Exp. Agric.* 35:997–1008.

Thornthwaite, C. W. 1948. An approach towards a rational classification of climate. *Geogr Ver.* 38:55–94.

Tisdall, J. M. and J. M. Oades. 1982. Organic matter and water-stable aggregates in soils. *J. Soil Sci.* 33:141–163.

Tripathi, S. C., K. D. Sayre, and J. N. Kaul. 2005. Planting systems on lodging behavior, yield components, and yield of irrigated spring bread wheat. *Crop Sci.* 45:1448–1455.

Truman, C. C., D. W. Reeves, J. N. Shaw, A. C. Motta, C. H. Burmester, R. L. Raper, and E. B. Schwab. 2003. Tillage impacts on soil property, runoff, and soil loss variations from a Rhodic Paleudult under simulated rainfall. *J. Soil Water Conserv.* 58:258–267.

Tsuda, M. 1986. Effects of water stress on the panicle emergence in rice and sorghum plants. *Japan. J. Crop Sci.* 55:196–200.

Tuong, T. P. and S. I. Bhuiyan. 1994. Innovations toward improving water use efficiency of rice. Paper presented at *the World Water Resources Seminar*, December, 13–15, 1994, Lansdowne Conference Resort, Virginia.

Turk, K. J. and A. E. Hall. 1980. Drought adaptation of cowpea. II. Influence of drought on seed yield. *Agron. J.* 72:421–427.

Turner, N. C. 2004. Agronomic options for improving rainfall-use efficiency of crops in dryland farming systems. *J. Exp. Bot.* 55:2413–2425.

Turner, N. C., J. C. O'Toole, R. T. Cruz, E. B. Yambao, S. Ahmad, O. S. Namuco, and M. Dingkuhn. 1986. Response of seven diverse rice cultivars to water deficits 2. Osmotic adjustment, leaf elasticity, leaf extension, leaf death, stomatal conductance and photosynthesis. *Field Crops Res.* 13:273–286.

Undersander, D. J. 1987. Alfalfa (*Medicago sativa* L.) growth response to water and temperature. *Irrig. Sci.* 8:23–33.

Unger, P. W. 1984. Tillage and residue effects on wheat, sorghum, and sunflower grown in rotation. *Soil Sci. Soc. Am. J.* 48:885–891.

Unger, P. A. 1994. Tillage effects on dryland wheat and sorghum production in the southern great plains. *Agron. J.* 86:310–314.

Unger, P. W. and R. L. Baumhardt. 1999. Factors related to dryland grain sorghum yield increases: 1939 to through 1997. *Agron. J.* 91:870–875.

Van Bavel, C. H. M. 1966. Potential evaporation: The combination concept and its experimental verification. *Water Resour. Res.* 2:455–467.

Van Bavel, C. H. M. 1967. Changes in canopy resistance to water loss from alfalfa induced by soil water depletion. *Agric. For. Meteorol.* 4:165–176.

Vieira, R. D., D. M. Tekrony, and D. B. Egli. 1992. Effect of drought stress on soybean seed germination and vigor. *J. Seed Technol.* 15:12–21.

Viets, F. G. Jr. 1972. Water deficits and nutrient availability. In: Water *Deficits and Plant Growth: Plant Responses and Control of Water Balance*, Vol. 3, ed., T. T. Kozlowski, pp. 217–239. New York: Academic Press.

Villalobos, F. J. and E. Fereres. 1990. Evaporation measurements beneath corn, cotton, and sunflower canopies. *Agron. J.* 82:1153–1159.

Virgona, J. M., K. T. Hubick, H. M. Rawson, G. D. Farquhar, and R. W. Downes. 1990. Genotypic variation in transpiration efficiency, carbon isotope discrimination, and carbon allocation during early growth in sunflower. *Aust. J. Plant Physiol.* 17:207–214.

Wang, F. H., Z. H. He, K. D. Sayre, S. D. Li, J. S. Si, B. Feng, and L. A. Kong. 2009. Wheat cropping systems and technologies in China. *Field Crops Res.* 111:181–188.

Wang, F. H., X. Q. Wang, and K. D. Sayre. 2004. Comparison of conventional, flood irrigated, flat planting with furrow irrigated, raised bed planting for winter wheat in China. *Field Crops Res.* 87:35–42.

Wani, S. P., W. B. McGill, K. L. Haugen-Kozyra, J. A. Robertson, and J. J. Thurston. 1994. Improved soil quality and barley yields with fababeans, manure, forages and crop rotation on a Gray Luvisol. *Can. J. Soil Sci.* 74:75–84.

Weinhold, B. J., T. P. Trooien, and G. Reichman. 1995. Yield and nitrogen use efficiency of irrigated corn in the northern Great Plains. *Agron. J.* 87:842–846.

Wells, R. 2002. Stem and root carbohydrate dynamics of two cotton cultivars bred fifty years apart. *Agron. J.* 94:876–882.

Westgate, M. E. 1994. Seed formation in maize during drought. In: *Physiology and Determination of Crop Yield*, eds., K. J. Boote, J. M. Bennett, T. R. Sinclair, and G. M. Paulsen, pp. 361–364. Madison, WI: ASA, CSSA, and SSSA.

Westgate, M. E. and J. S. Boyer. 1985. Carbohydrate reserves and reproductive development at low leaf water potentials in maize. *Crop Sci.* 25:762–769.

Westgate, M. E. and J. S. Boyer. 1986. Reproduction at low silk and pollen water potentials in maize. *Crop Sci.* 26:951–956.

Whitaker, J. R., G. L. Ritchie, C. W. Bednarz, J. E. Hook, and C. I. Mills. 2008. Subsurface drip and overhead irrigation in the Lower Coastal Plains: A comparison of water use and efficiency, maturity, yield and fiber quality. *Agron. J.* 100:1763–1768.

Wien, H. C., E. J. Littleton, and A. Ayanaba. 1979. Drought stress of cowpea and soybean under tropical conditions. In: *Stress Physiology of Crop Plants*, eds., H. Mussell and R. C. Staples. pp. 283–301, New York: Wiley Interscience.

Wilhelm, W. W., J. D. Doran, and J. F. Power. 1986. Corn and soybean yield response to crop residue management under no-tillage production systems. *Agron. J.* 78:184–189.

Wilson, J. H. H. and C. S. Allison. 1978. Effects of water stress on the growth of maize (*Zea mays L.*). *Rhod. J. Agric. Res.* 16:175–192.

Wood, S., K. Sebastian, and S. J. Scherr. 2001. *Pilot Analysis of Global Ecosystems: Agroecosystems*. Washington, DC: World Resource Institute.

Wraith, J. M. and C. K. Wright. 1998. Soil water and root growth. *Hort. Sci.* 33:951–959.

Wright, A. L. and F. M. Hons. 2004. Soil aggregation and carbon and nitrogen storage under soybean cropping sequences. *Soil Sci. Soc. Am. J.* 68:507–513.

Wright, G. C., R. C. N. Rao, and G. D. Farquhar. 1994. Water use efficiency and carbon isotope discrimination in peanut under water deficit conditions. *Crop Sci.* 34:92–97.

Wu, F., W. Bao, F. Li, and N. Wu. 2008. Effects of drought stress and N supply on the growth, biomass portioning and water use efficiency of *Sophora davidii* seedlings. *Environ. Exp. Bot.* 63:248–255.

Yang, J., J. Zhang, L. Liu, Z. Wang, and Q. Zhu. 2002. Carbon remobilization and grain filling in Japonica/Indica hybrid rice subjected to postanthesis water deficits. *Agron. J.* 94:102–109.

Zablotowicz, R. M., D. D. Focht, and G. H. Cannell. 1981. Nodulation and N fixation and droughty conditions. *Agron. J.* 73:9–12.

Zeiher, C., D. B. Egli, J. E. Leggett, and D. A. Reicosky. 1982. Cultivar differences in nitrogen redistribution in soybeans. *Agron. J.* 74:375–379.

Zhang, J., J. Sun, A. Duan, J. Wang, X. Shen, and X. Liu. 2007. Effects of different planting patterns on water use and yield performance of winter wheat in the Huang-Hai plain of China. *Agric. Water Manage.* 92:41–47.

Zhang, Z., S. Zhang, J. Yang, and J. Zhang. 2008. Yield, grain quality and water use efficiency of rice under non-flooded mulching cultivation. *Field Crops Res.* 108:71–81.

Zhu, X., H. Cong, G. Chen, S. Wang, and C. Zhang. 2005. Different solute levels in two spring wheat cultivars induced by progressive field water stress at different developmental stages. *J. Arid Environ.* 62:1–14.

4 Rhizosphere Chemistry

4.1 INTRODUCTION

Roots are among the most important organisms in the soil ecosystem, and the soil environment surrounding the roots is designated as the rhizosphere. The word rhizosphere has been derived from Greek, meaning the influence of a root on its surrounding (Pinton and Varanini, 2001). It was first introduced by the German scientist Hiltner in 1904 to describe the interaction between microorganisms and legume plant roots. Now the term includes all plants and is a topic of fundamental importance in crop production. The interaction between microorganisms and plant roots may be positive, negative, or neutral. Hiltner observed that microorganisms were much higher in the soil surrounding plant roots than that remote from the root. He designated this zone of soil in which microorganisms were affected by plant root as the rhizosphere (Rovira and Davey, 1974). Further, Rovira and Davey (1974) reported that colonization of the rhizosphere by microorganisms varied from plant species to species (Table 4.1). In the past decades, several definitions of the rhizosphere have been presented. According to the Soil Science Society of America (2008), the rhizosphere can be defined as the zone of soil immediately adjacent to plant roots in which the kinds, numbers, or activities of microorganisms differ from that of bulk soil. Bowen and Rovira (1999) defined the rhizosphere as the soil adjacent to roots with a different physical, chemical, and biological environment from bulk soil. Hinsinger (1998) and Hinsinger et al. (2009) defined the rhizosphere as the volume of soil influenced by root activity. As stressed by Hinsinger et al. (2005) and Gregory (2006), depending on the activity that one considers (exudation of reactive compounds, respiration, uptake of more or less mobile nutrients and water), the radial extension of the rhizosphere can range from submicrometer to supra-centimeter scales. Mengel et al. (2001) defined that the part of the soil that is directly influenced by roots and is within 1–3 mm from the root surface is called the rhizosphere. Neergaard and Magid (2001) reported that the measured extension of ryegrass (*Lolium perenne* L.) rhizosphere is 1–3 mm, whereas other researchers estimate that it ranges from 2 to 5 mm (Dijkstra et al., 1987; Youssef et al., 1989; Yeates and Darrah, 1991; Badalucco et al., 1996). Pepper and Bezdicek (1990) reported that the rhizosphere can extend to 20 mm as a series of gradients of organic substrate, microorganisms, pH O_2, CO_2, and H_2O. According to Darrah (1993), the inner boundary of the rhizosphere is not better defined.

Hinsinger et al. (2005) and Gregory (2006) reported that the rhizosphere is different from bulk soil due to a range of physical, chemical, and biological processes that occur as a consequence of root growth, water and nutrient uptake, respiration, and rhizodeposition. According to Darrah (1993), the rhizosphere is a zone of soil surrounding the root that is affected by it but its size differs spatially and temporally depending on the factor considered, ranging from a fraction of

TABLE 4.1
Presence of Bacterial Colony in the
Rhizosphere of Major Food Crops

Crop Species	Rhizosphere (Million g⁻¹ Soil)	Bulk Soil (Million g⁻¹ Soil)
Barley	505	140
Corn	614	184
Wheat	710	120
Oats	1090	184
Red clover	3255	134
Flax	1015	184

Source: Compiled from Rovira, D. and Davey, C.B., Biology of the rhizosphere, in: *The Plant Root and Its Environment*, ed., Carson, E.W., pp. 153–204, University Press of Virginia, Charlottesville, VA, 1974.

a millimeter for microbial populations and immobile nutrients, to tens of millimeters for mobile nutrients and water, to several tens of millimeters for volatile compounds and gases released from roots. This means that the interface between the root and the soil is complex, frequently ill defined, and heterogeneous in space and time (Gregory, 2006).

Bowen and Rovira (1999) reported that the extent of the rhizosphere is not precisely defined because organisms vary in their sensitivity to soluble and volatile substances coming from the root. The term "rhizosphere effect" describes the enhanced microbial growth and population densities in the rhizosphere, due to increased soluble C and nutrients, compared with the surrounding bulk soil (Elliott et al., 1984; Lupwayi and Kennedy, 2007). Furthermore, the extent of the rhizosphere effect into the soil will depend on the diffusion away from the root of the many compounds released from the root, their diffusion characteristics, and the water status of the soil (Bowen and Rovira, 1999). The rhizosphere is considered as an environment enriched in organic matter and often more acidic than bulk soil (Seguin et al., 2004).

When considering mobile nutrients such as NO_3^-, the size of the rhizosphere may extend several millimeters from the root (Darrah, 1993). However, for the immobile nutrient such as P, the extension of the rhizosphere is often limited to less than 1 mm (Hubel and Beck, 1993). The spatial extension of the rhizosphere also varies with plant species (Hinsinger, 1998). This is associated with difference in plant species for their root system development and also indirectly through mycorrhizal symbiosis (Clarkson, 1985; Bolan, 1991).

Pinton and Varanini (2001) suggested that the soil layer surrounding the roots should be termed as the ectorhizosphere and the root layer colonized by microorganisms should be designated as the endorhizosphere. The two areas are separated by the root surface known as the rhizoplane (Figure 4.1). Bowen and Rovira (1999) also

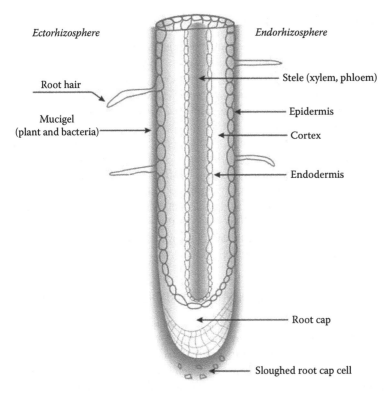

Ectorhizosphere   Endorhizosphere

Root hair

Mucigel
(plant and bacteria)

Stele (xylem, phloem)

Epidermis

Cortex

Endodermis

Root cap

Sloughed root cap cell

FIGURE 4.1   Root cross section showing ecto- and endorhizospheres. (From Fageria, N.K. and Stone, L.F., *J. Plant Nutr.*, 29, 1327, 2006.)

defined the rhizoplane as the actual root surface–soil interface. The rhizoplane or root surface also provides a highly favorable nutrient base for many species of bacteria and fungi, and these two zones together are often referred to as the soil–plant interface (Brimecombe et al., 2001). Brimecombe et al. (2001) defined endorhizosphere as the cell layers of the root itself and ectorhizosphere as the area surrounding the root. Kloepper et al. (1992) argued for the abolition of the term "endorhizosphere." Bowen and Rovira (1999) supported the argument of the Kloepper group and reported that organisms inside the root should be referred to as "internal root colonists."

The ability of roots to supply nutrients to the plant is influenced not only by their ability to absorb nutrients but also their ability to translocate the absorbed nutrient within root system and to the shoot (Atkinson, 1990). However, nutrient absorption by roots in adequate amount and proportion is the first step to meet plant nutrient requirements and higher productivity. The rhizosphere of plant roots ecologically links soil processes, plant growth, and microbial diversity and therefore plays diverse and essential roles in the earth's ecosystems (Luo et al., 2010). The objective of this chapter is to provide a comprehensive and updated review of the most recent advances in the oxidized rhizosphere and the physical, chemical, and biological changes and their association with nutrient availability.

4.2   EFFECT OF OXYGEN ON ROOT GROWTH

In the oxidized rhizosphere, plant growth is higher compared with the reduced rhizosphere. This may be associated with higher oxygen supply to plant roots. Plant roots grown in well-aerated soils are usually long, fibrous, and profusely branched with many root hairs and high root respiration (Huang and Nesmith, 1990). Bhattarai et al. (2005) reported that soybean, tomato, and cotton on a heavy clay soil consistently showed increased root weight, root length diameter, and soil respiration to oxygation. Soil respiration rate was almost doubled for soybean, and was proportionately even greater in cotton with oxygation. Oxygen deficiency not only reduced normal growth of roots but also induced formation of phytotoxic products of anoxic metabolism such as alcohols, Fe^{2+}, Mn^{2+}, and Al^{3+} and excessive sulfides and organic acids (acetic, propionic, butyric acid) and methane that inhibit root function and plant growth. Bhattarai et al. (2005) reported that in order to sustain root function, oxygen must be readily available to satisfy respiratory demand and to avoid building up of anaerobically induced phytotoxic products.

4.3   PHYSICAL, CHEMICAL, AND BIOLOGICAL CHANGES

Soil is a dynamic, three-phase (solid, liquids, and gas) system. Effective soil aeration ensures sufficient oxygen diffusion into the root zone for optimal crop production. The rhizosphere ecosystem is totally different than the rest of the bulk soil volume. In the rhizosphere, there are interactions between soil, plants, and microorganisms that bring physical, chemical, and biological changes. The overall change caused by the physical, chemical, and biological properties of the soil affects plant growth (Kang et al., 2011). Zhang et al. (2004) reported that managing the rhizosphere ecosystem and regulating the rhizosphere processes toward sustainable development may be an effective alternative approach to enhancing nutrient resource use efficiency and improving crop productivity in various cropping systems. Principal physical, chemical, and biological changes in the rhizosphere are synthesized in Figure 4.2, and a detailed discussion is given in the succeeding sections.

4.3.1   Physical Changes

Physical properties are the characteristics, processes, or reactions of a soil that are caused by physical forces and that can be described by or expressed in physical or chemical equations (Fageria et al., 2011). The soil physical properties most strongly influencing root growth and, consequently, nutrient uptake include temperature, water availability, and soil structure (Reisenauer, 1994). These physical changes are brought by the addition of organic matter through root residues and microorganisms associated with plant root systems. Further, these physical properties are interrelated and a change in one may cause a change in others that may be favorable or adverse. The soil physical properties largely determine rhizosphere extension due to their influence on root growth and transfer of ionic and molecular compounds (Nye, 1981; Hinsinger, 1998). Physical changes in the rhizosphere have been much less studied than chemical and biological changes, despite their potential consequences for

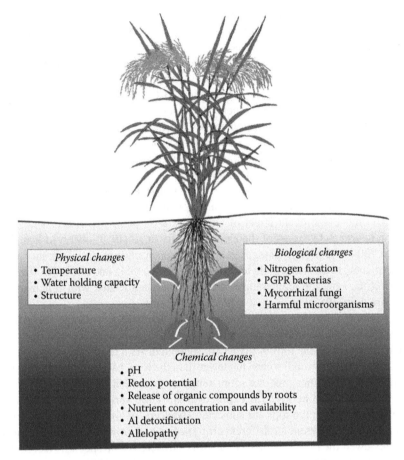

FIGURE 4.2   Major physical, chemical, and biological changes in the rhizosphere. (From Fageria, N.K. and Stone, L.F., *J. Plant Nutr.*, 29, 1327, 2006.)

the movement of water and solutes. Hence, limited data are available for physical changes in the rhizosphere.

4.3.1.1   Temperature

Soil temperature affects physical, chemical, and biological processes in the rhizosphere and nutrient availability. Lower as well as higher temperatures are detrimental for rhizosphere environment changes. The crops cultivated in the tropical climate such as corn grow well in the temperature range of 25°C–30°C, whereas, the crops in the temperate climate such as rye grow well in the temperature range of 12°C–18°C (Brady and Weil, 2002). Although there are exceptions, C_4 plants are generally more tolerant of high temperature than C_3 plants (Edwards et al., 1983). Many C_4 plants like sugarcane and corn are better able to grow under high temperatures than are C_3 plants such as wheat and barley (Fageria et al., 2011).

Root growth and microbial processes are also adversely affected by low as well as high temperatures. Temperature also influences the most important process, which is diffusive flow in the gaseous phase in any non-waterlogged soil. There is slow rate of O_2 depletion by roots and soil microorganisms at low temperature, compared to that at high temperature, when oxygen is less soluble (Bhattarai et al., 2005). When soil temperature is higher than 35°C and lower than 10°C, most of the biological processes are adversely affected. The optimum temperature for maximum production of root material for several species ranges from 20°C to 30°C (Voorhees et al., 1981).

The organic-matter-rich rhizosphere zone increases water content compared to bulk soil. The increase in water content is associated with a lowering of the thermal conductivity and an increase in the specific heat on a wet-weight basis. Further, higher organic matter content in the rhizosphere provides a buffer against sudden temperature fluctuations (Gupta et al., 1977).

Increased soil temperature favors absorption of iron and manganese; the effects on the Mn accumulation are relatively much greater than those on iron (Moraghan, 1985). The most significant effect of temperature on plant–Mn relationships is its effect on tolerance to excess of the element. Rufty et al. (1979) observed leaf tissue levels of 700–1200 mg kg^{-1} associated with the appearance of visual toxicity symptoms on plants grown at day/night temperature of 22°C/18°C, and 5000–8000 mg kg^{-1} at 30°C/26°C. Low soil temperatures are commonly associated with deficiencies of Mn (Moraghan, 1985).

Root temperatures are usually lower than air temperatures during the growing period, and variations in the temperatures of the root zone are less than that of the ambient air to which the tops of plants are subjected. As a result, the roots of plants have a temperature optimal somewhat lower than the tops, have become less adaptive to temperature extremes, and are more sensitive to sudden fluctuations. The roots of most plants would be killed if exposed to the same variations and durations of temperature to which the tops are subjected (Nielsen, 1974). Optimum root growth temperatures are different for different crop species and growing conditions. Nielsen (1974) reviewed the literature on root growth of different crop species. According to this author, barley root yields have been maximum at 6°C–13°C root temperature. In oats, optimum root temperature was 10°C–15°C. The optimum root temperature for wheat root growth was about 20°C, for corn 25°C–30°C; for cotton 28°C–30°C, for potato 10°C, for rice 25°C–30°C, and for dry bean about 28°C (Nielsen, 1974). Root zone temperature affects the morphology and distribution of roots. At optimum temperature, cell division is more rapid but of shorter duration than at lower temperatures. At cooler temperatures, roots are usually whiter, thicker in diameter, and less branched than at warmer temperatures (Nielsen, 1974).

4.3.1.2 Water-Holding Capacity

Plant roots add a substantial amount of organic matter to the rhizosphere after their decomposition. The increased organic matter content in the rhizosphere improves soil physical properties such as bulk density, porosity, and pore size distribution in favor of higher infiltration rate and water-holding capacity. The increase in the water retention capacity of soil with the addition of organic matter has been widely reported in the literature (McRae and Mehuys, 1985; Metzger and Yaron, 1987). Khaleel et al. (1981) analyzed experimental results obtained in a broad range of soils amended with various organic wastes. They noted that water retention (on weight basis) increases at both field capacity and wilting point when organic wastes are added and that relative increases in water retention capacity were greater for coarse-textured soils than for fine-textured soils. Soil organic matter also influences the biology of the soil by providing food to heterotrophic microorganisms. Root residues contain sugars, proteins, and starches that are quite readily metabolized by soil microbes. After their death these microorganisms improve soil physical and chemical properties in favor of better plant growth.

The release of root mucilage may also change the water relations of the rhizosphere. Passioura (1988) suggested that this might be particularly beneficial for water uptake at small values of matric potential (dry soils). Read and Gregory (1997) showed that mucilage reduced the surface tension of water and Read et al. (2003) confirmed that the addition of the surfactant component of mucilage can alter the relationship between water content and soil matric potential, making the soil drier at a given value of matric potential, especially at high matric potentials. Whalley et al. (2005) also found that the rhizosphere soil of corn and barley tended to be drier at a given matric potential than bulk soil but suggested that differences in wetting angle and pore connectivity were the likely explanation for these differences.

In comparison with bulk soil, root growth can alter the water release characteristic of the rhizosphere soil (Whalley et al., 2005) as well as its hydraulic conductivity (Hallett et al., 2003; Whalley et al., 2004). For a given matric potential, the rhizosphere soil is likely to be drier than bulk soil and has a smaller hydraulic conductivity (Gregory et al., 2010). This is not a general result, but it is the most frequent observation (Gregory, 2006). The explanation for these changes to soil hydraulic function in the rhizosphere has been related to changes in soil structure (Whalley et al., 2005) and differences in the contact angle between water and soil surface (Hallett et al., 2003; Read et al., 2003). Both these accounts show that different plant species can affect soil hydraulic properties differently (Gregory et al., 2010).

4.3.1.3 Structure

The binding of soil particles into aggregates results in structure. The size, arrangement, and stability of soil aggregates have a wide influence on soil physical and chemical process, root growth, and, consequently, nutrient uptakes. Soil aggregation is defined as a naturally occurring cluster or group of soil particles in which the forces holding the particles together are much stronger than forces between adjacent aggregates (Lynch and Bragg, 1985). The term soil structure and soil aggregation are often used synonymously, but the soil aggregates are the basic units of soil structure, rather than the whole. Soil aggregates are formed mainly by physical forces, while stabilizing is effected by several factors including organic materials, iron and aluminum oxides, and clay content (Lynch and Bragg, 1985). Organic matter is the main factor responsible for the water stability of soil aggregates. Aggregate stability is positively correlated with the soil organic carbon (Metzger and Yaron, 1987). When soil carbon is lower, soil compaction occurs, which adversely affects the chemical and biological processes in the soil. Soil compaction can be described as a loss of soil structure, resulting in an increase in bulk density and a decrease in porosity.

Plant roots have shown to increase the stability of surrounding aggregates (Tisdale and Oades, 1979; Reid and Gross, 1980; Young and Crawford, 2004). Roots change the soil structure by growing between aggregates and reshaping the spaces within the soil (Watt et al., 2006). Roots of lucerne are used to create biopores in deeper soil layers, which a subsequent crop can use (Cresswell and Kirkegaard, 1995; Davies and Peoples, 2003). The root system can be used to improve soil structure in the surface soil. Populous ryegrass roots and hairs can improve the structure of surface soil (Watt et al., 2006). Tisdall and Oades (1979) reported that the efficiency of ryegrass (*Lolium perenne* L.) in aggregate stabilizing was due to the large vesicular arbuscular mycorrhizal fungal population it supported, because length was related to aggregate stability, and they concluded that the organic materials that aggregate are decomposing roots and fungal hyphae. Roots themselves release large amounts of organic materials into soil, seen as slime layer around the root together with sloughed-off root cells and root hairs (Campbell and Porter, 1982). Grassland soils, generally, have a stable soil structure (Reid and Gross, 1980). This is because there is much greater root biomass and, hence, the associated rhizosphere microbial biomass produced under grass (Lynch, 1981). The root produces mucilage, and the associated microorganisms modify the products released by roots to produce their own mucilage, resulting in the formation of mucigel. These materials are both largely polysaccharide in nature and, hence, have the capacity to stabilize soil aggregates (Lynch and Bragg, 1985).

4.3.2   CHEMICAL CHANGES

Plant roots dramatically modify the chemical environment of the rhizosphere and thereby influence processes such as mineral weathering and the mobility of nutrients and pollutants (Kraemer et al., 2006). Several chemical changes take place in the rhizosphere due to plant roots and soil environmental interactions. Plants modify the chemical properties of their rhizosphere because of their metabolic activity, especially mineral nutrition. This activity induces many changes in ion equilibriums, the mobilization of mineral elements, and even the weathering of some primary minerals (Hinsinger, 1998). Among these changes, pH, oxidation potential, rhizodeposition, nutrient concentrations, and root exudates are prominent (Hinsinger et al., 2003; Fageria and Stone, 2006). These chemical changes in the rhizosphere significantly influence nutrient solubility and their uptake by plants. A detailed discussion of chemical changes in the rhizosphere is given by Marschner et al. (1986), Darrah (1993), Hinsinger (1998), and Hinsinger et al. (2009).

4.3.2.1   pH

Soil pH is one of the most important chemical properties that influences metal or nutrient solubility and, hence, their availability to plants. At lower pH (<5.5), the availability of most micronutrients is higher except Mo and decreases with increasing soil pH. This decrease is mostly associated with the absorption and precipitation process. Iron deficiency chlorosis is commonly associated with plants grown on high-pH calcareous soils (Chaney, 1988; Hansen et al., 2003). Solubility of Fe in the soil solution decreases 100-fold with each unit increase in pH and reaches the minimum between pH 7.5 and 8.5 (Rogovska et al., 2009). Similarly, Rogovska et al. (2009) reported that soybean grown on high-pH calcareous soils often suffers yield loss due to unavailability of Fe. Fageria et al. (2011) reported that upland rice suffered from iron deficiency when Oxisol pH was raised from 4.6 to 6.8. The Fe deficiency symptoms in upland rice when pH was raised were cultivar dependent. Cultivars that were bred under high acidity showed more severe Fe deficiency with increasing soil pH compared to cultivars that were bred in soils with liming. Availability of N as well as P is lower at lower pH and improves in a quadratic fashion with increasing pH until around 7.0. The increase in N availability is mainly associated with improved activity of N turnover bacteria. The availability of P is associated with the neutralizing of Al, Mn, and Fe compounds, which fix this element at lower soil pH (Fageria and Stone, 2006).

The rhizosphere pH may differ from the bulk soil pH by more than two units, depending on the form of N supply (NH_4^+, NO_3^-, symbiotic N_2 fixation), plant species, nutritional status of the plants, and pH buffering capacity of the soil (Marschner, 1991). The changes in the rhizosphere pH are associated with differences in cation/anion uptake and release of H^+ or OH^- (HCO_3^-) by plant roots. The release of protons or hydroxyl ions to neutralize the imbalance results in pH changes (Nye, 1981). When more cations are absorbed, H^+ is released by roots to keep ionic balance and pH decreases. Similarly, when more anions are absorbed, OH^- is released and pH increases. Nitrogen plays a prominent role in the cation–anion balance, because it is the nutrient that is taken up in maximum amount (some crops absorb N equal or lower than K) (Mengel et al., 2001; Fageria et al., 2011). In addition, it can be taken up as a cation (NH_4^+) or anion (NO_3^-), or even as an uncharged species (gaseous N_2 having neutral charge) in the case of N_2-fixing legume plants. Further, plants can directly use a significant proportion of N as amino acids that can either be positively charged, neutral, or negatively charged (Jones and Darrah, 1994; Hinsinger et al., 2003).

Nitrogen form is one of the main factors that affect cation or anion uptake and pH changes in the rhizosphere. When more N is available in the NH_4^+ form, H^+ is released and pH decreases. Similarly, when higher concentration of NO_3^- is present in the soil solution, its uptake releases OH^- ion and pH increases. Furthermore, nearly 70% of the cations or anions taken up by the plants are ammonium or nitrate (Van Beusichem et al., 1988). The NH_4^+/NO_3^- ion uptake can change the rhizosphere pH up to two units higher or lower compared with bulk soil (Mengel et al., 2001). Generally, in oxidized soils, NO_3^- is the dominant ion and under reduced soil conditions (flooded rice) NH_4^+ is the dominant ion.

Legumes are exceptions, which acidify the rhizosphere even after the supply of nitrate (Marschner and Romheld, 1983). Legumes fixing N_2 acidify their rhizosphere because of a relatively higher uptake of cations over anions (Aguilar and van Diest, 1981; Marschner and Romheld, 1983; Jarvis and Hatch, 1985; Marschner, 1995). However, some workers have reported that acidification of legume rhizosphere is associated with specific properties of legumes, which normally acidify the rhizosphere even under nitrate-fed conditions (Mraschner, 1995; Hinsinger, 1998). The pH also changes with the excretion of organic acids by roots and by microorganism activities in the rhizosphere. Further, the CO_2 produced by root and microorganism respiration can dissolve in soil solution and may form carbonic acid and lower the pH. Soil-buffering capacity (clay and organic matter content) and initial pH are the main parameters that determine changes in soil pH.

It is well known that acidification of the rhizosphere can solubilize several hardly soluble macronutrients (Riley and Braber, 1971) and micronutrients (Sarkar and Wyn Jones, 1982; Hinsinger and Gilkes, 1996). Root excretion of H^+ in the rhizosphere is an effective mechanism for improving uptake of micronutrients, except Mo (Fageria et al., 2002). Bar-Yosef et al. (1980) reported that root excretion of H^+ at root surface is an effective mechanism for enhancing Zn uptake compared to excretion of complexing agents. Enhanced reducing activity at root surfaces has been noted as root-induced responses to Fe deficiency in dicotyledonous and nongraminaceous monocotyledonous plants (Marschner, 1995).

When the rhizosphere pH is too low due to plant liberation of protons (H^+), it may have an adverse effect on uptake of some nutrients. For example, P is strongly sorbed on Fe and Al hydroxides in acid soils, the charge of these minerals being pH dependent. Further, too low pH can also induce Al toxicity and influence nutrient uptake and plant growth adversely. However, some studies have indicated that some plant species can increase the rhizosphere pH in acid conditions and decrease it in neutral or alkaline conditions, indicating the capacity of plants to adapt to adverse conditions (Youssef and Chino, 1988, 1989; Hinsinger, 1998). Whatever the origin of pH changes, modifications of 1–2 pH units have been commonly reported in the rhizosphere of various crop species (Riley and Barber, 1971; Marschner and Romheld, 1983; Hinsinger, 1998).

4.3.2.2 Oxidation–Reduction Potential

This is a chemical process in which electrons are given up or acquired. It is designated by Eh and measured in volts or millivolts. In this chemical process, the donor loses electrons and increases its oxidation number or is oxidized; the acceptor gains electrons and decreases its oxidation number or is reduced. Generally, there is a positive correlation between soil oxygen and Eh. In an oxidized soil, Eh ranges from +600 to +350 mV, whereas for most reduced (anaerobic) soils, the Eh varies from −300 to +350 mV (Masscheleyn et al., 1993). Root activity alters rhizosphere redox potential through respiratory oxygen consumption and ion uptake or exudation. In particular, root absorption and assimilation of NH_4^+ and NO_3^- consume 0.31 mol O_2 mol^{-1} NH_4^+ and 1.5 mol O_2 mol^{-1} NO_3^-, respectively (Bloom et al., 1992). Hence, when roots use NO_3^- as a nitrogen source, the rhizosphere redox potential declines more rapidly than when they use NH_4^+ (Bloom et al., 2003). The concentration of NH_4^+ and NO_3^- in the rhizosphere and rhizosphere redox potential may be partially responsible for the observed large fluctuations in the relative availability of soil NH_4^+ and NO_3^- and in root growth (Jackson and Bloom, 1990).

There are many other nutrient solubility or uptake processes in the rhizosphere which alter redox potential, such as redox reactions with various forms of Mn (Mn^{2+} and Mn^{4+}), Fe (Fe^{2+} and Fe^{3+}), and Cu (Cu^+ and Cu^{2+}) (Lindsay, 1979). However, the Fe and Mn redox reactions are considerably more important than Cu because of their higher concentrations in soil (Fageria et al., 2002). The primary source of electrons for biological redox reactions in soil is organic matter, but aeration, pH, and root and microbial activities also influence these reactions. Redox reactions in the rhizosphere can also be influenced by organic metabolites produced by roots and microorganisms.

The release of some acids like caffeic and malic from plant roots can also change redox potential of the rhizosphere (Romheld and Marschner, 1983). Bienfait et al. (1983) reported that reductant release in the rhizosphere by dry bean plants (*Phaseolus vulgaris* L.) contributed about 14% of the total amount of Fe reduced by roots of Fe-sufficient plants and less than 2% for Fe-deficient plants. It is also reported by Bienfait et al. (1983) and Romheld and Marschner (1983) that root-induced change in redox potential is likely to be restricted to the root–soil interface and not to extend into the rhizosphere.

4.3.2.3 Release of Organic Compounds

Plant roots not only absorb water and nutrients to support plant growth but also release into the rhizosphere organic and inorganic compounds. These compounds bring several chemical changes in the root environment, affecting microbial population and availability of nutrients (Neumann and Romheld, 2001). The release may occur as an active exudation, a passive leaking, the production of mucilage, or with the death and sloughing of root cells. Releases increase under a variety of conditions, particularly forms of abiotic and biotic stress (Marschner, 1995). Rovira et al. (1979) classified root-released organic compounds as follows: (1) exudates—compounds of low molecular weight that leak nonmetabolically from intact plant cells; (2) secretions—compounds metabolically released from active plant cells; (3) lysates—compounds released from the autolysis of older cells; (4) plant mucilages—polysaccharides from the root cap, root cap cells, primary cell wall, and other cells; and (5) mucigel—gelatinous material of plant and microbial origin.

The terms exudates and exudation are sometimes used collectively and perhaps incorrectly to include all of the organic compounds released from the roots and most if not all the mechanisms involved in the release of organic compounds (Pepper and Bezdicek, 1990). Roots release a wide range of organic compounds, although sugars and polysaccharides, organic and amino acids, peptides, and proteins constitute the bulk of the rhizodeposits (Bowen and Rovira, 1999). Much research has now demonstrated that other compounds released from roots may act as messengers that communicate and initiate root–root, root–microbe, and root–faunal interactions (Gregory, 2006). Plant roots can also release compounds that are able to disrupt communication between bacteria, thereby reducing their susceptibility to infection (Gregory, 2006). The important organic compounds released by roots are listed in Table 4.2. The major mechanisms of the release of these compounds are leakage and secretion. Leakage involves simple diffusion of these compounds due to the higher concentrations of compounds within the roots as compared to the soil (Pepper and Bezdicek, 1990). Secretion, on the other hand, requires metabolic energy because it occurs against concentration gradients. Sugars and amino acids provide energy for microorganisms in the rhizosphere that mineralize or solubilize many nutrients. Similarly, acids reduce the pH, and the availability of many micronutrients improves. Release of mucilages protects the root tips from injury, and desiccation also plays a role in nutrient uptake through its pH-dependent cation exchange

TABLE 4.2
Organic and Inorganic Compounds Released by the Roots in the Rhizosphere

Root Exudates	Compounds
Diffusitives	Sugars and polysaccharides (arabinose, fructose, galactose, glucose, maltose, mannose, oligosaccharides, ribose, sucrose, xylose)
	Organic acids (acetic, butyric, citric, oxalic, tartaric, succinic, propionic, malic, glycolic, benzoic)
	Amino acids (glutamine, glycine, serine, tryptophane, aspartic, cystine, cystathionine, α-alanine, β-alanine, γ-aminobutyric)
	Inorganic ions, oxygen
Secretives	Mucilage, protons, electrons, enzymes (amylase, invertase, peroxidase, phenolase, phosphatases, adenine, uridine/cytidine, nucleotides)
Excretives	CO_2, HCO_3, protons, electrons, ethylene
Root debris	Root cap cells, cell content

Source: Compiled from Neumann, G. and Romheld, V., The release of root exudates as affected by the plants physiological status, in *The Rhizosphere: Biochemistry and Organic Substances at the Soil-Plant Interface*, eds., Pinto, R., Varanini, Z., and Nannipieri, P., pp. 41–93, Marcel Dekker, New York, 2001; Uren, N.C., Types, amounts, and possible functions of compounds released into the rhizosphere by soil-grown plants, in *The Rhizosphere: Biochemistry and Organic Substances at the Soil-Plant Interface*, eds., Pinto, R., Varanini, Z., and Nannipieri, P., pp. 19–40, Marcel Dekker, New York, 2001; Bertin, C. et al., *Plant Soil*, 256, 67, 2003; Dakora, F.D. and Phillips, D.A., *Plant Soil*, 245, 35, 2002.

capacity (Jenny and Grossenbacher, 1963). The term mucilage is used to describe the range of carbohydrate-rich compounds and ionic compounds exuded by roots (Pojasok and Kay, 1990; Whalley et al., 2005). The quality and quantity of organic compound release is determined by plant species and genotypes within species, plant age, soil type, soil physical properties, and presence of microorganisms.

4.3.2.4 Nutrient Concentration and Availability

The nutrient mobility concept in the rhizosphere first proposed by Bray (1954) was a historic breakthrough for understanding nutrient uptake mechanisms in plants. According to this author, some nutrients move freely from bulk soil to the rhizosphere, such as N, and others do not move or move to a short distance, such as P. Based on this concept, the zone of competition for nutrients by roots differs depending upon the mobility of nutrients (Gregory, 2006). This change of thinking about the availability of nutrients to plants was paralleled by similar development regarding the movement of water toward roots (Gardner, 1960).

The nature and concentration of nutrients in the rhizosphere depend on soil type, soil fertility, and crop intensity. Similarly, nutrient absorption and utilization by plants are affected by many soil and plant factors and their interactions. The process of absorption of nutrients by plant roots changes ionic concentration in the rhizosphere. Much progress has been made in recent years in understanding nutrient concentrations and availability in the rhizosphere. With the release of organic compounds in the rhizosphere, concentration and availability of most nutrients increase (Claassen et al., 1986; Jianguo and Shuman, 1991). Some nutrients have larger or specific effect on their concentration and availability. Hence, availability of these nutrients is treated in this section.

4.3.2.4.1 Macronutrients

Macronutrients that are absorbed by the plants from the soil solution are nitrogen, phosphorus, potassium, calcium, magnesium, and sulfur. Their role in plant growth is already discussed in Chapter 3. However, their concentration and availability in the oxidized rhizosphere are discussed in the succeeding sections.

4.3.2.4.1.1 Nitrogen Symbiotic organisms such as rhizobia and, to a lesser extent, nonsymbiotic bacteria such as *Azospirillum* can both augment plant uptake of nitrogen via nitrogen fixation (Pepper and Bezdicek, 1990). Nonsymbiotic nitrogen fixation has also been well documented (Marschner, 1995; Mengel et al., 2001). Symbiotic N_2-fixing bacteria can fix a substantial amount of nitrogen, but free-living organisms must compete with all other rhizosphere organisms for substrate and, hence, are unlikely to fix large amounts of nitrogen. The quantity of symbiotic nitrogen fixation by legumes varies from species to species, and environmental conditions. However, contribution of symbiotic N_2 fixation by legumes of up to 300 kg ha^{-1} has been reported (Bezdicek et al., 1978). Free-living organisms can fix about 15–30 kg N ha^{-1} year^{-1} (Pepper and Bezdicek, 1990). Detailed discussion of nitrogen fixation by microorganisms is given in Section 4.3.3. In addition to supplying plants with fixed nitrogen, rhizosphere organisms greatly influence nitrogen cycle in the soil–plant system by mineralization, immobilization, nitrification, and denitrification.

Mengel et al. (1990) showed the ability of plants to promote the release of fixed, nonexchangeable ammonium in their rhizospheres. Scherer and Ahrens (1996) reported that depletion of NH_4^+ ion at the root surface contributes for the release of nonexchangeable NH_4^+ in the rhizosphere of ryegrass (*Lolium multiforum* L.) and red clover (*Trifolium pratense* L.). Furthermore, the mobilization of non-exchangeable NH_4^+-N in the soil–root interface may be increased by nitrifying and heterotrophic microorganisms with a higher activity in the rhizosphere, influencing the equilibrium between nonexchangeable NH_4^+ and NH_4^+ in the soil solution, thus favoring the release of NH_4^+ from interlayers of the clay minerals (Nommik and Vathras, 1982; Bottner et al., 1988; Scherer and Ahrens, 1996). According to Grinster et al. (1982) rape (*Brassica napus* L.) decreased the pH in the vicinity of the roots from 6.2 to 4.5. The relatively high replacing power of H^+ is presumably an important factor in mobilizing nonexchangeable NH_4^+, because H^+ leaves the crystal lattice in an expanded state and thus renders the nonexchangeable NH_4^+ more accessible to replacing cations. Jones et al. (2009) concluded that C and N flow in the rhizosphere is extremely complex, being highly plant and environmental dependent and varying both spatially and temporally along the root.

4.3.2.4.1.2 Phosphorus In almost all soils, P is often the most limiting mineral nutrient. Its bioavailability is low due to immobilization in the acid soils. It is also immobilized or fixed in calcareous soils due to high content of calcium.

Physical, chemical, and biological changes in the rhizosphere are associated with improved concentration of this element in the root vicinity and, consequently, uptake. Root exudates released in the rhizosphere contain phosphate radicals that will be subject to microbial modification. Furthermore, several microbial processes including mineralization, immobilization, and solubilization of inorganic phosphates also influence P availability to plants. Several studies have reported modification in concentration and uptake of P in the rhizosphere of crop plants. Jianguo and Shuman (1991) reported that in rice an uptake of P increased under low P conditions and this increase in P uptake was associated with decrease in the rhizosphere pH. Further, these authors reported that H^+ secretion by rice roots under low soil P conditions can be considered as a beneficial adaptation.

Under P deficiency, white lupin (*Lupinus albus* L.) increases the formation of cluster roots, release of H^+, and exudation of citrate (Dinkelaker et al., 1989; Lamont, 2003). For example, the enhanced release of H^+ from cluster roots acidified rhizosphere from pH 7.5 to 4.8 (Dinkelaker et al., 1989, 1995) and even as low as pH 3.6 in some cases (Li et al., 1997). Citrate concentration in root cluster rhizosphere can reach as high as $100\,\mu mol\ g^{-1}$ soil, and the amount of citrate exudation from roots represented up to 23% of the total plant dry weight (Dinkelaker et al., 1995; Shen et al., 2004). Large amounts of proton release and citrate exudation may facilitate P acquisition by white lupin from sparingly soluble P in soils (Marschner, 1995; Hinsinger, 1998, 2001).

There is overwhelming evidence now of the availability of unavailable soil P in some crop species (Jones, 1998). There are many plant-induced changes in the rhizosphere which make this phenomenon happen. This can include the manipulation of root morphology (hair length/density), enhanced root-to-shoot ratio, the extra provision of C for mycorrhizal exploitation of nonrhizosphere soil, the release of phosphatases to release organically bound soil P, and the release of organic acids and H^+ to solubilize inorganic P (Jones, 1998; Shen et al., 2004). Rape roots are capable of excreting large amounts of organic acids that contribute to the acidification and solubilization of soil P (Hoffland, 1992; Jones and Brassington, 1998; Neumann and Romheld, 1999). In contrast to rape, cereals such as wheat and barley do not show much acidification because anion–cation uptake is instantaneously compensated for by OH^- or HCO_3^- extrusion to maintain and balance the charge (Neumann and Romheld, 1999; Vong et al., 2004).

Enhanced secretion of acid phosphatases (APase) and phytases by plant roots and also by rhizosphere microorganisms under P-deficient conditions may contribute to P acquisition by hydrolysis of organic P esters in the rhizosphere (Neumann and Romheld, 2001). Some dicotyledonous plant roots, and especially nonmycorrihizal plants such as *Lupinas albus* and *Brassica napus* are capable of releasing large amounts of organic acids into the rhizosphere in response to P deficiency (Gerke, 1994; Jones, 1998). Mallic, citric, and tartaric appear to be the primary organic acids released by roots under P deficiency (Jones, 1998; Wang et al., 2000).

The depletion of soil solution phosphorus ions in the rhizosphere occurs as a consequence of uptake of P by roots (Hinsinger and Gilkes, 1996). This depletion gives rise to a replenishment of P from the solid phase (Morel and Hinsinger, 1999). However, the replenishment of depleted P depends on time and physicochemical conditions of the soil (Nye, 1981; Darrah, 1993; Hinsinger and Gilkes, 1996; Hinsinger, 1998). The P-deficiency conditions also induce the root exudation of phosphatases and RNAases in many plant species (Wasaki et al., 2003; Kochian et al., 2004). Because a significant fraction of soil P can be fixed in organic compounds, the activity of these enzymes could play roles in catalyzing the hydrolysis of organic P for uptake by roots (Kochian et al., 2004).

4.3.2.4.1.3 Potassium Potassium uptake is higher by crop plants and its availability to plants in sufficient amount is fundamental for higher productivity. It does not form any gases in the soil like N that could be lost to the atmosphere. Its fixation in the soil–plant system is also limited. It does not cause major environmental problems like N and P. The original source of K are primary minerals such as micas (biotite and muscovite), feldspar (orthoclase and microcline), and their weathering products. Availability of nonexchangeable K to plants has been reported to increase due to exchange reaction and mineral dissolution in the rhizosphere (Moritsuka et al., 2004). Increased exudation of sugars, organic acids, and amino acids has been detected in maize as a response to K limitation (Neuman and Romheld, 2001).

Hinsinger and Jaillard (1993) reported that release of interlayer K in phlogopite occurred in the rhizosphere of ryegrass (*Lolium perenne* L.) when the K concentration in the rhizosphere solution decreased below a threshold of about $80\,\mu M$ ($3.12\,mg\,kg^{-1}$). The release involves exchange of interlayer K by cation of high hydration energy and the consequent expansion of the interlayer space (Moritsuka et al., 2004). Similarly, Hinsinger et al. (1993) reported that dissolution of phlogopite structure occurred in the rhizosphere of brassica (*Brassica napus* L.) probably due to proton excretion by roots. The release of nonexchangeable K from feldspar has also contributed to exudation of acids like citric and oxalic (Song and Huang, 1988; Drever and Stillings, 1997; Wang et al., 2000). Other organic acids such as maleic, tartaric, succinic, formic, acetic, propionic, and butyric are also released by roots in the rhizosphere and can help in K dissolution from minerals (Moritsuka et al., 2004).

The release of nonexchangeable K from soil minerals requires very low concentration of K in the soil solution (Sparks, 1987; Fannings et al., 1989). The root-induced release of nonexchangeable K contributes up to 80% of the uptake of the plants in soils where the release of nonexchangeable K would have been expected to be negligible when considering the concentration of K in the bulk soil solution (Niebes et al., 1993; Hinsinger, 1998). Exchangeable Ca^{2+}, Mg^{2+}, and Na^+ are supplied to plant roots mainly by mass flow. Furthermore, due to their lower demand for plant growth than the supply by mass flow from the nonrhizosphere, these cations are often accumulated in the rhizosphere and adsorbed on the exchange sites of the solid phase (Yanai et al., 1996; Moritsuka et al., 2000, 2004). The adsorption of these cations on the exchange site in the rhizosphere may be an important process in the release of interlayer K (Mengel, 1985; Sparks, 1987). Moritsuka et al. (2004) reported that the release of nonexchangeable K in the rhizosphere by the adsorption of cations Ca^{2+}, Mg^{2+}, or Na^+ plays an important role for upland soils whose rhizosphere pH is usually above the range of intensive mineral dissolution as a result of the supply of nitrogen to plants mainly in the form of NO_3^-.

Cations such as Ca^{2+} and Mg^{2+} or Na^+ are supplied to plants mainly by mass flow. Due to their lower demand for plant growth than the supply by mass flow from nonrhizosphere, these cations are often accumulated in the rhizosphere and adsorbed on the exchange site of the soil solid phase (Yanai et al., 1996; Moritsuka et al., 2000). This adsorption of cations as Ca and Mg on the exchange site of the soil solid phase in the rhizosphere may be an important process in the release of interlayer K.

4.3.2.4.1.4   Calcium and Magnesium   Mass flow and diffusion theory predicts that Ca and Mg concentrations in the rhizosphere are in excess of plant needs (Barber, 1995). However, in tropical Oxisols and Ultisols, deficiency of Ca and Mg is frequently reported for plant growth and liming is an essential practice to raise the concentration of these two cations in the rhizosphere (Fageria and Baligar, 2003a; Fageria, 2009).

4.3.2.4.1.5   Sulfur   Sulfur deficiency is widely reported in many regions of the world (Scherer, 2001). The S deficiency is due to the use of high-analysis low S fertilizers, low S returns with organic manures, use of high-yielding cultivars that require high S, declining use of S-contaminated fungicides, and reduced atmospheric inputs caused by stricter emission regulations. Saito (2004) reported that S deficiency in agricultural areas in the world has been recently observed because emissions of S air pollutants in acid rain have been diminished from industrialized areas. An insufficient S supply can affect yield and quality of the crop caused by the S requirement for protein and enzyme synthesis as well it is a constituent of the amino acids methionine and cysteine (Scherer, 2001). It also takes part in the formation of vitamins and plant chlorophyll (Ceccotti, 1996). In addition to increasing crop yields and improving quality, it has been shown that on S-deficient soils application of S can considerably reduce the amount of N fertilizer required for a given level of production (Schnug et al., 1993).

Oxidation reactions in the rhizosphere environment of rape (*Brassica napus* L.) were found to be more intense than in bulk soil, due to the large presence of oxidizing bacteria (37%–70% of the total heterotrophic bacteria) harbored by the roots of this plant (Grayston and Germida, 1991). Protons released by oxidation reactions acidify the microenvironment around the roots and hydrolyze organic S into S-containing amino acids, which are then oxidized to produce sulfate S (Scherer, 2001). This indicates that root-derived C (including rhizodeposits) plays a crucial role in the cycling of nutrients such as sulfur (Vong et al., 2004). Deluca and Kenney (1994) reported that the immobilization of S was dependent on the quantity and quality of available C source rather than the total C.

Sulfate is the most common form of inorganic S and can be divided into SO_4^{2-} in soil solution, adsorbed SO_4^{2-}, and mineral sulfur (Barber, 1995). Generally, more than 95% of soil S is organic bonded and divided into sulfate ester S and carbon-bonded S. Although not readily available to plants, the large organic S fraction may potentially be an important source for the supply of S to plants in deficiency situations (Eriksen et al., 1998). The organic bonded S is liberated to plants with the mineralization process. In oxidized soils, S release is normal in the soil solution for plant uptake due to mineralization by the microorganisms or enzymes. In addition to sufficient oxygen supply, sulfatase activity is also associated with soil humus content and the activity decreases with depth in soil profile (Tabatabai and Bremner, 1970). Also soil temperature and soil moisture seem to play a regulatory role for the activity of arylsulfatase (Castellano and Dick, 1991). The release of sulfate from organic form also depends on C/S ratio. Sulfate is released from organic material when C/S ratio is less than 200 and is immobilized if C/S ratio is above 400. The C/S ratio between 200 and 400 can cause both mineralization and immobilization (Eriksen et al., 1998).

Once S is released into soil solution through mineralization, it can be adsorbed on soil colloids depending on pH of the equilibrium solution (Couto et al., 1979). The S adsorption is very strong at low soil pH and, therefore, negatively correlated with soil pH (Kparmwang et al., 1997). It becomes negligible at pH > 6.5 (Scherer, 2001). Soils may adsorb SO_4^{2-} by hydrous oxides of Fe and Al. Hence, S adsorption is higher in the Oxisols and Ultisols, which contain higher amounts of Fe and Al oxides. In acid soils, S adsorption can be minimized with the addition of lime. Application of lime and gypsum improves the uptake of SO_4^{2-} by plants. Sulfur can be applied through S-containing fertilizers to meet the crop demand for this element.

4.3.2.4.2   Micronutrients

Micronutrients that are essential for plant growth are zinc, iron, manganese, copper, boron, molybdenum, chlorine, and nickel. Their role in plant growth is already discussed in Chapter 3. However, their concentration changes in the oxidized rhizosphere and their availability to plants is discussed in the succeeding sections.

4.3.2.4.2.1   Iron   Iron deficiency is widespread in crop plants grown on high pH or calcareous soils. Plants have evolved different mechanisms for acquisition of iron from soils low in available iron. Plant acquisition of iron from low iron soils or adaptation to Fe-deficiency stress have been classified as strategy I and strategy II plants (Marschner, 1995). In dicotyledonous and nongraminaceous monocotyledonous plants, solubilization of Fe^{3+} is usually mediated by rhizosphere acidification, by complexation with chelating compounds, and by reduction to Fe^{2+}, which is taken up by the roots probably by a specific transporter for Fe^{2+} (Neumann and Romheld, 2001). These root responses are generally confined to subapical root zones and associated with distinct changes in root morphology, such as thickening of the root tips and formation of rhizodermal transfer cells. Rhizosphere acidification in response to Fe deficiency is most probably mediated by activation of the plasmalemma H^+-ATPase (Neumann and Romheld, 2001). These plants are known as strategy I plants.

Graminaceous species acquire iron from low iron growth medium by a mechanism known as strategy II. In response to Fe deficiency, graminaceous plants are able to release considerable amounts of nonprotein amino acids known as phytosiderophores (Neumann and Romheld, 2001). Phytosiderophores have been defined by Takagi et al. (1984) as a group of root exudates exhibiting strong complexing properties with respect to ferric iron and identified as nonproteinogenic amino acids, such as mugineic acid and its derivatives. In this respect they are analogues of microbial siderophores (Hinsinger, 1998). The synthesis and release of phytosiderophores in the rhizosphere are stimulated by Fe deficiency (Romheld, 1991).

The phytosiderophores are highly effective chelators for Fe^{3+} form of iron. The release of these phytosiderophores takes place predominantly in subapical root zone and these chelates are stable even at high soil pH (>7.0) (Neumann and Romheld, 2001). However, a major limitation of the efficiency of phytosiderophores is their degradation by rhizosphere microorganisms (Takagi et al., 1988; van Wiren et al., 1993). Takagi et al. (1984) reported that the release of phytosiderophores is a rhythmic phenomenon restricted to a period of 2–8 h after the onset of daylight. Tolerance to different graminaceous species to Fe deficiency is related to the amount of phytosiderophores released. The major graminaceous species are rated in the order of barley > wheat > oat > rye > corn > sorghum > rice (Romheld and Marschner, 1990; Neumann and Romheld, 2001). However, there is considerable difference among genotypes of same species.

In the rhizosphere, microorganisms utilize either organic acids or phytosiderophores to transport iron or produce their own low-molecular-weight metal chelators, called siderophores (Crowley, 2001). Since siderophores are produced only in response to iron deficiency, which is a function of crop relative growth rate, its mineral nutritional status and on soil chemical properties. Almost all microorganisms have been found to produce siderophores and can potentially compete with each other for iron, depending on their ability to utilize different siderophore types or based on the uptake kinetics of their siderophore transport systems (Crowley, 2001).

4.3.2.4.2.2 Zinc, Copper, and Manganese Soil pH is a major chemical property influencing solubility and availability of zinc, copper, and manganese. An increase in soil pH increases the adsorption of these micronutrients onto soil particles and thereby decreases the availability (Fageria et al., 2002). The availability of micronutrients to plants also depends on the capacity of different plant species to mobilize microelements via a range of rhizosphere processes. The major root-induced processes are (1) reduction in rhizosphere pH or decrease in redox potential and consequent dissolution of elements from soil exchange complex, (2) exudation of organic ligands and consequently complexation of metals in the soil solution, and (3) elemental uptake by roots and consequent desorption from the soil exchange complex (Hinsinger, 2001; Loosemore et al., 2004). However, all these changes vary in intensity and dynamics with both plant species and environmental conditions. Loosemore et al. (2004) reported increase in the bioavailability of zinc by tobacco genotypes due to decreased pH in the rhizosphere. Mobilization of Zn, Cu, and Mn in the rhizosphere has been reported due to rhizosphere acidification and to complexation with organic acids (citrate) in root exudates of various plant species (Neumann and Romheld, 2001).

The particular oxidation states of the metal micronutrients significantly control the biological activity and ultimately the availability of the element in living systems. An important consequence of redox state changes for Mn is the dramatic changes in solubility of this element at root surface. The reduction process or electron transfer (Mn^{4+} to Mn^{2+}) improves uptake of this element by plants. The reduction of Mn^{4+} to Mn^{2+} in the rhizosphere can occur through microbial activities, and is important in Mn supply to plants. Similar to Zn^{2+} it is assumed that the primary Mn^{2+} species absorbed by plants is the free metal (Fageria and Stone, 2006). Chaignon et al. (2002) showed that Fe deficiency resulted in enhanced release of phytosiderophores (complexing root exudates) by two genotypes of wheat. As a consequence of this, an enhanced mobilization of soil Zn was found in Zn-efficient genotypes, but not in Zn-inefficient genotypes (Loosemore et al., 2004).

4.3.2.4.2.3 Boron and Molybdenum Boron and molybdenum are required by crop plants in minimum amounts compared with other micronutrients (Fageria et al., 2002). Uptake of these two micronutrients is controlled by rhizosphere pH. As soil pH increases, Mo uptake increases but uptake of B decreases. Barber (1974) reported that uptake of B in the shoot of soybean decreased linearly when rhizosphere pH was increased from 5 to 7.5. The decrease in B uptake was also influenced by supply of NH_4^+ or NO_3^- forms of nitrogen. Boron uptake in the shoot of soybean was lower when N was supplied as NO_3^- compared to N supplied as NH_4^+. This may be due to a difference in pH change due to two forms of N.

4.3.2.4.2.3.1 Aluminum Detoxification: Aluminum toxicity is widespread in tropical as well as temperate acid soils. Aluminum inhibits root development and, consequently, nutrient water uptakes. When the soil pH drops below 5.0, Al^{3+} is solubilized into the soil solution, and this is the most important rhizotoxic Al species (Kochian et al., 2004). In addition, intense solubilization of mononuclear Al species strongly limits root growth by multiple cytotoxic effects mainly on root meristems (Neumann and Romheld, 2001). Aluminum toxicity significantly decreases crop productivity through inhibition of crop growth, nutrient uptake and utilization, and microbial activity (Yang et al., 2004). Organic acids released by plant roots in the rhizosphere can detoxify Al and reduce its toxicity. Dinkelaker et al. (1993) reported that root-induced exudates make complexation or chelate Al^{3+} in the rhizosphere and detoxify it or prevent its entry in the roots. Similarly, Neumann and Romheld (2001) reported that there is increasing evidence that Al complexation with carboxylates released in apical root zones in response to elevated external Al concentration is a widespread mechanism for Al exclusion in many plant species. The Al carboxylate complexes are less toxic than free ionic Al species and are not taken up by plant roots (Neumann and Romheld, 2001).

Hue et al. (1986) reported that the most efficient Al detoxifying acids are citric and oxalic. However, tartaric, malic, and malonic acids can also detoxify Al but to a lesser extent. The potential role of organic acid release in Al tolerance by crop species is discussed by Miyasaka et al. (1991) and Fuente-Martinez and Herrera-Estrella (1999). Miyasaka et al. (1991) reported that the root system of an Al-tolerant snap bean cultivar grown in Al-containing solutions released 70 times as much citrate as in the absence of Al and released 10 times as much citrate as an Al-sensitive cultivar grown in the presence of Al. Also, other constituents of root exudates have been implicated in the binding of Al and, thus, in Al detoxification in the rhizosphere. Examples are the release of Al-binding polypeptides and of phosphate anions in wheat. A high binding capacity for Al has also been demonstrated for mucilage released preferably in apical root zone (Neumann and Romheld, 2001). In several species, including snap bean, wheat, and buckwheat, Al resistance is correlated with the Al-induced secretion of citrate, oxalate, or malate (Ma et al., 1997; Li et al., 2002). Genetic analysis has also indicated that Al tolerance genes were multigenic and linked to several chromosome arms (Aniol, 1990).

4.3.2.4.2.3.2   Allelopathy: Originally, allelopathy was defined as the biochemical interactions between plants of all kinds including the microorganisms that are typically placed in the plant kingdom. Since then, the term has undergone several changes over time, and it is now defined as any direct or indirect harmful or beneficial effect by one plant on another through the production of chemical compounds that escape into the environment (Rice, 1974). The International Allelopathy Society (IAS) has defined allelopathy as any process involving secondary metabolites produced by plants, microorganisms, and viruses that influence the growth and development of agricultural and biological systems (Kruidhof, 2008). Bertin et al. (2003) defined allelopathy as the chemical interactions that occur among plants and are mediated by the release of allelochemicals into the rhizosphere. The organic compounds involved in allelopathy are collectively called allelochemicals (Olofsdotter, 2001) and chemistry of these various allelochemical compounds in terrestrial ecosystems has been extensively reported (Rice, 1979; Fageria and Baligar, 2003b). These allelochemicals include simple phenolic acids, aliphatic acids, coumarins, terpenoids, lactones, tannins, flavonoids, alkaloids, cyanogenic glycosides, and glucosinolates (Fageria and Baligar, 2003b). Phenolic acids have been identified in allelophathic rice germplasm (Rimando et al., 2001) and have previously been described as allelochemicals Inderjit (1996). Most are secondary metabolites released into the environment by leaching, volatilization, or exudation from shoots and roots. Many compounds are degradation products released during the decomposition of dead tissues.

Once these chemicals are released into the immediate environment, they must accumulate in sufficient quantity to affect plants, persist for some period of time, or be constantly released in order to have lasting effects (Putnam and Duke, 1978). Abiotic (physical and chemical) and biotic (microbial) factors can influence the phytotoxicity of chemicals in terms of quality and quantity required to cause injury (Inderjit, 2001). After entering the soil, allelochemicals encounter millions of soil microbes. The accumulation of chemicals at phytotoxic levels and their fate and persistence in soil are important determining factors for allelochemical interference. After entry into soil, all chemicals undergo processes such as retention, transport, and transformation, which, in turn, influence their phytotoxic levels (Cheng, 1995).

The production of allelochemicals may be considered as a chemical defense mechanism that is under both genetic and environmental control (Sene et al., 2001). For instance, herbivore attacks, insect damage, and reduced soil fertility will generally increase the synthesis of these compounds by the plants (Fageria and Baligar, 2003b). This has been partly explained by the balance between carbon (C) and nutrient availability (Bryant et al., 1983). In particular, N deficiency has been shown to strongly affect the synthesis of polyphenols (Koricheva et al., 1998), because it will affect growth more than photosynthesis, and thus N deficiency allows more carbohydrates to be available for phenolic synthesis.

In Brazilian Oxisols, upland rice yield under monoculture is significantly reduced after 2 or 3 years of consecutive planting on the same area and such reduction in yield is attributed to auto-allelopathy (Fageria and Baligar, 2003b). Similarly, experiments with rice in the Philippines have shown the residual effects of allelochemicals on the reduction of yield of subsequent rice crops (Olofsdotter, 2001). Fageria and Souza (1995) reported

TABLE 4.3

Response of Upland Rice to Chemical Fertilizers and Green Manure on Brazilian Oxisol

Fertility Level	Grain Yield (kg ha⁻¹)	Grain Yield (kg ha⁻¹)	Grain Yield (kg ha⁻¹)
	First crop	Second crop	Third crop
Low	2188	2383	480
Medium	2428	2795	1242
High	2330	2657	1324

Source: Fageria, N.K. and Souza, N.P., *Pesq. Agropec. Brás.*, 30, 359, 1995.

yield reduction of upland rice in the third year when grown in rotation with dry bean on a Brazilian Oxisol (Table 4.3). Allelochemical's adverse effects in the rhizosphere can be reduced by adopting appropriate soil and crop management practices. A detailed discussion of these management practices is given by Fageria and Baligar (2003b).

4.3.2.4.2.3.3 Greenhouse Gas Emission: Many greenhouse gases (GHGs) are produced in the oxidized and reduced rhizosphere due to chemical and biological changes and are released in the atmosphere. In addition, the release of GHGs in the atmosphere is also related to deforestation, biomass burning, and other land use changes. Figure 4.3 shows a degraded pasture in the state of Tocantins, north of Brazil. The land area occupied by degraded *Brachiaria* pastures in the tropical region of Brazil is estimated to be more than 25 million ha, an area greater than

FIGURE 4.3 Degraded pastures in the state of Tocantins, which is part of the Amazon forest in the north part of Brazil.

FIGURE 4.4   Burned vegetation with termite heaps in the state of Goiás, in the central part of Brazil.

Great Britain (Oliveira et al., 2004). Similarly, Figures 4.4 through 4.6 show burning of vegetation and forest in the state of Goiás and the Amazon forest of Brazil. Follett (2010) reported that land use practices such as cultivation, livestock grazing (degraded pasture), manure management, and fertilization have strongly contributed to the release of GHGs. Principal GHGs are carbon dioxide (CO_2), methane (CH_4), and nitrous oxide (N_2O). These gases' capacity to trap solar radiation within the earth's atmosphere, which is similar to that of the glass in a greenhouse, leads to their being called, collectively, greenhouse gases (Lal et al., 1998). The enrichment

FIGURE 4.5   Burned pasture in the state of Goiás, in the central part of Brazil.

FIGURE 4.6   Burned forest in the Amazon region of Brazil.

of atmosphere by GHGs can cause the greenhouse effect of creating a substantial increase in the mean global temperature. This change may bring significant changes in plant growth and shift in ecological zones.

The global warming potential (GWP) of N_2O is 296 times and CH_4 is 23 times that of CO_2 (Houghton et al., 2001; Forster et al., 2007; Johnson et al., 2010). The GWP of a GHG is defined as the ratio of the time-integrated radiative forcing from instantaneous O (Houghton et al., 2001). The atmospheric concentration of N_2O, like carbon dioxide, has been rising since the Industrial Revolution due to human activities. The concentration of N_2O in the atmosphere, estimated at 2.68×10^{-2} mL L^{-1} around 1750, has increased by about 17% as a result of human alterations of the global N cycle (Intergovernmental Panel on Climate Change, 2001). Global annual N_2O emissions from agricultural soils have been estimated to range between 1.9 and 4.2 Tg N, with about half arising from anthropogenic sources (Intergovernmental Panel on Climate Change, 2001). In addition, N_2O plays a role in the destruction of the earth's protective ozone layer. Nitrous oxide is responsible for about 6% of the current greenhouse effect (Intergovernmental Panel on Climate Change, 2007). Approximately, 70% of the anthropogenic N_2O emissions in the world is derived from agricultural activities (Kroeze et al., 1999; Guo et al., 2010; Lipps, 2010).

Nitrification and denitrification are the primary soil processes responsible for a majority of N_2O emissions from soils (Guo et al., 2010). These two processes are linked because denitrification is dependent on the products of nitrification (NO_2^- and NO_3^-) (Guo et al., 2010). Soil water content plays an important role in N_2O emissions because it influences the O_2 content of the soil. Nitrous oxide emissions generally increase with increasing soil water content, especially when the water-filled-pore space is >70% (Skiba and Ball, 2002; Drury et al., 2003; Khalil and Baggs, 2005). However, nitrification may play a significant role in N_2O emissions when the water-filled pore space is <70% (Stevens et al., 1997; Wolf and Russow, 2000; Adviento-Borbe et al., 2006; Guo et al., 2010).

Nitrous oxide is formed during enzymatic nitrification and denitrification processes and chemo-denitrification, with denitrification generally considered a major source of N_2O from agricultural soils (Conard, 1995; Venterea and Rolston, 2000; Follett, 2001). Legumes contribute to N_2O emissions because the atmospheric N_2 fixed by legume during *Rhizobium* symbiosis is nitrified and denitrified like fertilizer N. Also rhizobia in root nodules are capable of denitrification as well as N_2 fixation (Lupwayi and Kennedy, 2007). However, Lupwayi and Kennedy (2007) reported that legume crops are grown with little or no N fertilizer (soybean); emissions of N_2O are expected to be less in a legume crop than in a fertilized cereal crop even though legume root exudates may result in increased gas emissions. McSwiney and Robertson (2005) and Halvorson et al. (2008) reported that N fertilizer in excess of plant needs is more likely to result in N_2O flux. Agriculture contributes approximately 78% of the total N_2O emissions in the United States (USEPA, 2008). Similarly, Wagner-Riddle et al. (2007) noted the importance of matching N application (rate and timing) to crop needs as a strategy to reduce N_2O emission. Split application of N fertilizer significantly reduced N_2O emission in potato production (Burton et al., 2008; Hyatt et al., 2010).

Soil and crop management practices (e.g., crop type, fallow frequency, residue management, soil amendments, cover crops, rotations, tillage, irrigation, drainage, mulching, and fertilization) can play a major role in regulating GHG emissions and need to be optimized for minimizing GHG emissions (Collins et al., 1999; Doran et al., 1999; Johnson et al., 2010). Management of N fertilizer is one of the most critical factors affecting N_2O emissions (Mosier et al., 1998). Anhydrous ammonia and urea are two of the most commonly used fertilizers in the United States and the world (Venterea et al., 2010). Venterea et al. (2010) reported that N_2O emissions from corn fertilized with anhydrous ammonia were twice the emissions with urea fertilization. There has been less use of anhydrous ammonia compared to urea in the United States and other countries in the last few decades (Venterea et al., 2010). This trend of excessive use of urea compared to anhydrous ammonia may result in decrease in N_2O emission from the cropping systems worldwide.

As the atmospheric CO_2 concentration continues to rise at the rate of $1.7 \mu mol^{-1}$ year^{-1} (Tans, 2009), there is strong scientific interest in finding ways to slow or reverse this trend (Halpern et al., 2010). Adopting agricultural practices that increase the amount of C in the soil, thereby diverting it from the atmosphere, is a recognized mitigation strategy (Lal, 2008). Comprising >60% of the terrestrial C pool, soil organic matter is one of the principal factors regulating the global C cycle (Batjes, 1996; Bornemann et al., 2010). An important portion of GHGs' balance is the storage and release of C from soils under grasslands (MacDonald et al., 2010). Perennial grassland systems, whether natural prairies, pastures, or forage fields contain up to 1.5 times more C than annual production systems located on the same soil, depending on the land management history (Cambardella and Elliott, 1993; Chantigny et al., 1997; Six et al., 1998). There is, therefore, interest in converting current arable land to grassland systems due to their potential to sequester C (Soussana et al., 2004).

4.3.3 Biological Changes

Important biological changes associated with the rhizosphere are root colonization by many types of beneficial and harmful microorganisms (Bowen and Rovira,
1976). The plant beneficial microorganisms include dinitrogen (N_2)-fixing bacteria,
free-living nitrogen-fixing bacteria, plant-growth-promoting rhizobacteria, saprophytic microorganisms, bicontrol agents, and mycorrhizae fungi. The presence of
some deleterious rhizobacteria is also common and these bacteria may affect plant
growth adversely. Most of the soilborne pathogens are adapted to grow and survive
in bulk soil, but the rhizosphere is the playground and infection court where the
pathogen establishes a parasitic relationship with plants (Raaijmakers et al., 2009).
Raaijmakers et al. (2009) and Hartmann et al. (2009) reported that the rhizosphere
is a hot spot of microbial interactions as exudates released by plant roots are a main
food source for microorganisms and a driving force of their population density and
activity.

The plant rhizosphere is a dynamic environment in which many parameters may
influence the population structure, diversity, and activity of the microbial community. Two important factors determining the structure of microbial community present in the vicinity of plant roots are plant species and soil type (Garbeva et al., 2008).
Soil microorganisms play important roles in soil quality and plant productivity (Hill
et al., 2000). The microbial biomass in soil constitutes a pool of nutrients that have
a rapid turnover when compared with organic matter (Baath and Anderson, 2003).
Therefore, quantitative and qualitative changes in the composition of soil microbial
communities may serve as an important and sensitive indicator of both short- and
long-term changes in soil health (Hill et al., 2000). Soil microbial communities may
be strongly influenced by agricultural practices that change the soil environment
(Lundquist et al., 1999; Wang et al., 2008).

4.3.3.1 Dinitrogen Fixation

Dinitrogen (N_2) fixation is defined as the conversion of molecular nitrogen (N_2) to
ammonia and subsequently to organic nitrogen utilization in biological processes
(Soil Science Society of America, 2008). The N_2 fixation is considered one of the most
significant biological phenomena in food and fiber production for mankind. The total
terrestrial biological nitrogen fixation quantity is reported to be 175×10^6 Mg per year
as compared with 77×10^6 Mg per year produced by fertilizer industries (Brady and
Weil, 2002). The quantity of N_2 fixed by legumes depends on species and environmental conditions. Grain legumes in symbiosis with rhizobia fix up to 450 kg N_2 ha^{-1}
(Unkovich and Pate, 2000). Crop species that fix low amount of N_2 are chickpea
(0–141 kg ha^{-1}), dry bean (0–165 kg ha^{-1}), and lentil (5–191 kg ha^{-1}) (Lupwayi and
Kennedy, 2007). A higher amount of N_2 is fixed by lupin (19–327 kg ha^{-1}), faba
bean (12–330 kg ha^{-1}), and soybean (12–330 kg ha^{-1}) (Lupwayi and Kennedy, 2007).
Soybean, a legume planted on nearly 30 million ha annually in the United States, can
fulfill most of its nitrogen requirements via biological nitrogen fixation. Similarly,
Brazil is the second largest soybean-producing country in the world after the United
States, and in Brazil farmers do not apply chemical N fertilizers in soybean. Most of
the soybean N needs are met by inoculation with appropriate rhizobia.

The N_2 fixation is carried out by certain types of bacteria in association with plant roots. Microorganisms responsible for biological nitrogen fixation live symbiotically in the roots of higher plants and also in association with roots as well as free-living organisms in the rhizosphere. A major portion of biological nitrogen is fixed by species of *Rhizobium* and *Bradyrhizobium* living in symbiosis with legume roots. The genus *Rhizobium* contains fast-growing, acid-producing bacteria, while the genus *Bradyrhizobium* has slow-growing nonacid-producing bacteria (Marschner, 1995; Brady and Weil, 2002). These bacteria form root nodules that serve as site of N_2 fixation. In this symbiotic association, the plant provides the bacteria carbohydrates for energy and bacteria provide the plant fixed amonical nitrogen. Nodulation is host specific and a given *Rhizobium* or *Bradyrhizobium* species can infect certain plant species but not all. Table 4.4 provides names of genus/species that infect legume roots for nitrogen fixation. Before *Rhizobium* can form nodules in legume roots to fix N_2, the bacterium must be attracted to the roots through bidireaction host–bacterium communication (Lupwayi and Kennedy, 2007). Legume root exudates contain chemical compounds, including flavonoids, that attract rhizobia to roots (Dakora and Phillips, 2002). Nodulation and nitrogen fixation are complex processes and influenced by climatic, soil, and plant factors. Jackson et al. (2008) reported that temperature and P are the two most important factors affecting dinitrogen fixation in plants. Low as well as high temperature and low level of P in the soil decrease N_2 fixation.

Some nonlegume plant species also develop nodules and fix atmospheric nitrogen. The important angiosperms belong to the genus *Alnus*, *Ceanothus*, *Myrica*,

TABLE 4.4
Nitrogen-Fixing Bacteria and Their Host Plants

Genus/Species	Host Plants
Rhizobium leguminosarum	
biovar viceae	*Vicia* (vetch), *Pisum* (peas), *Lens* (lentils), *Lathyrus* (sweet pea)
biovar phaseoli	*Phaseolus* (common bean, runner bean)
biovar trifolii	*Trifolium* spp. (most clovers)
Rhizobium meliloti	Melilotus (sweet clover), Medicago (alfalfa), *Trigonella* (fenugreek)
Rhizobium loti	*Lotus* (trefoils), *Lupinus* (lupins), *Cicer* (chickpea), *Leucaena* (leucena)
Rhizobium fredii	*Glycine* (soybean)
Bradyrhizobium japonicum	*Glycine* (soybean)
Bradyrhizobium	*Lupinus* (lupins)
Bradyrhizobium arachis	*Arachis* (peanut)
Bradyrhizobium	*Vigna* (cowpea), *Cajanus* (pigeon pea), *Pueraria* (kudzu), *Crotolaria* (crotolaria)

Sources: Marschner, H., *Mineral Nutrition of Higher Plants*, 2nd edn., Academic Press, New York, 1995; Fageria, N.K. et al., *Growth and Mineral Nutrition of Field Crop*, 3rd edn., CRC Press, Boca Raton, FL, 2011; Brady, N.C. and Weil, R.R., *Nature and Properties of Soils*, 13th edn., Prentice Hall, Upper Saddle River, NJ, 2002.

Casuarina, Elaeagnus, and *Coriaria.* The roots of these angiosperms that are present in forests and wetlands are infected by actinomycetes of the genus *Frankia* (Brady and Weil, 2002). On the global level, nitrogen fixation by these angiosperms exceeds that fixed by agricultural legumes (Brady and Weil, 2002). Nitrogen fixation mechanisms by *Rhizobium* have been extensively reviewed by Denarie et al. (1996), Broughton and Perret (1999), Albrecht et al. (1999), and Lupwayi and Kennedy (2007).

Biological nitrogen fixation also changes rhizosphere pH. When soybean is infected with an effective *Bradyrhizobium japonicum* strain, symbiotic N_2 fixation results in H^+ release into the rhizosphere and a reducing environment in the root nodules (Gibson, 1980). When fertilizer N (NO_3^-) is added, nodulation and N_2 fixation are reduced (Gibson and Harper, 1985; Herridge and Brockwell, 1988; Wiersma and Orf, 1992) and rhizosphere pH and (OH^-) increase (Assmakopoulou, 2006; Zhao and Ling, 2007; Wiersma, 2010).

4.3.3.2 Nitrogen Fixation by Free-Living and Root-Associated Bacteria

Some bacteria in association with roots without nodule formation and also having free-living habit in the rhizosphere, fix atmospheric nitrogen. Asymbiotic nitrogen-fixing bacteria have been found in the rhizosphere of many plants in both tropical and temperate regions (Haynes and Goh, 1978). These bacteria release the fixed nitrogen into the soil upon microbial decomposition; hence, the soil factors affecting mineralization affect the forms of nitrogen released. One example of nonnodule nitrogen fixation by cyanobacteria is *Azolla-Anabaena.* These bacteria inhabit the leaves of the floating fern *Azolla* and flourish well in association with flooded rice in tropical and subtropical regions and fix a significant amount of nitrogen. Furthermore, in the rhizosphere of certain grasses and other nonlegumes, some bacteria such as *Azospirillum* (synonym *Spirillum*) and *Azotobacter* also fix atmospheric nitrogen. Plant root exudates provide energy to these microorganisms for their survival. Following the first reports of an *Azospirillium*–plant association (Dobereiner and Day, 1976), there has been considerable interest in the use of *Azospirillum* species as nitrogen-fixing bacteria in association with the roots of monocotyledonous plants, particularly grasses and grain crops. The original thesis was that *Azospirillum* invades cortical cells, forming a primitive symbiotic association that was suspected of fixing N_2 (Dobereiner and Day, 1976). *Azospirillum* is ubiquitous in the rhizosphere of grasses in the field. Usually two species are found: *Azospirillum lipoferum,* most commonly as a surface sterile isolate from the roots of C_4 grasses, and *Azospirillum brasilense* from C_3 grasses and sugarcane (C_4) in tropical and subtropical regions. Despite the fact that *Azospirillum* spp. are known to fix N_2, the quantity supplied to the plant is only of minor significance and nowhere nearly meets its N requirements.

In addition, some *Azotobacter* (in temperate zones) and *Beijerinckia* (in tropical zones) also fix nitrogen nonsymbiotically. Certain aerobic bacteria of the genus *Clostridium* are also able to fix nitrogen. These bacteria are heterotrophs and receive energy from plant residues available in the rhizosphere. Some autotrophs bacteria such as photosynthetic bacteria (*Klebsiella pneumoniae*) and cyanobacteria also fix nitrogen. In the presence of light, these organisms are able to fix carbon dioxide and nitrogen simultaneously (Brady and Weil, 2002).

4.3.3.3 Plant-Growth-Promoting Rhizobacteria

Colonization of the rhizosphere by microorganisms results in modifications in plant growth and development. Kloepper and Schroth (1978) introduced the term "plant growth–promoting rhizobacteria" (PGPR) to designate these bacteria. The PGPR have been divided into two classes according to whether they can affect plant growth directly or indirectly (Bashan and Holguin, 1998). Direct influence is related to increased solubilization and uptake of nutrients and production of phytohormones. It has long been known that soil microorganisms play an important role in the cycling of many soil nutrients including bacteria, yeasts, actinomycetes, and mycorrhizae fungi reported to cause increases in plant-available P in the soil (Whitelaw, 2000). Raghu and MacRae (1966), and Whipps and Lynch (1986) also reported that a high proportion of P-solubilizing microorganisms are concentrated in the rhizosphere of plants. Microorganisms in the rhizosphere obtain their nutrition from root exudates, plant mucigel, and root lysates. Rhizosphere microorganisms are normal heterotrophs but they live in an environment with high levels of nutrients, such as carbon and nitrogen, and tend to adapt rapidly to improvements of nutrient supply (Tinker, 1984; Whitelaw, 2000).

Indirect effect is associated with pathogen suppression, production of iron chelating siderophores and antibiotics, and the induction of plant resistance mechanisms (Persello-Cartieaux et al., 2003). The PGPR biosynthesize plant growth regulators, namely, auxins, gibberellins, cytokinins, ethylene, and abscisic acid (Arshad and Frankenberger, 1998). The detailed discussion regarding microorganisms involved and the quantity of growth-promoting hormones produced and functions are given by Arshad and Frankenberger (1998) and Persello-Cartieaux (2003). The rhizosphere of flooded rice is largely colonized by aerobic heterotrophic nitrogen-fixing bacteria including species of *Azospirillum, Herbaspirillum, Burkholderia, Azoarcus*, and *Pseudomonas* (Ladha and Reddy, 2003). Most species of the genus *Azospirillum* are known to act as plant growth–promoting rhizobacteria and stimulate plant growth directly either by synthesizing phytohormones or by promoting improved N nutrition through biological nitrogen fixation. Remans et al. (2008) reported differences in the responsiveness to PGPR between two genotypes of dry bean, suggesting genetic variation for this trait within germplasm. The properties of PGPR offer great promise for agronomic applications, but interactions with other bacteria and environmental factors are still a problem for sustainable application of PGPR (Persello-Cartieaux, 2003).

In recent decades, a wide variety of microbial strains have been isolated from the rhizosphere soil for the purpose of improving plant growth and health (Whipps, 2001; Bent, 2006). The most successfully used nonsymbiotic rhizosphere species in biological control of plant diseases are *Bacillus* spp. (Jacobsen et al., 2004; Schisler et al., 2004), *Pseudomonas* spp. (Weller, 2007), and *Trichoderma* spp. (Samuels, 2006). Environmentally friendly biocontrol materials may eventually replace some overused chemical pesticides in agriculture (Mathre et al., 1999; Conrath et al., 2006; Avis et al., 2008). Luo et al. (2010) reported that a *Bacillus subtilis* enhanced bio-organic fertilizer could effectively control cotton verticillium wilt, one of the most destructive diseases of cotton worldwide. That *Bacillus subtilis* can inhibit plant pathogens and promote plant growth has been proven and reviewed extensively (Stein, 2005; Nagorska et al., 2007; Earl et al., 2008).

4.3.3.4 Harmful Microorganisms

In the rhizosphere, some microorganisms may enhance plant diseases and reduce crop yields (Fageria, 1992). Soilborne pathogens such as actinomycetes, bacteria, fungi, and nematodes lead to pathogenic stress and change the morphology and physiology of roots and shoots (Lyda, 1981; Fageria, 1992, Fageria et al., 2011). Such changes reduce plant ability to absorb water and nutrients effectively. Plant diseases are also greatly influenced by nutritional deficiencies as well as toxicities (Huber, 1980). The severity of obligate and facultative parasites on plants is influenced by many micronutrients (Graham and Webb, 1991; Fageria et al., 2002).

The entry of root pathogens into susceptible tissue is generally controlled by the soil and rhizosphere environment, and disease progress after infection (pathogen in the plant tissue) is controlled by the plant-tissue environment. Conditions such as soil or plant water potentials, soil or plant temperature, gas exchange in the root zone or within the plant tissues, and nutrition of the host are all important in determining both the incidence of infection and symptom development, in that order. However, with many root, stem, and stalk diseases, the major influence of environment on disease development is expressed postinfection rather than as an aid to penetration of the host by the pathogen (Cook, 1990). Cook (1990) summarized control measures that can be adopted to avoid root infection by disease microorganisms. These measures include lowering the inoculum potential of the pathogen in soil below some economic threshold. This may be achieved by crop rotation, tillage, and possibly burial of crop residues which can be used by pathogen as a food source, soil fumigation, soil solarization, and planting clean seed or pathogen-free seeds. The second strategy mentioned by Cook (1990) to control disease pathogens is the application of fungicides. The most desirable and environmental strategy is to plant disease-resistant crop species or genotypes within species.

4.4 CONCLUSIONS

The rhizosphere is the soil zone adjacent to plant roots which is physically, chemically, and biologically different than bulk or nonrhizosphere soil. The rhizosphere has been also divided into ectorhizosphere (zone outside the root), rhizoplane (the root surface), and endorhizosphere (zone inside the root). During the evolution of existing plant species, a diversity of mechanisms has been adopted by plant communities to adapt to adverse environmental conditions. In the adaptability process, the soil area surrounding the roots is subject to many physical, chemical, and biological changes. The physical changes most prominent are root temperature, water holding capacity of roots, and soil structure. The chemical changes associated with plant adaptability are pH, redox potential, root exudates, nutrient concentrations, and organic matter content. Microbial association with root plays an important role in nutrient availability to plants. The microorganisms that are beneficial to plants are *Rhizobium* or *Franka* genera and mycorrhiza fungi, which are able to establish a symbiotic relation with their host plant. The availability of nutrients to plants increased with the association of these rhizobacteria or fungi. These changes are responsible for nutrient solubility and availability to plants. The magnitude of physical, chemical, and biological changes varied with plant species, soil type, and environmental factors and their interactions. Hence, rhizosphere changes are very complex in nature and dynamics, and to date their complete knowledge is not available and more studies are needed to understand or clarify these changes and their interactions with plant growth.

REFERENCES

Adviento-Borbe, M. A. A., J. W. Doran, R. A. Drijber, and A. Dobermann. 2006. Soil electrical conductivity and water content affect nitrous oxide and carbon dioxide emissions in intensively managed soils. *J. Environ. Qual.* 35:1999–2010.

Aguilar, S. A. and A. van Diest. 1981. Root-phosphate mobilization induced by the alkaline uptake pattern of legume utilizing symbiotically fixed nitrogen. *Plant Soil* 61:27–42.

Albrecht, C., R. Geurts, and T. Bisseling. 1999. Legume nodulation and mycorrhizae formation; two extremes in host specificity meet. *Eur. Mol. Biol. Organ. J.*, 18:281–288.

Aniol, A. 1990. Genetics of tolerance to aluminum in wheat (*Triticum aestivum* L.). *Plant Soil* 123:223–227.

Arshad, M. and W. T. Frankenberger, Jr. 1998. Plant growth-regulating substances in the rhizosphere: Microbial production and functions. *Adv. Agron.* 62:45–151.

Assmakopoulou, A. 2006. Effect of iron supply and nitrogen form on growth, nutritional status and ferric reducing activity of spinach in nutrient solution culture. *Sci. Hort.* 110:21–29.

Atkinson, D. 1990. Influence of root system morphology and development on the need for fertilizers and the efficiency of use. In: *Crops as Enhancer of Nutrient Use*, eds., V. C. Baligar and R. R. Duncan, pp. 411–451. San Diego, CA: Academic Press.

Avis, T. J., V. Gravel, H. Antoun, and R. J. Tweddell. 2008. Multifaceted beneficial effects of rhizosphere microorganisms on plant health and productivity. *Soil Biol. Biochem.* 40:1733–1740.

Baath, E. and T. H. Anderson. 2003. Comparison of soil fungal/bacterial ratios in a pH gradient using physiological and PLFA-based techniques. *Soil Biol. Biochem.* 35:955–963.

Badalucco, L., P. J. Kuikman, and P. Nannipieri. 1996. Protease and deaminase activities in wheat rhizosphere and their relation to bacterial and protozoan populations. *Biol. Fertil. Soils* 23:99–104.

Barber, S. A. 1974. Influence of the plant root on ion movement in soil. In: *The Plant Root and Its Environment*, ed., E. W. Carson, pp. 525–564. Charlottesville, VA: University Press of Virginia.

Barber, S. S. 1995. *Soil Nutrient Bioavailability*, 2nd edn. New York: John Wiley & Sons.

Bar-Yosef, B., S. Fishman, and H. Talpaz. 1980. A model of zinc movement to single roots in soils. *Soil Sci. Soc. Am.* 44:1272–1279.

Bashan, Y. and G. Holguin. 1998. Proposal for the division of plant growth-promoting rhizobacteria into two classifications: Biocontrol-PGPB (plant growth promoting bacteria) and PGPB. *Soil Biol. Biochem.* 30:1225–1228.

Batjes, N. H. 1996. Total carbon and nitrogen in the soils of the world. *Eur. J. Soil Sci.* 47:151–163.

Bent, E. 2006. Induced systemic resistance mediated by plant growth promoting rhizobacteria (PGPR) and fungi (PGPF). In: *Multigenic and Induced Systemic Resistance in Plants*, eds., S. Tuzun and E. Bent, pp. 225–258. New York: Springer Verlag.

Bertin, C., X. Yang, and L. A. Weston. 2003. The role of root exudates and allelochemicals in the rhizosphere. *Plant Soil* 256:67–83.

Bezdicek, D. F., D. W. Evans, B. Abebe, and R. W. Witters. 1978. Evaluation of peat and granular inoculum for soybean yield and N fixation under irrigation. *Agron. J.* 70:865–868.

Bhattarai, S. P., N. Su, and D. J. Midmore. 2005. Oxygation unlocks yield potentials of crops in oxygen-limited soil environment. *Adv. Agron.* 88:314–377.

Bienfait, H. F., R. J. Bino, A. M. van der Bliek, J. F. Duivenvoorden, and J. M. Fontaine. 1983. Characterization of ferric reducing activity in roots of Fe-deficient *Phaseolus vulgaris*. *Physiol. Plantrum*. 59:196–202.

Bloom, A. J., P. A. Meyerhoff, A. R. Taylor, and T. L. Rost. 2003. Root development and absorption of ammonium and nitrate from the rhizosphere. *J. Plant Growth Regul.* 21:416–431.

Bloom, A. J., S. S. Sukrapanna, and R. L. Warner. 1992. Root respiration associated with ammonium and nitrate absorption and assimilation by barley. *Plant Physiol.* 99:1294–1301.

Bolan, N. S. 1991. A critical review on the role of mycorrhiza fungi in the uptake of phosphorus by plants. *Plant Soil* 134:189–207.

Bornemann, L., G. Welp, and W. Amelung. 2010. Particulate organic matter at the field scale: Rapid acquisition using mid-infrared spectroscopy. *Soil Sci. Soc. Am. J.* 74:1147–1156.

Bottner, P., Z. Sallih, and G. Billes. 1988. Root activity and carbon metabolism in soils. *Biol. Fertil. Soils* 7:71–78.

Bowen, G. D. and A. D. Rovira. 1976. Microbial colonization of plant roots. *Annu. Rev. Phytopathol.* 14:121–144.

Bowen, G. D. and A. D. Rovira. 1999. The rhizosphere and its management to improve plant growth. *Adv. Agron.* 66:1–102.

Brady, N. C. and R. R. Weil. 2002. *Nature and Properties of Soils*, 13th edn. Upper Saddle River, NJ: Prentice Hall.

Bray, R. H. 1954. A nutrient mobility concept of soil-plant relationship. *Soil Sci.* 78:9–22.

Brimecombe, M. J., F. A. D. Leij, and J. M. Lynch. 2001. The effect of root exudates on rhizosphere microbial populations. In: *The Rhizosphere: Biochemical and Organic Substances at the Soil-Plant Interface*, eds., R. Pinton, Z. Varanini, and P. Nannipieri, pp. 95–140. New York: Marcel Dekker.

Broughton, W. J. and X. Perret. 1999. Genealogy of legume-*Rhizobium* symbiosis. *Curr. Opin. Plant Biol.* 2:305–311.

Bryant, J. P., F. S. Chapin, and D. R. Klein. 1983. Carbon/nutrient balance of boreal plants in relation to vertebrate herbivory. *Oikos* 40:357–368.

Burton, D. L., X. Li, and C. A. Grant. 2008. Influence of fertilizer nitrogen source and management practice on N_2O emissions from two Black Chernozemic soils. *Can. J. Soil Sci.* 88:219–227.

Cambardella, C. A. and E. T. Elliott. 1993. Carbon and nitrogen distribution in aggregates from cultivated and native grassland soils. *Soil Sci. Soc. Am. J.* 57:1071–1076.

Campbell, R. and R. Porter. 1982. Low temperature scanning electron microscopy of microorganisms in soil. *Soil Biol. Biochem.* 14:241–245.

Castellano, S. D. and R. P. Dick. 1991. Cropping and sulfur fertilization influence on sulfur transformations in soil. *Soil Sci. Soc. Am. J.* 55:114–121.

Ceccotti, S. P. 1996. Plant nutrient sulphur: A review of nutrient balance, environment impact and fertilizers. *Fertil. Res.* 43:117–125.

Chaignon, V., D. Malta, and P. Hinsinger. 2002. Fe-deficiency increases Cu acquisition by wheat cropped in a Cu-contaminated vineyard soil. *New Phytol.* 15:4121–4130.

Chaney, R. L. 1988. Recent progress and needed research in plant Fe nutrition. *J. Plant Nutr.* 11:1589–1603.

Chantigny, M. H., D. A. Angers, D. Prevost, L. P. Vezina, and F. P. Chalifour. 1997. Soil aggregation and fungal and bacterial biomass under annual and perennial cropping systems. *Soil Sci. Soc. Am. J.* 61:262–267.

Cheng, H. H. 1995. Characterization of the mechanisms of allelopathy: Modeling and experimental approaches. In: *Principles and Practices in Plant Ecology: Allelochemical Interactions*, ed., Inderjit, pp. 132–141. Boca Raton, FL: CRC Press.

Claassen, N., M. Syring, and A. Jungk. 1986. Verification of a mathematical model by simulating potassium uptake from soil. *Plant Soil* 95:209–220.

Clarkson, D. T. 1985. Factors affecting mineral nutrient acquisition by plants. *Annu. Rev. Plant Physiol.* 36:77–115.

Collins, H. P., R. L. Blevins, L. G. Bundy, D. R. Christensen, W. A. Dick, D. R. Huggins, and E. A. Paul. 1999. Soil carbon dynamics in corn-based agroecosystems: Results from carbon-13 natural abundance. *Soil Sci. Soc. Am. J.* 63:584–591.

Conard, R. 1995. Soil microbial processes involved in production and consumption of atmospheric trace gases. *Adv. Microb. Ecol.* 14:207–250.

Conrath, U., G. J. M. Beckers, V. Flors, P. Garcia-Agustin, G. Jakab, and F. Mauch. 2006. Priming: Getting ready for battle. *Mol. Plant Microbe Interact.* 19:1062–1071.

Cook, R. J. 1990. Diseases caused by root-infecting pathogens in dryland agriculture. *Adv. Soil Sci.* 13:215–239.

Couto, W., D. J. Lathwell, and D. R. Bouldin. 1979. Sulfate sorption by two Oxisols and Alfisols of the tropics. *Soil Sci.* 127:108–116.

Cresswell, H. P. and J. A. Kirkegaard. 1995. Subsoil amelioration by plant roots: The process and the evidence. *Aust. J. Soil Sci.* 33:221–239.

Crowley, D. 2001. Function of siderophores in the plant rhizosphere. In: *The Rhizosphere: Biochemistry and Organic Substances at the Soil-Plant Interface*, eds., R. Pinton, Z. Varanini, and P. Nannipieri, pp. 223–261. New York: Marcel Dekker.

Dakora, F. D. and D. A. Phillips. 2002. Root exudates as mediators of mineral acquisitional in low-nutrient environments. *Plant Soil* 245:35–47.

Darrah, P. R. 1993. The rhizosphere and plant nutrition: A quantitative approach. *Plant Soil* 156:3–22.

Davies, S. I. and M. B. Peoples. 2003. Identifying potential approaches to improve the reliability of teminating a lucerene pasture before cropping: A review. *Aust. J. Exp. Agric.* 43:429–447.

Deluca, T. H. and D. R. Keeney. 1994. Soluble carbon and nitrogen pools of prairie and cultivated soils: Seasonal variation. *Soil Sci. Soc. Am. J.* 58:835–840.

Denarie, J., F. Debelle, and J. C. Prome. 1996. Rhizobium lipochitooligosaccharide nodulation factors: Signalling molecules mediating recognition and morphogenesis. *Annu. Rev. Biochem.* 65:503–535.

Dijkstra, A. F., J. M. Govaert, G. H. N. Scholton, and J. D. van Elsas. 1987. A soil chamber for studying the bacterial distribution in the vicinity of roots. *Soil Biol. Biochem.* 19:351–352.

Dinkelaker, B., G. Hahn, and H. Marschner. 1993. Non-destructive methods for demonstrating chemical changes in the rhizosphere. II. Application of methods. *Plant Soil* 1155/156:73–76.

Dinkelaker, B., C. Hengeler, and H. Marschner. 1995. Distribution and function of proteoid roots and other root clusters. *Bot. Acta* 108:183–200.

Dinkelaker, B., V. Romheld, and H. Marschner. 1989. Citric acid excretion and precipitation of calcium in the rhizosphere of white lupin (*Lupinus albus* L.). *Plant Cell Environ.* 12:285–292.

Dobereiner, J. and J. M. Day. 1976. Associative symbiosis in tropical grasses: Characterization of microorganisms and dinitrogen fixing sites. In: *Proceedings of First International Symposium on N₂ Fixation*. Pullman, WA: Washington University Press.

Doran, J. W., A. J. Jones, M. A. Arshad, and J. E. Gilley. 1999. Determinants of soil quality and health. In: *Soil Quality and Soil Erosion*, ed., R. Lal, pp. 17–36. Boca Raton, FL: CRC Press.

Drever, J. I. and L. L. Stillings. 1997. The role of organic acids in mineral weathering. *Colloids Surf. A: Physicochem. Eng. Aspects* 120:167–181.

Drury, C. F., T. Q. Zhang, and B. D. Kay. 2003. The non-limiting and least limiting water ranges for soil nitrogen mineralization. *Soil Sci. Soc. Am. J.* 67:1388–1404.

Earl, A. M., R. Losick, and R. Kolter. 2008. Ecology and genomics of *Bacillus subtilis*. *Trends Microbiol.* 16:269–275.

Edwards, G. E., S. B. Ku, and J. G. Foster. 1983. Physiological constraints to maximum yield potential. In: *Challenging Problems in Plant Health*, eds., T. Kommedahl and P. H. Williams, pp. 105–119. St. Paul, MN: American Phytopatholgy Society.

Elliott, L. F., G. M. Gilmour, J. M. Lynch, and D. Tittemore. 1984. Bacterial colonization of plant roots. In: *Microbial Plant Interactions*, eds., R. L. Todd and J. E. Giddens, pp. 1–16. Madison, WI: SSSA.

Eriksen, J., M. D. Murphy, and E. Schnug. 1998. The soil sulphur cycle. In: *Sulphur in Agroecosystems*, ed., E. Schnug, pp. 39–73. Dordrecht, the Netherlands: Kluwer Academic Publishers.

Fageria, N. K. 1992. *Maximizing Crop Yields*, 274 pp. New York: Marcel Dekker.

Fageria, N. K. 2009. *The Nutrient Use of Crop Plant*. Boca Raton, FL: CRC Press.

Fageria, N. K. and V. C. Baligar. 2003a. Soil fertility management of tropical acid soils for sustainable crop production. In: *Handbook of Soil Acidity*, ed., Z. Rengel, pp. 359–385. New York: Marcel Dekker.

Fageria, N. K. and V. C. Baligar. 2003b. Upland rice and allelopathy. *Commun. Soil Sci. Plant Anal.* 34:1311–1329.

Fageria, N. K., V. C. Baligar, and R. B. Clark. 2002. Micronutrients in crop production. *Adv. Agron.* 77:185–268.

Fageria, N. K., V. C. Baligar, and C. A. Jones. 2011. *Growth and Mineral Nutrition of Field Crop*, 3rd edn. Boca Raton, FL: CRC Press.

Fageria, N. K. and N. P. Souza. 1995. Response of rice and common bean crops in succession to fertilization in cerrdo soil. *Pesq. Agropec. Brás.* 30:359–368.

Fageria, N. K. and L. F. Stone. 2006. Physical, chemical and biological changes in rhizosphere and nutrient availability. *J. Plant Nutr.* 29:1327–1356.

Fannings, D. S., V. Z. Keramidas, and M. A. El-Desoky. 1989. Micas. In: *Minerals in Soil Environment*, 2nd edn., eds., J. B. Dixon and S. B. Weed, pp. 551–634. Madison, WI: Soil Science Society of America.

Follett, R. F. 2001. Soil management concepts and carbon sequestration in cropland soils. *Soil Tillage Res.* 61:77–92.

Follett, R. F. 2010. Symposium: Soil carbon sequestration and greenhouse gas mitigation. *Soil Sci. Soc. Am. J.* 74:345–346.

Forster, P., V. Ramaswamy, P. Artaxo, T. Berntsen, R. Betts, and D. W. Fahey. 2007. Changes in atmospheric constituents and in radiative forcing. In: *Climate Change 2007: The Physical Science Basis*, ed., S. Solomon, pp. 129–234. Cambridge, U.K.: Cambridge University Press.

Fuente-Martínez, J. M. and L. H. Herrera-Estrella. 1999. Advances in the understanding of aluminum toxicity and the development of aluminum-tolerant transgenic plants. *Adv. Agron.* 66:103–120.

Garbeva, P., J. D. V. Elsas, and J. A. V. Veen. 2008. Rhizosphere microbial community and its response to plant species and soil history. *Plant Soil* 302:19–32.

Gardner, W. R. 1960. Dynamic aspects of water availability to plants. *Soil Sci.* 89:63–73.

Gerke, J. 1994. Kinetics of soil phosphate desorption as affected by citric acid. *Z. Pflanzenernahr Bodenk* 156:253–257.

Gibson, A. H. 1980. Methods for legumes in glasshouses and controlled environment cabinets. In: *Methods for Evaluating Biological Nitrogen Fixation*, ed., F. J. Bergersen, pp. 139–148. New York: John Wiley & Sons.

Gibson, A. H. and J. E. Harper. 1985. Nitrate effect on nodulation of soybean by *Bradyrhizobium japonicum*. *Crop Sci.* 25:497–501.

Graham, R. D. and M. J. Webb. 1991. Micronutrients and disease resistance and tolerance in plants. In: *Micronutrient in Agriculture*, 2nd edn., eds., J. J. Mortvedt, F. R. Fox, L. M. Shuman, and R. M. Welch, pp. 329–370. Madison, WI: Soil Science Society of America.

Grayston, S. J. and J. J. Germida. 1991. Sulfur-oxidizing bacteria as plant growth promoting rhizobacteria for canola. *Can. J. Microbiol.* 37:521–529.

Gregory, P. J. 2006. Roots, rhizosphere and soil: The route to a better understanding of soil science? *Eur. J. Soil Sci.* 57:2–12.

Gregory, A. S., C. P. Webster, C. W. Watts, W. R. Whalley, C. J. A. Macleod, A. Joynes et al. 2010. Soil management and grass species effects on the hydraulic properties of shrinking soils. *Soil Sci. Soc. Am. J.* 74:753–761.

Grinster, M. J., M. J. Hedley, R. E. White, and P. H. Nye. 1982. Plant-induced changes in the rhizosphere of rape seedlings. I. pH change and the increase in P concentration in the soil solution. *New Phytol.* 91:19–29.

Guo, X., C. F. Drury, X. Yang, and R. Zhang. 2010. Influence of constant and fluctuating water contents on nitrous oxide emissions from soils under varying crop rotations. *Soil Sci. Soc. Am. J.* 74:2077–2085.

Gupta, S. C., R. H. Dowdy, and W. E. Larson. 1977. Hydraulic and thermal properties of a sandy soil as influenced by incorporation of sewage sludge. *Soil Sci. Soc. Am. J.* 41:601–605.

Hallett, P. D., D. C. Gordon, and A. G. Bengough. 2003. Plant influence on rhizosphere hydraulic properties: Direct measurements using a miniaturized infiltrometer. *New Phytol.* 157:597–603.

Halpern, M. T., J. K. Whalen, and C. A. Madramootoo. 2010. Long-term tillage and residue management influences soil carbon and nitrogen dynamics. *Soil Sci. Soc. Am. J.* 74:1211–1217.

Halvorson, A. D., S. J. Del Grosso, and C. A. Reule. 2008. Nitrogen, tillage and crop rotation effects on nitrous oxide emissions from irrigated cropping systems. *J. Environ. Qual.* 37:1337–1344.

Hansen, N. C., M. A. Schitt, J. E. Anderson, and J. S. Strock. 2003. Iron deficiency of soybean in the upper Midwest and associated soil properties. *Soil Sci. Soc. Am. J.* 95:1595–1601.

Harrier, L. A. and C. A. Watson. 2003. The role of arbuscular mycorrhiza fungi in sustainable cropping systems. *Adv. Agron.* 20:185–225.

Hartmann, A., M. Schmid, D. V. Tuinen, and G. Berg. 2009. Plant driven selection of microbes. *Plant Soil* 321:235–257.

Haynes, R. J. and K. M. Goh. 1978. Ammonium and nitrate nutrition of plants. *Biol. Rev.* 53:465–510.

Herridge, D. F. and J. Brockwell. 1988. Contributions of fixed nitrogen and soil nitrate to the nitrogen economy of irrigated soybean. *Soil Biol. Biochem.* 20:711–717.

Hill, G. T., N. A. Mitkowski, L. Aldrich-Wolfe, L. R. Emele, D. D. Jurkonie, A. Ficke, S. Maldonado-Ramirez, S. T. Lynch, and E. B. Nelsona. 2000. Methods of assessing the composition and diversity of soil microbial communities. *Appl. Soil Ecol.* 15:25–36.

Hinsinger, P. 1998. How do plant roots acquire mineral nutrients? Chemical processes involved in the rhizosphere. *Adv. Agron.* 64:225–265.

Hinsinger, P. 2001. Bioavailability of trace elements as related to root-induced chemical changes in the rhizosphere. *Plant Soil* 237:173–195.

Hinsinger, P., A. G. Bengough, D. Vetterlein, and I. M. Young. 2009. Rhizosphere: Biophysics, biogeochemistry and ecological relevance. *Plant Soil* 321:117–152.

Hinsinger, P., F. Elsass, B. Jaillard, and M. Robert. 1993. Root-induced irreversible transformation of a trioctahedral mica in the rhizosphere of rape. *J. Soil Sci.* 44:535–545.

Hinsinger, P. and R. J. Gilkes. 1996. Mobilization of phosphate from phosphate rock and alumina-sorbed phosphate by the roots of ryegrass and clover as related to rhizosphere pH. *Eur. J. Soil Sci.* 47:533–544.

Hinsinger, P., G. R. Gobran, P. J. Gregory, and W. W. Wenzel. 2005. Rhizosphere geometry and heterogeneity arising from root-mediated physical and chemical processes. *New Phytol.* 168:293–303.

Hinsinger, P. and B. Jaillard. 1993. Root induced release of interlayer potassium and vermiculation of phogopite as related to potassium depletion in the rhizosphere of ryegrass. *J. Soil Sci.* 44:525–534.

Hinsinger, P., C. Plassard, C. Tang, and B. Jaillard. 2003. Origin of root-mediated pH changes in the rhizosphere and their responses to environmental constraints: A review. *Plant Soil* 248:43–59.

Hoffland, E. 1992. Quantitative evaluation of the role of organic acid exudation in the mobilization of rock phosphate by rape. *Plant Soil* 140:279–289.

Houghton, J. T., Y. Ding, D. J. Griggs, M. Noguer, P. J. van der Linden, X. Dai, K. Maskell, and C. A. Johnson. 2001. *Climate Change 2001: The Scientific Basis.* Cambridge, U.K.: Cambridge University Press.

Huang, B. R. and D. S. Nesmith. 1990. Soil aeration effects on root growth and activity. *Acta Hortic.* 504:41–49.

Hubel, F. and E. Beck. 1993. In-situ determination of the P-relations around the primary root of maize with respect to inorganic and phytate-P. *Plant Soil* 157:1–9.

Huber, D. M. 1980. The role of mineral nutrition in defense. In: *Plant Pathology: An Advanced Treatise*, eds., J. G. Horsfall and E. B. Cowling, pp. 381–406. New York: Academic Press.

Hue, N. V., G. R. Craddock, and F. Adams. 1986. Effect of organic acids on aluminum toxicity in subsoils. *Soil Sci. Soc. Am. J.* 50:28–34.

Hyatt, C. R., R. T. Venterea, C. J. Rosen, M. McNearney, M. L. Wilson, and M. S. Dolan. 2010. Polymer-coated urea maintains potato yields and reduces nitrous oxide emissions in a Minnesota loamy sand. *Soil Sci. Soc. Am. J.* 74:419–428.

Inderjit. 1996. Plant phenolics in allelopathy. *Bot. Rev.* 62:186–202.

Inderjit. 2001. Soil: Environmental effects on allelochemical activity. *Agron. J.* 93:79–84.

Intergovernmental Panel on Climate Change (IPCC). 2001. *Climate Change 2001: The Scientific Basis.* Contribution of working group I to the third assessment report of the intergovernmental panel on climate change. Cambridge, U.K.: Cambridge University Press.

Intergovernmental Panel on Climate Change (IPCC). 2007. *Climate Change 2007: The Physical Science Basis.* Cambridge, U.K.: Cambridge University Press.

Jackson, L. E., M. Berger, and T. R. Cavagnaro. 2008. Root, nitrogen transformations, and ecosystem services. *Annu. Rev. Plant Biol.* 59:341–363.

Jackson, L. E. and A. J. Bloom. 1990. Root distribution in relation to soil nitrogen availability in field grown tomatoes. *Plant Soil* 128:115–126.

Jacobsen, B. J., N. K. Zidack, and B. J. Larson. 2004. The role of Bacillus-based biological control agents in integrated pest management systems: Plant diseases. *Phytopathology* 94:1272–1275.

Jarvis, S. C. and D. J. Hatch. 1985. Rates of hydrogen ion efflux by nodulated legumes grown in flowing solution culture with continuous pH monitoring and adjustment. *Ann. Bot.* 55:41–51.

Jenny, H. and K. Grossenbacher. 1963. Root-soil boundary zones as seen in the electron microscope. *Soil Sci. Soc. Proc.* 27:273–277.

Jianguo, H. and L. M. Shuman. 1991. Phosphorus status and utilization in the rhizosphere. *Soil Sci.* 152:360–364.

Johnson, J. M. F., D. Archer, and N. Barbour. 2010. Greenhouse gas emission from contrasting management scenarios in the northern corn belt. *Soil Sci. Soc. Am. J.* 74:396–406.

Jones, D. L. 1998. Organic acids in the rhizosphere-a critical review. *Plant Soil* 205:25–44.

Jones, D. L. and D. S. Brassington. 1998. Sorption of organic acids in acid soil and its implication in the rhizosphere. *Eur. J. Soil Sci.* 49:447–455.

Jones, D. L. and P. R. Darrah. 1994. Amino-acid influx and efflux at the soil-plant interface of *Zea mays* L. and its implications in the rhizosphere. *Plant Soil* 163:1–12.

Jones, D. L., C. Nguyen, and R. D. Finlay. 2009. Carbon flow in the rhizosphere: Carbon trading at the soil-root interface. *Plant Soil* 321:5–33.

Kang, Y. I., J. M. Park, S. H. Kim, N. J. Kang, K. S. Park, S. Y. Lee, and B. R. Jeong. 2011. Effects of root zone pH and nutrient concentration on the growth and nutrient uptake of tomato seedlings. *J. Plant Nutr.* 34:640–652.

Khaleel, R., K. R. Reddy, and M. R. Overcash. 1981. Changes in soil physical properties due to organic waste applications: A review. *J. Environ. Qual.* 10:133–141.

Khalil, M. I. and E. M. Baggs. 2005. CH_4 oxidation and N_2O emissions at varied soil water-filled pore space and headspace CH_4 concentrations. *Soil Biol. Biochem.* 37:1785–1794.

Kloepper, J. W., B. Schippers, and P. A. H. M. Bakker. 1992. Proposed elimination of the term "endorhizosphere." *Phytopathology* 82:726–727.

Kloepper, J. W. and M. N. Schroth. 1978. Plant growth-promoting rhizobacteria on radishes. In: *Proceedings of the Fourth International Conference on Plant Pathogenic Bacteria*, Vol. 2, pp. 879–882. Tours, France: Station de Pathologie Vegetale et de Phytobacteriologie.

Kochian, L. V., O. A. Hoekenga, and M. A. Pineros. 2004. How do crop plants tolerate acid soils? Mechanisms of aluminum tolerance and phosphorus deficiency. *Annu. Rev. Plant Biol.* 55:459–493.

Koricheva, J., S. Larrson, E. Haukioja, and M. Keinanen. 1998. Regulation of woody plant secondary metabolism by resource availability: Hypothesis testing by means of meta-analysis. *Oikos* 83:212–226.

Kparmwang, T., I. E. Esu, and V. O. Chude. 1997. Sulphate adsorption-desorption characteristics of three Ultisols and an Alfisol developed on basalts in the Nigerian savanna. *Discov. Innov.* 9:197–204.

Kraemer, S. M., D. E. Crowley, and R. Kretzschmar. 2006. Geochemical aspects of phytosiderophore promoted iron acquisition by plants. *Adv. Agron.* 91:1–46.

Kroeze, C., A. Mosier, and L. Bouwman. 1999. Closing the global N_2O budget: A retrospective analysis 1500–1994. *Glob. Biogeochem. Cycles* 13:1–8.

Kruidhof, H. M. 2008. Cover crops-based ecological weed management: Exploitation and optimization. PhD thesis. Wageningen, the Netherlands: Wageningen University.

Ladha, J. K. and P. M. Reddy. 2003. Nitrogen fixation in rice systems: State of knowledge and future prospects. *Plant Soil* 252:151–167.

Lal, R. 2008. Carbon sequestration. *Philos. Trans. R. Soc. Lond. Ser.* 78:131–141.

Lal, R., J. M. Kimble, R. F. Follett, and C. V. Cole. 1998. *The Potential of U. S. Cropland to Sequester Carbon and Mitigate the Greenhouse Effect.* Boca Raton, FL: CRC Press.

Lamont, B. B. 2003. Structure, ecology and physiology of root clusters: A review. *Plant Soil* 248:1–19.

Li, X. F., J. F. Ma, and H. Matsumoto. 2002. Aluminum induced secretion of both citrate and malate in rye. *Plant Soil* 242:235–243.

Li, M., T. Shinano, and T. Tadano. 1997. Distribution of exudates of lupin roots in the rhizosphere under phosphorus deficient conditions. *Soil Sci. Plant Nutr.* 43:237–245.

Lindsay, W. L. 1979. *Chemical Equilibrium in Soils.* New York: Wiley.

Lipps, L. 2010. Examining the role of soils in emissions of the potent greenhouse gas. *C. S. A. News* 55:4–9.

Loosemore, N., A. Straczek, P. Hinsinger, and B. Jaillard. 2004. Zinc mobilization from a contaminated soil by three genotypes of tobacco as affected by soil and rhizosphere pH. *Plant Soil* 260:19–32.

Lundquist, E. J., K. M. Scow, L. E. Jackson, S. L. Uesugi, and C. R. Johnson. 1999. Rapid response of soil microbial communities from conventional, low input, and organic farming systems to a wet/dry cycle. *Soil Biol. Biochem.* 31:1661–1675.

Luo, J., W. Ran, J. Hu, X. Yang, Y. Xu, and Q. Shen. 2010. Application of bio-organic fertilizer significantly affected fungal diversity of soils. *Soil Sci. Soc. Am. J.* 74:2029–2048.

Lupwayi, N. Z. and A. C. Kennedy. 2007. Grain legumes in northern Great Plains: Impacts on selected biological processes. *Agron. J.* 99:1700–1709.

Lyda, S. D. 1981. Alleviating pathogen stress. In: *Modifying the Root Environment to Reduce Crop Stress*, eds., G. F. Arkin and H. M. Taylor, pp. 195–214. St. Joseph, MI: American Society of Agricultural Engineering Monograph No. 4.

Lynch, J. M. 1981. Interactions between bacteria and plants in the root environment. In: *Bacteria and Plants*, eds., M. E. Rodes-Roberts and F. A. Skinner, pp. 1–23. London, U.K.: Academic Press.

Lynch, J. M. and E. Bragg. 1985. Microorganisms and soil aggregates stability. *Adv. Soil Sci.* 2:133–171.

Ma, J. F., S. J. Zheng, S. Hiradate, and H. Matsumoto. 1997. Detoxifying aluminum with buckwheat. *Nature* 390:569–570.

MacDonald, J. D., D. A. Angers, P. Rochette, M. H. Chantigny, I. Royer, and M. O. Gasser. 2010. Plowing a poorly drained grassland reduces soil respiration. *Soil Sci. Soc. Am. J.* 74:2067–2076.

Marschner, H. 1991. Mechanisms of adaptation of plants to acid soils. *Plant Soil* 134:1–20.

Marschner, H. 1995. *Mineral Nutrition of Higher Plants*, 2nd edn. New York: Academic Press.

Marschner, H. and V. Romheld. 1983. In vivo measurement of root-induced pH changes at soil-root interface. Effect of plant species and nitrogen source. *Z. Pflanzenphysiol.* 111:241–251.

Marschner, H., V. Romheld, W. J. Horst, and P. Martin. 1986. Root-induced changes in the rhizosphere: Importance for the mineral nutrition of plants. *Z. Pflanzenern. Bodenk.* 149:441–456.

Masscheleyn, P. H., R. D. Laune, and W. H. P. Jr Patrick. 1993. Methane and nitrous oxide emissions from laboratory measurements of rice soil suspension. Effects of soil oxidation-reduction status. *Chemosphere* 26:251–260.

Mathre, D. E., R. J. Cook, and N. W. Callan. 1999. From discovery to use: Traversing the world of commercializing biocontrol agents for plant disease control. *Plant Dis.* 83:972–983.

McRae, R. J. and G. R. Mehuys. 1985. The effect of green manuring on the physical properties of temperate-area soils. *Adv. Soil Sci.* 3:71–94.

McSwiney, C. P. and G. P. Robertson. 2005. Nonlinear response of N_2O flux to incremental fertilizer addition in a continuous maize (*Zea mays* L.) cropping system. *Global Change Biol.* 11:1712–1719.

Mengel, K. 1985. Dynamics and availability of major nutrients in soils. *Adv. Soil Sci.* 2:65–131.

Mengel, K., D. Horn, and H. Tributh. 1990. Availability of interlayer ammonium as related to root vicinity and mineral type. *Soil Sci.* 149:131–137.

Mengel, K., E. A. Kirkby, H. Kosegarten, and T. Appel. 2001. *Principles of Plant Nutrition*, 5th edn. Dordrecht, the Netherlands: Kluwer Academic Publishers.

Metzger, L. and B. Yaron. 1987. Influence of sludge organic matter on soil physical properties. *Adv. Soil Sci.* 7:141–162.

Miyasaka, S. C., J. G. Buta, R. K. Howell, and C. D. Foy. 1991. Mechanism of aluminum tolerance in snapbean. *Plant Physiol.* 96:737–743.

Moraghan, J. T. 1985. Manganese deficiency in soybeans as affected by FeEDDHA and low soil temperature. *Soil Sci. Soc. Am. J.* 49:1584–1985.

Morel, C. and P. Hinsinger. 1999. Root induced modifications of the exchange of phosphate ion between soil solution and solid phase. *Plant Soil* 211:103–110.

Moritsuka, N., J. Yanai, and T. Kosaki. 2000. Effect of plant growth on the distribution and forms of soil nutrients in the rhizosphere. *Soil Sci. Plant Nutr.* 46:439–447.

Moritsuka, N., J. Yanai, and T. Kosaki. 2004. Possible processes releasing nonexchangeable potassium from the rhizosphere of maize. *Plant Soil* 258:261–268.

Mosier, A., C. Kroeze, C. Nevison, O. Oenema, S. Seitzinger, and O. Van Cleempt. 1998. Closing the global N_2O budget: Nitrous oxide emissions through the agricultural nitrogen cycle. *Nutr. Cycling Agroecosyst.* 52:225–248.

Nagorska, K., M. Bikowski, and M. Obuchowski. 2007. Multicellular behavior and production of a wide variety of toxic substances support usage of *Bacillus subtilis* as a powerful biocontrol agent. *Acta Biochim. Pol.* 54:495–508.

Neergaard, A. D. and J. Magid. 2001. Influence of the rhizosphere on microbial biomass and recently formed organic matter. *Eur. J. Soil Sci.* 52:377–384.

Neumann, G. and V. Romheld. 1999. Root excretion of carboxylic acids and protons in phosphorus deficient plants. *Plant Soil* 211:121–130.

Neumann, G. and V. Romheld. 2001. The release of root exudates as affected by the plants physiological status. In: *The Rhizosphere: Biochemistry and Organic Substances at the Soil-Plant Interface*, eds., R. Pinto, Z. Varanini, and P. Nannipieri, pp. 41–93. New York: Marcel Dekker.

Niebes, J. F., J. E. Dufey, B. Jaillard, and P. Hinsinger. 1993. Release of nonexchangeable potassium from different size fractions of two highly K-fertilized soils in the rhizosphere of rape (*Brassica napus* cv. Drakkar). *Plant Soil* 155/156:403–406.

Nielsen, K. F. 1974. Roots and root temperatures. In: *The Plant Root and Its Environment*, ed., E. W. Carson, pp. 293–333. Charlottesville, VA: University Press of Virginia.

Nommik, H. and K. Vathras. 1982. Retention and fixation of ammonium in soils. In: *Nitrogen in Agricultural Soils*, ed., F. J. Stevenson, pp. 123–171. Madison, WI: American Society of Agronomy, Monograph 22.

Nye, P. H. 1981. Changes of pH across the rhizosphere induced by roots. *Plant Soil* 61:7–26.

Oliveira, O. C., I. P. Oliveira, B. J. R. Alves, S. Urquiaga, and R. M. Boddey. 2004. Chemical and biological indicators of decline/degradation of *Brachiaria* pastures in the Brazilian cerrado. *Agric. Ecosyst. Environ.* 103:289–300.

Olofsdotter, M. 2001. Rice-a step toward use of allelopathy. *Agron. J.* 93:3–8.

Passioura, J. B. 1988. Water transport in and to roots. *Annu. Rev. Plant Physiol. Plant Mol. Biol.* 39:245–265.

Pepper, I. L. and D. F. Bezdicek. 1990. Root microbial interactions and rhizosphere nutrient dynamics. In: *Crops as Enhancer of Nutrient Use*, eds., V. C. Baligar and R. R. Duncan, pp. 375–410. San Diego, CA: Academic Press.

Persello-Cartieaux, E., L. Nussaume, and C. Robaglla. 2003. Tales from the underground: Molecular plant-rhizobacteria interactions. *Plant Cell Environ.* 26:189–199.

Pinton, R. and Z. Varanini. 2001. The rhizosphere as a site of biochemical interactions among soil components, plants and microorganisms. In: *The Rhizosphere: Biochemical and Organic Substances at the Soil-Plant Interface*, eds., R. Pinto, Z. Varanini, and P. Nannipieri, pp. 1–17. New York: Marcel Dekker.

Pojasok, T. and B. D. Kay. 1990. Effect of root exudates from corn and bromegrass on soil structural stability. *Can. J. Soil Sci.* 70:351–362.

Putnam, A. R. and W. B. Duke. 1978. Allelopathy in agroecosystem. *Annu. Rev. Phytopathol.* 16:431–451.

Raaijmakers, J. M., T. C. Paulitz, C. Steinberg, C. Alabouvette, and Y. Moenne-Loccoz. 2009. The rhizosphere: A playground and battlefield for soilborne pathogens and beneficial microorganisms. *Plant Soil* 321:341–361.

Raghu, K. and I. C. MacRae. 1966. Occurrence of phosphate-dissolving micro-organisms in the rhizosphere of rice plants and in submerged soils. *J. Appl. Bacteriol.* 29:582–586.

Read, D. B., A. G. Bengough, P. J. Gregory, J. W. Crawford, D. Robinson, C. M. Scrimgeour, I. M. Young, K. Zhang, and X. Zhang. 2003. Plant roots release phospholipids surfactants that modify the physical and chemical properties of soil. *New Phytol.* 157:315–326.

Read, D. B. and P. J. Gregory. 1997. Surface tension and viscosity of axenic maize and lupin mucilages. *New Phytol.* 137:623–628.

Reid, J. B. and M. J. Gross. 1980. Changes in the aggregate stability of a sandy loam soil affected by growing roots of a perennial ryegrass (*Lolium perenne* L.). *J. Sci. Food Agric.* 31:325–328.

Reisenauer, H. M. 1994. The interactions of manganese and iron. In: *Biochemistry of Metal Micronutrients in the Rhizosphere*, eds., J. A. Manthey, D. E. Crowley, and D. G. Luster, pp. 147–164. Boca Raton, FL: Lewis Publishers.

Remans, R., S. Beebe, M. Bliar, G. Manrique, E. Tovar, I. Rao et al. 2008. Physiological and genetic analysis of root responsiveness to auxin-producing plant growth-promoting bacteria in common bean (*Phaseolus vulgaris* L.). *Plant Soil* 302:149–161.

Rice, E. L. 1974. *Allelopathy*, 353 pp. New York: Academic Press.

Rice, E. L. 1979. Allelopathy: An update. *Bot. Rev.* 62:186–202.

Riley, D. and S. A. Barber. 1971. Effect of ammonium and nitrate fertilization on phosphorus uptake as related to root-induced pH changes at the root soil-interface. *Soil Sci. Soc. Am. Proc.* 35:301–306.

Rimando, A. M., M. Olofsdotter, F. E. Dayan, and S. O. Duke. 2001. Searching for rice allelochemicals: An example of bioassay-guided isolation. *Agron. J.* 93:16–20.

Rogovska, N. P., A. M. Blackmer, and G. L. Tylka. 2009. Soybean yield and soybean cyst nematode densities related to soil pH, soil carbonate concentrations, and alkalinity stress index. *Agron. J.* 101:1019–1026.

Romheld, V. 1991. The role of phytosiderophores in acquisition of iron and other micronutrients in graminaceous species: An ecological approach. *Plant Soil* 130:127–134.

Romheld, V. and H. Marschner. 1983. Mechanism of iron uptake by peanut plants. I. Fe(III) reduction, chelate splitting, and release of phenolics. *Plant Physiol.* 71:949–954.

Romheld, V. and H. Marschner. 1990. Genotypic differences among graminaceous species in release of phytosiderophores and uptake of iron phytosiderophores. *Plant Soil* 123:147–153.

Rovira, D. and C. B. Davey. 1974. Biology of the rhizosphere. In: *The Plant Root and Its Environment*, ed., E. W. Carson, pp. 153–204. Charlottesville, VA: University Press of Virginia.

Rovira, A. D., R. C. Foster, and J. K. Martin. 1979. Note on terminology: Origin, nature and nomenclature of the organic materials in the rhizosphere. In: *The Soil-Root Interface*, eds., J. L. Harley and R. Scott Russell, pp. 1–4. London, U.K.: Academic Press.

Rufty, T. W., G. S. Miner, and C. D. Raper Jr. 1979. Temperature effects on growth and manganese tolerance in tobacco. *Agron. J.* 71:638–644.

Saito, K. 2004. Sulfur assimilatory metabolism. The long and smelling road. *Plant Physiol.* 136:2443–2450.

Samuels, G. J. 2006. *Trichoderma*: Systematics, the sexual state, and ecology. *Phytopathology* 96:195–206.

Sarkar, A. N. and R. G. Wyn Jones. 1982. Effect of rhizosphere pH on availability and uptake of Fe, Mn, and Zn. *Plant Soil* 66:361–372.

Scherer, H. W. 2001. Sulphur in crop production. *Eur. J. Agron.* 14:81–111.

Scherer, H. W. and G. Ahrens. 1996. Depletion of non-exchangeable NH_4^+-N in the soil-root interface in relation to clay mineral composition and plant species. *Eur. J. Agron.* 5:1–7.

Schisler, D. A., P. J. Slininger, R. W. Behle, and M. A. Jackson. 2004. Formulation of Bacillus spp. For biological control of plant diseases. *Phytopathology* 94:1267–1271.

Schnug, E., S. Haneklaus, and D. Murphy. 1993. Impact of sulfur fertilization on fertilizer nitrogen efficiency. *Sulphur Agric.* 17:3012.

Seguin, V., C. Gagnon, and F. Courchesne. 2004. Changes in water extractable metals, pH and organic carbon concentrations at the soil-root interface of forested soils. *Plant Soil.* 260:1–17.

Sene, M., T. Dore, and C. Gallet. 2001. Relationships between biomass and phenolic production in grain sorghum under different conditions. *Agron. J.* 93:49–54.

Shen, J., C. Tang, Z. Rengel, and F. Zhang. 2004. Root induced acidification and excess cation uptake by N_2 fixing *Lupinas albus* grown in phosphorus deficient soil. *Plant Soil* 260:69–77.

Six, J., E. T. Elliott, K. Paustian, and J. W. Doran. 1998. Aggregation and soil organic matter accumulation in cultivated and native grassland soils. *Soil Sci. Soc. Am. J.* 62:1367–1377.

Skiba, U. and B. Ball. 2002. The effect of soil texture and soil drainage on emissions of nitric oxide and nitrous oxide. *Soil Use Manage.* 18:56–60.

Soil Science Society of America. 2008. *Glossary of Soil Science Terms*, ed., Soil Science Society of America, 134 pp. Madison, WI: SSSA.

Song, S. K. and P. M. Huang. 1988. Dynamics of potassium release from potassium-bearing minerals as influenced by oxalic and citric acid. *Soil Sci. Soc. Am. J.* 52:383–390.

Soussana, J. F., P. Loiseau, N. Vuichard, E. Ceschia, J. Balesdent, T. Chevallier, and D. Arrouays. 2004. Carbon cycling and sequestration opportunities in temperate grasslands. *Soil Use Manage.* 20:219–230.

Sparks, D. L. 1987. Potassium dynamics in soils. *Adv. Soil Sci.* 6:1–63.

Stein, T. 2005. *Bacillus subtilis* antibiotics: Structures, syntheses and specific functions. *Mol. Microbiol.* 56:845–857.

Stevens, R. J., R. J. Laughlin, L. C. Burns, J. R. M. Arah, and R. C. Hood. 1997. Measuring the contribution of nitrification and denitrification to the flux of nitrous oxide from soil. *Soil Biol. Biochem.* 29:139–151.

Tabatabai, M. A. and J. M. Bremner. 1970. Arylsulfatase activity of soils. *Soil Sci. Soc. Am. Proc.* 34:225–229.

Takagi, S., S. Kamei, and M. H. Yu. 1988. Efficiency of iron extraction from soil by mugineic acid family phytosiderophores. *J. Plant Nutr.* 11:643–651.

Takagi, S., K. Nomoto, and T. Takemoto. 1984. Physiological aspects of mugineic acid, a possible phytosiderophore of graminaceous plants. *J. Plant Nutr.* 7:469–477.

Tans, P. 2009. Trends in atmospheric carbon dioxide: Recent Manual Loa CO_2. Available at www.esrl.noaa.gov/gmd/ccgg/trends/ (verified May 18, 2010). Boulder, CO: Earth System Research Laboratory.

Tinker, P. B. 1984. The role of microorganisms in mediating and facilitating the uptake of plant nutrients from soil. *Plant Soil* 76:77–91.

Tisdall, J. M. and J. M. Oades. 1979. Stabilization of soil aggregates by the root systems of ryegrass. *Aust. J. Soil Res.* 17:429–441.

Unkovich, M. J. and J. S. Pate. 2000. An appraisal of recent field measurements of symbiotic N_2 fixation by annual legumes. *Field Crops Res.* 65:211–228.

Uren, N. C. 2001. Types, amounts, and possible functions of compounds released into the rhizosphere by soil-grown plants. In: *The Rhizosphere: Biochemistry and Organic Substances at the Soil-Plant Interface*, eds., R. Pinto, Z. Varanini, and P. Nannipieri, pp. 19–40. New York: Marcel Dekker.

USEPA. 2008. Inventory of U. S. greenhouse gas emissions and sinks: 1990–2006. EPA 430-R-08-005. available at www.epa.gov/climatechange/emissions/downloads/08_CR.pdf (verified December 12, 2009). Washington. DC: USEPA.

Van Beusichem, M. L., E. A. Kirkby, and R. Baas. 1988. Influence of nitrate and ammonium nutrition and the uptake, assimilation, and distribution of nutrients in *Ricinus communis*. *Plant Physiol.* 86:914–921.

Venterea, R. T., M. S. Dolan, and T. E. Ochsner. 2010. Urea decreases nitrous oxide emissions compared with anhydrous ammonia in a Minnesota corn cropping system. *Soil Sci. Soc. Am. J.* 74:407–418.

Venterea, R. T. and D. E. Rolston. 2000. Mechanisms and kinetics of nitric and nitrous oxide production during nitrification in agricultural soil. *Glob. Change Biol.* 6:303–316.

Vong, P. C., O. Dedourge, and A. Guckert. 2004. Immobilization and mobilization of labeled sulphur in relations to soil arysulphatase activity in rhizosphere soil of field grown rape, barley, and fallow. *Plant Soil* 258:227–239.

Voorhees, W. B., R. R. Allmaras, and C. E. Johnson. 1981. Alleviating temperature stress. In: *Modifying the Root Environment to Reduce Crop Stress*, G. F. Arkin and H. Taylor, eds., pp. 217–266. St. Joseph, MI: American Society of Agricultural Engineering, Monograph 4.

Wagner-Riddle, C., A. Furon, N. L. McLaughlin, I. Lee, J. Barbeau, S. Jayasundara, G. Parkin, P. von Bertoldi, and J. O. N. Warland. 2007. Intensive measurement of nitrous oxide emissions from a corn-soybean-wheat rotation under two contrasting management systems over 5 years. *Glob. Change Biol.* 13:1722–1736.

Wang, J., S. Kang, F. Li, F. Zhang, Z. Li, and J. Zhang. 2008. Effects of alternate partial root-zone irrigation on soil microorganisms and maize root. *Plant Soil* 302:45–52.

Wang, J. G., F. S. Zhang, X. L. Zhang, and Y. P. Cao. 2000. Release of potassium from K-bearing minerals: Effect of plant roots under P deficiency. *Nutr. Cycling Agroecosyst.* 56:45–52.

Wasaki, J., T. Yamamura, T. Shinano, and M. Osaki. 2003. Secreted acid phosphatase is expressed in cluster roots of lupin in response to phosphorus deficiency. *Plant Soil* 248:129–136.

Watt, M., J. A. Kirkegaard, and J. B. Passioura. 2006. Rhizosphere biology and crop productivity: A review. *Aust. J. Soil Res.* 44:299–317.

Weller, D. M. 2007. Pseudomonas biocontrol agents of soilborne pathogens: Looking back over 30 years. *Phytopathology* 97:250–256.

Whalley, W. R., P. B. Leeds-Harrison, P. K. Leech, B. A. Risely, and N. R. A. Bird. 2004. The hydraulic properties of soil at root-soil interface. *Soil Sci.* 169:90–99.

Whalley, W. R., B. Riseley, P. B. Leeds-Harrison, N. R. A. Bird, P. K. Leech, and W. P. Adderley. 2005. Structural differences between bulk and rhizosphere soil. *Eur. J. Soil Sci.* 56:353–360.

Whipps, J. M. 2001. Microbial interactions and biocontrol in the rhizosphere. *J. Exp. Bot.* 52:487–511.

Whipps, J. M. and J. M. Lynch. 1986. The influence of the rhizosphere on crop productivity. *Adv. Microbial. Ecol.* 9:187–244.

Whitelaw, M. A. 2000. Growth promotion of plants inoculated with phosphate-solubilizing fungi. *Adv. Agron.* 69:99–151.

Wiersma, J. V. 2010. Nitrate-induced iron deficiency in soybean varieties with varying iron-stress responses. *Agron. J.* 102:1738–1744.

Wiersma, J. V. and J. H. Orf. 1992. Early maturing soybean nodulation and performance selected *Bradyrhizobium japonicum* strains. *Agron. J.* 84:449–459.

von Wiren, N., V. Romheld, J. L. Morel, A. Guckert, and H. Marschner. 1993. Influence of microorganisms on iron acquisition in maize. *Soil Biol. Biochem.* 25:371–376.

Wolf, I. and R. Russow. 2000. Different pathways of formation of N_2O, N_2 and NO in black earth soil. *Soil Biol. Biochem.* 32:229–239.

Yanai, J., D. J. Linehan, D. Robinson, I. M. Young, C. A. Hackett, K. Kyuma, and T. Kosaki. 1996. Effects of inorganic nitrogen application on the dynamics of the soil solution composition in the root zone of maize. *Plant Soil* 180:1–9.

Yang, X., W. Wang, Z. Ye, Z. He, and V. C. Baligar. 2004. Physiological and genetic aspects of crop plant adaptation to elemental stresses in acid soils. In: *The Red Soils of China*, eds., M. J. Wilson, Z. He, and X. Yang, pp. 171–218. Dordrecht, the Netherlands: Kluwer Academic Publishers.

Yeates, G. and P. R. Darrah. 1991. Microbial changes in a model rhizosphere. *Soil Biol. Biochem.* 23:963–971.

Young, I. M. and J. W. Crawford. 2004. Interactions and self-organization in the soil-microbe complex. *Science* 304:1634–1637.

Youssef, R. A. and M. Chino. 1988. Development of a new rhizobox system to study the nutrient status in the rhizosphere. *Soil Sci. Plant Nutr.* 34:461–465.

Youssef, R. A. and M. Chino. 1989. Root-induced changes in the rhizosphere of plants. I. pH changes in relation to the bulk soil. *Soil Sci. Plant Nutr.* 35:461–468.

Youssef, R. A., S. Kanazawa, and M. Chino. 1989. Distribution of the microbial biomass across the rhizosphere of barley (*Hordeum vulgare* L.) in soils. *Biol. Fertil. Soils* 7:341–345.

Zhang, F., J. Shen, L. Li, and X. Liu. 2004. An overview of rhizosphere processes related with plant nutrition in major cropping systems in China. *Plant Soil* 260:89–99.

Zhao, T. and H. Q. Ling. 2007. Effects of pH and nitrogen forms on expression profiles of genes involved in iron homeostasis in tomato. *Plant Cell Environ.* 30:518–527.

5 The Rhizosphere Chemistry of Flooded Rice

5.1 INTRODUCTION

Rice (*Oryza sativa* L.) is an important food crop which was grown on a land area of more than 155 million hectares worldwide in 2008 (USDA, 2009). Rice is one of the oldest cultivated crops on earth. It also is probably the world's most versatile crop that grows at more than 3000 m elevation in the Himalayas and at sea level in the deltas of the great rivers of Asia. Floating cultivars grow in water as deep as 4 m in Thailand, and in Brazil, rice is grown as a dryland crop much like wheat or corn. In West Africa, rice is grown in mangrove swamps. Rice, like barley (*Hordeum vulgare* L.), oats (*Avena sativa* L.), rye (*Secale cereale* L.), and wheat (*Triticum aestivum* L.), belongs to the Gramineae family. It was first domesticated about 10,000 years ago; however, its exact origin of domestication is not known. The domestication of rice could have occurred independently at several places in a broad belt from the foothills of the Himalayas to Vietnam and southern China (Chang, 1975). The geographical dispersal and the selection pressures of farming led to a large number of varieties of *O. sativa*, the Asian species. Another species, *Oryza glaberrima*, was later domesticated in western Africa (Hargrove, 1988). Rice is the staple food crop in the diet of about one-half of the world's population; it provides 35%–60% of the dietary calories consumed by nearly 3 billion people (Fageria et al., 2003). More than 90% of the world's rice is grown and consumed in Asia, where about 60% of the world's people live. Rice is also a staple food in Latin America, parts of Africa, and the Middle East. China and India are the largest producers and consumers of rice in the world.

Under normal conditions, lowland rice fields are flooded with water about 3–4 weeks after sowing. A water level of about 10–15 cm is maintained during the crop growth cycle and drained before harvest. Due to flooding, lowland rice suffers less from diseases, insects, and weeds compared to upland rice. These factors also contribute to higher yield of lowland rice compared to upland rice. Flooding or waterlogging eliminates oxygen from the rhizosphere and causes changes in the physical, biological, and chemical properties of soil. These changes are associated with physical reactions between the soil and water and also because of biological processes set in motion as a result of excess water or oxygen deficiency (Patrick and Mahapatra, 1968; Gao et al., 2004). Oxygen deficiency or exclusion in submerged soils can occur within a day after flooding. The oxygen movement through the flooding water is usually much slower than the rate at which oxygen can be reduced in the soil. This situation may result in the formation of two distinctly different layers in a waterlogged soil: on the top is an oxidized or aerobic surface layer where oxygen is present and at the bottom is a reduced or anaerobic layer in which no free oxygen is present.

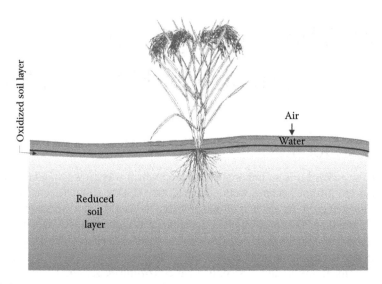

FIGURE 5.1   Oxidized and reduced soil layer in submerged rice soil.

Illustrated in Figure 5.1 is the thin oxidized layer (usually 1–20 mm in thickness) normally found at the interface between water and soil (Bouldin, 1986). The thickness of the oxidized surface layer is determined by the net effect of the oxygen consumption rate in the soil and the oxygen supply rate through the floodwater; a high consumption rate results in a thin oxidized surface layer, while a low consumption rate results in a thicker oxidized layer. Because of the higher demand for oxygen, a soil with an appreciable supply of readily decomposable organic matter usually has a thin oxidized layer (Patrick and Mahapatra, 1968). In addition, flooding also has major effects on the availability of macro- and micronutrients. Some nutrients are increased in availability to the crop, whereas others are subject to greater fixation or loss from the soil as a result of flooding (Patrick and Mikkelsen, 1971).

The presence of floodwater for part or all of the growing season requires that the rice root system is adapted to largely anaerobic soil conditions. The rice plant has adapted to this environment by transporting oxygen from the aerial portions of the plant to the root system via aerenchyma tissues (Yoshida, 1981; Fageria et al., 2003). Rice plants are genetically furnished with aerenchyma. Ando et al. (1983), using an oxygen electrode, showed that oxygen released from rice roots was taken from the atmosphere and released within 5 min through the stems and leaves into the rhizosphere. Horiguchi (1995) reported that in an anatomical study on the aerenchyma of the stem of rice and other graminaceous plants, it was found that the development of aerenchyma in rice varied depending on the conditions of the cultivation environment. The development of aerenchyma serves two purposes: one, it reduces the diffusive resistance to axial transport of gases between shoots and roots and, second, it enhances the oxygen supply to the root tips by reducing the oxygen demand per unit volume of tissue (Armstrong, 1979; Armstrong et al., 1991).

In flooded soils, rice roots developed more lysigenous intercellular space compared to roots grown in oxidized soils. Horiguchi (1995) also reported differences among rice cultivars in the development of aerenchyma in the stems and lysigenous intercellular space in the roots. A secondary adaptive mechanism is the development of an extensive lateral, fibrous root system located in the surface 1–2 mm of oxidized soil at the soil–water interface (Fageria et al., 2003). Oxygen diffusing through the water layer allows this zone of soil to remain oxidized. For these reasons, flooded rice normally has a shallow, fibrous root system (Wells et al., 1993). Figures 5.2 and 5.3

FIGURE 5.2 Lowland rice root growth at low N (0 mg kg^{-1}) (left) and at high N rate (300 mg kg^{-1}) (right).

FIGURE 5.3 Root growth of upland rice at low N (0 mg N kg^{-1}) (left) and at high N rate (300 mg N kg^{-1}) (right).

compare the root growth of lowland and upland rice at two N rates. The root growth of upland rice or aerobic rice was much longer at low as well as at high N rates compared to lowland or flooded rice. The aquatic environment not only influences the development of the root system but also alters the availability of several essential nutrients, affects nutrient uptake and use efficiency, fertilization practices, and makes rice especially unique among crop production systems. The enhancement of rice production and its sustainability are important features of grain production to benefit the world's 3.5 billion people who depend on rice for their livelihood and as their basic food. The objective of this chapter is to discuss the physical, biological, and chemical changes in the rhizosphere of lowland or flooded rice soils that may help in better nutrient management and, consequently, higher yields.

5.2   RICE ECOSYSTEMS

The ecosystem deals with the environmental conditions under which a crop is grown. Description of a crop ecosystem is important for adopting or improving production practices for higher yields. For rice ecosystem, the important environmental parameters are water regime and types of soils including texture, drainage, topography, and temperature. Rice is an ancient crop that has evolved along with man's knowledge of land, water, and soils, and is now grown in a multitude of environments (Garrity, 1984). It grows on dry and flooded soil, in low- and high-temperature environments, and in a number of soils. There are three principal rice ecosystems. These are classified as upland, lowland, and deepwater or floating cultivation. However, these ecosystems have been further classified into several subgroups based on an area's or country's rice production systems (De Datta, 1981; Garrity, 1984; Khush, 1984). Among the factors that will determine the relative productivity of the different ecosystems are technology used by the farmers, which is determined by their economic situation, the availability of credits to buy good quality seeds and fertilizers, and the availability of water. In addition, the research effort directed to the system, public and private investment in production and infrastructure, and the actual and predicted rice prices also determine the productivity of each ecosystem.

5.2.1   UPLAND ECOSYSTEM

Rice grown on well-drained sloppy or level lands that are not bunded, without water accumulation, and depend on rainfall for moisture is called upland rice (Fageria, 2001). This type of rice culture is also known as aerobic rice. This type of rice ecosystem is most common in Asia, Latin America, and Africa (De Datta, 1981). Latin America and Africa offer more areas for future upland rice expansion of the world's rice land. The yield of lowland rice is much higher compared to upland rice due to low biotic and abiotic stresses, an assured water supply, and the use of high inputs by farmers. For example, in Brazil, upland rice average yield is about 2.2 Mg ha^{-1}, whereas lowland rice yield is more than 5 Mg ha^{-1}. The lower

FIGURE 5.4   Upland rice crop grown on Oxisol, in the central part of Brazil.

FIGURE 5.5   Lowland rice crops in the state of Tocantins, in the central part of Brazil.

yield of upland rice is associated with biotic and abiotic stresses (Fageria, 2001). Upland rice has lower yield compared to lowland rice; however, its cost of production also is lower. Due to lower cost and lack of irrigation facilities, upland rice will continue to be an important component of cropping systems in South America, Africa, and Asia. Figures 5.4 and 5.5 show upland and lowland rice ecosystems or growth, respectively, in the central part of Brazil.

5.2.2   LOWLAND ECOSYSTEM

The lowland rice ecosystem is also known as flooded rice. Lowland rice is grown in rain-fed lowlands and in irrigated lowlands. Among rice ecosystems, the lowland ecosystem is most important because a large amount of rice produced worldwide comes from this ecosystem. About 76% of the global rice is produced from irrigated-lowland rice systems (Fageria et al., 2003). Lowland rice is grown on lands having sufficient water to flood during the major part of the growth cycle. In directly seeded rice, crop is generally flooded 25–30 days after sowing and water is drained about 1 week before harvesting. Generally, the water level is maintained at about 10–15 cm during crop growth. However, the actual water level depends on land leveling and water control system. From a yield sustainability standpoint, traditionally, wetland rice cultivation has been extremely successful. Moderate but stable yields have been maintained for thousands of years without deterioration of the environment (Roger, 1996). But maintaining the sustainability of rice-producing environments in the face of increased demands will require new concepts and agricultural practices. In Brazil, this ecosystem is broadly classified as irrigated rice. Irrigated rice is an important crop in Brazil due to economical and social reasons. Irrigated rice systems account for about 70% of the total rice production in Brazil.

5.2.3 DEEPWATER OR FLOATING RICE ECOSYSTEM

Deepwater rice is a general term used for rice culture or the variety that is planted when the standing water for a certain period of time is more than 50 cm. Floating rice is grown where maximum water depth ranges between 1 and 6 m for more than half of the growth duration (De Datta, 1981). Deepwater and floating rice are mostly grown in India, Bangladesh, Thailand, Myanmar, Vietnam, Kampuchea, Mali, Nigeria, and Indonesia (De Datta, 1981; Vergara, 1985). This deepwater rice covers more than 10 million hectares and millions of farmers depend on it for survival and security (Vergara, 1985). Although excess water is common to all deepwater rice, growth and development vary according to the crop establishment practice, and the onset, depth, speed, and duration of flooding. As floodwater rises, plants elongate, produce nodal roots, and may produce nodal tillers. The water level may reach 100 cm near the end of the monsoon season, when photoperiod-sensitive rice varieties flower. As the water recedes, grain ripens. Crop duration may be 300 days, and harvest may be on dry or wet fields (Vergara, 1985). This type of rice culture requires varietal and cultural practice improvements to increase yields.

The major distinguishing features of floating rice appear to be a semi-prostrate appearance near the base of the plant, even in the early stages of growth under shallow water, the ability to elongate rapidly under rising water conditions (up to 10 cm day^{-1}), the formation of adventitious roots at the higher nodes, and a distinct photoperiodic type of flowering behavior (Jackson et al., 1972). Under deepwater conditions, the leaves appear to float on the surface. When the water recedes, and flowering occurs, a tangled mass of stems results. However, the upper portion of the stem usually exhibits phototropism, which reduces the damage to the panicle caused by water and mud and which also facilitates harvesting (Jackson et al., 1972).

5.3   TYPE OF SOILS USED FOR LOWLAND RICE CULTIVATION

Lowland rice is produced on a variety of soils in different agroecological regions of the world. Due to the heterogeneity of agroecological regions, the pedogenetic and morphological characteristics of soils used to grow rice also vary considerably. The soils used for rice production worldwide are distributed over the 10 soil orders (Moormann, 1978; Hudnall, 1991). Moorman (1978) summarized that rice is grown worldwide on all soil orders identified in the soil classification system (USDA, 1975). Worldwide, the wide array of soils used to produce rice results in an equally diverse assortment of management practices implemented for successful rice production on these soils. Murthy (1978) reported that the soils on which rice grows in India are so extraordinarily varied that there is hardly a type of soil, including salt-affected soils, on which it cannot be grown with some degree of success. In Brazil, flooded rice is mainly grown on Alfisols, Vertisols, Inceptisols, Histosols, and Entisols (Moraes, 1999). In Sri Lanka, rice is grown on Alfisols, Ultisols, Entisols, Inceptisols, and Histosols (Panabokke, 1978). In Indonesia, the main rice soils are Entisols, Inceptisols, Vertisols, Ultisols, and Alfisols (Soepraptohardjo and Suhardjo, 1978). Raymundo (1978) reported that in the Philippines, the soils used for wetland rice production are mainly Entisols, Inceptisols, Alfisols, and Vertisols. In Europe, rice is planted in limited areas in Albania, Bulgaria, France, Greece, Hungary, Italy, Portugal, Romania, Spain, and Yugoslavia where the predominant soil orders are Inceptisols, Entisols, and Vertisols (Matsuo et al., 1978). In the United States, rice is grown primarily on Alfisols, Inceptisols, Mollisols, and Vertisols (Flach and Slusher, 1978). However, in Florida, a small hectarage of rice is produced on Histosols. Most of the soils used for rice production in the United States and some other geographic areas have properties that make them ideally suited for flood-irrigated rice. The soils are relatively young, contain significant amounts of weatherable minerals, and have relatively high base saturations despite the fact that some are in areas of high precipitation (Flach and Slusher, 1978).

Soil parameters for optimum rice yields are optimum soil depth, compact subsoil horizon, good soil moisture retention, good internal drainage, good fertility, and a favorable soil structure (Fageria et al., 2003). Clayey to loamy clay texture soils are appropriate for lowland rice production. Permeable, coarse-textured soils are less suitable for flood-irrigated rice production because they have low water- or nutrient-holding capacities. A sufficient supply of nutrients is one of the key factors required to improve crop yields and maintain sustainable agricultural production on these soils. Flood-irrigated rice is an important crop that needs to be included in the cropping system of these poorly drained areas during the rainy seasons. During dry periods, other crops can be planted in rotation, provided there is proper drainage. These soils generally have an adequate natural water supply throughout the year, but are acidic and require routine applications of lime if legumes are grown in rotation with rice. The physical and chemical properties of Várzea soils of Brazil are presented in Tables 5.1 through 5.4. Data in these tables show that chemical and physical properties varied largely from state to state and from municipality to municipality within states.

TABLE 5.1

Chemical Properties of Várzea (Lowland) Soils of Some States of Brazil

State	Mo (g kg⁻¹)	pH in H₂O	P (mg kg⁻¹)	K (mg kg⁻¹)	Ca (cmol_c kg⁻¹)	Mg (cmol_c kg⁻¹)	Al (cmol_c kg⁻¹)
Goiás	42	5.2	15.2	85	4.7	2.6	1.5
Mato Grosso	16	5.1	6.9	68	2.5	1.4	1.3
Mato Grosso do Sul	69	5.3	21.7	75	7.8	3.4	1.1
Paraná	138	4.3	36.4	84	2.6	1.8	4.4
Minas Gerais	25	5.0	17.7	133	3.9	1.6	0.5
Rio Grande do Norte	25	7.1	45.1	168	10.4	6.6	0.1
Piauí	10	5.6	13.6	115	10.3	6.7	0.7
Maranhão	8	4.8	1.9	82	6.7	10.7	1.5
Average	42	5.3	19.8	101	6.1	4.4	1.5

Sources: Fageria, N.K. et al., *Commun. Soil Sci. Plant Anal.*, 22, 1631, 1991; Fageria, N.K. et al., *Pesq. Agropec. Bras.*, 29, 267, 1994; Fageria, N.K. et al., *Commun. Soil Sci. Plant Anal.*, 28, 37, 1997. Values are 0–20 cm soil depth and lowland rice is generally grown on these soils during rainy season.

TABLE 5.2

Micronutrient Concentrations, Cation Exchange Capacity (CTC), Base Saturation (V), and Aluminum Saturation (M) of Várzea (Lowland) Soils of Some States of Brazil

State	Cu (mg kg⁻¹)	Zn (mg kg⁻¹)	Fe (mg kg⁻¹)	Mn (mg kg⁻¹)	CTC (cmol_c kg⁻¹)	V (cmol_c kg⁻¹)	M (cmol_c kg⁻¹)
Goiás	7.4	3.0	436	42	27	33	16
Mato Grosso	1.3	1.4	263	33	12	33	32
Mato Grosso do Sul	11.9	2.5	193	23	26	42	18
Paraná	6.3	1.5	65	12	52	22	29
Minas Gerais	2.9	7.9	627	98	15	42	10
Rio Grande do Norte	1.9	2.0	307	163	39	95	1
Piauí	3.4	3.2	382	61	30	81	3
Maranhão	0.9	3.7	320	43	29	70	7
Média	4.5	3.2	324	59	29	54	15

Sources: Fageria, N.K. et al., *Commun. Soil Sci. Plant Anal.*, 22, 1631, 1991; Fageria, N.K. et al., *Pesq. Agropec. Bras.*, 29, 267, 1994; Fageria, N.K. et al., *Commun. Soil Sci. Plant Anal.*, 28, 37, 1997. Values are 0–20 cm soil depth and lowland rice is generally grown on these soils during rainy season.

TABLE 5.3
Textural Analysis of Várzea (Lowland) Soils of Some States of Brazil

State	Sand (g kg⁻¹)	Silt (g kg⁻¹)	Clay (g kg⁻¹)
Goiás	350	220	422
Mato Grosso	408	282	310
Mato Grosso do Sul	394	250	356
Paraná	600	187	213
Minas Gerais	223	184	593
Rio Grande do Norte	431	308	261
Piauí	301	335	364
Maranhão	118	410	472
Average	354	272	374

Sources: Fageria, N.K. et al., *Commun. Soil Sci. Plant Anal.,* 22, 1631, 1991; Fageria, N.K. et al., *Pesq. Agropec. Bras.,* 29, 267, 1994; Fageria, N.K. et al., *Commun. Soil Sci. Plant Anal.,* 28, 37, 1997. Values are 0–20 cm soil depth and lowland rice is generally grown on these soils during rainy season.

5.4 PHYSICAL, BIOLOGICAL, AND CHEMICAL CHANGES IN THE FLOODED OR SUBMERGED SOILS

With the omission of oxygen from the large part of the soil profile, physical, biological, and chemical changes occur in the submerged or flooded rice soils. These changes vary with the type of soil, the presence of microbial biomass, the quality and quantity of organic matter, the cultivar planted, and the level of soil fertility. In addition, these changes affect the availability of essential plant nutrients and, consequently, plant growth and yield. Furthermore, the percolation rate decreases with flooding because of physical and chemical changes such as swelling, dispersion, and disintegration of soil aggregates; the reduction of soil pores by microbial activity, and organic matter decomposition that reduces the binding effect of aggregates and causes the soil to seal off (Wickham and Singh, 1978).

5.4.1 Physical Changes

Submerging or waterlogging rice-growing soils create conditions markedly different from those of a well-drained aerobic soil. As soon as soils of lowland rice are flooded, oxygen level begins to decline. The rate of decline is very fast, and within 6–10h after flooding the O_2 level drops to near zero (Patrick and Mikkelsen, 1971). The rapid decline of O_2 from the soil is accompanied by an increase of other gases produced through microbial respiration. The major gases that accumulate in the flooded soils are carbon dioxide (CO_2), methane (CH_4), nitrogen (N_2), and hydrogen (H_2). Patrick and Mikkelsen (1971) reported that the composition of these gases may vary from 1%

TABLE 5.4

**Chemical and Textural Properties of Várzea (Lowland) Soils of State
of Rio Grande do Sul of Brazil**

Location/Municipality	Mo (g kg⁻¹)	pH in H_2O	Ca+Mg (cmol$_c$ kg⁻¹)	K (mg kg⁻¹)	Al (cmol$_c$ kg⁻¹)
Palmares	20	5.0	0.5	20	0.6
Vacacaí	119	5.0	1.3	20	1.7
Pelotas	9	5.4	3.1	27	1.1
Meleiro	8	4.8	0.9	39	2.9
Colégio	130	4.5	18.6	234	1.2
Jundaí	32	4.4	3.9	39	8.8
Jacinto Machado	10	5.5	10.5	78	0.2
Curumim	40	5.5	1.2	27	3.8
Average	41	5.0	5.0	61	2.5

Location/Municipality	V (%)	M (%)	Sand (g kg⁻¹)	Silt (g kg⁻¹)	Clay (g kg⁻¹)
Palmares	15	70	840	60	100
Vacacaí	23	53	640	260	100
Pelotas	52	24	440	360	190
Meleiro	9	73	370	280	350
Colégio	44	6	240	270	490
Jundaí	12	67	20	310	670
Jacinto Machado	74	1	420	270	280
Curumim	7	73	790	130	80
Average	28	46	470	246	284

Source: Klamt, E. et al., Várzea soils of state of Rio Grande do Sul, Faculty of Agronomy, Department of
Soils, Technical bulletin 4, Porto Alegre, Brazil, Federal University of Rio Grande do Sul, 1985.
Values are 0–20 cm soil depth and lowland rice is generally grown on these soils during rainy season. Rio
Grande do Sul is the largest lowland rice-producing state in Brazil. V (base saturation) and M (alumi-
num saturation).

to 20% CO_2, 10% to 95% N_2, 15% to 75% CH_4, and 0% to 10% H_2. This variation may
be associated with the presence of microbial biomass, the presence of organic matter
and inorganic substances, and also influenced by the cultivar planted.

Flooding may also alter the soil temperature and may disintegrate soil structure.
Patrick and Mahapatra (1968) reported that the structure of the soil, which is considered to be of greater importance in well-drained soils, is destroyed by a minimum of
manipulation due to the weakening effect of water on the bonds holding the soil particles together as stable aggregates. At a given soil moisture content and as bulk density increases, the thermal conductivity increases (Ghildyal and Tripathi, 1971). As
the thermal conductivity of soil particles is higher than that of air, increased density
decreases the volume of gases and increases thermal contact between the soil particles.
As a result, the thermal conductivity increases (Ghildyal, 1978). Permeability to water
may be reduced by the clogging of soil pores due to physical, chemical, and biological
changes. This may help to reduce percolation of water and the leaching of nutrients.

FIGURE 5.6 Puddling is done in lowland rice plots in the state of Para in the Amazon basin Brazil (earlier project of Jari, funded by D. K. Ludwig).

In the lowland rice production system, the subsoil layer is compacted with the help of a roller and the process is known as puddling. According to the Soil Science Society of America (2008), puddling is defined as any process involving both shearing and compactive forces that destroy natural structure and result in a condition of greatly reduced pore space. Ghildyal (1978) defined puddling as the mixing of soil with water to render it impervious. Intensive tillage by repeated plowing of a wet soil breaks down coarse aggregates and the mean particle size decreases. Soil compaction affects water retention characteristics, water intake rates, and the exchange of gases. In compacted soil, bulk density, microvoids, thermal conductivity and diffusivity, and nutrient mobility increase, and macrovoids, hydraulic conductivity, and water intake rates decrease. Medium-textured soils are most susceptible to compaction.

Puddling is very common in Asian rice-producing countries. Puddling in intensive wetland cultivation breaks the natural aggregates to finer fractions. It decreases the apparent specific volume and hydraulic conductivity, creates an anaerobic environment, and affects Eh and pH (Ghildyal, 1978). Ghildyal (1978) reported that rice root growth, nutrient uptake, and water use are favorably affected by moderate compaction of a flooded soil where the soil strength is low. Gathala et al. (2011) reported that although puddling is known to be beneficial for rice growth, it can adversely affect the growth and yield of a subsequent upland crop because of its adverse effects on the physical properties of soil, which include poor soil structure, suboptimal permeability in the subsurface layer, poor soil aeration, and soil compaction. Figure 5.6 shows that puddling is done in lowland rice plots before sowing the pregerminated seeds of rice. In many Asian countries, rice is transplanted in the puddle fields by small farmers.

5.4.2 Biological Changes

In waterlogged or flooded rice soils, aerobic microorganisms become quiescent or die, and facultative and obligate anaerobic bacteria proliferate. These new microorganisms bring many biological changes in the reduced soil environment. In the absence of oxygen, many facultative and obligate anaerobes oxidized organic compounds with the release of energy in a process called "anaerobic fermentation" (Patrick and Mikkelsen, 1971). Anaerobic fermentation usually produces lactic acid as a first product. This is subsequently converted to acetic, formic, and butyric acids. Among aerobic organisms, oxygen serves as the electron acceptor, but in anaerobic forms, either an organic metabolic by-product or some inorganic substance must substitute for oxygen (Patrick and Mikkelsen, 1971). In the flooded soils, organic matter decomposition is retarded due to the lower carbon assimilation rate of anaerobic bacteria. In a submerged soil, the facultative and obligate anaerobic organisms utilize NO_3^-, Mn^{4+}, Fe^{3+}, SO_4^{2-}, dissimilation products of organic matter, CO_2, and H^+ ions as electron acceptors in their respiration, reducing NO_3^- to N_2, Mn^{4+} to Mn^{2+}, Fe^{3+} to Fe^{2+}, SO_4^{2-} to S^{2-}, CO_2 to CH_4, and H^+ to H_2 gas (Patrick and Reddy, 1978).

5.4.3 Chemical Changes

The most important chemical changes that occur in flooded or submerged rice soils are pH, redox potential, and ionic strength or electrical conductivity. In addition, various reducing compounds such as hydrogen sulfide (H_2S), ferrous ions (Fe^{2+}), and nitrite ions (NO_2^-) tend to be readily produced (Horiguchi, 1995). These changes occur due to oxygen depletion in the waterlogged soils.

5.4.3.1 pH

Soil pH is an important chemical property due to its influence on soil microorganisms, metal dissolution chemistry, and availability of nutrients to plants. As a single measurement, pH describes more than relative acidity and alkalinity. It is determined by a pH meter using a glass electrode and in a specific soil–solution ratio. Usually, distilled water or 0.01 M $CaCl_2$ or 1 M KCl solution is used for soil pH determination. Schofield and Taylor (1955) found that the addition of 0.01 mol L^{-1} $CaCl_2$ to the soil suspension decreased the pH by approximately 0.50 pH unit from that of H_2O. Seasonal rainfall variation could, therefore, be expected to affect pH readings that are performed in water because soluble salts will vary from season to season (Miller and Kissel, 2010). The standard measurement of soil pH in North America is a 1:1 soil/water suspension, which is performed in approximately 55% of the soil analysis laboratories enrolled in North America Proficiency Testing (Miller and Kissel, 2010). In Brazil, most laboratories use 1:2.5 soil/water suspensions in routine soil analyses for making fertilizer and lime recommendations.

Soil pH indicates acidity, alkalinity, or neutrality of a soil. Soil pH 7.0 is a neutral value and above this pH soils are designated as alkaline and below this pH soils are acidic in reaction. The pH of acid soils increases and alkaline soils decreases due to flooding. Overall, pH of most soils tends to change toward neutral after flooding. An equilibrium pH in the range of 6.5–7.5 is usually attained (Patrick and Reddy, 1978). A majority of oxidation–reduction reactions in flooded soils involve either consumption or production of H^+/OH^- ions (Ponnamperuma, 1972). The increase in pH of acid soils is mainly determined by the reduction of iron and manganese oxides, which consume H^+ ions. Actually, in all the reduction processes encountered in soils, the consumption of hydrogen ions is involved (Yu, 1985). These reduction processes are shown by the following equations:

$$Fe_2O_3 + 6H^+ + 2e^- \leftrightarrow 2Fe^{2+} + 3H_2O$$

$$MnO_2 + 4H^+ + 2e^- \leftrightarrow Mn^{2+} + 2H_2O$$

However, the rate and extent of change in pH are determined by a variety of factors. Among these factors, the presence of decomposable organic matter plays a prominent role. It is the production of organic reducing substances that causes the development of reduction processes of inorganic compounds and, thus, the consumption of hydrogen ions. On the other hand, organic acids and CO_2 produced during the decomposition of organic matter decrease the pH (Yu, 1991).

The decrease in the pH of alkaline soils is associated with the microbial decomposition of organic matter, which produces CO_2, and the produced CO_2 reacts with H_2O to form carbonic acid, which dissociates into H^+ and HCO_3^- ions. Patrick and Reddy (1978) reported that the decrease in the pH of alkaline and calcareous submerged soils is associated with Na_2CO_3–H_2O–CO_2 and $CaCO_3$–H_2O–CO_2 systems, respectively. Figure 5.7 shows changes in soil pH of lowland rice collected from four locations of the state of Rio Grande do Sul, Brazil. It can be seen from Figure 5.7 that soil pH increases with the flooding period and stabilizes around 56 days after flooding in all the soils. However, the magnitude of pH change differs from soil to soil. The pH of most agricultural soils is in the range of 4–9 (Fageria, 2009). The most suitable pH for the growth of annual crops like soybean, corn, dry bean, and wheat in Brazilian lowland soils is around 6.5 (Fageria and Baligar, 1999).

The direct effect of soil pH on rice growth is not very important, because the rice plant can adapt to a wide range of pH. Yu (1991) reported that the growth of three

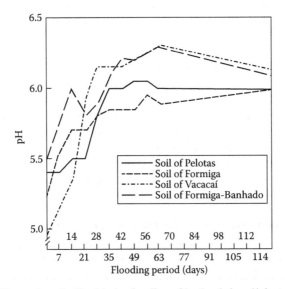

FIGURE 5.7 Change in soil pH with the flooding of lowland rice. (Adapted from Moraes, J.F.V. and Freira, C.J.S., *Pesq. Agropec. Bras.*, 9, 45, 1974, EMBRAPA.)

lowland rice cultivars was not affected significantly in the pH range of 5–7. However, the change in the pH affects a series of other properties of the soil that may directly or indirectly influence rice growth.

5.4.3.2 Oxidation–Reduction Potential

Oxidation–reduction or the redox potential has a significant influence on the chemistry of iron and other nutrients in the submerged soils. It is the best single indicator of the degree of anaerobiosis in the flooded soil, and allows reasonable predictions to be made concerning the behavior of several essential plant nutrients (Patrick and Mikkelsen, 1971). Oxidation is the donation and reduction is the acceptance of electrons from other substances. Oxidizing agents accept electrons from other substances and thereby reduce themselves. Reducing agents donate electrons to other substances. For example, iron (II) is an electron donor or a reducing agent when it oxidizes to iron (III). Hydrogen peroxide (H_2O_2) is an oxidizing agent when it accepts electrons from organic matter and oxidizes them to CO_2 (Bohn et al., 1979).

Oxidation–reduction potential is measured in millivolts and the symbol used for this chemical change in flooded soil is Eh. Oxidized soils have a redox potential in the range of +400 to +700 mV, whereas waterlogged soils' redox potential is generally in the range of −250 to −300 mV (Patrick and Mahapatra, 1968). Important oxidation–reduction processes that occur in waterlogged soils are presented in Table 5.5. Some of the oxidized soil components that undergo reduction after oxygen is depleted are reduced sequentially; that is, all of the oxidized components of one system will be reduced before any of the oxidized components of another system begin to be reduced. Others overlap during reduction (Patrick and Reddy, 1978). As the O_2 depletes from the waterlogged soils, reduction processes occur in sequence. Nitrate and manganese compounds are reduced first, then ferric compounds are reduced to the ferrous form, and, last, sulfate is reduced to sulfide. The redox potential or oxidation–reduction potential decreases with the flooding of rice soils (Figure 5.8).

A rapid decline in the redox potential is characteristic of soils with low contents of reducing iron and manganese and high organic matter content. Iron and manganese compounds serve as buffers against the development of reducing conditions in the soil (Patrick and Mahapatra, 1968). The critical redox potentials for Fe reduction and consequent dissolution are between +300 and +100 mV at pH 6 and 7, and −100 mV at pH 8, while at pH 5, appreciable reductions occur at +300 mV (Gotoh and Patrick, 1974). Oxidation–reduction or potential reduction values for oxidized and submerged soils and reduction processes are given in Table 5.6.

5.4.3.3 Soil Redox Status versus Straw Incorporation

Soil organic N is the largest source of plant-available N for rice, representing 50%–80% of total N assimilated by the crop (Mikkelsen, 1987; Eagle et al., 2001). This means that maintenance of or increase in soil organic matter is fundamental for adequate N supply to rice. Crop residue is an important source of soil organic matter. It is widely reported in the literature that crop residue management practices

TABLE 5.5
Thermodynamic Sequence of Reduction Processes in the Submerged Soils

Reaction	Redox Potential $E_0{}^a$ (V)
$O_2 + 4H^+ + 4e^- \leftrightarrow 2H_2O$	0.81
$2NO_3^- + 12H^+ + 10e^- \leftrightarrow N_2 + 6H_2O$	0.74
$MnO_2 + 4H^+ + 2e^- \leftrightarrow Mn^{2+} + 2H_2O$	0.40
$CH_3COCOOH + 2H^+ + 2e^- \leftrightarrow CH_3CHOHCOOH$	−0.16
$Fe(OH)_3 + 3H^+ + e^- \leftrightarrow Fe^{2+} + 3H_2O$	−0.19
$SO_4^{2-} + 10H^+ + 8e^- \leftrightarrow H_2S + 4H_2O$	−0.21
$CO_2 + 8H^+ + 8e^- \leftrightarrow CH_4 + 2H_2O$	−0.24
$N_2 + 8H^+ + 6e^- \leftrightarrow 2NH_4^+$	−0.28
$NADP^+ + 2H^+ + 2e^- \leftrightarrow NADPH$	−0.32
$NAD^+ + 2H^+ + 2e^- \leftrightarrow NADH$	−0.33
$2H^+ + 2e^- \leftrightarrow H_2$	−0.41
Ferredoxin (ox) + $e^- \leftrightarrow$ Ferrodoxin (red)	−0.43

Sources: Ponnamperuma, F.N., *Adv. Agron.*, 24, 29, 1972; Ponnamperuma, F.N., Physicochemical properties of submerged soils in relation to fertility, in *The Fertility of Paddy Soils and Fertilizer Application for Rice*, ed., Food and Fertilizer Technology Center, pp. 1–27, Taipei City, Taiwan, Food and Fertilizer Technology Center, 1976; Patrick, W.H. Jr. and Reddy, C.N., Chemical changes in rice soils, in *Soils and Rice*, ed., IRRI, pp. 361–379, Los Banos, Philippines, IRRI, 1978.

[a] E_0 corrected to pH 7.0.

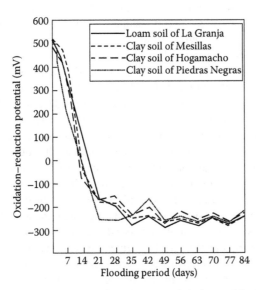

FIGURE 5.8 Influence of flooding on the redox potential of some Mexican soils. (Adapted from Moraes, J.F.V. and Freira, C.J.S., *Pesq. Agropec. Bras.,* 9, 45, 1974.)

TABLE 5.6

Range of Oxidation–Reduction Potential Values in Oxidized and Submerged Soils at which Reduction Processes Occur

Soil Moisture/Reduction Processes	Redox Potential (mV)
Well-oxidized soils	+700 to +500
Moderately reduced soils	+400 to +200
Reduced soils	+100 to −100
Highly reduced soils	−100 to −300
NO_3^- to N_2	+280 to +220
Mn^{4+} to Mn^{2+}	+280 to +220
Fe^{3+} to Fe^{2+}	+180 to +150
SO_4^{2-} to S^{2-}	−120 to −180
CO_2 to CH_4	−200 to −280
O_2 to H_2O	+380 to +320
Absence of free O_2	+350

Sources: Adapted from Patrick, W.H. Jr., *Nature,* 212, 1278, 1966; Patrick, W.H. Jr. and Reddy, C.N., Chemical changes in rice soils, in *Soils and Rice,* ed., IRRI, pp. 361–379, Los Banos, Philippines, IRRI, 1978; Marschner, H., *Mineral Nutrition of Higher Plants,* 2nd edn., New York, Academic Press, 1995; Fageria, N.K. et al., *J. Plant Nutr.,* 31, 1676, 2008.

can affect N immobilization and stabilization processes important to the efficient utilization of N from fertilizers, crop residues, and soil organic matter (Bird et al., 2002). When one considers agricultural production on a global basis, greater than 50% of the production is crop residues or straw (Lafond et al., 2009). Crop residue production is a function of the intensity of cropping, the quantities of inorganic fertilizers used, mainly N and P, organic amendments, conservation tillage, crop rotation, and inherent soil fertility (Lafond et al., 2009). Figures 5.9 and 5.10 show increased straw production of lowland rice with the addition of N fertilizer. Plant residues, roots, and root exudates are the major sources of C input into terrestrial ecosystems, with animal manures representing a secondary source, although indirectly derived from plant materials. The rate of C input is the most important factor in determining whether C can be maintained or increased in the soil. Straw incorporation is an important strategy in improving organic matter content of the soils and recycling nutrients. To have a positive effect on plant growth, the added straw should be decomposed. The decomposition of straw is a dynamic and complex process, involving physical, chemical, and biological reactions (Shechter et al., 2010).

Previous work from long-term rice management studies in tropical (Cassman et al., 1996; Bellakki et al., 1998) and temperate (Eagli et al., 2000; Bird et al., 2001) climates indicate increased plant-available soil N supply after 5–10 years of straw incorporation. In addition, after many years of straw incorporation, a sustained,

FIGURE 5.9   Shoot dry matter yield of genotype BRS Tropical at different N rate: (left to right) 0, 50, 100, 150, and 200 kg N ha^{-1}.

FIGURE 5.10   Shoot dry matter yield of genotype BRS Jaçanã at different N rate: (left to right) 0, 50, 100, 150, and 200 kg N ha^{-1}.

greater soil microbial biomass (SMB) C and N was reported (Powlson et al., 1987; Bird et al., 2001). An increase in SMB can affect C and N sequestration rates of fertilizer and crop residues through greater immobilization of and conversion to stable soil organic matter as well as through greater mineralization of stabilized soil organic matter and N (Bird et al., 2001). Furthermore, crop residue management has affected the utilization of N fertilizer in rice (Huang and Broadbent, 1989; Bird et al., 2001). These studies indicate that long-term straw management in lowland rice can affect the size and stability of the soil N supply (Bird et al., 2002).

There is no single indicator of straw quality that can predict residue decomposition (Kumar and Goh, 2000). Moisture, aeration, and temperature are principal soil factors determining the rate of organic residue decomposition. Below a certain critical moisture level biological processes are arrested, while at high moisture levels anaerobic conditions result (Linquist et al., 2006). Decomposition under anaerobic conditions is thought to be slower than under aerobic conditions (Tate, 1979). In a lab study, Pal and Broadbent (1975) reported that 14% more straw C was lost from soil at 60% water-holding capacity compared to 150% water-holding capacity after 4 months of incubation. The potential for N immobilization under flooded soil conditions is thought to be less than in aerobic soils (Broadbent and Nakashima, 1970). However, in pot and field studies, N immobilization has resulted in N deficiencies in rice (Bacon et al., 1989; Becker et al., 1994).

The structural components of residue, C and N content, and C/N ratio are known to be indicators of residue decomposition and N mineralization (Baggie et al., 2004; Goh and Tutuna, 2004; Stubbs et al., 2009). High hemicellulose is linked to rapid decomposition, and high lignin content, high C/N ratio, and low total N are associated with slower breakdown (Stubbs et al., 2009). Williams et al. (1968) reported that a straw N concentration of 0.54% N was the critical level of straw N, determining whether or not N immobilization would affect yield response in single growing seasons. The initial stage of degradation is a rapid loss of readily decomposable fractions such as polysaccharides and protein, followed by a slower stage of degradation and transformation of the refractory plant components, such as lignin and aliphatic biopolymers (Nierop et al., 2001). The extent of decomposition and transformation depends on the amount, composition, and properties of the plant materials as well as on microbial activity and soil properties (Stevenson, 1994).

Since few commercial uses of rice straw have been found, incorporating rice straw into the soil is an alternative to burning to improve air quality as well as the physical and chemical properties of soil (Gao et al., 2004). Gao et al. (2002) reported that straw incorporation in flooded rice soils did enhance reducing conditions. The reducing reactions occur in sequence, that is, O_2, NO_3^-, Mn (III, IV), Fe (III), SO_4^{2-}, and CO_2. However, sometimes significant overlapping of these reactions may occur (Patrick and Jugsujinda, 1992; Gao et al., 2002). A number of studies have sought to determine the intensity of reduction at which each of these oxidized electron acceptors is reduced following the curtailment of O_2 (Gotoh and Patrick, 1972, 1974). These redox processes yield products such as fatty acids and hydrogen sulfide that may affect rice yield.

Gao et al. (2004) reported that observations in the field plots revealed that straw incorporation into the soil was not likely to cause an adverse effect on rice plants on a large scale. However, sulfide toxicity symptoms on rice in randomly localized sites were observed mostly in the drain outlet ends of the plots. Sulfide toxicity symptoms were observed with blackened roots, much shorter plant height, and fewer number of standing plants in addition to the rotten-egg odor of collected soil samples in the affected area, compared with the healthy plant areas. Other literature reported that symptoms for sulfide toxicity on rice included gray-green-colored leaves, sterile florets and subsequent reduced grain yield, and death in severe cases (Allam and Hollis, 1972; Kuo and Mikkelsen, 1981). Gao et al. (2004) reported that in California the average straw return rate is about 4 g straw kg^{-1} soil and with this amount of straw sulfide toxicity is not expected on a large scale. Hence, it can be concluded that the adverse effect of straw incorporation in flooded rice soils depends on the quantity of rice straw incorporated.

5.4.3.4 Ionic Strength

Ionic strength is defined as the measure of the electrical environment of ions in a solution. Ionic strength can be calculated by using the following formula (Fageria et al., 2008):

$$\text{Ionic strength} = \frac{1}{2} \sum M_i Z_i^2$$

where
  M is the molarity of the ion
  Z_i is the total charge of the ion (regardless of sign)
  $\sum$ is a symbol meaning the "sum of"

Sodium-affected soils may lose their fertility due to clay dispersion, structure collapse, or reduction in hydraulic conductivity (Sumner, 1993; Saejiew et al., 2004). In rice fields, flooding with freshwater may lower the ionic strength of the soil solution to critical levels where clay dispersion can occur. There are extended areas of Na-affected rice fields over the world, like in West Africa (Ceuppens and Wopereis, 1999), Australia (Chartres, 1993; Naidu et al., 1993), and Asia (Qadir et al., 1998). Concentration of ions in the soil solution is measured by electrical conductivity. Ionic strength of submerged soil increases with the release of macro- and micronutrients in the soil solution (Patrick and Mikkelsen, 1971; Figure 5.11).

5.4.3.5 Nutrient Availability

Reducing condition in flooded rice soils changes the concentration and forms of applied as well as native soil nutrients. Hence, availability of essential macro- and micronutrients is significantly influenced in flooded rice soils (Patrick and Wyatt, 1964; Patrick et al., 1986).

5.4.3.5.1 Nitrogen

Nitrogen is a key nutrient in improving growth and yield of crop plants in all agro-ecosystems. Its main role is in increasing the photosynthesis process in plants, which is associated with improving grain yield. Presently, 50% of the human population relies on N fertilizer for food production. Since the green revolution in the 1960s, the application of chemical N fertilizers on fertilizer-responsive and lodging-resistant modern crop cultivars of rice has boosted food output by about 260% (Ladha et al., 2005). Figure 5.12 shows the influence of N on shoot or straw growth of three lowland rice cultivars. Similarly, the influence of N on grain yield of three lowland rice genotypes to N fertilization is presented in Figure 5.13. Shoot dry weight increased linearly with the increasing N rate in the range of 0–200 kg ha^{-1}. Similarly, grain yield of two genotypes increased linearly and one increased quadratically with the increasing N rate in the 0–200 kg ha^{-1}

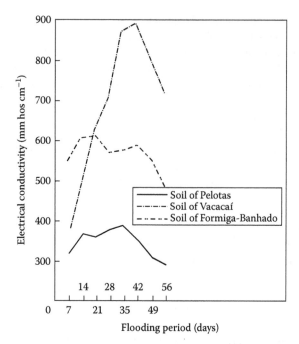

FIGURE 5.11 Influence of flooding on electrical conductivity of some Mexican soils. (Adapted from Moraes, J.F.V. and Freira, C.J.S., *Pesq. Agropec. Bras.*, 9, 45, 1974.)

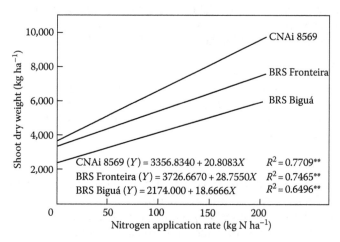

FIGURE 5.12 Influence of N on shoot dry weight of three lowland rice genotypes. *Significant at the 5% probability level and **Significant at the 1% probability level.

(Figure 5.13). Nitrogen is responsible for increasing yield components like panicles or heads in cereals and pods in legumes (Fageria, 2009). A high-yielding rice crop takes up 150–200 kg N ha^{-1} and soil N mineralization accounts for at least half of this amount (Cassman et al., 1996). Table 5.7 shows straw and grain yield and N uptake by straw and grain of lowland rice grown on a Brazilian Inceptisol.

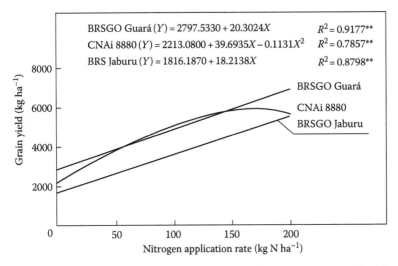

FIGURE 5.13 Influence of N on grain yield of three lowland rice genotypes. *Significant at the 5% probability level and **Significant at the 1% probability level.

TABLE 5.7
Nitrogen Uptake by Straw and Grain of Lowland Rice Grown on a Brazilian Inceptisol under Different N Timing Treatments

Treatment	Straw Yield (kg ha⁻¹)	Grain Yield (kg ha⁻¹)	N Uptake in Straw (kg ha⁻¹)	N Uptake in Grain (kg ha⁻¹)
T_1	9216	8909ab	57	101ab
T_2	9352	9674a	58	121a
T_3	9671	9570a	55	111ab
T_4	9161	9153a	57	110ab
T_5	10.089	9437a	61	111ab
T_6	8774	9176a	59	105ab
T_7	8993	8240b	52	90b
Average	9322	9166	57	107
F-test	NS	**	NS	**
CV (%)	9	6	19	9

Source: Adapted from Fageria, N.K. and Prabhu, A.S., *Pesq. Agropec. Bras.*, 39, 123, 2004.

T_1, all N applied at sowing; T_2, 1/3N applied at sowing + 1/3N applied at active tillering + 1/3N applied at primordium initiation; T_3, 1/2N applied at sowing + 1/2N applied at active tillering; T_4, 1/2N applied at sowing + 1/2N applied at primordium initiation; T_5, 2/3N applied at sowing + 1/3N applied at active tillering; T_6, 2/3N applied at sowing + 1/3N applied at promordium initiation; T_7 = 1/3N applied at sowing + 2/3N applied at initiation of tillering. Means followed by the same letter in the same column are statistically not different at 5% probability level by Tukey's test.

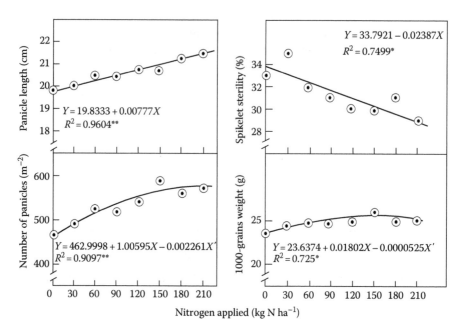

FIGURE 5.14   Influence of N on number of panicles, panicle length, 1000-grain weight and spikelet sterility of lowland rice. *Significant at the 5% probability level and **Significant at the 1% probability level. (From Fageria, N.K. and Baligar, V.C., *Commun. Soil Sci. Plant Anal.*, 32, 1405, 2001.)

Nitrogen also improves grain weight and reduces grain sterility. Figure 5.14 shows the influence of nitrogen on yield components of lowland rice. Grain yield in rice is a function of panicles per unit area, number of spikelets per panicle, 1000-grain weight, and spikelet sterility or filled spikelets (Fageria, 2007). Therefore, it is very important to understand the management practices that influence yield components and, consequently, grain yield. Nitrogen application up to 210 kg ha^{-1} influenced panicle length significantly ($P < 0.01$) and the relationship between the N applied and the panicle length was linear (Figure 5.14). The number of panicles m^{-2} and 1000-grain weight were also increased significantly and quadratically with the application of N fertilizer. Spikelet sterility, however, decreased significantly and linearly with increasing N rates. Nitrogen treatment accounted for about 96% variation in panicle length, about 91% variation in panicles m^{-2}, about 75% variation in spikelet sterility, and about 73% variation in 1000-grain weight. Fageria (2007) also reported that panicles per unit area, filled spikelet percentage, and 1000-grain weight were major contributors to increased grain yield in modern high-yielding rice varieties.

A major part of N in flooded rice soils is lost through leaching and denitrification (Fageria and Baligar, 2005). The major biological reaction involving nitrate in flooded soil is denitrification. Denitrification is the biological process that reduces nitrate to nitrogen gas or nitrous oxide, or both. Patrick and Mikkelsen (1971) reported that denitrification losses of 50% or more of applied N are common in flooded rice soils. Frequent fluctuations in the moisture content of a field as a result of flooding and drainage create ideal conditions for denitrification. Nitrogen converted to the nitrate form during the period when the soil is drained is lost through denitrification when the soil is flooded. The repeated submergence and drying of rice soils has been shown to increase N losses compared with losses in continuously submerged soils (Patrick and Wyatt, 1964; Kundu and Ladha, 1999). The deep placement of N in flooded rice reduces N lost through denitrification. Nitrate produced in the surface oxidized layer of a waterlogged soil can easily move downward by diffusion and percolate into the underlying reduced layer, where it is rapidly denitrified (Patrick and Mahapatra, 1968).

Even with best management practices like adequate rate, forms, methods, and timing of application, the utilization of added nitrogen is generally poorer in flooded rice soils. Fageria and Baligar (2001) and Fageria et al. (2007) studied N recovery efficiency of lowland rice grown on Brazilian Inceptisols (Table 5.8). Average efficiency under different rates was 39%, whereas average N recovery efficiency of five genotypes was 29%. Hence, a large part of applied N is lost in soil–plant systems. Patrick and Mahapatra (1968) reported that in Japan 30%–40% applied N is recovered by lowland rice as compared to an availability of 50%–60% when applied to upland crops.

In aerated soils, most of the N is in the form of NO_3^- due to nitrification process. In waterlogged soils, the absence of O_2 inhibits the activity of *the Nitrosomonas* microorganisms that oxidize NH_4^+, and, therefore, nitrogen mineralization stops at the NH_4^+ form. The accumulation of NH_4^+ in waterlogged soils would mean that the nitrogen is not lost from the soil–plant system, as is the case in denitrification.

TABLE 5.8
Nitrogen Recovery Efficiency in Lowland Rice as Influenced by N Rate and Genotypes

N Rate (kg ha^{-1})	N Recovery Efficiency (%)	Lowland Rice Genotype	N Recovery Efficiency (%)
30	49	CNAi 8886	37
60	50	CNAi 8569	29
90	37	BRSGO Guará	29
120	38	BRS Jaburu	26
150	34	BRS Giguá	23
180	33	Average	29
210	32		
Average	39		
R^2	0.82**		

Sources: Adapted from Fageria, N.K. and Baligar, V.C., *Commun. Soil Sci. Plant Anal.*, 32, 1405, 2001; Fageria, N.K. et al., *Pesq. Agropec. Brás.*, 42, 1029, 2007.

N recovery efficiency (%)

$$= \frac{\text{N uptake by plants in N fertilized plot} - \text{N uptake by plants in control treatment}}{\text{Quantity of N applied}}$$

**Significant at the 1% probability level.

This may only happen if rice fields constantly remain flooded during the crop growing cycle. If the availability of water is not under farmers' control due to lack of rainfall or storage facility, the situation may change in regard to the transformation and availability of N to plants. Hence, if N is applied in the reduced soil layer and water level is maintained in the rice field constantly, N uptake may improve in flooded rice.

5.4.3.5.2 Phosphorus

Phosphorus plays an important role in the growth and development of crop plants. Its role is well documented in many physiological and biological processes in the plants (Fageria, 2009). Phosphorus deficiency is one of the most important yield-limiting factors in annual crops grown on highly weathered acid soils of the tropics (Sanchez and Salinas, 1981; Dobermann et al., 2002; Fageria and Barbosa Filho, 2007). Saleque et al. (1998) reviewed the extent of P deficiency in lowland rice soils of tropical Asian countries. They found that most of the soils of Kampuchea, Vietnam, Thailand, Indonesia, Myanmar, Malaysia, Bangladesh, and Sri Lanka were low in available P. Similarly, Goswami and Banerjee (1978) reported P deficiency in Oxisols and coastal alluvial soils of Indonesia, in acid sulfate soils of the Republic of Korea, in Andosols of Japan, and in the black, red, and lateritic soils of India. Fageria et al. (2003, 2011d) reported a highly significant response of lowland rice grown on a Brazilian Inceptisol in the central part of Brazil. The P deficiency is associated with low natural P as well as with the high P-fixation capacity of these soils. Added soluble P is usually rapidly adsorbed on the surfaces of Fe and Al oxides, which is followed by immobilization in other forms and within soil particles (Hedley et al., 1994; Linquist et al., 1997). Data in Table 5.9 show that shoot dry weight and yield components of lowland rice were significantly improved with the addition of P in a Brazilian Inceptisol.

Depending on the soil mineralogy, the oxidation–reduction changes can significantly alter P solubility and sorption mechanisms (Reddy et al., 1998). The term "P sorption" refers to an instantaneous physical reaction of P with the soil (Havlin et al., 1999). Phosphorus sorption is closely related to the characteristics of the soil, such as its surface area (Olsen and Watanabe, 1957), $CaCO_3$ content (Richardson and Vaithiyanathan, 1995), and pH (Khalid et al., 1977; Reddy et al., 1988; Janardhanan and Daroub, 2010). Phosphorus sorption is also related to the hydrous Fe and Al oxyhydroxide content of the soil (Khalid et al., 1977; Reddy et al., 1998; Villapando and Graetz, 2001). Oxilate-extracted Fe and Al have been shown to be reliable predictors of the P-sorption capacity of peat soils in the Everglades (Reddy et al., 1998; Giesler et al., 2005). Several researchers have also reported a positive relationship between P sorption and the organic matter content of the soil (Richardson and Vaithiyanathan, 1995; Villapando and Graetz, 2001).

Some researchers reported that reduced soils released more soluble P (Ann et al., 2000; Pant and Reddy, 2001). Phosphorus availability is increased in flooded soils due to the reduction of ferric phosphate to the more soluble ferrous form and the hydrolysis of phosphate compounds (Patrick et al., 1973; Gotoh and Patrick, 1974; Holford and Patrick, 1981). This may be more pronounced in acid soils where P is immobilized by iron and aluminum oxides. Similarly, P uptake in flooded alkaline soils also improves due to the liberation of P from Ca and $CaCO_3$ due to decrease

TABLE 5.9
Dry Matter Yield of Shoot, Panicle Number, Panicle Length, 1000-Grain Weight, Spikelet Sterility, and GHI as Influenced by Phosphate Treatments

P Rate (kg ha^{-1})	Shoot Dry Weight (kg ha^{-1})	Panicle Number (m^{-2})	Panicle Length (cm)	1000-Grain Weight (g)	Spikelet Sterility (%)	Grain Harvest Index
0	3930.3	264.3	19.0	23.2	17.9	0.23
131	7088.7	365.0	20.1	25.1	21.2	0.29
262	7753.5	432.0	21.3	26.3	18.2	0.37
393	7664.3	412.2	20.9	26.5	17.9	0.40
524	8093.3	417.3	20.8	26.0	16.9	0.37
655	7021.0	419.2	22.3	26.6	13.3	0.43
F-test						
Year (Y)	**	**	**	*	NS	NS
P rate (P)	**	**	**	**	NS	**
Y × P	NS	NS	NS	NS	NS	NS
CV (%)	22	10	5	5	22	17

Regression Analysis

P rate (X) vs. shoot dry weight (Y) = 4297.5420 + 19.0682X − 0.0229X^2, R^2 = 0.6250**
P rate (X) vs. panicle number (Y) = 276.1075 + 0.7040X − 0.00077X^2, R^2 = 0.7497**
P rate (X) vs. panicle length (Y) = 19.3166 + 0.0059X − 0.0000031X^2, R^2 = 0.5632**
P rate (X) vs. 1000-grain weight (Y) = 23.4333 + 0.0135X − 0.000014X^2, R^2 = 0.5944**
P rate (X) vs. spikelet sterility (Y) = 18.5500 + 0.0104X − 0.0000027X^2, R^2 = 0.3845*
P rate (X) vs. grain harvest index (Y) = 0.2314 + 0.00061X − 0.00000051X^2, R^2 = 0.8193**
Shoot dry weight (X) vs. grain yield (Y) = −10332.13 + 3.9319X − 0.00025X^2, R^2 = 0.8654**
Panicle number (X) vs. grain yield (Y) = −15604.45 + 87.1577X − 0.0919X^2, R^2 = 0.8654**
Panicle length (X) vs. grain yield (Y) = −122564.60 + 11227.76X − 246.516X^2, R^2 = 0.7768**
1000-grain weight (X) vs. grain yield (Y) = −30651.68 + 1822.5990X − 18.3971X^2, R^2 = 0.8107**
Spikelet sterility (X) vs. grain yield (Y) = 7529.74 − 244.5112X + 2.2283X^2, R^2 = 0.1493NS
Grain harvest index (X) vs. grain yield (Y) = −10743.00 + 69155.70X − 74449.30X^2, R^2 = 0.9440**

Source: Fageria, N.K. and Santos, A.B., *Commun. Soil Sci. Plant Anal.*, 39, 873, 2008.
Values are averaged across 2 years.
NS, nonsignificant.
*, **Significant at the 5% and 1% probability levels, respectively.

in pH. The formation of insoluble tricalcium phosphate is favored at a high pH. Patrick and Mahapatra (1968) reported a marked increase in the extractable P with the lowering of redox potential below +200 mV (at pH 5.7). Extractable P increased from 10 to 35 mg kg^{-1} between the potentials of 200 and −200 mV. The fact that at +200 mV iron also began to be reduced to the ferrous form tends to confirm that this increase in P came from the conversion of ferric phosphate to more soluble ferrous phosphate (Patrick and Mahapatra, 1968). Fageria and Baligar (1999) determined P in an Inceptisol after harvest of flooded rice to be influenced by pH (Table 5.10).

TABLE 5.10
Influence of pH on Mehlich
1-Extractable P after Harvest
of Lowland Rice

Soil pH in H_2O	P mg kg^{-1}
4.9	54.3
5.9	71.9
6.4	58.4
6.7	43.3
7.0	29.8
R^2	0.72**

Source: Adapted from Fageria, N.K. and
Baligar, V.C., *J. Plant Nutr.,* 22,
1495, 1999.
**Significant at the 1% probability level.

The Mehlich 1-extractable P increased from 54.3 mg P kg^{-1} at pH 4.9 to 71.9 mg P kg^{-1} at pH 5.9. When soil pH was raised higher than 5.9, P concentration decreased. These results have special significance in managing acid soils, because P deficiency is one of the most important yield-limiting nutrients in crop production in the acid lowland soils of Brazil (Fageria and Baligar, 1999).

Huguenin-Elie et al. (2003) reported that flooding the soil tends to increase the availability of phosphorus to plants, both because of faster diffusion to roots and because P becomes more soluble as a result of reductive dissolutions of iron oxides. Kirk and Saleque (1995) showed that in spite of increased P solubility and mobility under flooded conditions, rice plants often rely upon root-induced solubilization by acidification for the bulk of their P. If the soil remains aerobic, the P only slowly reverts to its preflooding state over a period of months (Willett, 1991). Yu (1991) reported that a moderate reduction of acid soils favors the availability of P, directly through the release of phosphates associated with iron oxides and indirectly through the reduction in phosphate adsorption as a result of the rise in pH. Tanaka et al. (1969) reported that the pH changes associated with the reduction of ferric to ferrous iron are of greater importance in liberating P than the release of P by reductive dissolution.

Zhang et al. (2010) studied P sorption of an organic-rich silt loam under air-dried, field wet, and reduced conditions. Additionally, the influence of farm wastewater on soil P sorption was also studied. Major results indicated that soil reduction increased the maximum amount of P that can be sorbed (S_{max}) and decreased the aqueous P concentration at which P sorption and desorption are equal (EPC_0), both determined from a modified Langmuir model. The slightly reduced field wet soils had no significant differences in S_{max} due to limited soil reduction. Using the dilute wastewater as the sorption solution matrices instead of 0.01 mol L^{-1} KCl solution, the soils generally exhibited greater S_{max} and lower EPC_0, except for the EPC_0 of a reduced surface

TABLE 5.11
Grain Yield of 12 Genotypes as Influenced by Phosphorus Fertilization

	Phosphorus Rate (kg P ha⁻¹)				
Genotype	0	22	44	66	88
BRS Jaçnã	932.9	5903.2	7373.3	6750.0	5681.2
CNAi 8860	1153.4	588.8	6361.2	5470.7	4874.7
CNAi 8879	1174.9	5517.8	6500.3	4736.2	5038.8
BRS Fronteira	1259.0	5185.2	6190.5	4979.3	4182.0
CNAi 8880	1738.0	5346.7	4794.7	5477.8	4994.7
CNAi 8886	1698.3	5377.8	6017.0	5377.1	4318.0
CNAi 8885	1289.0	5139.8	5853.0	5046.3	4854.2
CNAi 8569	58.0	3482.8	3684.0	3402.7	1565.0
BRS Guará	248.9	5100.2	6163.3	5303.3	4826.2
BRS Alvorada	64.2	4435.7	4802.8	4255.2	3764.7
BRS Jaburu	41.6	3779.8	5384.7	4704.3	4441.5
BRS Biguá	128.6	3844.3	5717.5	5026.7	5005.5

F-test	
Year (Y)	**
P rate (P)	**
Y × P	NS
Genotype (G)	**
Y × G	**
CV (%)	24.3

Source: Fageria, N.K. et al., *J. Plant Nutr.*, 34, 1087, 2011d.
NS, nonsignificant.
**Significant at the 1% probability level, respectively.

soil, implying more complex P sorption in the field. These authors reported that the transformation of Fe compounds during soil reduction is primarily responsible for the changes in soil P sorption. Vadas and Sims (1998, 1999) reported that reduced soils amended with poultry litter released less soluble P.

In addition to the use of adequate rate of P, planting P-efficient genotypes is an important strategy in improving lowland rice yield in P-deficient soils. Fageria et al. (2011d) studied the response of 12 lowland rice in a P-deficient, Brazilian lowland (Table 5.11). Grain yield increased significantly in a quadratic fashion of 12 genotypes with increasing P rate in the range of 0–88 kg P ha⁻¹ (Table 5.12). The increase in grain yield of genotypes was associated with increase in panicle number and shoot dry weight with increasing P rate (Fageria et al., 2011d). Although the response of 12 genotypes to P fertilization was similar, the magnitude of grain yield was different (Table 5.11). At lowest P rate (control treatment), genotype BRS Jaburu produced lowest grain yield (42 kg ha⁻¹) and genotype CNAi 8880 produced maximum grain yield (1738 kg ha⁻¹). Similarly, at highest P rate (88 kg P ha⁻¹), genotype CNAi 8569

TABLE 5.12

Regression Equations Showing Relationship between Phosphorus Rate (X) and Grain Yield (Y) of 12 Lowland Rice Genotypes

Genotype	Regression Equation	R^2	P Rate for Maximum Grain Yield (kg ha⁻¹)
BRS Jaçnã	$Y = 1234.94 + 231.06X - 2.09X^2$	0.87**	55
CNAi 8860	$Y = 1655.52 + 182.48X - 1.68X^2$	0.82**	54
CNAi 8879	$Y = 1657.66 + 172.19X - 1.60X^2$	0.73**	54
BRS Fronteira	$Y = 1564.98 + 177.11X - 1.72X^2$	0.80**	51
CNAi 8880	$Y = 2148.82 + 120.44X - 1.03X^2$	0.67**	58
CNAi 8886	$Y = 1973.31 + 163.50X - 1.16X^2$	0.74**	70
CNAi 8885	$Y = 1656.87 + 156.74X - 1.42X^2$	0.73**	55
CNAi 8569	$Y = 279.20 + 156.29X - 1.62X^2$	0.88**	48
BRS Guará	$Y = 659.75 + 205.91X - 1.86X^2$	0.89**	55
BRS Alvorada	$Y = 500.62 + 170.98X - 1.57X^2$	0.82**	54
BRS Jaburu	$Y = 255.94 + 177.80X - 1.48X^2$	0.88**	60
BRS Biguá	$Y = 323.35 + 180.07X - 1.48X^2$	0.86**	61
Average of 12 genotypes	$Y = 1156.88 + 175.02X - 1.61X^2$	0.90**	54

Source: Fageria, N.K. et al., *J. Plant Nutr.*, 34, 1087, 2011d.
Values are averages of 2 year field trial.
**Significant at the 1% probability level.

produced minimum grain yield of 1565 kg ha⁻¹ and genotype BRA Jaçanã produced maximum grain yield of 5681 kg ha⁻¹. At an intermediate P rate (44 kg P ha⁻¹), genotypes CNAi 8569 produced minimum grain yield (3684 kg ha⁻¹) and genotype BRS Jaçanã produced maximum grain yield (7373 kg ha⁻¹). Hence, rice genotypes varied in response to magnitude at low, medium, and high P rates.

5.4.3.5.3 Potassium

Potassium plays an important role in many plant physiological and biochemical processes. It helps in osmotic and ionic regulation, functions as a cofactor or activator for many enzymes of carbohydrate and protein metabolism. The role of K in controlling a number of diseases is widely reported (Fageria et al., 2010). Potassium is absorbed in greater quantities by rice, especially by high-yielding cultivars, than any other essential nutrient (Fageria et al., 2011a). A single lowland rice crop producing 9.8 Mg ha⁻¹ of grain in about 115 days took up 218 kg N ha⁻¹, 31 kg P ha⁻¹, 258 kg K ha⁻¹, and 9 kg S ha⁻¹ (De Datta and Mikkelsen, 1985). As grain yield increases, the demand for plant nutrients, particularly K, is known to increase (Fageria et al., 2011a).

Data in Table 5.13 show the response of lowland rice to K fertilization. The results of K soil test calibration to applied K are presented in Figure 5.15. There was a quadratic response of increasing soil-extractable K in the range of 14–86 mg K kg⁻¹

TABLE 5.13

Grain Yield of Rice as Influenced by Potassium Fertilization

K Rate (kg ha⁻¹)	Grain Yield (kg ha⁻¹)		
	1st Year	2nd Year	Average
0	2961	2631	2796
125	5880	6244	6062
250	5526	5590	5558
375	4853	5092	4973
500	4962	5076	5019
625	5091	5329	5210
F-test			
Year (Y)	**		
K rate (K)	**		
Y X K	**		
CV (%)	6.8		

Regression Analysis

K rate (X) vs. 1st year grain yield (Y)
$= 3641.72 + 10.3063X - 0.0138X^2, R^2 = 0.4466^{**}$

K rate (X) vs. 2nd year grain yield (Y)
$= 3467.83 + 12.3452 X - 0.0163X^2, R^2 = 0.4567^{**}$

K rate (X) vs. average grain yield (Y)
$= 3555.04 + 11.3227X - 0.0151X^2, R^2 = 0.4674^{**}$

Source: Fageria, N.K. et al., *Commun. Soil Sci. Plant Anal.* 41, 1, 2010.
**Significant at the 1% probability level.

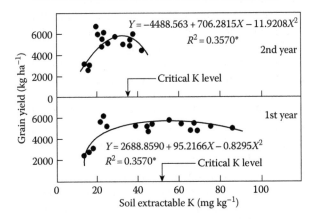

FIGURE 5.15 Relationship between soil-extractable K and grain yield of lowland rice. *Significant at the 5% probability level and **Significant at the 1% probability level. (From Fageria N.K. et al., *Commun. Soil Sci. Plant Anal.*, 42, 1, 2011b.)

Mehlich 1-extractable soil K in the first year. In the second year also, rice yield increased in the quadratic fashion when K soil test was in the range of 15–42 mg kg^{-1}. In the first year, based on regression equation, maximum grain yield was obtained at 57 mg K kg^{-1} of soil. In the second year, maximum grain yield was obtained at 30 mg kg^{-1} extractable soil K. Fageria et al. (1990) reported that the Mehlich 1-extractable soil K level associated with maximum rice yield was 59 mg kg^{-1} for the first crop and 34 mg kg^{-1} for the second crop in a Brazilian Inceptisol. The results of the present study fall in this K level range. The maximum grain yield obtained at the low K level (30 mg kg^{-1}) in the second year indicates that there was a significant reduction in K content of the soil during 2 years of rice cultivation. It also suggests that in the second year nonexchangeable K might have played an important role in the supply of K to the rice crop. Sparks (1987) and Hinsinger (1998) reported that the availability of nonexchangeable K to plants is associated with very low level of K in the soil and this may be due to chemical changes in the rhizosphere.

The forms of K in soil in the order of their availability to plants and microbes are solution, exchangeable, nonexchangeable, and mineral. Soil solution K is the form taken up directly by plants and microbes and is also subject to leaching (Sparks, 1987). The influence of flooding is less on the chemistry of K as compared to N and P. The reducing conditions caused by flooding result in a larger fraction of the K ions being displaced from the exchange complex into the soil solution. The release of a relatively large amount of ferrous and manganese ions and the production of ammonium ions result in the displacement of some of the K ions from the exchange complex to the soil solution. This may lead to higher availability of K to rice in flooded soils (Patrick and Mikkelsen, 1971).

5.4.3.5.4 Sulfur

In flooded soils, SO_4^{2-} ion is reduced to H_2S by anaerobic microbial activities. Furthermore, in flooded soils, Fe^{3+} reduction to Fe^{2+} precedes SO_4^{2-} reduction; Fe^{2+} will always be present in the soil solution by the time H_2S is produced, so that H_2S will be converted to insoluble FeS. This reaction protects microorganisms and higher plants from the toxic effects of H_2S (Patrick and Reddy, 1978). Sulfate reduction began to occur very early, immediately after flooding, and decreased linearly to significant level within 3 weeks after flooding in the straw treatment and 6 weeks after flooding when no straw was added, indicating sulfate reduction started and finished within 3–6 weeks after flooding (Gao et al., 2004).

Overall, the availability of S is reduced in flooded soils due to the formation of insoluble FeS. The hydrogen sulfide (H_2S) may occur under the following circumstances: (1) if soils containing a high amount of fresh organic matter are submerged at a high temperature, the initial formation of Fe^{2+} and Mn^{2+} may be insufficient to precipitate all the newly formed sulfide ions; (2) in strongly leached, acid, sandy soils with a high content of organic matter, the amount of reducible ion oxide may be too low; and (3) if acid sulfate soils are planted with rice immediately after submergence, H_2S can be present in high concentrations, due to the low pH of the medium (Yu, 1991). Yu (1991) reported that H_2S toxicity may occur in lowland d rice when concentration of H_2S is 0.68 mg kg^{-1}. A higher concentration of Fe^{2+} and Mn^{2+} decreased H_2S toxicity in lowland rice.

5.4.3.5.5 Calcium and Magnesium

Calcium and magnesium deficiency are rare in lowland rice. Rice is highly tolerant to soil acidity. Optimum soil pH for lowland rice grown on a Brazilian Inceptisol was reported to be 4.9 (Fageria and Baligar, 1999). In highly acidic soils, dolomitic lime can be added to supply Ca and Mg. Only small amounts of these elements are removed in the grain, and unless the straw is removed from the field, the total removal is small. Changes in Ca and Mg concentrations are at minimum in flooded soils. However, Gao et al. (2010) reported that Ca concentration significantly increased in flooded soil compared to aerobic soil (Table 5.14).

5.4.3.5.6 Micronutrients

The availability of micronutrients to plants is governed by their concentration in soil solution, which is modified by hydrolysis, complexation, and redox potential. The availability of micronutrient cations is particularly sensitive to changes in the soil environment. Part of the sensitivity to these changes is related directly to the performance of the root system in exploring the soil volume for these immobile elements (Zn, Cu, Fe, and Mn), and part is related to the pools or bonding of the elements in the soil (Fageria, 2009; Fageria et al., 2011c).

5.4.3.5.6.1 Iron and Manganese The Fe^{3+} reduces to Fe^{2+} and Mn^{4+} reduces to Mn^{2+}; hence, the uptake of these elements increased in the flooded rice soils (Tables 5.15 and 5.16). The reduction processes of iron and manganese are shown under the heading pH changes. The concentration of Fe^{2+} and Mn^{2+} reached a peak, then decreased, and later stabilized. After reaching maximum concentrations, the decrease in Fe^{2+} and Mn^{2+} concentrations is considered to reflect the formation of precipitates such as with sulfide, produced from sulfate reduction, and possibly carbonate too (Gao et al., 2004). Gao et al. (2004) reported that the maximum concentration of Mn^{2+} was observed 3 weeks after flooding with straw incorporation and 6 weeks after flooding when no straw was added, indicating Mn reduction started and finished within 3–6 weeks after flooding. These authors also reported that the iron reduction showed the same trend but was more strongly affected by the straw

---

TABLE 5.14

Change in Soil pH and Ca Concentration in Pore Water Collected from the Controlled Experiment at 5, 42, and 148 Days after Application of Water Treatment in Oxidized (Unflooded) and Reduced (Flooded) Soils

Time in Days	Oxidized Soil			Reduced Soil		
	RP (mV)	pH	Ca mg L⁻¹	RP (mV)	pH	Ca mg L⁻¹
5	425	6.7	85	200	6.1	205
42	—	6.9	74	0	6.7	479
148	433	6.8	79	−170	6.5	370

Source: Adapted from Gao, X. et al., *Soil Sci. Soc. Am. J.,* 74, 301, 2010.
RP stands for redox potential.

TABLE 5.15

Change in Soil pH and Fe Concentration in Pore Water Collected from the Controlled Experiment at 5, 42, and 148 Days after Application of Water Treatment in Oxidized (Unflooded) and Reduced (Flooded) Soils

Time in Days	Oxidized Soil			Reduced Soil		
	RP (mV)	pH	Fe mg L^{-1}	RP (mV)	pH	Fe mg L^{-1}
5	425	6.7	0.1	200	6.1	4.2
42	—	6.9	0.1	0	6.7	1.1
148	433	6.8	0.01	−170	6.5	0.9

Source: Adapted from Gao, X. et al., *Soil Sci. Soc. Am. J.*, 74, 301, 2010.
RP stands for redox potential.

TABLE 5.16

Change in Soil pH and Mn Concentration in Pore Water Collected from the Controlled Experiment at 5, 42, and 148 Days after Application of Water Treatment in Oxidized (Unflooded) and Reduced (Flooded) Soils

Time in Days	Oxidized Soil			Reduced Soil		
	RP (mV)	pH	Mn mg L^{-1}	RP (mV)	pH	Mn mg L^{-1}
5	425	6.7	2.4	200	6.1	6.7
42	—	6.9	1.1	0	6.7	19.0
148	433	6.8	0.2	−170	6.5	16.0

Source: Adapted from Gao, X. et al., *Soil Sci. Soc. Am. J.*, 74, 301, 2010.
RP stands for redox potential.

treatment. A higher concentration of iron may play an important role in reducing sulfide toxicity by forming FeS. On the other hand, a higher amount of straw incorporation could deplete the Fe source within a relatively short time and, thus, may stimulate sulfide toxicity through sulfate reduction (Gao et al., 2004).

Reported concentration values of Fe^{2+} in soil solution of flooded paddy fields range from 0.7 to 20 mmol L^{-1} or 39 to 1117 mg L^{-1} (Ponnamperuma, 1972; Genon et al., 1994; Boivin et al., 2002; Saejiew et al., 2004). The higher concentration of Fe^{2+} (>300 mg L^{-1}) may be toxic to rice plants under certain conditions (Fageria, 1984; Fageria et al., 2008). Sims and Johnson (1991) reported that for most crops the critical deficiency soil Fe concentration range was 2.5–5.0 mg kg^{-1} of DTPA-extractable (diethylene-triamine penta-acetic acid) Fe, but is also influenced by soil pH. In both field and pot experiments the degree of bronzing in a given variety showed a highly significant correlation ($r = 0.90**$) with yield (Breemen and Moormann, 1978). Iron toxicity in rice plants as indicated by the bronzing of leaves was reported when soluble iron in soil solution was more than 300–500 mg kg^{-1} (Ponnamperuma et al., 1955; Tanaka et al., 1966). By DTPA-extracting solution. However, Breemen and Moorman (1978) reported that the bronzing symptoms appear generally when iron concentration in soil solution is in the range of 300–400 mg kg^{-1} by DTPA-extracting solution. Barbosa Filho et al. (1983) reported that iron toxicity in lowland rice occurred when Mehlich 1-extracting iron in the soil was in the range of 420–730 mg kg^{-1}. This means that iron toxicity level in the soil also depends on the extracting solution used to extract the iron from the soil. Values for the Mehlich 1-extracting solution are higher compared to DTPA extraction.

The ability of flooded rice plants to release oxygen into the rhizosphere can result in the formation of a coating or plaque around roots composed primarily of iron and possibly some manganese (Bacha and Hossner, 1977; Chen et al., 1980a,b; Crowder and Macfie, 1986). Iron coatings are restricted to older roots of rice and they were not found near root tips or on the surfaces of young secondary roots (Chen et al., 1980b). The coatings filled exposed cavities within the epidermal cells of rice (Chen et al., 1980b). In addition to the relationship between oxygen release by roots and the formation of iron coating, extractable iron, pH, organic matter, inorganic carbonates, and microorganisms have also been suggested as possible factors influencing plaque formation.

Iron toxicity in lowland rice has been reported in South America, Asia, and Africa (Sahu, 1968; Barbosa Filho et al., 1983; Fageria, 1984; Fageria and Rabelo, 1987; Fageria et al., 2003; Sahrawat, 2004). Metal toxicity in crop plants can be expressed in two ways. One is when metal is absorbed in higher amounts and becomes lethal to the plant cells. This is known as direct toxicity of metals. Another metal toxicity is associated with the inhibition of uptake and the utilization of essential nutrients by plants. This is known as indirect metal toxicity. Indirect toxicity creates nutrient imbalance in plants. This type of iron toxicity is more common in lowland rice compared to direct toxicity (Fageria et al., 1990a, 2006). The most important nutrient deficiencies observed in irrigated or flooded rice in Brazil are P, K, and Zn (Barbosa Filho et al., 1983). The yield reduction of rice cultivars due to Fe toxicity depends on the tolerance or susceptibility of cultivars to toxicity. Ikehashi and Ponnamperuma (1978) reported that the reduction of yield on an iron toxic soil ranged from a mean of 29% for five moderately tolerant lines to a mean of 74% for five susceptible lines.

Effective measures to ameliorate Fe^{2+} toxicity include periodic surface drainage to oxidize reduced Fe^{2+}, liming of acid soils, use of adequate amounts of essential nutrients, and planting of iron toxic cultivars/genotypes. The Fe-excluding ability of rice plants is lowered by deficiencies of P, K, Ca, and Mg (Obata, 1995). In particular, K deficiency readily induces Fe toxicity (Fageria et al., 2003). Silicon fertilization may be useful to alleviate Fe toxicity in lowland rice. The application of 2 Mg ha^{-1} to soil with a high concentration of Fe decreases the content of this element in rice plants without producing deficiency symptoms. This could be due to an increase in the oxidizing capacity of rice roots supplied with adequate amounts of Si (Okuda and Takahashi, 1965). According to Ponnamperuma (1965), sufficient Si supply facilitates oxygen transport more efficiently from the plant tops to the roots through the enlargement or rigidity of gas channels and, as a result, increases oxidation and the subsequent deposition of iron and manganese on the root surface, thus excluding the iron from absorption.

Among these management practices, the use of tolerant cultivars is most economical and environmentally sound practice. Genetic variability in rice cultivars to Fe^{2+} toxicity has been reported (Gunawarkena et al., 1982; Fageria et al., 1990; Fageria and Rabelo, 1987). Tadano (1976) reported that there might be three mechanisms that are responsible for iron toxicity variability in rice cultivars. These mechanisms are (1) oxidation of Fe^{2+} in the rhizosphere, (2) exclusion of Fe^{2+} at the root surface, and (3) retention of Fe in the root tissues, which prevents translocation of Fe from the root to the shoot. The introduction of modern high-yielding cultivars has often led to an increase in the incidence of bronzing, apparently because the traditional cultivars had better tolerance for iron toxicity and other adverse soil conditions (Breemen and Moormann, 1978).

5.4.3.5.6.2 Zinc Zinc was one of the first minerals known to be essential for plants, animals, and man (Welch, 1993; Kabata-Pendias, 2000), and yet, in spite of this knowledge Zn deficiencies still plague us today (Obrador et al., 2003). Zinc deficiency is the most widespread micronutrient disorder among different crops (Romheld and Marschner, 1991). Zinc deficiency in rice has been reported in lowland rice in Brazil (Fageria, 2001a; Fageria et al., 2011d), in India (Mandal et al., 2000; Qadar, 2002), and in the Philippines (De Datta, 1981). A global study by FAO showed that about 30% of the cultivated soils of the world are Zn deficient (Sillanpaa, 1982). Additionally, about 50% of the soils used worldwide for cereal production contain low levels of plant-available Zn (Graham et al., 1992; Welch, 1993). Zinc deficiency in crop plants reduces not only grain yield, but also the nutritional quality of the grain. Zinc is a cofactor for several enzymes that are involved with N metabolism (e.g., glutamate dehydrogenase) and anaerobic metabolism (e.g., alcohol dehydrogenase).

Zinc application recommendations generally varied from 5 to 10 kg Zn ha^{-1} for lowland rice, depending on soil type, cultivar planted, and soil pH (Fageria and Barbosa Filho, 1994; Slaton et al., 2001; Fageria et al., 2011a). Zinc bioavailability in soil is mostly regulated by adsorption–desorption reactions and solubility relations between the solution and solid phases (Lindsay, 1991; Catlett et al., 2002). The behavior of Zn in aerobic soils differs significantly from anaerobic soils (Dutta et al., 1989). Soil properties including the pH, redox potential, organic matter, pedogenic oxide, and soil S contents exert the most significant influence on the adsorption–desorption reactions of Zn in soils and, thus, regulate the amount of Zn dissolved in the soil solution (Guadalix and Pardo, 1995). Zinc concentrations generally decreased after the flooding of rice soils. The decrease in concentration with the flooding may be associated with increase in soil pH after the flooding. In addition, Carbonell-Barrachina et al. (2000) reported that extensive flooding periods may lead to a considerable decrease in the concentration of Zn if the formation of insoluble ZnS occurs, as is often observed experimentally in anaerobic sediments (Schroder et al., 2008). Dutta et al. (1989) reported that DTPA-extractable Zn in 26 acidic to near neutral alluvial soils on submergence was decreased. The percentage of decrease was significantly correlated with initial pH, organic carbon, clay, cation exchange capacity and initial DTPA-extractable Zn. These authors also reported that the pH seemed to play the most important role in influencing the magnitude of decrease. The results of path coefficient analysis revealed that increase in pH directly contributed 37% of the percentage of decrease in DTPA-extractable Zn in soil on submergence. Additionally, more, multiple regression equation showed that 40% variability in the magnitude of decrease of Zn was accounted by some of the initial soil properties. Gao et al. (2010) reported that Zn availability was higher in anaerobic soil compared to aerobic soil but soil pH of aerobic soil was higher compared to anaerobic soil. Increasing soil pH decreases micronutrients availability except Mo (Fageria et al., 2002).

Mehlich 1 and DTPA are two principal extractants for determining Zn levels in the soils and Zn availability to plants (Abreu et al., 1998; Fageria et al., 2011a). Mehlich 1 (Mehlich, 1953)- as well as DTPA (Lindsay and Norvell, 1978)-extractable Zn increased significantly with increasing Zn rate in the soil (Table 5.17). However, Mehlich 1-extractable soil Zn was much higher as compared to Mehlich 1-extractable soil Zn. Fageria (2000) also reported higher Zn extraction by Mehlich 1-extracting solution as compared to DTPA-extracting solution in a Brazilian Oxisol. Linear correlation coefficients were determined between Zn extracted by Mehlich 1 and DTPA extractants and Zn uptake in grain and shoot of rice (Table 5.18). Zinc extracted by two extractants had highly significant correlation. Similarly, Zn uptake in the shoot as well as in the grain was also highly correlated with the Zn extracted by Mehlich 1 as well as DTPA extractants. This means both the extractants can be used to evaluate Zn availability for lowland rice in an Inceptisol of Brazil. However, in Brazil most of the soil testing laboratories use Mehlich 1-extracting solution not only for the extraction of Zn, Cu, Mn, and Fe but also for the extraction of P and K (Abreu et al., 1998). Hence, Mehlich 1-extracting solution has an economic advantage over the DTPA-extracting solution.

Nutrient use efficiency in the literature is defined in five different ways (Fageria and Baligar, 2001). All five Zn use efficiencies were calculated and are presented in Table 5.19. Agronomic efficiency (AE), physiological efficiency (PE), agrophysiological

TABLE 5.17
Mehlich 1- and DTPA-Extractable Soil Zn under Different Zn Treatments at Harvest of Lowland Rice

Zn Rate (mg kg^{-1})	Mehlich 1 Zn (mg kg^{-1})	DTPA Zn (mg kg^{-1})
0	1.16	0.56
5	3.20	1.21
10	5.07	1.91
20	11.73	4.84
40	20.90	8.43
80	44.36	16.21
120	96.66	33.14
F-test	**	**

Regression Analysis
Zn rate vs. Mehlich 1-extractable Zn $(Y) = -3.32 + 0.75X$, $R^2 = 0.96$**
Zn rate vs. DTPA-extractable Zn $(Y) = -0.67 + 0.26X$, $R^2 = 0.97$**

Source: Fageria, N.K. et al., *Commun. Soil Sci. Plant Anal.*, 42, 1719, 2011c.
**Significant at the 1% probability level.

TABLE 5.18

Linear Correlation Coefficient (*r*) between Mehlich 1- and DTPA-Extractable Zn in the Soil and Zn Uptake in Straw and Grain of Lowland Rice

Variable	Value of *r*
Mehlich 1-extractable soil Zn vs. DTPA-extractable Zn	0.98**
Mehlich 1-extractable soil Zn vs. Zn uptake in grain	0.57**
DTPA-extractable soil Zn vs. Zn uptake in grain	0.59**
Mehlich 1-extractable soil Zn vs. Zn uptake in straw	0.91**
DTPA-extractable soil Zn vs. Zn uptake in straw	0.96**

Source: Fageria, N.K. et al., *Commun. Soil Sci. Plant Anal.*, 42, 1719, 2011c.

**Significant at the 1% probability level.

efficiency (APE), apparent recovery efficiency (ARE), and utilization efficiency (UE) significantly decreased with increasing Zn rate in the soil. However, agrophysiological (APE) had significant quadratic increase with increasing Zn levels. Higher nutrient use efficiency at lower level is common due to the efficient utilization of nutrients at lower level (Fageria, 1992). The decrease in nutrient use efficiency with increasing nutrient rate is also related to progressive decrease in dry matter or grain yield with increasing nutrient rate in nutrient-deficient soils or where crop response to applied nutrient is obtained (Fageria and Baligar, 2001). Genc et al. (2002) also reported higher Zn use efficiency at the lowest Zn level in barley genotypes. The apparent recovery efficiency of Zn across the Zn rates was about 6%. In the literature, information on Zn use efficiency is limited, and, hence, we cannot compare our results with the published work. However, Mortvedt (1994) reported that crop recovery of micronutrients is relatively low (5%–10%) compared to that of macronutrients (10%–50%) because of poor distribution from low rates applied, fertilizer reactions with soil to form unavailable products, and low mobility in soil.

5.4.3.5.6.3 Copper Copper is a redox-sensitive transition metal and is found as a trace element in most soils at an average concentration of 30 mg kg^{-1} and in a range of 2–100 mg kg^{-1} (Bigalke et al., 2010). It occurs predominantly as Cu^{2+} while also Cu^+ species and elemental Cu may be present (Grybos et al., 2007; Weber et al., 2009). In soils, Cu is mostly complexed by organic matter, incorporated and adsorbed to Fe oxides and primary or secondary silicates. Copper uptake may increase or decrease depending on biogeochemical processes. At low redox potentials in flooded soils, Fe oxides are reduced and associated metals including Cu are released; however, frequently changing redox conditions may also cause an enhanced fixation of Cu in Fe oxy(hydr)oxides (Contin et al., 2007). In the presence of Fe^{2+}, Cu^{2+} can be transformed to Cu^+ in reducing environments (Matocha et al., 2005), which may also alter its binding type and its mobility (Bigalke et al., 2010). One possible effect of Cu reduction may also be precipitation as Cu(s) (Grybos et al., 2007). The increase in

TABLE 5.19

Zinc Use Efficiency in Lowland Rice as Influenced by Zn Application Rate

Zn Rate (mg kg⁻¹)	AE (μg μg⁻¹)	PE (μg μg⁻¹)	APE (μg μg⁻¹)	ARE (%)	UE (μg μg⁻¹)
5	815	10326	4881	16.5	1742
10	450	12756	6457	6.9	867
20	266	13072	6440	44.4	535
40	109	10276	4519	2.4	245
80	58	6739	3985	1.5	98
120	36	2106	1033	3.5	74
Average	289	9213	4553	5.9	594
F-test	**	*	NS	**	**

Source: Fageria, N.K. et al., *Commun. Soil Sci. Plant Anal.*, 42, 1719, 2011c.

AE, agronomic efficiency; PE, physiological efficiency; APE, agrophysiological efficiency; ARE, apparent recovery efficiency; and UE, utilization efficiency.

NS, nonsignificant.

$AE\ (\mu g\ g^{-1}) = \dfrac{GY_1 - GY_2}{\text{Quantity of Zn applied}}$, where GY_1 = grain yield of Zn-fertilized pots, GY_2 = grain yield of Zn-unfertilized pots.

$PE\ (\mu g\ g^{-1}) = \dfrac{GY_1 - GY_2}{Zn_1 - Zn_2}$, where GY_1 = grain plus straw yield of Zn-fertilized pots, GY_2 = grain plus straw yield of Zn-unfertilized pots, Zn_1 = Zn uptake in grain and straw of Zn-fertilized pots and Zn_2 = Zn uptake in grain and straw of Zn-unfertilized pots.

$APE\ (\mu g\ g^{-1}) = \dfrac{GY_1 - GY_2}{Zn_1 - Zn_2}$, where GY_1 = grain yield of Zn-fertilized pots, GY_2 = grain yield of Zn-unfertilized pots, Zn_1 = Zn uptake in grain and straw of Zn-fertilized pots and Zn_2 = Zn uptake in grain and straw of Zn-unfertilized pots.

$ARE\ (\%) = \dfrac{Zn_1 - Zn_2}{\text{Quantity of Zn applied}}$, where Zn_1 = Zn uptake in grain and straw of Zn-fertilized pots, Zn_2 = Zn uptake in grain and straw of Zn-unfertilized pots.

$UE\ (\mu g\ g^{-1}) = PE \times ARE$.

*, **Significant at the 1% and 5% probability levels, respectively.

pH during chemical reduction reactions may cause the release of dissolved organic matter, which enhances metal mobility (Grybos et al., 2007). Weber et al. (2009) reported the formation of elemental Cu and Cu_xS nanoparticles in flooded soils, which can enhance Cu mobility. But in unflooded soils, Cu^{2+} might also be reduced to elemental Cu nanoparticles by plants. The formation of Cu nanoparticles by plants is thought to be a defense mechanism against Cu toxicity and may only occur in polluted soils (Manceau et al., 2008). Dutta et al. (1989) reported that DTPA-extractable Cu in 26 acidic to near-neutral alluvial soils decreased on submergence, and an increase in pH was one of the major factors for this decrease. In conclusion, when pH is increased due to flooding, the availability of Cu decreases.

5.4.3.5.6.4 Boron, Molybdenum, and Silica Boron is an essential micronutrient required by plants, and a toxicity of this element may occur at elevated concentration in the soil solution (Goldberg et al., 2004). The range between B deficiency and toxicity is narrow, typically 0.028–0.093 mmol L^{-1} or 0.30–1.0 mg kg^{-1} for sensitive crops, and 0.37–1.39 mmol L^{-1} or 4–15 mg kg^{-1} for tolerant crops (Keren and Bingham, 1985). Little is known about the behavior of B and Mo in submerged soils. A variety of factors such as pH, organic matter, clay minerals, Fe and Al oxides, carbonates, and tillage management may change the content of extractable B and transformation among different soil B fractions (Jin et al., 1987; Mandal et al., 1993). The content of water-soluble B in soils tends to increase with soil pH, but not always in a consistent manner, probably because B adsorption by soil components also increases with the increase of pH and reaches a maximum in the alkaline pH range (Gu and Lowe, 1990; Goldberg et al., 1993, 1996a). In practice, liming soils may result in a significant decrease of B uptake presumably because of increased B sorption (Lehto and Malkonen, 1994). Boron concentration seems to remain more or less constant after the submergence of rice soils (Ponnamperuma, 1975).

Molybdenum deficiency has been reported for many agronomic crops throughout the world (Murphy and Walsh, 1972). The availability of Mo to plants is affected by a variety of factors including soil solution pH, soil texture, soil moisture, temperature, oxide content, organic matter content, and clay mineralogy (Reisenauer et al., 1973; Goldberg et al., 2002). The dominant Mo adsorption surfaces in soil are oxides, clay minerals, and organic matter (Goldberg et al., 1996). Molybdenum concentration in rice soils was found to increase after submergence (Ponnamperuma, 1975), possibly because of the increased pH. In flooded soils, Si generally tends to increase after submergence. This increase is probably due to the release of adsorbed and occluded Si from oxyhydroxides of Fe and Al as well as to the effect of the increased pH resulting from submergence. Decomposing rice straw with its high silica content may also contribute to the increased Si content of the soil solution of flooded soils (Patrick and Reddy, 1978).

5.4.3.6 Methane Production

Methane (CH_4), following CO_2, is the second-most abundant carbon compound in the atmosphere. Globally, CH_4 budget is dominated by biogenic source, natural wetlands (23%), and rice fields (21%), accounting for almost half of the total budget (Conrad, 2007). Extensive research has shown that wetlands dominated by graminoids are large sources for CH_4 emissions (Whalen, 2005), in part because air spaces within plants act as a conduit for CH_4 to move from the soil to the atmosphere, thereby passing microbial oxidation in oxic parts of the soil. Although plant species effects on CH_4 emission to the atmosphere are well known (Joabsson et al., 1999), much less is known about how specific plant types control methane production (Williams and Yavitt, 2010). In flooded rice soils when rice plants are growing, organic compounds are derived from recent photosynthesis leaks from roots and fuel microorganisms living close to the root surface, including anaerobic CH_4 producers (methanogens) (Williams and Yavitt, 2010). When rice straw is incorporated in the soil after crop harvesting, the decomposition of plant straw also fuels methanogenesis (Conrad, 2007; Williams and Yavitt, 2010).

Conrad (2007) suggested that methane emission from rice fields can be mitigated by water management, nutrient management, and crop management. Water management is probably the most effective management tool for CH_4 mitigation. Mid-season drainage or frequent intermittent drainage generally results in a drastic reduction of CH_4 production and emission. The application of ferric iron or sulfate to the soil also suppresses CH_4 emission. The effect is based on the outcompetition of methanogens by iron-reducing bacteria that utilize the common substrates H_2 and acetate more effectively (Conrad, 2007). Conrad (2007) also reported that the use of urea as an N fertilizer also has CH_4 suppression effect. The suppression of CH_4 emission by urea may be due to the stimulation of CH_4 oxidation or the suppression of CH_4 production.

5.5 CONCLUSIONS

Rice is the staple food crop for about 50% of the world's population. Rice ecosystems are classified into several groups based on water regime, soil type, drainage, topography, and temperature. However, major world production comes from two ecosystems known as upland and lowland or flooded. Flooded rice contributes about 76% of the global rice production. The anaerobic soil environment created by flood irrigation of lowland rice brings several physical, chemical, and biological changes in the rice rhizosphere that may influence growth and development and, consequently, yield. The main changes that occur in flooded or waterlogging rice soils decrease in oxidation–reduction or redox potential and increase in Fe^{2+} and Mn^{2+} concentrations due to reduction of Fe^{3+} to Fe^{2+} and Mn^{4+} to Mn^{2+}. The pH of acid soils increase and alkaline soils decrease due to flooding. Reduction of NO_3^- and NO_2^- to N_2 and N_2O, reduction of SO_4^{2-} to S^{2-}, reduction of CO_2 to CH_4, improvement in the concentration and availability of P, Ca, Mg, Fe, Mn, Mo, and Si, and decrease in concentration and availability of Zn, Cu, and S. The uptake of N may increase if properly managed or applied in the reduced soil layer. The chemical changes occur because of physical reactions between the soil and water and also because of biological activities of anaerobic microorganisms. The magnitude of these chemical changes is determined by soil type, soil organic matter content, soil fertility, cultivars, and microbial activities. The exclusion of O_2 from flooded soils is accompanied by an increase of other gases (CO_2, CH_4, and H_2), produced largely through processes of microbial respiration. The knowledge of the chemistry of lowland rice soils is important for fertility management and maximizing rice yield. This chapter has discussed physical, biological, and chemical changes in flooded or lowland rice soils.

REFERENCES

Abreu, C. A., M. F. Abreu, J. C. Andrade, and B. V. Raij. 1998. Restrictions in the use of correlation coefficients in comparing methods for the determination of the micronutrients in soils. *Commun. Soil Sci. Plant Anal.* 29:1961–1972.
Allam, A. I. and J. P. Hollis. 1972. Sulfide inhibition of oxidases in rice roots. *Phytopathology* 62:634–639.
Ando, T., S. Yoshida, and I. Niiyama. 1983. Nature of oxidizing power of rice roots. *Plant Soil* 72:57–71.
Ann, Y., K. R. Reddy, and J. J. Delfino. 2000. Influence of redox potential on phosphorus solubility in chemically amended wetland organic soils. *Ecol. Eng.* 14:169–180.
Armstrong, W. 1979. Aeration in higher plants. *Adv. Bot. Res.* 7:225–332.

Armstrong, W., S. H. F. M. Justin, P. M. Beckett, and S. Lythe. 1991. Root adaption to soil waterlogging. *Aquat. Bot.* 39:57–73.

Bacha, R. E. and L. R. Hossner. 1977. Characteristics of coatings formed on rice roots as affected by iron and manganese additions. *Soil Sci. Soc. Am. J.* 41:931–935.

Bacon, P. E., L. G. Lewin, J. W. McGarity, E. H. Hoult, and D. Alter. 1989. The effect of stubble management and N fertilization practices on the nitrogen economy under intensive rice cropping. *Aust. J. Soil Res.* 27:685–698.

Baggie, I., D. L. Rowell, J. S. Robinson, and G. P. Warren. 2004. Decomposition and phosphorus release from organic residues as affected by residue quality and added inorganic phosphorus. *Agrofor. Syst.* 63:125–131.

Barbosa Filho, M. P., N. K. Fageria, and L. F. Stone. 1983. Water management and liming in relation to grain yield and iron toxicity. *Pesq. Agrope. Bras.* 18:903–910.

Becker, M., J. K. Ladha, I. C. Simpson, and J. C. G. Ottow. 1994. Parameters affecting residue nitrogen mineralization in flooded soils. *Soil Sci. Soc. Am. J.* 58:1666–1671.

Bellakki, M. A., V. P. Badanur, and R. A. Setty. 1998. Effect of long term integrated nutrient management on important properties of a Vertisol. *J. Indian Soc. Soil Sci.* 46:176–180.

Bigalke, M., S. Weyer, and W. Wilcke. 2010. Stable copper isotopes: A novel tool to trace copper behavior in hydromorphic soils. *Soil Sci. Soc. Am. J.* 74:60–73.

Bird, J. A., W. R. Horwath, A. J. Eagle, and C. V. Kessel. 2001. Immobilization of fertilizer N in rice; effects of straw management practices. *Soil Sci. Soc. Am. J.* 65:1143–1152.

Bird, J. A., C. V. Kessel, and W. R. Horwath. 2002. Nitrogen dynamics in humic fractions under alternative straw management in temperate rice. *Soil Sci. Soc. Am. J.* 66:478–488.

Bohn, H. L., B. L. McNeal, and G. A. O'Connor. 1979. *Soil Chemistry*. New York: John Wiley & Sons.

Boivin, P., F. Favre, C. Hammecker, J. L. Maeght, J. Delariviere, J. C. Poussin, and M. C. S. Wopereis. 2002. Processes driving soil solution chemistry in a flooded rice-cropped vertisol: Analysis of long time monitoring data. *Geoderma* 10:199–206.

Bouldin, D. R. 1986. The chemistry and biology of flooded soils in relations to the nitrogen economy in rice fields. *Fertil. Res.* 9:1–14.

Breemen, N. V. and F. R. Moormann. 1978. Iron-toxic soils. In: *Soils and Rice*, ed., IRRI, pp. 781–800. Los Bãnos, Philippines: IRRI.

Broadbent, F. E. and T. Nakashima. 1970. Nitrogen immobilization in flooded soils. *Soil Sci. Soc. Am. Proc.* 35:922–926.

Carbonell-Barrachina, A. A., A. Jugsujinda, F. Burlo, R. D. Delaune, and W. H. Patrick. 2000. Arsenic chemistry in municipal sewage sludge as affected by redox potential and pH. *Water Res.* 34:216–224.

Cassman, K. G., G. C. Gines, M. A. Dizon, M. I. Samson, and J. M. Alcantara. 1996. Nitrogen use efficiency in tropical lowland rice systems: Contribution from indigenous and applied nitrogen. *Field Crops Res.* 47:1–12.

Catlett, K. M., D. M. Heil, W. L. Lindsay, and M. H. Ebinger. 2002. Soil chemical properties controlling Zn^{2+} activity in 18 Colorado soils. *Soil Sci. Soc. Am. J.* 66:1182–1189.

Ceuppens, J. and M. C. S. Wopereis. 1999. Impact of non-drained irrigated rice cropping on soil salinization in the Senegal River Delta. *Georderma* 92:125–140.

Chang, T. T. 1975. The origin, evolution, cultivation, dissemination, and diversification of Asian and African rices. *Euphytica* 25:425–441.

Chartres, C. J. 1993. Sodic soils: An introduction to their formation and distribution in Australia. *Aust. J. Soil Res.* 31:751–760.

Chen, C. C., J. B. Dixon, and F. T. Turner. 1980a. Iron coatings on rice roots: Mineralogy and quantity influencing factors. *Soil Sci. Soc. Am. J.* 44:635–639.

Chen, C. C., J. B. Dixon, and F. T. Turner. 1980b. Iron coatings on rice roots: Morphology and models of development. *Soil Sci. Soc. Am. J.* 44:1113–1119.

Conrad, R. 2007. Microbial ecology of methanogens and methanotrophs. *Adv. Agron.* 96:1–63.

Contin, M., C. Mondini, L. Leita, and M. D. Nobili. 2007. Enhanced soil toxic metal fixation in iron (hydr)oxides by redox cycles. *Geoderma* 140:164–175.

Crowder, A. A. and S. M. Macfie. 1986. Seasonal deposition of ferric hydroxide plaque on roots of wetland plants. *Can. J. Bot.* 64:2120–2124.

De Datta, S. K. 1981. *Principles and Practices of Rice Production*. New York: John Wiley & Sons.

De Datta, S. K. and D. S. Mikkelsen. 1985. Potassium nutrition of rice. In: *Potassium in Agriculture*, ed., R. D. Munson, pp. 665–699. Madison, WI: American Society of Agronomy.

Dobermann, A., T. George, and N. Thevs. 2002. Phosphorus fertilizer effects on soil phosphorus pools in acid upland soils. *Soil Sci. Soc. Am. J.* 66:652–660.

Dutta, D., B. Mandal, and L. N. Mandal. 1989. Decrease in availability of zinc and copper in acidic to near neutral soils on submergence. *Soil Sci.* 147:187–195.

Eagle, A. J., J. A. Bird, J. E. Hill, W. R. Horwath, and C. V. Kessel. 2001. Nitrogen dynamics and fertilizer use efficiency in rice following straw incorporation and winter flooding. *Agron. J.* 93:1346–1354.

Eagle, A. J., J. A. Bird, W. R. Horwath, B. A. Lindquist, S. M. Brouder, J. E. Hill, and C.V. Kessel. 2000. Rice yield and nitrogen efficiency under alternative straw management practices. *Agron. J.* 92:1096–1103.

Fageria, N. K. 1984. *Fertilization and Mineral Nutrition of Rice*. Goiânia/Rio de Janeiro, Brazil: EMBRAPA-CNPAF/Editora Campus.

Fageria, N. K. 1992. *Maximizing Crop Yields*. New York: Marcel Dekker.

Fageria, N. K. 2000. Adequate and toxic levels of zinc for rice, common bean, corn, soybean and wheat production in cerrado soil. *Rev. Bras. Eng. Agri. Ambien.* 4:390–395.

Fageria, N. K. 2001a. Screening method of lowland rice genotypes for zinc uptake efficiency. *Scientia Agricola* 58:623–626.

Fageria, N. K. 2001b. Nutrient management for improving upland rice productivity and sustainability. *Commun. Soil Sci. Plant Anal.* 32:2603–2629.

Fageria, N. K. 2007. Yield physiology of rice. *J. Plant Nutr.* 30:843–879.

Fageria, N. K. 2009. *The Use of Nutrients in Crop Plants*. Boca Raton, FL: CRC Press.

Fageria, N. K. and V. C. Baligar. 1999. Growth and nutrient concentrations of common bean, lowland rice, corn, soybean, and wheat at different soil pH on an Inceptisol. *J. Plant Nutr.* 22:1495–1507.

Fageria, N. K. and V. C. Baligar. 2001. Lowland rice response to nitrogen fertilization. *Commun. Soil Sci. Plant Anal.* 32:1405–1429.

Fageria, N. K. and V. C. Baligar. 2005. Enhancing nitrogen use efficiency in crop plants. *Adv. Agron.* 88:97–185.

Fageria, N. K. and V. C. Baligar. 2006. Nutrient efficient plants in improving crop yields in the twenty first century. Paper presented at *the 18th World Soil Science Congress*, July 9–15, 2006, Philadelphia, PA.

Fageria, N. K., V. C. Baligar, and R. B. Clark. 2002. Micronutrients in crop production. *Adv. Agron.* 77:185–268.

Fageria, N. K., V. C. Baligar, and R. B. Clark. 2006. *Physiology of Crop Production*. New York: The Haworth Press.

Fageria, N. K., V. C. Baligar, and C. A. Jones. 2011a. *Growth and Mineral Nutrition of Field Crops*, 3rd edn. Boca Raton, FL: CRC Press.

Fageria, N. K., V. C. Baligar, and R. J. Wright. 1990a. Iron nutrition of plants: An overview on the chemistry and physiology of its deficiency and toxicity. *Pesq. Agropec. Bras.* 25:553–570.

Fageria, N. K., V. C. Baligar, R. J. Wright, and J. R. P. Carvalho. 1990b. Lowland rice response to potassium fertilization and its effect on N and P uptake. *Fertil. Res.* 21:157–162.

Fageria, N. K. and M. P. Barbosa Filho. 1994. *Nutritional Deficiency in Rice: Identification and Correction*, 36 pp. Goiania, Brazil: Embrapa Arroz e Feijão.

Fageria, N. K. and M. P. Barbosa Filho. 2007. Dry-matter and grain yield, nutrient uptake, and phosphorus use-efficiency of lowland rice as influenced by phosphorus fertilization. *Commun. Soil Sci. Plant Anal.* 38:1289–1297.

Fageria, N. K., M. P. Barbosa Filho, and F. J. P. Zimmermann. 1994. Chemical and physical characterization of varzea soils of some States of Brazil. *Pesq. Agropec. Bras.* 29:267–274.

Fageria, N. K., G. D. Carvalho, A. B. Santos, E. P. B. Ferreira, and A. M. Knupp. 2011b. Chemistry of lowland rice soils and nutrient availability. *Commun. Soil Sci. Plant Anal.* 42:1–21.

Fageria, N. K. and A. S. Prabhu. 2004. Blast control and nitrogen management in lowland rice cultivation. *Pesq. Agropec. Bras.* 39:123–129.

Fageria, N. K. and N. A. Rabelo. 1987. Tolerance of rice cultivars to iron toxicity. *J. Plant Nutr.* 10:653–661.

Fageria, N. K. and A. B. Santos. 2008. Lowland rice response to thermophosphate fertilization. *Commun. Soil Sci. Plant Anal.* 39:873–889.

Fageria, N. K., A. B. Santos, M. P. Barbosa Filho, and C. M. Guimarães. 2008. Iron toxicity in lowland rice. *J. Plant Nutr.* 31:1676–1697.

Fageria, N. K., A. B. Santos, and T. Cobucci. 2011c. Zinc nutrition of lowland rice. *Commun. Soil Sci. Plant Anal.* 42:1719–1727.

Fageria, N. K., A. B. Santos, and V. A. Cutrim. 2007. Yield and nitrogen use efficiency of lowland rice genotypes as influenced by nitrogen fertilization. *Pesq. Agropec. Brás.* 42:1029–1034.

Fageria, N. K., A. B. Santos, and A. B. Heinemann. 2011d. Lowland rice genotypes evaluation for phosphorus use efficiency in tropical lowland. *J. Plant Nutr.* 34:1087–1095.

Fageria, N. K., A. B. Santos, I. D. G. Lins, and S. L. Camargo. 1997. Characterization of fertility and particle size of varzea soils of Mato Grosso and Mato Grosso do Sul States of Brazil. *Commun. Soil Sci. Plant Anal.* 28:37–47.

Fageria, N. K., A. B. Santos, A. Moreira, and M. F. Moraes. 2010. Potassium soil test calibration for lowland rice on an Inceptisol. *Commun. Soil Sci. Plant Anal.* 41:1–7.

Fageria, N. K., N. A. Slaton, and V. C. Baligar. 2003. Nutrient management for improving lowland rice productivity and sustainability. *Adv. Agron.* 80:63–152.

Fageria, N. K., R. J. Wright, V. C. Baligar, and C. M. Sousa. 1991. Characterization of physical and chemical properties of varzea soils of Goias State of Brazil. *Commun. Soil Sci. Plant Anal.* 22:1631–1646.

Flach, K. W. and D. F. Slusher. 1978. Soils used for rice culture in the United States. In: *Soils and Rice*, ed., IRRI, pp. 199–215. Los Banos, Philippines: IRRI.

Gao, S., T. J. Schroder, E. Hoffland, C. Zou, F. Zhang, and S. E. A. T. M. van der Zee. 2010. Geochemical modeling of zinc bioavailability for rice. *Soil Sci. Soc. Am. J.* 74:301–309.

Gao, S., K. K. Tanji, and S. C. Scardaci. 2004. Impact of rice incorporation on soil redox status and sulfide toxicity. *Agron. J.* 96:70–76.

Gao, S., K. K. Tanji, S. C. Scardaci, and A. T. Chow. 2002. Comparison of redox indicators in a paddy soil during rice growing season. *Soil Sci. Soc. Am. J.* 66:805–817.

Garrity, D. P. 1984. Rice environmental classification: A comparative review. In: *Terminology for Rice Growing Environments*, ed., IRRI, pp. 11–26. Los Banos, Philippines: IRRI.

Gathala, M. K., J. K. Ladha, V. Kumar, Y. S. Saharawat, V. Kumar, P. K. Sharma, S. Sharma, and H. Pathak. 2011. Tillage and crop establishment affects sustainability of South Asia rice-wheat system. *Agron. J.* 103:961–971.

Genc, Y., G. K. McDonald, and R. D. Graham. 2002. Critical deficiency concentration of zinc in barley genotypes differing in zinc efficiency and its relations to growth responses. *J. Plant Nutr.* 25:545–560.

Genon, J. G., N. Hepcee, B. Delvaux, J. E. Dufey, and P. Hennebert. 1994. Redox conditions and iron chemistry in highland swamps of Burundi. *Plant Soil* 166:165–171.

Ghildyal, B. P. 1978. Effects of compaction and puddling on soil physical properties and rice growth. In: *Soils and Rice*, ed., IRRI, pp. 317–336. Los Banos, Philippines: IRRI.

Ghildyal, B. P. and R. P. Tripathi. 1971. Effect of varying bulk densities on the thermal characteristics of lateritic sandy clay loam soil. *J. Indian Soc. Soil Sci.* 19:5–10.

Giesler, R., T. Andersson, L. Lovgren, and P. Persson. 2005. Phosphorus sorption in aluminum and iron rich humus soils. *Soil Sci. Soc. Am. J.* 69:77–86.

Goh, K. M. and S. S. Tutuna. 2004. Effects of organic and plant residue quality and orchard management practices on decomposition rates of residues. *Commun. Soil Sci. Plant Anal.* 35:441–460.

Goldberg, S., H. S. Forster, and C. L. Godfrey. 1996a. Molybdenum adsorption on oxides, clay minerals, and soils. *Soil Sci. Soc. Am. J.* 60:425–432.

Goldberg, S., H. S. Forster, S. M. Lesch, and E. L. Heick. 1993. Boron adsorption mechanisms on oxides, clay minerals, and soils inferred from ionic strength effects. *Soil Sci. Soc. Am. J.* 57:704–708.

Goldberg, S., H. S. Forester, S. M. Lesch, and E. L. Heick. 1996b. Influence of anion competition on boron adsorption by clays and soils. *Soil Sci.* 161:99–103.

Goldberg, S., S. M. Lesch, and D. L. Suarez. 2002. Predicting molybdenum adsorption by soils using soil chemical parameters in the constant capacitance model. *Soil Sci. Soc. Am. J.* 66:1836–1842.

Goldberg, S., D. L. Suarez, N. T. Basta, and S. M. Lesch. 2004. Predicting boron adsorption isotherms by Midwestern soils using the constant capacitance model. *Soil Sci. Soc. Am. J.* 68:795–801.

Goswami, N. N. and N. K. Banerjee. 1978. Phosphorus, potassium and other micronutrients. In: *Soils and Rice*, ed. IRRI, pp. 561–580, 317–336. Los Banos, Philippines: IRRI.

Gotoh, S. and W. H. Patrick, Jr. 1972. Transformation of manganese in a waterlogged soil as affected by redox potential and pH. *Soil Sci. Soc. Am. Proc.* 36:738–742.

Gotoh, S. and W. H. Patrick, Jr. 1974. Transformation of iron in a waterlogged soil as influenced by redox potential and pH. *Soil Sci. Soc. Am. Proc.* 38:66–71.

Graham, R. D., J. S. Ascher, and S. C. Hynes. 1992. Selecting zinc efficient cereal genotypes for soils of low zinc status. *Plant Soil* 146:241–250.

Grybos, M., M. Davranche, G. Grauau, and P. Petitjean. 2007. Is trace metal release in wetland soils controlled by organic matter mobility or Fe-oxyhydroxides reduction? *J. Colloid Interface Sci.* 314:490–501.

Gu, B. and L. E. Lowe. 1990. Studies on the adsorption of boron on humic acids. *Can. J. Soil Sci.* 70:305–311.

Guadalix, M. E. and M. T. Pardo. 1995. Zinc sorption by acid tropical soils as affected by cultivation. *Eur. J. Soil Sci.* 46:317–322.

Gunawarkena, I., S. S. Virmani, and F. J. Sumo. 1982. Breeding rice for tolerance to iron toxicity. *Oryza* 19:5–12.

Hargrove, T. R. 1988. Rice production leaps forward. *Span* 30:114–115.

Havlin, J. L., J. D. Beaton, S. L. Tisdale, and W. L. Nelson. 1999. *Soil Fertility and Fertilizers: An Introduction to Nutrient Management*, 6th edn. Upper Saddle River, NJ: Prentice Hall.

Hedley, M. J., G. J. D. Kirk, and M. B. Santos. 1994. Phosphorus efficiency and the forms of soil phosphorus utilized by upland rice cultivars. *Plant Soil* 158:53–62.

Hinsinger, P. 1998. How do plant roots acquire mineral nutrients? Chemical processes involved in the rhizosphere. *Adv. Agron.* 64:225–265.

Holford, I. A. R. and W. H. Patrick, Jr. 1981. Effects of duration of anaerobiosis and reoxidation on phosphate sorption characteristics of an acid soil. *Aust. J. Soil Res.* 19:69–78.

Horiguchi, T. 1995. Rhizosphere and root functions. In: *Science of the Rice Plant: Physiology*, Vol. 2, eds., T. Matsuo, K. Kumazawa, R. Ishii, K. Ishihara, and H. Hirata, pp. 221–248. Tokyo, Japan: Food and Agriculture Policy Research Center.

Huang, Z. W. and F. E. Broadbent. 1989. The influence of organic residues on utilization of urea N by rice. *Fert. Res.* 18:213–220.

Hudnall, W. H. 1991. Taxonomy of acid rice growing soils of the tropics. In: *Rice Production on Acid Soils of the Tropics*, eds., P. Deturck and F. N. Ponnamperuma, pp. 3–8. Kandy, Sri Lanka: Institute of Fundamental Studies.

Huguenin-Elie, O., G. P. D. Kirk, and E. Frossard. 2003. Phosphorus uptake by rice from soil that is flooded, drained or flooded then drained. *Eur. J. Soil Sci.* 54:77–90.

Ikehashi, H. and F. N. Ponnamperuma. 1978. Varietal tolerance of rice for adverse soils. In: *Soils and Rice*, ed., IRRI, pp. 801–825. Los Baños, Philippines: IRRI.

Jackson, B. R., A. Yantasast, C. Prechachart, M. A. Chowdhry, and S. M. H. Zaman. 1972. Breeding rice for deep water areas. In: *Rice Breeding*, ed., IRRI, pp. 515–528. Los Banos, Philippines: IRRI.

Janardhanan, L. and S. H. Daroub. 2010. Phosphorus sorption in organic soils in South Florida. *Soil Sci. Soc. Am. J.* 74:1597–1606.

Jin, J., D. C. Martens, and L. W. Zelazny. 1987. Distribution and plant availability of soil boron fractions. *Soil Sci. Soc. Am. J.* 51:1228–1231.

Joabsson, A., T. R. Christensen, and B. Wallen. 1999. Vascular plant controls on methane emission from northern peat forming wetlands. *Trends Ecol. Evol.* 14:385–388.

Kabata-Pendias, A. 2000. *Trace Elements in Soils and Plants*, 3rd edn. Boca Raton, FL: CRC Press.

Keren, R. and F. T. Bingham. 1985. Boron in water, soils, and plants. *Adv. Soil Sci.* 1:229–276.

Khalid, R. A., W. H. Patrick, Jr., and R. D. DeLaune. 1977. Phosphorus sorption characteristics of flooded soil. *Soil Sci. Soc. Am. J.* 41:305–310.

Khush, G. S. 1984. Terminology for rice growing environments. In: *Terminology for Rice Growing Environments*, ed., IRRI, pp. 5–10. Los Banos, Philippines: IRRI.

Kirk, G. J. D. and M. A. Saleque. 1995. Solubilization of phosphate by rice plants growing in reduced soil prediction of the amount solubilized and the resultant increase in uptake. *Eur. J. Soil Sci.* 46:247–255.

Klamt, E., N. Kampf, and P. Schneider. 1985. Varzea soils of state of Rio Grande do sul. Faculty of Agronomy, Department of Soils, Technical bulletin 4, Porto Alegre, Brazil: Federal University of Rio Grande do Sul.

Kumar, K. and K. M. Goh. 2000. Crop residues and management practices: Effects on soil quality, soil nitrogen dynamics, crop yield, and nitrogen recovery. *Adv. Agron.* 68:197–319.

Kundu, D. K. and J. K. Ladha. 1999. Sustaining productivity of lowland rice soils: Issues and options related to N availability. In: *Resource Management in Rice Systems: Nutrients*, eds., V. Balasubramanian, J. K. Ladhs, and G. L. Denning, pp. 27–44. Dordrecht, the Netherlands: Kluwer Academic Publishers.

Kuo, S. and D. S. Mikkelsen. 1981. The effects of straw and sulfate amendments and temperature on sulfide production in two flooded soils. *Soil Sci.* 132:353–357.

Ladha, J. K., H. Pathak, T. J. Krupnik, J. Six, and C. V. Kessel. 2005. Efficiency of fertilizer nitrogen in cereal production: Retrospects and prospects. *Adv. Agron.* 87:85–156.

Lafond, G. P., M. Stumborg, R. Lemke, W. E. May, C. B. Holzapfel, and C. A. Campbell. 2009. Quantifying straw removal through baling and measuring the long-term impact on soil quality and wheat production. *Agron. J.* 101:529–537.

Lehto, T. and E. Malkonen. 1994. Effects of liming and boron fertilization on boron uptake of *Picea abies. Plant Soil* 163:55–64.

Lindsay, W. L. 1991. Inorganic equilibrium affecting micronutrients in soils. In: *Micronutrients in Agriculture*, 2nd edn., eds., J. J. Mortvedt, F. R. Fox, L. M. Shuman, and R. L. Welch, pp. 477–521. Madison, WI: SSSA.

Lindsay, W. L. and W. A. Norvell. 1978. Development of a DTPA soil test for zinc, iron, manganese, and copper. *Soil Sci. Soc. Am. J.* 42:421–428.

Linquist, B. A., S. M. Brouder, and J. E. Hill. 2006. Winter straw and water management effects on soil nitrogen dynamics in California rice systems. *Agron. J.* 98:1050–1059.

Linquist, B. A., P. W. Singleton, R. S. Yost, and K. G. Cassman. 1997. Aggregate size effects on the sorption and release of phosphorus in an Ultisol. *Soil Sci. Soc. Am. J.* 61:160–166.

Manceau, A., K. L. Nagy, M. A. Marcus, M. Lanson, N. Geoffroy, T. Jacquer, and T. Kirpichtchikova. 2008. Formation of metallic copper nanoparticles at the soil-root interface. *Environ. Sci. Technol.* 42:1766–1772.

Mandal, B., T. K. Adhikari, and D. K. De. 1993. Effect of lime and organic matter application on the availability of added boron in acidic alluvial soils. *Commun. Soil Sci. Plant Anal.* 24:1925–1935.

Mandal, B., G. C. Hazra, and L. N. Mandal. 2000. Soil management influences on zinc desorption for rice and maize nutrition. *Soil Sci. Soc. Am. J.* 64:1699–1705.

Marschner, H. 1995. *Mineral Nutrition of Higher Plants*, 2nd edn. New York: Academic Press.

Matocha, C. J., A. D. Karathanasis, S. Rakshit, and K. M. Wagner. 2005. Reduction of copper (II) by iron (II). *J. Environ. Qual.* 34:1539–1546.

Matsuo, H., A. J. Pecrot, and J. Riquier. 1978. Rice soils of Europe. In: *Soils and Rice*, ed., IRRI, pp. 191–198. Los Banos, Philippines: IRRI.

Mehlich, A. 1953. *Determination of P, Ca, Mg, K and NH₄ (Mimeograph)*. Raleigh, NC: North Carolina Soil Test Division.

Mikkelson, D. S. 1987. Nitrogen budgets in flooded soils used for rice production. *Plant Soil* 100:71–97.

Miller, R. O. and D. E. Kissel. 2010. Comparison of soil pH methods on soils of North America. *Soil Sci. Soc. Am. J.* 74:310–316.

Moormann, F. R. 1978. Morphology and classification of soils on which rice is grown. In: *Soils and Rice*, ed., IRRI, pp. 255–272. Los Banos, Philippines: IRRI.

Moraes, J. F. V. 1999. Soils. In: *Rice Culture in Brazil*, eds., N. R. A. Vieira, A. B. Santos, and E. P. Santana, pp. 88–115. Santo Antônio de Goiás, Brazil: Embrapa Arroz e Feijão.

Moraes, J. F. V. and C. J. S. Freira. 1974. Influence of flooding water depth on growth and yield of rice. *Pesq. Agropec. Bras.* 9:45–48.

Mortvedt, J. J. 1994. Needs for controlled availability of micronutrient fertilizers. *Fertil. Res.* 38:213–221.

Murphy, L. S. and L. M. Walsh. 1972. Correction of micronutrient deficiencies with fertilizers. In: *Micronutrients in Agriculture*, eds., J. J. Mortvedt, P. M. Giordano, and W. L. Lindsay, pp. 347–387. Madison, WI: Soil Science Society of America.

Murthy, R. S. 1978. Rice soils of India. In: *Soils and Rice*, ed., IRRI, pp. 3–17. Los Banos, Philippines; IRRI.

Naidu, R., R. H. Merry, G. J. Churchman, M. J. Wright, R. S. Murray, R. W. Fitzpatrick, and B. A. Zarcinas. 1993. Sodicity in South Australia: A review. *Aust. J. Soil Sci.* 31:911–930.

Nierop, K. G. J., B. van Lagen, and P. Buurman. 2001. Composition of plant tissues and soil organic matter in the first stages of a vegetation succession. *Geoderma* 29:1009–1016.

Obata, H. 1995. Micro essential elements. In: *Science of the Rice Plant: Physiology*, Vol. 2, eds., T. Matsuo, K. Kumazawa, R. Ishii, K. Ishihara, and H. Hirata, pp. 402–419. Tokyo, Japan: Food and Agriculture Policy Research Center.

Obrador, A., J. Novillo, and J. M. Alvarez. 2003. Mobility and availability to plants of two zinc sources applied to a calcareous soil. *Soil Sci. Soc. Am. J.* 67:564–572.

Okuda, A. and E. Takahashi. 1965. The role of silicon. In: *The Mineral Nutrition of the Rice Plant*, ed., IRRI, pp. 123–146. Baltimore, MD: Johns Hopkins Press.

Olsen, S. R. and F. S. Watanabe. 1957. A method to determine a phosphorus adsorption maximum of soils as measured by the Langmuir isotherm. *Soil Sci. Soc. Am. Proc.* 21:144–149.

Pal, D. and F. E. Broadbent. 1975. Influence of moisture o rice straw decomposition in soils. *Soil Sci. Soc. Am. Proc.* 39:59–63.

Panabokke, C. R. 1978. Rice soils of Sri Lanka. In: *Soils and Rice*, ed., IRRI, pp. 19–33. Los Banos, Philippines: IRRI.

Pant, H. K. and K. R. Reddy. 2001. Phosphorus sorption characteristics of estuarine sediments under different redox conditions. *J. Environ. Qual.* 30:1474–1480.

Patrick, W. H. Jr. 1966. Apparatus for controlling the oxidation–reduction potential of water-logged soils. *Nature* 212:1278–1279.

Patrick, W. H. Jr., S. Gotoh, and B. G. Williams. 1973. Strengite dissolution in flooded soils and sediments. *Science* 179:564–565.

Patrick, W. H. Jr. and A. Jugsujinda. 1992. Sequential reduction and oxidation of inorganic nitrogen, manganese, and iron in flooded soil. *Soil Sci. Soc. Am. J.* 56:1071–1073.

Patrick, W. H. Jr. and I. C. Mahapatra. 1968. Transformation and availability to rice of nitrogen and phosphorus in waterlogged soils. *Adv. Agron.* 20:323–356.

Patrick, W. H. Jr. and D. S. Mikkelsen. 1971. Plant nutrient behavior in flooded soil. In: *Fertilizer Technology and Use*, 2nd edn., ed., R. A. Olson, pp. 187–215. Madison, WI: Soil Science Society of America.

Patrick, W. H. Jr., D. S. Mikkelsen, and B. R. Wells. 1986. Plant nutrient behavior in flooded soils. In: *Fertilizer Technology and Use*, 3rd edn., ed., O. P. Englested, pp. 197–228. Madison, WI: SSSA.

Patrick, W. H. Jr. and C. N. Reddy. 1978. Chemical changes in rice soils. In: *Soils and Rice*, ed., IRRI, pp. 361–379. Los Banos, Philippines: IRRI.

Patrick, W. H. Jr. and R. Wyatt. 1964. Soil nitrogen loss as a result of alternate submergence and drying. *Soil Sci. Soc. Am.* 28:647–653.

Ponnamperuma, F. N. 1965. Dynamic aspects of flooded soils and the nutrition of the rice plant. In: *The Mineral Nutrition of the Rice Plant*, ed., IRRI, pp. 295–298. Baltimore, MD: Johns Hopkins Press.

Ponnamperuma, F. N. 1972. The chemistry of submerged soils. *Adv. Agron.* 24:29–96.

Ponnamperuma, F. N. 1975. Micronutrient limitations in acid tropical rice soils. In: *Soil Management in Tropical America*, eds., E. Bornemisza and A. Alvarado, pp. 330–347. Ralegigh, NC: North Carolina State University.

Ponnamperuma, F. N. 1976. Physicochemical properties of submerged soils in relation to fertility. In: *The Fertility of Paddy Soils and Fertilizer Application for Rice*, ed., Food and Fertilizer Technology Center, pp. 1–27. Taipei City, Taiwan: Food and Fertilizer Technology Center.

Ponnamperuma, F. N., R. Bradfield, and M. Peech. 1955. Physiological disease of rice attributable to iron toxicity. *Nature* 175:265.

Powlson, D. S., P. C. Brooks, and B. T. Christensen. 1987. Measurement of soil microbial biomass provides an early indication of changes in total soil organic matter due to straw incorporation. *Soil Biol. Biochem.* 19:159–164.

Qadar, A. 2002. Selecting rice genotypes tolerant to zinc deficiency and sodicity stresses. I. Differences in zinc, iron, manganese, copper, phosphorus concentrations, and phosphorus/zinc ratio in their leaves. *J. Plant Nutr.* 25:457–473.

Qadir, M., R. H. Qureshi, and N. Ahmad. 1998. Horizontal flushing: A promising ameliorative technology for hard saline-sodic and sodic soils. *Soil Tillage Res.* 45:119–131.

Raymundo, M. E. 1978. Rice soils of the Philippines. In: *Soils and Rice*, ed., IRRI, pp. 115–133. Los Banos, Philippines: IRRI.

Reddy, K. R., G. A. O'Conner, and P. M. Gale. 1998. Phosphorus sorption capacities of wetland soils and stream soils impacted by dairy effluent. *J. Environ. Qual.* 27:438–447.

Reisenauer, H. M., L. M. Walsh, and R. G. Hoeft. 1973. Testing soils for sulphur, boron, molybdenum and chlorine. In: *Soil Testing and Plant Analysis*, eds., L. M. Walsh and J. D. Beaton, pp. 173–200. Madison, WI: American Society of Soil Science.

Richardson, C. J. and P. Vaithiyanathan. 1995. Phosphorus sorption characteristics of Everglades soils along a eutrophication gradient. *Soil Sci. Soc. Am. J.* 59:1782–1788.

Roger, P. A. 1996. *Biology and Management of the Floodwater Ecosystem in Rice Fields*. Los Banos, Philippines: International Rice Research Institute.

Romheld, V. and H. Marschner. 1991. Function of micronutrients in plants. In: *Micronutrients in Agriculture*, eds., J. J. Mortvedt, F. R. Cox, L. M. Shuman, and R. M. Welch, pp. 297–328. Madison, WI: Soil Science Society of America.

Saejiew, A., O. Grunberger, S. Arunin, F. Favre, D. Tessier, and P. Boivin. 2004. Critical coagulation concentration of paddy soil clays in sodium-ferrous iron electrolyte. *Am. Soc. Soil Sci. J.* 68:789–794.

Sahrawat, K. L. 2004. Iron toxicity in wetland rice and its role of other nutrients. *J. Plant Nutr.* 27:1471–1504.

Sahu, B. N. 1968. Bronzing disease of rice in Orissa as influenced by soil types and manuring and its control. *J. Indian Soc. Soil Sci.* 16:41–54.

Saleque, M. A., M. J. Abedin, G. M. Panaullah, and N. I. Bhuiyan. 1998. Yield and phosphorus efficiency of some lowland rice varieties at different levels of soil available phosphorus. *Commun. Soil Sci. Plant Anal.* 29:2905–2916.

Sanchez, P. A. and J. G. Salinas. 1981. Low-input technology for managing Oxisols and Ultisols in Tropical America. *Adv. Agron.* 34:279–406.

Schofield, R. K. and A. W. Taylor. 1955. The measurement of soil pH. *Soil Sci. Soc. Am. Proc.* 19:164–167.

Schroder, T. J., W. H. van Riemsdijk, S. E. A. T. M. van der Zee, and J. P. M. Vink. 2008. Monitoring and modeling of the solid-solution partitioning of metals and As in a river floodplain redox sequence. *Appl. Geochem.* 23:2350–2363.

Shechter, M., B. Xing, and B. Chefetz. 2010. Cutin and cutan biopolymers: Their role as natural sorbents. *Soil Sci. Soc. Am. J.* 74:1139–1146.

Sillanpaa, M. 1982. *Micronutrients and Nutrient Status of Soils: A Global Study*. FAO Soils bulletins, 48, Rome: FAO.

Sims, J. T. and G. V. Johnson. 1991. Micronutrient soil tests. In: *Micronutrient in Agriculture*, eds., J. J. Mortvedt, F. R. Fox, L. M. Shuman, and R. M. Welch, pp. 427–476. Madison, WI: SSSA.

Slaton, N. A., C. E. Wilson, Jr., S. Ntamatungiro, R. J. Norman, and D. L. Boothe. 2001. Evaluation of zinc seed treatments for rice. *Agron. J.* 93:152–157.

Soepraptohardjo, M. and H. Suhardjo. 1978. Rice soils of Indonesia. In: *Soils and Rice*, ed., IRRI, pp. 99–113. Los Banos, Philippines: IRRI.

Soil Science Society of America. 2008. *Glossary of Soil Science Terms*. Madison, WI: Soil Science Society of America.

Sparks, D. L. 1987. Potassium dynamics in soils. *Adv. Soil Sci.* 6:1–63.

Stevenson, F. J. 1994. *Humus Chemistry: Genesis, Composition, Reactions*, 2nd edn. New York: John Wiley & Sons.

Stubbs, T. L., A. C. Kennedy, P. E. Reisenauer, and J. W. Burns. 2009. Chemical composition of residue from cereal crops and cultivars in dryland ecosystems. *Agron. J.* 101:538–545.

Sumner, M. E. 1993. Sodic soils: New perspectives. *Aust. J. Soil Res.* 31:683–750.

Tadano, T. 1976. Studies on the methods to prevent iron toxicity in lowland rice. *Mem. Fac. Agric.* 10:22–88.

Tanaka, A., R. Loe, and S. A. Navasero. 1966. Some mechanisms involved in the development of iron toxicity symptoms in the rice plant. *Soil Sci. Plant Nutr.* 12:32–38.

Tanaka, A., N. Watanabe, and Y. Ishizuka. 1969. A critical study of the phosphorus concentration in the soil solution of submerged soils. *J. Soil Sci. Manure, Jpn.* 40:406–414.

Tate, R. L. 1979. Effect of flooding on microbial activity in organic soils: Carbon metabolism. *Soil Sci.* 128:267–273.

USDA (United States Department of Agriculture). 1975. *Soil Taxonomy: A Basic System of Soil Classification for Making and Interpreting Soil* Surveys, 754 pp. USDA Agriculture Handbook 436. Washington, DC: U.S. Government Printing Office.

USDA (United States Department of Agriculture). 2009. Rough rice area, by country and geographical region, 1960/61-2008/09. Available at http://beta.irri.org/solutions/images/stories/wrs/wrs_jun09_2009_table02_usda_area.xls (accessed November 16, 2009; verified January 8, 2010). Washington, DC: USDA.

Vadas, P. A. and J. T. Sims. 1998. Redox status, poultry litter, and phosphorus solubility in Atlantic Coastal Plain soils. *Soil Sci. Soc. Am. J.* 62:1025–1034.

Vadas, P. A. and J. T. Sims. 1999. Phosphorus sorption in manured Atlantic Coastal Plain soils under flooded and drained conditions. *J. Environ. Qual.* 28:1870–1877.

Vergara, B. S. 1985. *Growth and Development of the Deep Water Rice Plant.* International Rice Research Institute paper Series Number 103, Los Banos, Philippines. 38 pp.

Villapando, R. R. and D. A. Graetz. 2001. Phosphorus sorption and desorption properties of the spodic horizon from selected *Florida spodosols. Soil Sci. Soc. Am. J.* 65:331–339.

Weber, F. A., A. Voegelin, R. Kaegi, and R. Kretzschmar. 2009. Contaminant mobilization by metallic copper and metal sulphide colloids in flooded soils. *Nat. Geosci.* 2:267–271.

Welch, R. M. 1993. Zinc concentrations and forms in plants for humans and animals. In: *Zinc in Soils and Plants*, ed., A. D. Robson, pp. 183–195. Dordrecht, the Netherlands: Kluwer Academic Publishers.

Wells, B. R., B. A. Huey, R. J. Norman, and R. S. Helms. 1993. Rice. In: *Nutrient Deficiencies and Toxicities in Crop Plants*, ed., W. F. Bennett, pp. 15–19. St. Paul, MN: The American Phytopathological Society.

Whalen, S. C. 2005. Biogeochemistry of methane exchange between natural wetlands and the atmosphere. *Environ. Eng. Sci.* 22:73–94.

Wickham, T. H. and V. P. Singh. 1978. Water movement through wet soils. In: *Soils and Rice*, ed., IRRI, pp. 337–358. Los Banos, Philippines: IRRI.

Willett, I. R. 1991. Phosphorus dynamics in acidic soils that undergo alternate flooding and drying. In: *Rice Production on Acid Soils of the Tropics*, eds., P. Deturck and F. N. Ponnamperuma, pp. 43–49. Kandy, Sri Lanka: Institute of Fundamental Studies.

Williams, W. A., D. S. Mikkelsen, K. E. Mueller, and J. E. Ruckman. 1968. Nitrogen immobilization by rice straw incorporated in lowland rice production. *Plant Soil* 28:49–60.

Williams, C. J. and J. B. Yavitt. 2010. Temperate wetland methanogenesis: The importance of vegetation type and root ethanol production. *Soil Sci. Soc. Am. J.* 74:317–325.

Yoshida, S. 1981. *Fundamentals of Rice Crop Science.* Los Banos, Philippines: International Rice Research Institute.

Yu, T. R. 1985. *Physical Chemistry of Paddy Soils.* Beijing, Berlin: Springer Verlag.

Yu, T. R. 1991. Physico-chemical properties of acid soils of the tropics in relation to rice growth. In: *Rice Production on Acid Soils of the Tropics*, eds., P. Deturck and F. N. Ponnamperuma, pp. 33–42. Kandy, Sri Lanka: Institute of Fundamental Studies.

Zhang, W., J. W. Faulkner, S. K. Giri, L. D. Geohring, and T. S. Steenhuis. 2010. Effect of soil reduction on phosphorus of an organic-rich silt loam. *Soil Sci. Soc. Am. J.* 74:240–249.

6 Mycorrhizal Associations in the Rhizosphere

6.1 INTRODUCTION

The association between certain fungi and the roots of higher plants called mycorrhizae has long been proved. Mycorrhizae are presumably an important contributor to plant growth in most ecosystems (Troeh and Loynachan, 2009; Ortas and Akpinar, 2011; Shen et al., 2011). Brady and Weil (2002) reported that mycorrhizal structures have been found in fossils of plants that lived some 400 million years ago, indicating that mycorrhizal infection may have played a role in the evolutionary adaptation of plants to the land environment. However, the topic of microbial colonization of plant roots has received special attention from agricultural scientists in the recent years. This has happened due to the increased cost of agricultural inputs for crop production and concern for environmental pollution. In nature, the roots of most plants are colonized by bacteria, fungi, and other organisms for mutual or individual benefits. The root–microbial association varies with plant species and is influenced by environmental conditions. Plant roots that are invaded by fungus may transform into mycorrhizal or fungus roots. The term "mycorrhiza" was first used by the German scientist A. B. Frank in 1885 to describe the fungal hyphae closely associated with plant roots (Mengel et al., 2002). "Mycorrhizal" is a Greek word, which means "fungus" and "root." The singular form of this word is mycorrhiza and plural is mycorrhizae. When roots are colonized by mycorrhizae, the root morphology is modified; however, as long as a balanced relationship is maintained, there are no pathological symptoms (Gerdemann, 1974). Marschner (1995) reported that as a rule the fungus is partially or wholly dependent on the higher plant, whereas the plant may or may not benefit from the former. In some instances, mycorrhizae are essential.

A large number of monocots (about 79%), dicots (83%), and all gymnosperms normally form mycorrhizal associations (Smith et al., 1997; Mukerji et al., 2000; Epstein and Bloom, 2005). On the other hand, plant roots from some families like Brassicaceae, that is, cabbage; Chenopodiaceae, that is, spinach; and Proteaceae, that is, macadamia nuts rarely form this association (Epstein and Bloom, 2005). Marschner (1995) also reported that mycorrhizae are rarely found or are even absent in certain families such as Cruciferae and Chenopodiaceae. Epstein and Bloom (2005) reported that mycorrhizal association did not occur under adverse environmental conditions like high salt concentration, flooding, dry conditions, and soils having very high or low fertility. However, there may be some exceptions (Gupta and Kumar, 2000). For example, Brown and Bledsoe (1996) reported that the roots of a tidal salt marsh halophyte were colonized by arbuscular mycorrhizal fungi (AMF).

Mycorrhizal association with crop plants has special importance in modern agriculture in reducing the cost of crop production, since it improves nutrient and water availability. Mycorrhizae also impart other benefits to plants including the enhanced activity of enzymes such as endoglucanase, endopolymethylgalacturonase, and endoxyloglucanase (Adriano-Anaya et al., 2006); the production of certain secondary metabolites (Schliemann et al., 2008); the increased rate of photosynthesis (Wu and Xia, 2006); the enhancement of nitrogen fixation by symbiotic or association N_2-fixing bacteria (Meghvansi et al., 2008; Javaid, 2009); increased resistance to pests (Khaosaad et al., 2007); tolerance to various abiotic stress factors (Javaid, 2008); improved soil aggregation (Rillig and Mummey, 2006; Wu et al., 2008); and, thus, improved soil physical properties and stability. These fungi also play an important role in heavy metal phytoremediation (Gohre and Paszkowski, 2006). In addition, it is also important because this technology does not degrade natural resources such as soil and water. The objective of this chapter is to describe the basic features of the association of mycorrhizal fungi with crop plants and also to discuss the benefits of the fungi to plants and soil to show the importance of integrating their management into technology for improving crop production.

6.2 TAXONOMY AND BIOLOGY OF MYCORRHIZAL FUNGI

There are several distinct kinds of mycorrhizae (Gerdemann, 1974). However, mycorrhizal fungi are currently grouped into three major groups. These are known as ericoid mycorrhizae, ectomycorrhizae, and endomycorrhizae (Paul and Clark, 1996; Mengel et al., 2002). Among these mycorrhizae groups, endomycorrhizae, the most common, represent the arbuscular mycorrhiza (AM) for economic crop plants. Mycorrhizal fungi are composed of fine, tubular filaments called hyphae. The mycelium is the mass of hyphae that forms the subterranean body of a fungus (Epstein and Bloom, 2005).

6.2.1 ERICOID MYCORRHIZAE

The ericoid mycorrhizae have mainly a symbiotic association with plant species living on poor acid humic soils, especially heather plants such as *Calluna* and *Vaccinium* species. This group of fungi can survive at a soil pH as low as 3.5–4.2. Present in the hyphae are proteases and polyphenol oxidases, enzymes that degrade organic soil matter with a high C/N ratio (>100) (Mengel et al., 2002).

6.2.2 ECTOMYCORRHIZAE

There are many species of fungi that produce ectomycorrhizae. The ectomycorrhiza mostly makes association with roots of trees and shrubs of the temperate zone (Hogberg and Nylum, 1981). It is found on the species of the Pinaceae, Betulaceae, Fagaceae, and a few other plant families. Ectomycorrhizae are formed by a range of fungi with a number of tree species, the latter usually originating in the temperate zone and including oak, beech, pine, eucalyptus, birch, and larch (Harley, 1969;

Marks and Kozlowski, 1973; Tinker, 1980). The formation of ectomycorrhizae is easily observed, both by the presence of the mantle and from changes in the root morphology (Harley, 1969). A loose weft of hyphae forms round the root first, which thickens to produce a sheath or mantle, perhaps some 20–40 μM thick around the roots (Tinker, 1980). This is continuous with a network of hyphae lying between the cortical cells (Hartig net), but which never penetrates the endodermis. The exterior of the sheath carries mycelial strands, rhizomorphs, and a relatively sparse growth of external hyphae that penetrate the surrounding soil (Tinker, 1980). The exchange of materials occurs primarily in the Hartig net, where there is extensive contact between fungus and host cells.

There is sufficient published information that ectomycorrhizal (ECM) association increases nutrient absorption and improves plant growth (Harley, 1969). This may happen due to the increase in the absorbing surface that is in contact with the soil. In addition, mycorrhizal roots are stimulated to branch, and the diameters of individual mycorrhizae are greater than those of comparable nonmycorrhizal roots. This fungus also prolongs the life of roots, thus greatly increasing the total amount of actively absorbing surface area roots (Gerdemann, 1974). ECM fungi also produce auxins that are responsible for some of the important morphological differences between mycorrhizal and nonmycorrhizal roots. Further, fungi-infected root tissue is protected from attack by pathogens by the formation of a mechanical barrier and also by producing antibiotics (Zak, 1964).

The intensity of ECM infection tends to be higher when plants are grown in soils with moderate fertility or unbalanced nutrient supply. The host infection with these fungi may be reduced either in extremely infertile soils or under conditions of higher balanced mineral nutrition (Harley, 1969). Higher levels of infection also occur in raw humus forest soils rather than in agricultural soils that are high in mineral nutrients but low in humus (Gerdemann, 1974). These fungi utilize only relatively simple carbohydrates. They cannot decompose lignin, and only a few have been shown to utilize cellulose. Therefore, it is unlikely that these fungi are capable of much decomposition of plant debris or making appreciable growth in the soil apart from living roots (Harley, 1969; Gerdemann, 1974). Ecomycorrhizal fungi generally are not native in soils where their hosts do not occur, and there are numerous reports of failure of pines to grow when introduced into regions where pines are not native (Zak, 1964). The inoculation of plant roots with these fungi is possible using pure cultures. Mikola (1970) has discussed in detail the importance and techniques of inoculation with ECM fungi.

6.2.3 ENDOMYCORRHIZAE

There are several groups of endomycorrhizae, but the most representative group of these fungi is AM, previously called vesicular arbuscular mycorrhiza (VAM). It is also most important because it is found in many economically important plant species (Mukerji et al., 2000; Javaid, 2009). The AMF live in the roots of up to 90% of land plants, including woody plants (Beck et al., 2007). These fungi have been shown to benefit plants, but spatial and temporal distribution, composition, density, and cost–benefit analyses in ecosystems vary in response to soil properties and management practices (Walker et al., 1982; Merryweather and Fitter, 1998). The host plant type and plant density often determine the abundance and diversity of AMF in the soil (Troeh and Loynachan, 2009). Different AM species were found to be associated with corn than with soybean after 3 years of continuous cropping (Troeh and Loynachan, 2003). Other researchers have reported that the roots of different plants host different AM fungal species (Daniell et al., 2001; Vandenkoornhuyse et al., 2002).

Soil properties are also important in the distribution of AM in the roots of plants. Higher spore counts were reported in poorly drained soils than in well-drained soils (Khalil and Loynachan, 1994). Spores have been detected at a depth of 2.2 m in the soil, but the top 45 cm layer may harbor 70%–85% of the spores (Douds and Millner, 1999). Certain AMF, such as *Glomus etunicatum*, *Glomus intraradices*, *Glomus constrictum*, and *Glomus mosseae*, are often found in cultivated fields (Khalil et al., 1992; Helgason et al., 1998). Troeh and Loynachan (2009) studied the population diversity of AMF in well-drained and poorly drained soils of four Iowa soybean fields. These authors concluded that the *Glomus* species dominates in field soils used to produce two important grain crops, corn and soybean, in Midwestern United States.

The importance and functional significance of AM in terrestrial ecosystem have been well documented (Wang et al., 2010). In addition, it occurs throughout the world from the tropics to the Arctic, and there are very few plant associations that do not contain some species with AM (Gerdemann, 1974; Janos, 1980; Allen, 1991; Smith and Read, 1997, 2008; Muthukumar and Udaiyan, 2000; Shi et al., 2006; Wang et al., 2010). Groups of mycorrhizae colonize plant roots as diverse as orchids, trees, shrubs, and agriculturally important plants. The group that colonizes grasses and many crop plants, AMF is also the largest group of mycorrhiza (Wright, 2005). The presence of AMF in wetland ecosystems has been recently reported (Kumar and Ghose, 2008; Wang et al., 2010).

These ubiquitous fungi function as extensions of plant roots that influence both plant growth and soil quality (Wright, 2005). AM infections produce very little change in external root morphology, and because of this they are frequently overlooked. The fungus is a member of the family *Endogonaceae*, the genus *Glomus*, and is considered to be the most abundant of all soil fungi (Lamont, 1982; Marschner, 1995). It is an obligate symbiotic fungus and is not very host specific; AM are found on the majority of the world's vegetation. However, they are not formed in plants belonging to families such as Cruciferae (rape seed, cabbage, radish, mustard, etc.), Chenopodiaceae (sugar beet, spinach, etc.), Polygonaceae (buck wheat), and Amaranthaceae (amaranthus) (Arihara and Karasawa, 2001).

The basic structures of AMF are spores, hyphae, and arbuscule. Spores germinate and colonize plant roots by using hair-like projections called hyphae. These fungi grow both intercellularly and intracellularly in the root cortex and do not invade the endodermis, stele, or root meristem. The fungal mantle extends well into the soil (Marschner, 1995). The hyphae form complex coils and loops within cells. Within the cells, the hyphae typically form oval structures called vesicles and branched structures called arbuscules. Hence, the term vesicular arbuscular mycorrhiza) is used. The arbuscules appears to be sites of nutrient transfer between the fungus and the host plant (Epstein and Bloom, 2005). Hyphal length in soils is an indicator of the activity of AMF. Reported hyphal length densities range from 1 to 50 m g^{-1} with most values in the 5–15 m g^{-1} range (Jacobsen et al., 1992; Sylvia, 1992; Wright, 2005).

The life cycle of AMF is complex for a soilborne fungus and impacts the ecology and function of organisms in the field (Wright, 2005). It has limited saprophytic ability (Warner and Mosse, 1980) and depends on the colonization of plant roots to survive and propagate (Smith and Gianinazzi-Pearson, 1988). Also, there are large differences in the rates of colonization by different taxonomic groups of AMF (Hart and Reader, 2002).

The quantification of AM hyphae and biomass is difficult because there are no simple and reliable assays that differentiate AMF from other fungi. However, DNA technologies are emerging as a tool to identify this fungus under field conditions (Millner and Wright, 2002; Vandenkoornhuyse et al., 2002; Wright, 2005). Molecular and cellular features of AM have been discussed by Harrison (1999). Arihara and Karasawa (2001) reported that inoculating AMF can improve AM associations of crops but it is not yet practical for widespread adoption in field crop production. The utilization of indigenous AMF is, therefore, more appropriate to enhance the growth and uptake of nutrients.

6.3 FACTORS AFFECTING MYCORRHIZAL COLONIZATION

Host, climate, and soil are important factors that affect the mycorrhizal colonization of plant roots. If these factors are favorable, the colonization of plant roots with mycorrhizae may be high and vice versa. In addition, interactions among these factors also determine the colonization of plant roots with mycorrhizae.

6.3.1 Host Factors

Hosts are important factors for mycorrhizal fungi symbiosis. Although, fungus infects large number of plants, the rate of mycorrhiza formation during the periods of rapid growth may be crucial to plant growth. All the fungi are obligate symbionts and acquire carbon from their host plants to complete their life cycles (Bago et al., 2000). In return, the fungus provides multiple benefits for the plant, including enhanced mineral nutrition and tolerance to abiotic and biotic stresses (Sawers et al., 2008; Smith and Read, 2008). Different hosts growing in the same soil show widely differing susceptibility to fungus infection, with cereals usually being more heavily infected than potatoes (Hayman, 1975; Tinker, 1980). It is unknown whether this is due to a genuinely different root susceptibility to differing amounts of external mycelium that carries the infection further, or to differences in root density or distribution (Tinker, 1980).

It is generally agreed that among crop species that have extensive highly branched root systems such as cereals like wheat and grasses, the growth and P uptake by AMF is very limited. The situation is inverse in crop plants with coarse root systems and not highly branched, such as many woody species and root species such as cassava (Howeler et al., 1982; Marschner, 1995). In legumes, growth stimulation by AMF can improve P uptake and the growth of nodules and atmospheric nitrogen fixation.

The different responses of *Stylosanthes* and *Lotus* to AMF are another example of both the root morphology in general and root hair length in particular in the uptake of P from the deficient soils (Marschner, 1995).

Old and modern cultivars of wheat are significantly different in AMF responsiveness (Zhu et al., 2001). Modern cultivars generally have lower response to AMF and AMF association decreases with increasing P utilization efficiency. Hence, the economic benefit of growing modern P-efficient crop cultivars comes from lower cost inputs, but the soil may suffer from a lack of inputs by AMF (Wright, 2005). More research work is needed to get information on AMF inoculation and performance under field conditions using appropriate crop rotation.

6.3.2 CLIMATIC FACTORS

Climatic factors like temperature and light significantly influence mycorrhizal infection in plant roots. Unfavorable climatic conditions, such as shading, can decrease AM infection in crop plants. Low-temperature stress is one of the main abiotic stresses that limits the growth and natural distribution of plants. In most plants, various physiological processes such as water status and photosynthesis will change when plants are exposed to low temperature (Zhu et al., 2010). Mycorrhizal root colonization changes under low temperature (Hayman, 1974; Liu et al., 2004). The mycorrhizal colonization could affect the leaf water potential of bean plants (El-Tohamy et al., 1999).

The optimum temperature for the infection of AMF is reported to be in the range of 20°C–25°C (Hayman, 1974; Marschner, 1995). However, Bowen (1987) reported that AM colonization usually increases up to about 30°C, but some plant–fungus combinations develop normally up to 35°C. Such variation may represent adaptations to different climates. Zhu et al. (2010) studied the effect of temperature on root colonization with AMF. These authors grew corn plants in a sand and soil mixture at 25°C for 7 weeks, and then subjected them to 5°C, 15°C, and 25°C for 1 week. Low temperature (5°C and 15°C) stress decreased AM root colonization as well as root-and-shoot dry weight, compared to 25°C temperature treatments. Similar results were also reported by Anderson et al. (1987), Volkmar and Woodbury (1989), and Gavito et al. (2003). Maximum plant growth and root colonization with AMF occurred at 25°C. The AM colonization of soybean was strongly decreased at a root zone temperature of 15°C (Zhang et al., 1995). Similarly, AM colonization of sorghum was significantly reduced at a root temperature of 15°C for 10 weeks and was almost completely inhibited at 10°C (Liu et al., 2004). The positive effect of AM symbiosis on corn growth and root dry weight under different temperatures was quite in agreement with many greenhouse studies on other plant species (Volkmar and Woodbury, 1989; Gavito et al., 2003).

Unfavorable environmental conditions such as shading and defoliation depress mycorrhizal growth (Same et al., 1983). Harley (1969) reported that the rate of mycorrhizal infection depends upon the availability of free carbohydrates in the host roots. From this hypothesis it could be expected that more light will increase mycorrhizal infection by increasing the photosynthesis.

6.3.3   Soil Factors

Soil pH and fertility are important factors affecting mycorrhizae association with plant roots (Mosse, 1972; Graw, 1979). Slightly acid pH is considered as an optimal soil pH for mycorrhizal infection of plant roots (Marschner, 1995). Mosse (1972) reported that *G. mosseae* infected hosts only in soils with pH greater than 5.5. However, Tinker (1980) reported that mycorrhizal infection of grasses occurs at lower pH. The infection rate is very low in N-deficient plants and increases with increasing nitrate supply and nitrogen content of roots (Hepper, 1983).

Generally, high soil fertility, especially P, leads to lower infection with both ectotrophic and AM fungus (Sanders and Tinker, 1971, 1973; Marx et al., 1977). A moderate deficiency of a nutrient such as P tends to promote infection (Epstein and Bloom, 2005). Plants with abundant nutrients tend to suppress mycorrhizal infection. High P level in the soil decreases the density of arbuscular development (Smith and Gianinazzi-Pearson, 1988), the amount of external mycelium (Abbott et al., 1984), and the number of AM hyphal entry points in roots (Amijee, 1989). In well-fertilized soils, the association between fungi and plants may shift from beneficial to parasitic. Under these situations, fungi may receive carbohydrates from the host plant but host plant is not benefited from the fungi symbiotic association in relation to nutrient uptake. Brundrett (1991), Marschner (1995), and Epstein and Bloom (2005) reported that under these conditions, the host plant may treat mycorrhizal fungi as it does bacterial or fungal pathogens, and the host plant may initiate resistance responses that release antimicrobial substance that proceed to isolate tissues.

However, high Bray P values for Iowa soils did not suppress AMF colonization of soybean (Khalil et al., 1992), but extraradical hyphae were not measured to determine whether the production of hyphal network was suppressed (Wright, 2005). There are other examples of the complexity of field-based studies of the direct effects of P on AMF and plants. There is evidence that mycorrhizal plants can use organic P sources more effectively than nonmycorrhizal plants (Jayachandran et al., 1992). Microbial activity can suppress hyphal P transferred between plants by AM fungal hyphae (Heap and Newman, 1980; Chiariello et al., 1982; Wright, 2005). An indirect effect on P transfer may be collembolans that graze on hyphae (McGonigle and Fitter, 1988). However, effects on plants double-inoculated with collembolans and AMF under field or under greenhouse conditions are not linear (Harris and Boerner, 1990; Lussenhop, 1996; Wright, 2005).

Soil moisture content is also an important factor that affects mycorrhizal infection of plant roots. In general, waterlogged soils tend to have poorly infected roots (Tinker, 1980). Similarly, dry soils also decrease mycorrhizal infection (Meyer, 1974). Arihara and Karasawa (2001) reported that increased soil moisture (approximately at field capacity) improved the efficiency of AM colonization in corn roots, thus promoting AM formation, which in turn might stimulate P uptake and increase plant growth. Jasper et al. (1993) reported that AM spores were drought resistant but the hyphae did not survive very long in dry soil as they lost the vigor of colonization within 6 weeks. This suggests that adequate soil moisture status might improve the vitality of hyphae grown from spores and enhance the efficiency of AM colonization even in soils with a few spores following the cultivation of nonmycorrhizal crops in the previous seasons. Soil moisture also directly affects root growth and P uptake by crops, as low moisture content reduces P diffusion through the soil to the root surface (Olsen et al., 1965).

6.3.4 TILLAGE AND AGROCHEMICALS

Tillage and the use of agrochemicals (fertilizers, fungicides, insecticides, and herbicides) are important cultural practices in modern crop production systems. Ploughing soils or disturbance reduces AM infection in corn and wheat plants (Evans and Miller, 1990). Garcia et al. (2007) reported that root colonization by AMF was reduced by one-time tillage of continuous no-till, maybe partly because of increased root P concentration with one-time tillage. Soil AMF was also reduced by one-time tillage and had not recovered 2 years after the tillage event (Wortmann et al., 2008). Tillage disruption of the hyphal network can reduce AMF growth and infectivity of hyphal fragments (Evans and Miller, 1990; Johnson et al., 2001; Goss and Varennes, 2002). Channels formed by previous root growth often have high AMF, which may be reduced by the destruction of these channels (Evans and Miller, 1990). Wortmann et al. (2008, 2010) reported that bacteria, fungi, and actinomycete microbial groups were either increased or decreased by one-time tillage but returned to no-till levels in 1–3 years.

Brady and Weil (2002) reported that soil tillage disrupts hyphal networks, and, therefore, minimum tillage may increase the effectiveness of native mycorrhizae. Heavy N fertilization can adversely affect the colonization of mycorrhizae due to its direct effects on the fungus rather than simply the result of changes in root growth (Arihara and Karasawa, 2001). Nitrogen added as ammonium sulfate decreases AM colonization more than nitrate ions added as sodium nitrate (Chambers et al., 1980). The effect of pesticides on mycorrhiza varies with the type of chemical. Soil fumigants such as methyl bromide are toxic (Kleinschmidt and Gerdemann, 1972), whereas herbicides have little adverse effect on most AMF. Paraquat and simazine, however, showed an adverse effect on *G. etunicatum* (Nemec and Tucker, 1983). The influence of insecticides varied depending on the type of insecticides. For example, parathion decreased the formation of AM but carbaryl and endosulfan at 5 mg L^{-1} had negligible effect except for a delay in AM infection (Parvathi et al., 1985). Fungicides such as chloroneb, metalaxyl, and captan (at low concentrations) had no adverse effects on mycorrhizal infection of sour orange (Nemec, 1980). Zak and McMichael (2001) and Wright (2005) reported that fungicides, pesticides, and herbicides, when properly applied, probably have little impact on AM colonization.

6.4 BENEFITS OF MYCORRHIZAL FUNGI

Mycorrhizal fungus has many benefits for the growth of plants. These benefits are improvement in the uptake of nutrients and water and pathogen protection. Many woody plants are dependent on ECM fungi for their growth and survival (Stack and Sinclair, 1975; Perry et al., 1989; Horton et al., 1999; Nara, 2006; Kazantseva et al., 2009). Mycorrhizae usually incur the greatest benefit under conditions of high stress (Dickie et al., 2002, 2005; Dickie and Reich, 2005; Kazantseva et al., 2009), especially those where soil nutrients are limited (Treseder and Allen, 2002; Jones and Smith, 2004; Johnson et al., 2006), depending on the plant and fungal species involved (Johnson et al., 1997; Klironomos, 2003; Jones and Smith, 2004). In addition, colonization with AMF improves salinity tolerance in crop plants. Further, it improves soil structure and also helps in carbon sequestration in the soil.

6.4.1   Improvement in Nutrient Uptake

Nutrient supply in adequate amount and proportion is an important factor for improving plant growth. The transport of nutrients from soil to plant roots is a major function of mycorrhizal fungi. In general, the hyphae of fungi explore a large volume of soil than roots and transfer nutrients via the hyphal network to plant roots. Improvement in the uptake of N, P, K, Ca, Mg, Fe, Mn, Cu, and Zn has been reported by mycorrhizal colonization (Chu, 1999; Cardoso and Kuyper, 2006; Meding and Zasoski, 2008; Javaid, 2009). However, results are generally variable (Clark, 2002). The improvement in N uptake by plant roots infected by mycorrhizal colonization has been reported by Ames et al. (1983), Mader et al. (2000), and Javaid (2009). Generally, mycorrhizal symbiosis more influences on N uptake and translocation if NH_4^+ rather than NO_3^- is the nitrogen source (Smith, 1980). Consequently, plant growth response to mycorrhizal colonization may be greater in the presence of NH_4^+ rather than NO_3^- (Chambers et al., 1980). Since NO_3^- is the dominant form of N in most oxidized agricultural soils, being highly mobile they reduce the importance of mycorrhizal fungi in such soils (Harley, 1989). However, NO_3^- mobility is severely restricted by drought due to its low concentration and diffusion rate (Azcon et al., 1996). Thus, under such conditions, the role of mycorrhizae in NO_3^- transport to the root surface may be significant (Subramanian and Charest, 1999).

Mycorrhizal-infected plant roots have been reported to be involved in the absorption of N from organic compounds (Javaid, 2009). The AMF have been found to be proliferate in decomposing organic residues (St. John et al., 1983) and, therefore, are likely to enhance residue decomposition by stimulating the activity of saprophytic microbes (Hodge et al., 2001). Because of their small size, AMF hyphae are better able than plant roots to penetrate decomposing organic material and are, therefore, better competitors for recently mineralized N (Hodge, 2003). By capturing simple organic N compounds, AMF can short-circuit the N cycle (Hawkins et al., 2000; Hodge et al., 2001). Rains and Bledsoe (2007) reported that AMF are especially important in the acquisition of organic N in plants growing in areas where net N mineralization in soils is low. The mycorrhizal fungi have been reported to improve N_2 fixation in legumes (Javaid, 2009). They are known to enhance nodulation and the nitrogen-fixing ability of symbiotic nitrogen fixers of legumes (Patreze and Cordeiro, 2004; Kuster et al., 2007). Mycorrhizal colonization is also believed to stimulate the number and activity of free-living nitrogen fixers such as *Azotobacter*, *Pseudomonas*, and *Beijerinckia* in the soil (Barea et al., 2002; Raimam et al., 2007).

Nye and Tinker (1977) reported that P uptake in plants infected with mycorrhizal fungi may be four times higher compared to nonmycorrhizal plants. Phosphate is relatively immobile in soil and diffuses only slowly to the plant roots, which results in the formation of a depletion zone around the root, and, consequently, limits the supply of phosphorus to the plant (Smith and Read, 1997). To cope with such adverse conditions in the soil, plants have evolved elaborate mechanisms to facilitate phosphorus uptake, including the formation of symbiotic associations with mycorrhizal fungi (Javaid, 2009). It is now a well-established fact that AM colonization can enhance the uptake of P by plant roots (Janouskova et al., 2007; Medina et al., 2007).

Studies on P uptake from isotopically labeled media have shown that the external fungal hyphae are able to absorb P directly from the soluble P pools in the soil and translocate it to the host root (Wang et al., 2002). Hyphal inflow of P can differ with fungal species and this may in part explain the differences in the efficiency of various fungi in promoting plant growth. AMF, however, may be more than just an extension of the plant's root system (Javaid, 2009). Besides hyphae that extend beyond the root depletion zone, various other mechanisms have been proposed to explain P uptake by mycorrhizal fungi, such as the kinetics of P uptake into hyphae that differ from those of roots either through a higher affinity (lower K_m) or a lower threshold concentration at which influx equals efflux (C_{min}) (Joner and Jakobsen, 1995). Plant roots' infection with mycorrhizal colonization depends on the original soil P level. Soils with lower P concentration are subject to increased mycorrhizal colonization in roots (Covacevich et al., 2007). In addition, roots infected with mycorrhizae may take P from nonlabile sources such as iron and aluminum phosphates (insoluble inorganic P sources) in the soil through interaction between the roots and hyphae (Shibata and Yano, 2003).

In addition to P, N, Zn, and Cu, uptake also improves in plants infected with mycorrhizal fungi. Read and Perez-Moreno (2003) have reported that mycorrhizal fungi may improve N and P uptake by the mineralization of organic N and P from the organic complexes in the growth medium. Wang et al. (2010) reported that the inoculation of AMF significantly improved growth, resulting in greater plant height, diameter at ground level, and biomass, as well as increased absorption of N, P, and K by mangrove plant species. These authors also concluded that AMF play an important role in mangrove ecosystems. Recent studies show that mycorrhizal fungi would have access to rock phosphate through localized alterations of pH and/or by the production of organic acid anions that may act as a chelating agent (Javaid, 2009). Generally, mycorrhizal symbiosis has more influence on nitrogen uptake and translocation if ammonium rather than nitrate is the nitrogen source.

Neumann et al. (2009) reported that the concentrations of P, Zn, and Cu in the shoots of AM-inoculated sweet potato plants were in an optimum range, whereas in shoots of non-AM plants, the concentrations of these elements were lower than optimum, indicating that AM colonization had contributed significantly to the uptake of these nutrients. Arihara and Karasawa (2001) reported that the improving effect of AMF colonization on crop growth is also clear when crops are grown in soils with high P-fixing capacity such as Andosol. These authors further reported that to establish cropping systems efficient in P utilization, arranging crop sequence from the point of view of AMF symbiosis is important for environmentally sound and sustainable crop production. Similarly, Mosse (1973) reported that the improvement of P uptake and plant growth by AMF, one of the most common symbiotic systems, is well known in many crops. According to Subramanian and Charest (1997) mycorrhizal corn plants exhibited significantly higher K, Mg, Mn, and Zn in grains than nonmycorrhizal plants under drought stress. Sulfur uptake by mycorrhizal plants and hyphal translocation of S have also been demonstrated (Cooper and Tinker, 1978; Banerjee et al., 1999).

6.4.2 IMPROVEMENT IN WATER UPTAKE AND USE EFFICIENCY

Water is essential for many physiological and biochemical processes in the plants. Mycorrhizal and nonmycorrhizal plants often display different water status (Auge, 2001). Relative water content, leaf water potential, and water conservation capacity are important characteristics that influence plant water status (Zhu et al., 2010). Water conservation of detached leaves is commonly used to characterize the water-holding capacity of plant leaves (Zhu et al., 2010). The higher water conservation indicated that the plant had lower water loss (Bai et al., 2008). Zhu et al. (2010) reported that mycorrhizal corn plants had higher water conservation and water use efficiency at different temperatures (5°C, 15°C, and 25°C). El-Tohamy et al. (1999) reported that mycorrhizal bean plants had higher water potential during chilling stress. Based on these results it can be concluded that AM symbiosis could improve plant water status. Better water status in mycorrhizal plant leaves might be due to an extraction of soil water contributed by AM fungal hypha (Faber et al., 1991), higher activity, and hydraulic conductivity of roots (Auge and Stodola, 1990). Better water status in mycorrhizal plants was beneficial for stomatal opening in leaves and water flow through the plants to the evaporating surfaces in the leaves (Nelson and Safir, 1982). Furthermore, the improvement of water status via AM symbiosis could play an indirect role in enhancing nutrient uptake, osmotic adjustment, the capacity of gas exchange, and the efficiency of photochemistry of photosystem II (Zhu et al., 2010).

Auge (2001) and Neumann et al. (2009) reported that AM plants have been shown to better withstand periods of drought compared to non-AM control and to maintain higher biomass production under soil conditions. Frequently, these effects could be attributed to an improved mineral nutrition status of AM plants (Kwapata and Hall, 1985; Al-Karaki and Clark, 1998). A higher internal water use efficiency of AM compared to non-AM plants, irrespective of the plant nutritional status, has also been reported (Porcel and Ruiz-Lozano, 2004; Pinior et al., 2005), but the precise physiological reasons behind these observations are not yet completely understood (Neumann et al., 2009). Auge et al. (2003, 2007) reported that the extension of AM fungal mycelium in the soil is an important determinant for the contribution of the AM symbiosis to plant performance under drought stress. Faber et al. (1991) reported that AM fungal hyphae have access to fine soil pores that remain filled with soil solution even under low soil moisture regimes. This may render the AM fungal colony less sensitive toward decreasing soil water potentials compared to non-AM plant roots (Neumann et al., 2009).

6.4.3 PROTECTION FROM PATHOGENS

Infection of mycorrhizal fungus can protect plants from soilborne pathogens (Azcon-Aguilar and Barea, 1996; Borowicz, 2001). A mycorrhizosphere similar to the plant rhizosphere is established around hyphae where substrates are enriched and microbial activity is affected (Linderman, 1992). It is, therefore, reasonable to expect that increased microbial activity mediated by fungi could be antagonistic to root pathogen (Wright, 2005). In addition, improved nutrition and water use efficiency in fungi-infected plants may also control diseases.

ECM fungi have been postulated as a biological option to prevent *Phytophthora cambivora* (ink disease of sweet chestnut) infection in new plantation and nursery stock. Numerous studies have highlighted the inhibitory ability of certain ECM fungal species, decreasing the virulence *of Phytophthora* in tree species like *Pinus echinata* Mill, *Pinus taeda* L. (Marx and Davey, 1969; Marx, 1973; Barham et al., 1974; Blom et al., 2009), and *Eucalyptus marginata* Donn ex Sm. (Malajczuk, 1988). Multiple mechanisms are hypothesized by which ECM fungi could confer protection against root pathogens, by providing a physical barrier to penetration, by secreting antibiotics inhibitory to pathogens, by utilizing surplus carbohydrates, by favoring protective rhizosphere microorganisms, and by inducing in the host inhibitors to the pathogen (Zak, 1964; Marx, 1969, 1972; Blom et al., 2009).

6.4.4 Improvement in Chlorophyll Concentration

Mycorrhizal infection increases chlorophyll concentration in the leaves of the infected plants. Chlorophyll concentration controls the photosynthetic activity of a plant and, consequently, its yield. Zhu et al. (2010) reported that AM colonization increased the concentration of chlorophyll a, chlorophyll b, and chlorophyll a + b in the corn plants. These authors also reported that the maximal fluorescence, the maximum quantum efficiency of photosystem II primary photochemistry, and the potential photochemical efficiency were higher in AM plants compared to non-AM plants. Chlorophyll fluorescence is a very powerful tool to probe and elucidate the function of the photosynthetic apparatus and is now widespread in physiological and ecophysiological studies (Fracheboud et al., 1999; Baker, 2008). These authors also reported that AM inoculation notably increased the net photosynthetic rate and the transpiration rate of corn plants.

6.4.5 Improvement in Salinity Tolerance

Salinity is a serious problem in arid and semiarid regions and is also increasing in irrigated areas in various parts of the world. Munns and Tester (2008) reported that more than 6% of the world's total land area is salt affected. AMF widely occur in saline soils (Aliasgharzadeh et al., 2001). Salinity negatively affects the formation and function of mycorrhizal symbiosis (Juniper and Abbott, 1993). However, several studies have demonstrated that inoculation with AM improves growth and productivity on salt-affected soils in both glycophytes and halophytes (Asghari et al., 2005; Sannazzaro et al., 2006; Giri et al., 2007).

Hajiboland et al. (2010) reported that tomato plants inoculated with AM produced higher yields compared to noninoculated plants under control ($0.63\,dS\ m^{-1}$), low ($EC = 5\,dS\ m^{-1}$) or high ($EC = 10\,dS\ m^{-1}$) salinity levels. Mycorrhization alleviated salt-induced reduction of P, Ca, and K uptake. The Ca/Na and K/Na ratios were also better in AM-inoculated treatments compared to noninoculated treatments. A marked effect of AM on the uptake of P, Ca, and K was also observed even in the control plants. The enhancement of P (Giri et al., 2007) and K (Mohammad et al., 2003) uptake by AM has been reported and was considered one of the main reasons for the amelioration of growth in salt-affected plants colonized by AM (Ruiz-Lozano and Azcon, 2000). There is evidence that AM enhances the ability of plants to cope with salt stress by improving the uptake of nutrients other than P (Hajiboland et al., 2010). Improved ion balance, stimulation of the activity of protecting enzymes, higher water uptake, the activity of protecting enzymes, and changes in the rhizosphere microbial community have been suggested as possible mechanisms for AM-induced amelioration of salt stress (Asghari et al., 2005). Improvements in physiological processes like photosynthetic activity or water use efficiency have also been observed in mycorrhizal plants growing on salt-affected soils (Hajiboland et al., 2010).

Like other abiotic stresses, salinity also induces oxidative stress in plants (Santos et al., 2001; Hajiboland and Joudmand, 2009; Hajiboland et al., 2010). Plant cells contain an array of protection mechanisms and repair systems that can minimize the occurrence of oxidative stress caused by reactive oxygen species (Creissen and Mullineaux, 2002). The induction of reactive oxygen species' enzymes, such as superoxide dismutase, peroxidase, catalase, and ascorbate peroxidase, is the most common mechanism for detoxifying reactive oxygen species synthesized during stress response (Hajiboland et al., 2010). Proline is one of the most common compatible solutes in plants. Its function in the plants is osmotic adjustment of the cells and as a protectant of enzymes and cellular structures (Rhodes et al., 2004). The AM increases proline accumulation in the plants and reduces salinity stress (Santos et al., 2001).

6.4.6 CARBON SEQUESTRATION AND SOIL AGGREGATE STABILIZATION

The activity of mycorrhizal fungi in the plant rhizosphere is responsible for carbon sequestration and also the stabilizing of soil aggregates. Sequestration of C in vegetation and soil is recognized as a mechanism that can mitigate atmospheric CO_2 accumulation (Janzen, 2004; Maillard et al., 2010). Soil aggregation is one of the important characteristics that mediates many chemical, physical, and biological properties of the soil and improves soil quality and the sustainability of cropping systems (Mikha et al., 2010). Glomalin is a C storage molecule in soil directly related to the activity of AMF (Rillig et al., 1999; Rillig and Steinberg, 2002). Improved microbial activity in the rhizosphere lowers CO_2 emissions (Curtin et al., 2000). Golmalin is a sugar–protein complex secreted by AMF in the plant rhizosphere that is thought to contribute to soil aggregation (Soil Science Society of America, 2008). Golmalin can amplify C storage by stabilizing aggregates that contain C from roots and other microorganisms and their by-products. Hyphae of fungi also add C to terrestrial ecosystems (Wright, 2005).

6.5 USE OF MYCORRHIZAE IN THE CROPPING SYSTEMS

Knowledge about the benefits of mycorrhizal inoculation in cropping systems that involve different crop species is important for the practical application of this technology. Arihara and Karasawa (2001) reported the results of a series of experiments involving various crops in the cropping systems. Highly mycorrhizal crops (sunflower, corn, soybean, adzuki bean, and kidney bean), moderately mycorrhizal crops (potato and spring wheat), and nonmycorrhizal crops (buckwheat, sugar beet, mustard, and Japanese radish) were included in the studies. The effects of previous crops varied widely with the degree of mycorrhizal association of succeeding crops. In highly mycorrhizal crops, the stronger the association of previous crops with AM, the better was the yield of the following crop. Moderately mycorrhizal crops were also affected by preceding crops but the magnitude of effect was less than in highly mycorrhizal crops. By contrast, yields of nonmycorrhizal crops were only slightly affected. These authors concluded that highly mycorrhizal crops may be preferred as previous crops for cultivation any crop in the following seasons, irrespective of its mycorrhizal association. However, in the case of highly mycorrhizal crops, cultivation after highly mycorrhizal crops gives best results. Moderately mycorrhizal crops grown in the previous season may have beneficial effects on high to moderate mycorrhizal crops, but the effects may not be clear in the case of nonmycorrhizal crops. Nonmycorrhizal preceding crops were not suitable for most crops irrespective of their mycorrhizal association, although negative effects were generally nonmycorrhizal crops. The use of effective species as cover crops or in crop rotation may favor the buildup of effective mycorrhizae in soils (Brady and Weil, 2002).

6.6 CONCLUSIONS

"Mycorrhizal" is a Greek word which means "fungus" and "root." The term was first introduced by a German scientist A. B. Frank in 1885 to describe fungal hyphae symbiosis with plant roots. There are three major types or groups of mycorrhizae known as ericoid mycorrhizae, ectomycorrhizae (ECM), and endomycorrhizae or AM. The ericoid group of mycorrhizae infect mainly plants of poor acid humic soils, ECM infect mainly forest trees and shrubs, and AM infect plants of economic importance, like food crops (cereals and legumes) and pasture grasses and legumes. In addition, AM is particularly important to legume species since its presence can enhance N_2 fixation. This is a mutually beneficial symbiosis in which the higher plants provide organic carbon to the fungus, mainly as carbohydrates and the fungus exploits the soil volume, providing the root system with additional water and nutrients. Mycorrhiza-infected roots live longer than noninfected ones and may be responsible for the absorption of water and nutrients during the longer period of plant growth cycle and, consequently, higher yield. Host, climate, and soil factors influence mycorrhizal association with plant roots. Among soil factors, temperature is most important, and optimal temperature is reported to be in the range of 20°C–25°C. Soil pH and fertility are important factors affecting the mycorrhizal infection of plant roots. Poor soil fertility is considered to be a prerequisite for better infection of plant roots with mycorrhiza. ECM fungi can be multiplied on synthetic culture media for the inoculation of tree species. However, the AMF needs a host plant for producing inoculum in sufficient quantity. More research work is needed to fully exploit mycorrhizal technology in agricultural production.

REFERENCES

Abbott, L. K., A. D. Robson, and G. D. Boer. 1984. The effect of phosphorus on the formation of hyphae in soil by the vesicular-arbuscular mycorrhizal fungus, *Glomus fasciculatum*. *New Phytol.* 97:437–446.

Adriano-Anaya, M. L., M. Salvador-Figueroa, J. A. Ocampo, and I. Garcia-Romera. 2006. Hydrolytic enzyme activities in maize *(Zea mays)* and sorghum (*Sorghum bicolor*) roots inoculated with *Gluconacetobacter diazotrophicus* and *Glomus intraradices*. *Soil Biol. Biochem.* 38:879–886.

Aliasgharzadeh, N., N. S. Rastin, H. Towfighi, and A. Alizadeh. 2001. Occurrence of arbuscular mycorrhizal fungi in saline soils of the Tabriz plain of Iran in relation to some physical and chemical properties of soil. *Mycorrhiza* 11:119–122.

Al-Karaki, G. N. and R. B. Clark. 1998. Growth, mineral acquisition, and water use by mycorrhizal wheat grown under water stress. *J. Plant Nutr.* 21:263–276.

Allen, M. F. 1991. *The Ecology of Mycorrhizae*. Cambridge, U.K.: Cambridge University Press.

Ames, R. N., C. P. P. Read, and L. K. Porter. 1983. Hyphal uptake and transport of nitrogen from two ^{15}N-labelled sources by *Glomus mosseae*, a vesicular arbuscular mycorrhizal fungus. *New Phytol.* 95:381–396.

Amijee, F. 1989. Colonization of roots by vesicular-arbuscular mycorrhizal fungi in relation to phosphorus and carbon nutrition. *Aspect Appl. Biol.* 22:219–226.

Anderson, C. P., E. I. Sucoff, and R. K. Dixon. 1987. The influence of low soil temperature on the growth of vesicular-arbuscular mycorrhizal *Fraxinus pennsylvanica*. *Can. J. For. Res.* 17:951–956.

Arihara, J. and T. Karasawa. 2001. Phosphorus nutrition in cropping systems through arbuscular mycorrhizal management. In: *Plant Nutrient Acquisition: New Perspectives*, eds., A. Ae, K. Okada, and A. Srinivasan, pp. 319–337. New York: Springer Verlag.

Asghari, H. R., P. Marschner, S. E. Smith, and F. A. Smith. 2005. Growth response of *Atriplex nummularia* to inoculation with arbuscular mycorrhizal fungi at different salinity levels. *Plant Soil* 273:245–256.

Auge, R. M. 2001. Water relations, drought and vesicular-arbuscular mycorrhizal symbiosis. *Mycorrhiza* 11:3–42.

Auge, R. M., J. L. Moore, J. C. Stutz, D. M. Sylvia, A. Al-Agely, and A. M. Saxton. 2003. Relating dehydration resistance of mycorrhizal *Phaseolus vulgaris* L. to soil and root colonization by hyphae. *J. Plant Physiol.* 160:1147–1156.

Auge, R. M. and J. W. Stodola. 1990. An apparent increase in symplastic water contributes to greater turgor in mycorrhizal roots of droughted *Rosa* plants. *New Phytol.* 115:285–295.

Auge, R. M., H. D. Toler, J. L. Moore, K. Cho, and A. M. Saxton. 2007. Comparing contributions of soil versus root colonization to variations in stomatal behaviour and soil drying in mycorrhizal *Sorghum biocolor* and *Cucurbita pepo*. *J. Plant Physiol.* 164:1289–1299.

Azcon, R., M. Gomez, and R. Tobar. 1996. Physiological and nutritional responses by *Lactuca sativa* to nitrogen sources and mycorrhizal fungi under drought. *Biol. Fertil. Soils* 22:156–161.

Azcon-Aguilar, C. and J. M. Barea. 1996. Arbuscular mycorrhizas and biological control of soil-borne plant pathogens—An overview of the mechanisms involved. *Mycorrhiza* 6:457–464.

Bago, B., P. E. Pfeffer, and Y. Shachar-Hill. 2000. Carbon metabolism and transport in arbuscular mycorrhizas. *Plant Physiol.* 124:949–958.

Bai, X. F., H. Han, F. Y. Zhou, S. J. Yang, B. C. Liu, Z. Y. Lei, and Z. Jin. 2008. Analysis on dynamic change of relative water content and needle water conservation of *Pinus sylvestris* var. *Mongolica* on sandy land. *Prot. For. Sci. Technol.* 3:51–54.

Baker, N. R. 2008. Chlorophyll fluorescence: A probe of photosynthesis in vivo. *Annu. Rev. Plant Biol.* 59:89–113.

Banerjee, M. R., S. J. Chapman, and K. Killham. 1999. Uptake of fertilizer sulfur by maize from soils of low sulfur status as affected by vesicular arbuscular mycorrhizae. *Can. J. Soil Sci.* 79:557–559.

Barea, J. M., R. Azcon, and C. Azcon-Aguilar. 2002. Mycorrhizosphere interactions to improve plant fitness and soil quality. *Antonie Van Leeuwenkoek* 81:343–351.

Barham, R. O., D. H. Marx, and J. L. Ruehle. 1974. Infection of ectomycorrhizal and non-mycorrhizal roots of shortleaf pine by nematodes and *Phytophthora cinnamomi*. *Phytopathology* 64:1260–1264.

Beck, A., I. Haug, F. Obserwinkler, and I. Kottke. 2007. Structural characterization and molecular identification of mycorrhiza morphotypes of *Akzatea verticillata* (Alzateaceae), a prominent tree in the tropical mountain rainforest of South Equador. *Mycorrhiza* 17:607–625.

Blom, J. M., A. Vannini, A. M. Vettraino, M. D. Hale, and D. L. Godbold. 2009. Ectomycorrhizal community structure in a healthy and a *Phytophthora*-infected chestnut (*Castanea sativa* Mill.) stand in central Italy. *Mycorrhiza* 20:25–38.

Borowicz, V. A. 2001. Do arbuscular mycorrhizal fungi alter plant-pathogen relations? *Ecology* 82:3057–3068.

Bowen, G. D. 1987. The biology and physiology of infection and its development. In: *Ecophysiology of VA Mycorrhizal Plants*, ed., G. R. Sifir, pp. 27–70. Boca Raton, FL: CRC Press.

Brady, N. C. and R. R. Weil. 2002. *The Nature and Properties of Soils*, 13th edn. Upper Saddle River, NJ: Prentice Hall.

Brown, A. M. and C. Bledsoe. 1996. Spatial and temporal dynamics of mycorrhizas in *Jaumea carnosa*, a tidal saltmarsh halophytes. *J. Ecol.* 84:703–715.

Brundrett, M. C. 1991. Mycorrhizas in natural ecosystems. *Adv. Ecol. Res.* 21:171–313.

Calvet, C., J. Pinochet, A. H. Dorrego, V. Estaun, and A. Camprubi. 2001. Field microplot performance of the peach-almond hybrid FG-677 after inoculation with AM fungi in a replant soil infested with root-knot nematodes. *Mycorrhiza* 10:295–300.

Cardoso, I. M. and T. W. Kuyper. 2006. Mycorrhizas and tropical soil fertility. *Agric. Ecosyst. Environ.* 116:72–84.

Chambers, C. A., S. E. Smith, and F. A. Smith. 1980. Effects of ammonium and nitrate ions on mycorrhizal infection, nodulation and growth of *Trifolium subterraneum*. *New Phytol.* 85:47–62.

Chiariello, N., J. C. Hickman, and H. A. Mooney. 1982. Endomycorrhizal role for interspecific transfer of phosphorus in a community of annual plants. *Science* 217:941–943.

Chu, E. Y. 1999. The effects of arbuscular mycorrhizal fungi inoculation on *Euterpe oleracea* Mart. Seedlings. *Pesq. Agropec. Bras.* 34:1019–1024.

Clark, R. B. 2002. Differences among mycorrhizal fungi for mineral uptake per root length of switch grass grown in acidic soil. *J. Plant Nutr.* 25:1753–1772.

Cooper, K. M. and P. B. Tinker. 1978. Translocation and transfer of nutrients in vesicular arbuscular mycorrhizae. II. Uptake and translocation of phosphorus, zinc and sulphur. *New Phytol.* 81:43–54.

Covacevich, F., H. E. Echeverria, and L. A. N. Aguirrezabal. 2007. Soil available phosphorus status determines indigenous mycorrhizal colonization of field and glasshouse grown spring wheat from Argentina. *Appl. Soil Ecol.* 35:109.

Creissen, G. P. and P. M. Mullineaux. 2002. The molecular biology of the ascorbate-glutathione cycle in higher plants. In: *Oxidative Stress in Plants*, eds., D. Inze and M. V. Montgan, pp. 247–270. Boca Raton, FL: CRC Press.

Curtin, D., H. Wang, F. Selles, B. G. McConkey, and C. A. Campbell. 2000. Tillage effects on carbon fluxes in continuous wheat and fallow-wheat rotations. *Soil Sci. Soc. Am. J.* 64:2080–2086.

Daniell, T. J., R. Husband, A. H. Fitter, and J. P. W. Young. 2001. Molecular diversity of arbuscular mycorrhizal fungi colonizing arable crops. *FEMS Microbiol Ecol.* 36:203–209.

Dickie, I. A., R. T. Koide, and K. C. Steiner. 2002. Influences of established trees on mycorrhizas, nutrition, and growth of *Quercus rubra* seedlings. *Ecol. Monogr.* 72:505–521.

Dickie, I. A. and P. B. Reich. 2005. Ectomycorrhizal fungal communities at forest edges. *J. Ecol.* 93:244–255.

Dickie, I. A., S. A. Schnitzer, P. B. Reich, and S. E. Hobbie. 2005. Spatially disjunct effects of co-occurring competition and facilitation. *Ecol. Lett.* 8:1191–1200.

Douds, D. D. and P. D. Millner. 1999. Biodiversity of arbuscular-mycorrhizal fungi in agroecosystems. *Agric. Ecosyst. Environ.* 74:77–93.

El-Tohamy, W., W. H. Schnitzler, U. EI-Behairy, and M. S. EI-Beltagy. 1999. Effect of VA mycorrhiza on improving drought and chilling tolerance of bean plants (*Phaseolus vulgaris* L.). *J. Appl. Bot.* 73:178–183.

Epstein, M. E. and A. J. Bloom. 2005. *Mineral Nutrition of Plants: Principles and Perspectives*, 2nd edn. Sunderland, MA: Sinauer Associates, Inc. Publishers.

Evans, D. G. and M. H. Miller. 1990. The role of external mycelial network in the effect of soil disturbance upon vesicular arbuscular mycorrhizal colonization of maize. *New Phytol.* 114:65–71.

Faber, B. A., R. J. Zasoski, D. N. Munns, and K. Shackel. 1991. A method for measuring hyphal nutrient and water uptake in mycorrhizal plants. *Can. J. Bot.* 69:87–94.

Fracheboud, Y., P. Haldimann, J. Leipner, and P. Stamp. 1999. Chlorophyll fluorescence as a selection tool for cold tolerance of photosynthesis in maize (*Zea mays* L.). *J. Exp. Bot.* 50:1533–1540.

Garcia, J. P., C. S. Wortmann, M. Mamo, R. A. Drijber, J. A. Quincke, and D. Tarkalson. 2007. One-time tillage of no-till: Effects on nutrients, mycorrhizae, and phosphorus uptake. *Agron. J.* 99:1093–1103.

Gavito, M. E., P. Schweiger, and I. Jakobsen. 2003. P uptake by arbuscular mycorrhizal hyphae: Effect of soil temperature and atmospheric CO_2 enrichment. *Glob. Change Biol.* 9:106–116.

Gerdemann, J. W. 1974. Mycorrhizae. In: *The Plant Root and Its Environment*, ed., E. W. Carson, pp. 205–217. Charlottesville, VA: University Press of Virginia.

Giri, B., R. Kapoor, and K. G. Mukerji. 2007. Improved tolerance of *Acacia nilotica* to salt stress by arbuscular mycorrhizae, *Glomus fasciculatum* may be partly related to elevated K/Na ratios in root and shoot tissues. *Microb. Ecol.* 54:753–760.

Gohre, V. and U. Paszkowski. 2006. Contribution of arbuscular mycorrhizal symbiosis to heavy metal phytoremediation. *Planta* 223:1115–1122.

Goss, M. J. and A. Varennes. 2002. Soil disturbance reduces the efficacy of mycorrhizal associations for early soybean growth and N_2 fixation. *Soil Biol. Biochem.* 34:1167–1173.

Graw, D. 1979. The influence of pH on the efficiency of vesicular-arbuscular mycorrhiza. *New Phytol.* 82:687–695.

Gupta, R. K. and P. Kumar. 2000. Mycorrhizal plants in response to adverse environmental conditions. In: *Mycorrhizal Biology*, eds., K. G. Mukerji, B. P. Chamola, and J. Singh, pp. 67–84. New York: Plenum Publishers.

Hajiboland, R., N. Aliasgharzadeh, S. F. Laiegh, and C. Poschenrieder. 2010. Colonization with arbuscular mycorrhizal fungi improves salinity tolerance of tomato (*Solanum lycopersicum* L.) plants. *Plant Soil* 331:313–327.

Hajiboland, R. and A. Joudmand. 2009. The K/Na replacement and function of antioxidant defense system in sugar beet (*Beta vulgaris* L.) cultivars. *Acta Agric. Scand. Sect. B Soil Plant Sci.* 59:246–259.

Harley, J. L. 1969. *The Biology of Mycorrhiza*. London, U.K.: Leonard Hill.

Harley, J. L. 1989. The significance of mycorrhiza. *Mycol. Res.* 92:129–139.

Harris, K. K. and R. E. J. Boerner. 1990. Effects of belowground grazing by collembolan on growth, mycorrhizal infection, and P uptake of *Geraniium robertanium*. *Plant Soil* 129:203–210.

Harrison, M. J. 1999. Molecular and cellular aspects of the arbuscular mycorrhizal symbiosis. *Annu. Rev. Plant Physiol. Plant Mol. Biol.* 50:361–389.

Hart, M. M. and R. J. Reader. 2002. Taxonomic basis for variation in the colonization strategy of arbuscular mycorrhizal fungi. *New Phytol.* 153:335–344.

Hawkins, H. J., A. Johansen, and E. George. 2000. Uptake and transport of organic and inorganic nitrogen by arbuscular mycorrhizal fungi. *Plant Soil* 226:275–285.

Hayman, D. S. 1974. Plant growth responses to vesicular-arbuscular mycorrhiza. VI. Effects of light and temperature. *New Phytol.* 73:71–80.

Hayman, D. S. 1975. Phosphorus cycling by soil microorganisms and plant roots. In: *Soil Microbiology*, ed., N. Walker, pp. 67–91. London, U.K.: Butterworths.

Heap, A. J. and E. I. Newman. 1980. The influence of vesicular-arbuscular mycorrhizas on phosphorus transfer between plants. *New Phytol.* 85:173–179.

Helgason, T., T. J. Daniell, R. Husband, A. H. Fitter, and J. W. P. Young. 1998. Ploughing up the wood-wide web? *Nature* 394:431.

Hepper, C. M. 1983. The effect of nitrate and phosphate on the vesicular-arbuscular mycorrhizal infection of lettuce. *New Phytol.* 93:389–399.

Hodge, A. 2003. Plant nitrogen capture from organic matter as affected by spatial dispersion, interspecific competition and mycorrhizal colonization. *New Phytol.* 157:303–314.

Hodge, A., C. D. Campbell, and A. H. Fitter. 2001. An arbuscular mycorrhizal fungus accelerates decomposition and acquires nitrogen directly from organic material. *Nature* 413:297–288.

Hogberg, P. and J. E. Nylum. 1981. Ectomycorrhiae in coastal miombo woodland of Tanzania. *Plant Soil* 63:283–289.

Horton, T. R., T. D. Bruns, and V. T. Parker. 1999. Ectomycorrhizal fungi associated with *Arctostaphylos* contribute to *Pseudotsuga menziesii* establishment. *Can. J. Bot.* 77:93–102.

Howeler, R. H., L. F. Cadavid, and E. Burckhardt. 1982. Response of cassava to VA mycorrhizal inoculation and phosphorus application in greenhouse and field experiments. *Plant Soil* 69:327–339.

Jacobsen, I., L. K. Abbott, and A. D. Robson. 1992. External hyphae of vesicular-arbuscular mycorrhizal fungi associated with *Trifolium subterraneum* L. I. Spread of hyphae and phosphorus inflow into roots. *New Phytol.* 120:371–380.

Janos, D. P. 1980. Vesicular-arbuscular mycorrhizae affect lowland tropical rain forest growth. *Ecology* 61:151–162.

Janouskova, M., M. Vosatka, L. Rossi, and N. Lugon-Moulin. 2007. Effects of arbuscular mycorrhizal inoculation on cadmium accumulation by different tobacco (*Nicotiana tabacum* L.) types. *Appl. Soil Ecol.* 35:502–510.

Janzen, H. H. 2004. Carbon cycling in earth systems: A soil science perspective. *Agric. Ecosyst. Environ.* 104:399–417.

Jasper, D. A., L. K. Abbott, and A. D. Robson. 1993. The survival of ineffective hyphae of vesicular-arbuscular mycorrhizal fungi in dry soils: An interaction with sporulation. *New Phytol.* 124:473–479.

Javaid, A. 2008. Allelopathy in mycorrhizal symbiosis in the Poaceae family. *Allelopathy J.* 21:207–218.

Javaid, A. 2009. Arbuscular mycorrhizal mediated nutrition in plants. *J. Plant Nutr.* 32:1595–1618.

Jayachandran, K., A. P. Schwab, and B. A. D. Hetrick. 1992. Mineralization of organic phosphorus by vesicular-arbuscular mycorrhizal fungi. *Soil Biol. Biochem.* 24:897–903.

Johnson, N. C., J. H. Graham, and F. A. Smith. 1997. Functioning of mycorrhizal associations along the mutualism-parasitism continuum. *New Phytol.* 135:575–586.

Johnson, N. C., J. D. Hoeksema, J. D. Bever, V. B. Chaudhary, C. Gehring, J. Klironomos et al. 2006. From Lilliput to Brobdingnag: Extending models of mycorrhizal function across scales. *Bioscience* 56:889–900.

Johnson, D., J. R. Leake, and D. J. Read. 2001. Novel in-growth core system enables functional studies of grassland mycorrhizal mycelial networks. *New Phytol.* 152:555–562.

Joner, E. J. and I. Jakobsen. 1995. Growth and extracellular phosphatase activity of arbuscular mycorrhizal hyphae as influenced by soil organic matter. *Soil Biol. Biochem.* 27:1153–1159.

Jones, M. D. and S. E. Smith. 2004. Exploring functional definition of mycorrhizas: Are mycorrhizas always mutualisms? *Can. J. Bot.* 82:1089–1109.

Juniper, S. and L. Abbott. 1993. Vesicular arbuscular mycorrhizas and soil salinity. *Mycorrhiza* 4:45–57.

Kazantseva, O., M. Bingham, S. W. Simard, and S. M. Berch. 2009. Effects of growth medium, nutrients, water, and aeration on mycorrhization and biomass allocation of greenhouse-grown interior Douglas-fir seedlings. *Mycorrhiza* 20:51–66.

Khalil, S. and T. E. Loynachan. 1994. Soil drainage and distribution of VAM fungi in two toposequences. *Soil Biol. Biochem.* 26:929–934.

Khalil, S., T. E. Loynachan, and H. S. McNabb, Jr. 1992. Colonization of soybean by mycorrhizal fungi and spore populations in Iowa soils. *Agron. J.* 84:832–836.

Khaosaad, T., J. M. Garcia-Garrido, S. Steinkellner, and H. Vierheilig. 2007. Take-all disease is systematically reduced in roots of mycorrhizal barley plants. *Soil Biol. Biochem.* 39:727–734.

Kleinschmidt, G. D. and J. W. Gerdemann. 1972. Stunting of citrus seedlings in the fumigated nursery soil related to the absence of endomycorrhizae. *Phytopathology* 62:1447–1452.

Klironomos, J. N. 2003. Variation in plant response to native and exotic mycorrhizal fungi. *Ecology* 84:2292–2301.

Kumar, T. and M. Ghose. 2008. Status of arbuscular mycorrhizal fungi (AMF) in the Sunderbans of India in relation to tidal inundation and chemical properties of soil. *Wetlands Ecol. Manage.* 16:471–483.

Kuster, H., M. F. Vieweg, K. Manthey, M. C. Baier, N. Hohnjec, and A. M. Perlick. 2007. Identification and expression regulation of symbiotically activated legume genes. *Phytochemistry* 68:8–18.

Kwapata, M. B. and A. E. Hall. 1985. Effects of moisture regime and phosphorus on mycorrhizal infection, nutrient uptake and growth of cowpea (*Vigna unguiculata* L. WALP.). *Field Crops Res.* 12:241–250.

Lamont, B. 1982. Mechanisms for enhancing nutrient uptake, with particular reference to Mediterranean South Africa and Western Australia. *Bot. Rev.* 48:597–689.

Linderman, R. G. 1992. Vesicular-arbuscular mycorrhizae and soil microbial interactions. In: *Mycorrhizae in Sustainable Agriculture*, eds., G. J. Bethlenfalvay and R. G. Linderman, pp. 45–70. Madison, WI: ASA.

Liu, A., B. Wang, and C. Hamel. 2004. Arbuscular mycorrhiza colonization and development at suboptimal root zone temperature. *Mycorrhiza* 14:93–101.

Lussenhop, J. 1996. Collembolan as mediators of microbial symbiont effects upon soybean. *Soil Biol. Biochem.* 28:363–369.

Mader, P., H. Vierheilig, and R. Streitwolf-Engel. 2000. Transport of ^{15}N from a soil compartment separated by a polytetrafluorethylene membrane to plant roots via the hyphae of arbuscular mycorrhizal fungi. *New Phytol.* 146:155–161.

Maillard, E., D. Pare, and A. D. Munson. 2010. Soil carbon stock and carbon stability in a twenty-year old temperate plantation. *Soil Sci. Soc. Am. J.* 74:1775–1785.

Malajczuk, N. 1988. Interaction between *Phytophthora cinnamomi* zoospores and microorganisms on non-mycorrhizal and ectomycorrhizal roots of *Eucalyptus marginata*. *Trans. Br. Mycol. Soc.* 90:375–382.

Marks, G. C. and T. T. Kozlowski. 1973. *Ectomycorrhizae*. New York: Academic Press.

Marschner, H. 1995. *Mineral Nutrition of Higher Plants*, 2nd edn. New York: Academic Press.

Marx, D. H. 1969. The influence of ectotrophic mycorrhizal fungi on the resistance to pathogenic infections. I. Antagonism of mycorrhizal fungi to pathogenic fungi and soil bacteria. *Phytopathology* 59:153–163.

Marx, D. H. 1972. Ectomycorrhizae as biological deterrents. *Annu. Rev. Phytopathol.* 10:429–454.

Marx, D. H. 1973. Growth of ectomycorrhizal and nonmycorrhizal shortleaf pine seedlings in soil with *Phytophthora cinnamomi*. *Phytopathology* 63:18–23.

Marx, D. H. and C. B. Davey. 1969. The influence of ectotrophic mycorrhizal fungi on the resistance of pine roots to pathogenic interactions. III. Resistance of naturally occurring mycorrhizae to infection by *Phytophthora cinnamomi*. *Phytopathology* 59:559–565.

Marx, D. H., A. B. Hatch, and J. F. Mendicino. 1977. High soil fertility decreases sucrose content and susceptibility of loblolly pine roots to ectomycorrhizal infection by *Pisolithus tinctorius*. *Can. J. Bot.* 55:1569–1574.

McGonigle, T. P. and A. H. Fitter. 1988. Growth and phosphorus inflows of *Trifolium repens* L. with a range of indigenous vesicular-arbuscular mycorrhizal infection levels under field conditions. *New Phytol.* 108:59–65.

Medina, A., I. Jakobsen, N. Vassilev, and A. R. Larsen. 2007. Fermentation of sugar beet waste by *Aspergillus niger* facilitates growth and P uptake of external mycelium of mixed populations of arbuscular mycorrhizal fungi. *Soil Biol. Biochem.* 39:485–492.

Meding, S. M. and R. J. Zasoski. 2008. Hyphal-mediated transfer of nitrate, arsenic, cesium, rubidium, and strontium between arbuscular mycorrhizal forbs and grasses from a California oak woodland. *Soil Biol. Biochem.* 40:126–134.

Meghvansi, M. K., K. Prasda, D. Harwani, and S. K. Mahna. 2008. Response of soybean cultivars toward inoculation with three arbuscular mycorrhizal fungi and *Bradyrhizobium japonicum* in the alluvial soil. *Eur. J. Soil Biol.* 44:316–323.

Mengel, K., E. A. Kirkby, H. Kosegarten, and T. Appel. 2002. *Principles of Plant Nutrition*, 5th edn. Dordrecht, the Netherlands: Kluwer Academic Publishers.

Merryweather, J. and A. Fitter. 1998. The arbuscular mycorrhizal fungi of *Hyacinthoides nonscripta*: I. Diversity of fungal taxa. *New Phytol.* 138:117–129.

Meyer, F. 1974. Physiology of mycorrhizae. *Annu. Rev. Plant Physiol.* 25:567–586.

Mikha, M. M., J. G. Benjamin, M. F. Vigil, and D. C. Nielson. 2010. Cropping intensity impacts on soil aggregation and carbon sequestration in the central Great Plains. *Soil Sci. Soc. Am. J.* 74:1712–1719.

Mikola, P. 1970. Mycorrhizal inoculation in afforestation. *Int. Rev. For. Res.* 3:123–196.

Millner, P. D. and S. F. Wright. 2002. Tools for support of ecological research on arbuscular mycorrhizal fungi. *Symbiosis* 33:101–123.

Mohammad, M. J., H. I. Malkawi, and R. Shibli. 2003. Effects of arbuscular mycorrhizal fungi and phosphorus fertilization on growth and nutrient uptake of barley grown on soils with different levels of salts. *J. Plant Nutr.* 26:125–137.

Mosse, B. 1972. Effect of different Endogone strain on the growth of *Paspalum notatum*. *Nature* 239:221–223.

Mosse, B. 1973. Advances in the study of vesicular-arbuscular mycorrhiza. *Annu. Rev. Phytopathol.* 11:171–196.

Mukerji. K. G., B. P. Chamola, and J. Singh. 2000. *Mycorrhizal Biology*. New York: Plenum Publishers.

Munns, R. and M. Tester. 2008. Mechanisms of salinity tolerance. *Annu. Rev. Plant Biol.* 59:651–681.

Muthukumar, T. and K. Udaiyan. 2000. Arbuscular mycorrhizas of plants growing in the western Ghats region, southern India. *Mycorrhiza* 9:297–313.

Nara, K. 2006. Ectomycorrhizal networks and seedling establishment during early primary succession. *New Phytol.* 169:169–178.

Nelson, C. E. and G. R. Safir. 1982. The water relations of well-watered mycorrhizal and non-mycorrhizal onion plants. *J. Am. Hort. Sci.* 107:271–274.

Nemec, S. 1980. Effects of 11 fungicides on endomycorrhizal development in sour orange. *Can. J. Bot.* 58:522–526.

Nemec, S. and D. Tucker. 1983. Effects of herbicides on endomycorrhizal fungi in Florida USA citrus (*Citrus spp.*) soils. *Weed Sci.* 31:427–431.

Neumann, E., B. Schmid, V. Romheld, and E. George. 2009. Extraradical development and contribution to plant performance of an arbuscular mycorrhizal symbiosis exposed to complete or partial rootzone drying. *Mycorrhiza* 20:13–23.

Nye, P. H. and P. B. Tinker. 1977. *Solute Movement in the Soil-Root System.* Berkeley, CA: University of California.

Olsen, S. R., W. D. Kemper, and J. C. V. Schalik. 1965. Self diffusion coefficients of phosphorus in soil measured by transient and steady-state methods. *Soil Sci. Soc. Am. Proc.* 29:154–158.

Ortas, I. and C. Akpinar. 2011. Response of maize genotypes to several mycorrhizal inoculums in terms of plant growth, nutrient uptake and spore production. *J. Plant Nutr.* 34:970–987.

Parvathi, K., K. Venkateswarlu, and A. S. Rao. 1985. Effects of pesticides on development of *Glomus mosseae* in groundnut. *Trans. Br. Mycol. Soc.* 84:29–33.

Patreze, C. M. and L. Cordeiro. 2004. Nitrogen fixing and vesicular arbuscular mycorrhizal symbiosis in some tropical legume trees of tribe Mimoseae. *For. Ecol. Manage.* 196:275–285.

Paul, E. A. and F. E. Clark. 1996. *Soil Microbiology and Biochemistry.* London, U.K.: Academic Press.

Perry, D. A., M. P. Amaranthus, J. G. Borchers, S. L. Borchers, and R. E. Brainerd. 1989. Bootstrapping in ecosystems. *Bioscience* 39:230–237.

Pinior, A., G. Grunewaldt-Stocker, H. V. Alten, and R. J. Strasser. 2005. Mycorrhizal impact on drought stress tolerance of rose plants probed by chlorophyll a fluorescence, proline content and visual scoring. *Mycorrhiza* 15:596–605.

Porcel, R. and J. M. Ruiz-Lozano. 2004. Arbuscular mycorrhizal influence on leaf water potential, solute accumulation, and oxidative stress in soybean plants subjected to drought stress. *J. Exp. Bot.* 55:1743–1750.

Raimam, M. P., P. U. Albino, M. F. Cruz, G. M. Lovato, F. Spago, T. P. Ferracin et al. 2007. Interaction among free-living N-fixing bacteria isolated from *Drosera villosa* var. *Villosa* and AM fungi (*Glomus clarum*) in rice (*Oryza sativa*). *Appl. Soil Ecol.* 35:25–34.

Rains, K. C. and C. S. Bledsoe. 2007. Rapid uptake of ^{15}N ammonium and glycine- [13]C, ^{15}N by arbuscular and ericoid mycorrhizal plant native to a northern California coastal pygmy forest. *Soil Biol. Biochem.* 39:1078–1086.

Read, D. J. and J. Perez-Moreno. 2003. Mycorrhizas and nutrient cycling in ecosystems: A journey towards relevance? *New Phytol.* 157:475–492.

Rhodes, D., A. Nadolska-Orczyk, and P. J. Rich. 2004. Salinity, osmolytes and compatible solutes. In: *Salinity: Environment-Plants-Molecules*, eds., A. Lauchli and U. Luttge, pp. 181–204. Dordrecht, the Netherlands: Springer Verlag.

Rillig, M. C. and D. L. Mummey. 2006. Mycorrhizas and soil structure. *New Phytol.* 171:41–53.

Rillig, M. C. and P. D. Steinberg. 2002. Glomalin production by an arbuscular mycorrhizal fungus: A mechanisms of habitat modification? *Soil Biol. Biochem.* 34:1371–1374.

Rillig, M. C., S. F. Wright, M. F. Allen, and C. B. Field. 1999. Rise in carbon dioxide changes soil structure. *Nature* 400:628.

Ruiz-Lozano, J. M. and R. Azcon. 2000. Symbiotic efficiency and infectivity of an autochthonous arbuscular mycorrhizal *Glomus sp.* from saline soils and *Glomus deserticola* under salinity. *Mycorrhiza* 10:137–143.

Same, B. I., A. D. Robson, and L. K. Abbott. 1983. Phosphorus, soluble carbohydrates and endomycorrhizal infection. *Soil Biol. Biochem.* 15:593–597.

Sanders, F. E. and P. B. Tinker. 1971. Mechanism of absorption of phosphate from soil by *Endogone* mycorrhizas. *Nature* 233:278–279.

Sanders, F. E. and P. B. Tinker. 1973. Phosphate flow into mycorrhizal roots. *Pestic. Sci.* 4:385–395.

Sannazzaro, A. I., O. A. Ruiz, E. O. Alberto, and A. B. Menendez. 2006. Alleviation of salt stress in *Lótus glaber* by *Glomus intraraduces*. *Plant Soil* 285:279–287.

Santos, C. L. V., A. Campos, H. Azevedo, and G. Caldeira. 2001. In situ and in vitro senescence induced by KCl stress: Nutritional imbalance, lipid peroxidation and antioxidant mechanisms. *J. Exp. Bot.* 52:351–360.

Sawers, R. J. H., C. Gutjahr, and U. Paszkowski. 2008. Cereal mycorrhiza: An ancient symbiosis in modern agriculture. *Trends Plant Sci.* 13:93–97.

Schliemann, W., C. Ammer, and D. Strack. 2008. Metabolite profiling of mycorrhizal roots of *Medicago truncatula*. *Phytochemistry* 69:112–146.

Shen, H., H. Yang, and T. Guo. 2011. Influence of arbuscular mycorrhizal fungi and ammonium: Nitrate ratios on growth and pungency of spring onion plants. *J. Plant Nutr.* 34:743–752.

Shi, Z. Y., Y. L. Chen, G. Feng, R. J. Liu, P. Christie, and X. L. Li. 2006. Arbuscular mycorrhizal fungi associated with the Meliaceae on Hainan Island, China. *Mycorrhiza* 16:81–87.

Shibata, R. and K. Yano. 2003. Phosphorus acquisition from non-labile sources in peanut and pegeonpea with mycorrhizal interaction. *Appl. Soil Ecol.* 24:133–141.

Smith, S. E. 1980. Mycorrhizas of autotrophic higher plants. *Biol. Rev.* 55:475–510.

Smith, S. E. and V. Gianinazzi-Pearson. 1988. Physiological interactions between symbionts in vesicular-arbuscular mycorrhizal plants. *Annu. Rev. Plant Physiol. Plant Mol. Biol.* 39:221–244.

Smith, S. E. and D. J. Read. 1997. *Mycorrhizal Symbiosis*. San Diego, CA: Academic Press.

Smith, S. E. and D. J. Read. 2008. *Mycorrhizal Symbiosis*, 3rd edn. London, U.K.: Academic Press.

Smith, S. E., D. J. Read, and J. L. Harley. 1997. *Mycorrhizal Symbiosis*. San Diego, CA: Academic Press.

Soil Science Society of America. 2008. *Glossary of Soil Science Terms*. Madison, WI: SSSA.

Stack, R. W. and W. A. Sinclair. 1975. Protection of Douglas-fir seedlings against *Fusarium* root rot by a mycorrhizal fungus in the absence of mycorrhiza formation. *Phytopathology* 65:468–472.

St. John, T. V., D. C. Coleman, and C. P. P. Reid. 1983. Association of vesicular-arbuscular mycorrhizal hyphae with soil organic particles. *Ecology* 64:957–959.

Subramanian, K. S. and C. Charest. 1997. Nutritional, growth and reproductive response of maize to arbuscular mycorrhizal inoculation during and after drought stress at tasseling. *Mycorrhiza* 7:25–32.

Subramanian, K. S. and C. Charest. 1999. Acquisition of N by external hyphae of an arbuscular mycorrhizal fungus and its impact on physiological responses in maize under drought-stressed and well-watered conditions. *Mycorrhiza* 9:69–75.

Sylvia, D. M. 1992. Quantification of external hyphae of vesicular-arbuscular mycorrhizal fungi. *Meth. Microbiol.* 24:53–65.

Tinker, P. B. 1980. Role of rhizosphere microorganisms in phosphorus uptake by plants. In: *The Role of Phosphorus in Agriculture*, ed., R. C. Dinauer, pp. 617–654. Madison, WI: ASA, CSSA, and SSSA.

Treseder, K. K. and M. F. Allen. 2002. Direct nitrogen and phosphorus limitations of arbuscular mycorrhizal fungi: A model and field test. *New Phytol.* 155:507–515.

Troeh, Z. I. and T. E. Loynachan. 2003. Endomycorrhizal fungal survival in continuous corn, soybean, and fallow. *Agron. J.* 95:224–230.

Troeh, Z. I. and T. E. Loynachan. 2009. Diversity of arbuscular mycorrhizal fungal species in soils of cultivated soybean fields. *Agron. J.* 101:1453–1462.

Vandenkoornhuyse, P., R. Husband, T. J. Daniell, I. J. Watson, J. M. Duck, A. H. Fitter, and J. P. W. Young. 2002. Arbuscular mycorrhizal community composition associated with two plant species in a grassland ecosystem. *Mol. Ecol.* 11:1555–1564.

Volkmar, K. M. and W. Woodbury. 1989. Effects of soil temperatures and depth on colonization and root and shoot growth of barley inoculated with vesicular-arbuscular mycorrhizae indigenous to Canadian prairie soil. *Can. J. Bot.* 67:1702–1707.

Walker, C., C. W. Mize, and H. S. McNabb, Jr. 1982. Populations of endogonaceous fungi at two locations in central Iowa. *Can. J. Bot.* 60:2518–2529.

Wang, B., D. M. Funakoshi, Y. Dalpe, and C. Hamel. 2002. Phosphorus 32 absorption and translocation to host plants by arbuscular mycorrhizal fungi at low root-zone temperature. *Mycorrhiza* 12:93–96.

Wang, Y., Q. Qiu, Z. Yang, Z. Hu, N. F. Y. Tam, and G. Xin. 2010. Arbuscular mycorrhizal fungi in two mangroves in south China. *Plant Soil* 331:181–191.

Warner, A. and B. Mosse. 1980. Independent spread of vesicular-arbuscular mycorrhizal fungi in soil. *Trans. Br. Mycol. Soc.* 74:407–446.

Wortmann, C. S., R. A. Drijber, and T. G. Franti. 2010. One-time tillage of no-till crop land five years post-tillage. *Agron. J.* 102:1302–1307.

Wortmann, C. S., J. A. Quincke, R. A. Drijber, M. Mamo, and T. Franti. 2008. Soil microbial change and recovery after one-time tillage of continuous no-till. *Agron. J.* 100:1681–1686.

Wright, S. F. 2005. Management of arbuscular mycorrhizal fungi. In: *Roots and Soil Management: Interactions between Roots and the Soil*, eds., R. W. Zobel and S. F. Wright, pp. 183–197. Madison, WI: ASA, CSSA, and SSSA.

Wu, O. S. and R. X. Xia. 2006. Arbuscular mycorrhizal fungi influence growth, osmotic adjustment and photosynthesis of citrus under well-watered and water stress conditions. *J. Plant Physiol.* 163:417–425.

Wu, O. S., R. X. Xia, and Y. N. Zou. 2008. Improved soil structure and citrus growth after inoculation with three arbuscular mycorrhizal fungi under drought stress. *Eur. J. Soil Biol.* 44:122–128.

Zak, B. 1964. The role of mycorrhizae in root disease. *Annu. Rev. Phytopathol.* 2:377–392.

Zak, J. C. and B. McMichael. 2001. Agroecology of arbuscular mycorrhizal activity. In: *Structure and Function in Agroecosystem Design and Management*, eds., M. Shiyomi and H. Koizumi, pp. 145–166. Boca Raton, FL: CRC Press

Zhang, F., C. Hamel, H. Kianmehr, and D. L. Smith. 1995. Root zone temperature and soybean vesicular-arbuscular mycorrhizae: Development and interactions with the nitrogen fixing symbiosis. *Environ. Exp. Bot.* 35:287–298.

Zhu, Y. G., S. E. Smith, A. R. Barritt, and F. A. Smith. 2001. Phosphorus efficiencies and mycorrhizal responsiveness of old and modern wheat cultivars. *Plant Soil* 237:249–255.

Zhu, X. C., F. B. Song, and H. W. Xu. 2010. Arbuscular mycorrhizae improve low temperatures stress in maize via alterations in host water status and photosynthesis. *Plant Soil* 331:129–137.

7 Effects of Mineral Nutrition on Root Growth of Crop Plants

7.1 INTRODUCTION

The importance of essential mineral nutrition on plant growth and development is indisputable. Plant productivity and yield quality depend to a great degree on plant nutrition (Marschner, 1995; Fageria, 2009; Fageria et al., 2011a). In nutrient-deficient soils (like Brazilian Oxisols and Ultisols), the supply of an adequate amount of essential nutrients increases crop yields of important crops like rice, dry bean, wheat, corn, and soybean by about 40% (Fageria, 2009; Fageria et al., 2011), provided other factors of production (cultivar, soil moisture, control of diseases, insects, and weeds) are at an optimal level. Figure 7.1 shows upland rice response to N fertilization grown on a Brazilian Oxisol. It is clear from Figure 7.1 that nitrogen-deficient plants had yellow leaves without tillering. Plant growth was also reduced compared with plants that received adequate rate of nitrogen (right). Similarly, upland rice response to phosphorus is shown in Figure 7.2. Growth and tillering of upland rice was significantly increased with increasing P rate from 0 to 175 mg kg^{-1} soil. In Figure 7.3, the root growth of upland rice also improved with the addition of phosphorus to a Brazilian Oxisol. Figures 7.4 and 7.5 show the response of lowland rice and dry bean, respectively, to K fertilization grown on a Brazilian Inceptisol. The growth of lowland rice as well as dry bean was significantly improved in a quadratic fashion with the addition of K. Similarly, root growth of upland rice was also improved with the addition of potassium compared with control treatment (Figure 7.6). These figures show the importance of nitrogen, phosphorus, and potassium fertilization for the growth of rice and dry bean crops grown on a Brazilian Oxisol and Inceptisol, respectively.

There are 17 essential nutrients for plant growth. These are hydrogen (H), oxygen (O), carbon (C), nitrogen (N), phosphorus (P), potassium (K), calcium (Ca), magnesium (Mg), sulfur (S), zinc (Zn), copper (Cu), iron (Fe), manganese (Mn), boron (B), molybdenum (Mo), chlorine (Cl), and nickel (Ni). For vigorous root growth and development, plants need these essential nutrients in adequate amount and proper balance. The deficiency and/or excess of any essential nutrient in the growth medium will affect the growth of plant roots adversely. Root is an important organ for supplying water and nutrient to plants. Hence, root growth is directly related to plant productivity. One of the most commonly observed effects of nutrient deficiency on plant development is a decrease in the shoot–root ratio, particularly in fast-growing

FIGURE 7.1   Upland rice plants without N (left) and with N (right) grown on a Brazilian Oxisol.

FIGURE 7.2   Growth of upland rice plants at 0, 50, and 175 mg P kg^{-1} applied to a Brazilian Oxisol.

species adapted to sites of high fertility (Chapin, 1980). Deficiencies of nutrients like N, P, or S, all result in a shift in dry matter allocation in favor of root growth (Ericsson, 1995).

The extension of the root system into new soil volumes results in a continued absorption of nutrients to meet the demand of plants. Okajima (2001) reported that nutrient availability and nutrient uptake are primarily the functions of root extension as well as root activity. A number of studies have pointed out that nutrient uptake is closely related to morphological properties such as root length,

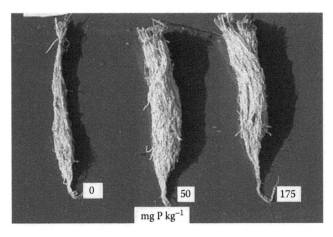

FIGURE 7.3 Root growth of upland rice at different P levels applied to a Brazilian Oxisol.

FIGURE 7.4 Lowland rice response to K fertilization grown on a Brazilian lowland soil, known locally as "Varzea soil."

surface area, fineness, and root hair. Barber et al. (1963) have estimated the significance of root extension quantitatively. Barley (1970), who reviewed the subject in detail, stressed the significance of these root properties in nutrient uptake. Root growth varied among crop species and even among the genotypes of same species under similar environmental conditions (Bonser et al., 1996; Forde and Lorenzo, 2001; Fageria et al., 2008). The root growth of crop plants is genetically controlled. However, it is also influenced by environmental factors (Russelle and Lamb, 2011). These include the physical, chemical, and biological activities of the soil. The objective of this chapter is to discuss the latest advances on the role of mineral nutrition on the root growth of crop plants. Most of the discussion was supported by experimental results and root growth pictures to make the topic as practical as possible.

FIGURE 7.5 Response of dry bean to K fertilization grown on a Brazilian lowland soil known locally as "Varzea soil."

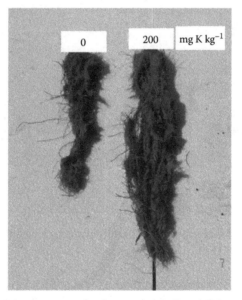

FIGURE 7.6 Upland rice root growth at 0 mg K kg⁻¹ (left) and 200 mg K kg⁻¹ (right).

7.2 MACRONUTRIENTS

Macronutrients are required by crop plants in larger amounts compared to micronutrients. The reason for this is that most of the micronutrients are involved in the structure or protoplasm formation of the plants. Figure 7.7 shows the uptake of macro- and micronutrients in dry bean plants at flowering. The nutrient uptake was in the order of $K > N > Ca > Mg > P$ for macronutrients and $Fe > Mn > Z > Cu$ for micronutrients. The uptake of macronutrients was in several kilograms per hectare, whereas the updake of micronutrients was only in grams per hectare. Root development is

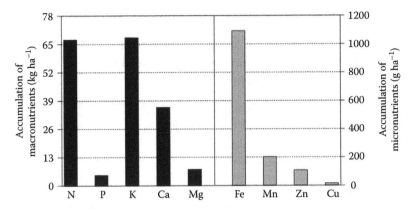

FIGURE 7.7 Accumulation of macro- and micronutrients by dry bean plants at flowering. (From Fageria, N.K. et al., *Growth and Mineral Nutrition of Field Crops*, 3rd edn., CRC Press, Boca Raton, FL, 2011a.)

remarkably sensitive to variations in the supply and distribution of inorganic nutrients in the soil. Macronutrients such as N, P, and K can affect development processes such as root branching, root hair production, root diameter, root growth angle, nodulation, and proteoid root formation (Forde and Lorenzo, 2001). Forde and Lorenzo also reported that the nutrient supply can affect root development either directly, as a result of changes in the external concentration of the nutrient, or indirectly through changes in the internal nutrient status of the plant. Root branching and lateral root length often increase in zones of higher soil fertility when the nutrient in question is otherwise in short supply (Hodge, 2004; Denton et al., 2006; Russelle and Lamb, 2011). This plastic response of plants to resource heterogeneity is well known, and species differ in their nutrient acquisition strategies (Vance et al., 2003; Lynch, 2007).

7.2.1 NITROGEN

Nitrogen influences the root growth of crop plants significantly when applied in an adequate amount. It influences root length, root dry weight, as well as root hairs formation. The effects of added N as fertilizer on root growth is more pronounced when the organic matter content of the soil is lower compared to soils with higher organic matter content (Fageria and Moreira, 2011). Data in Table 7.1 show the maximum root length and root dry weight of 12 lowland rice genotypes grown on a Brazilian Inceptisol. These two traits were significantly influenced by N rate and genotype treatments. Maximum root growth was significantly influenced by N rate and genotype treatments. However, N × G interaction was not significant for this growth parameter, indicating that each of the 12 genotypes reacted similarly to changes in N rates. Maximum root length varied from 21.17 to 29.00 cm, with an average value of 24.84 cm. Root length was significantly higher at low N rate compared to high N rate. When there is deficiency of a determined nutrient, roots try to grow longer to take nutrients from lower soil depths (Fageria and Moreira, 2011). Fageria (2010) studied the influence of N on root length and root dry weight of 20 upland rice

TABLE 7.1

Maximum Root Length and Root Dry Weight of 12 Lowland Rice Genotypes as Influenced by N and Genotype Treatments

N Rate/Genotype	Maximum Root Length (cm)	Root Dry Weight (g Plant^{-1})	
		N (0 mg kg^{-1})	N (300 mg kg^{-1})
0 mg N kg^{-1}	26.58a		
300 mg N kg^{-1}	23.61b		
BRS Tropical	29.00a	6.04a	5.59d
BRS Jaçanã	24.00ab	4.85abc	2.67e
BRA 02654	25.83ab	5.88ab	8.56bc
BRA 051077	23.83ab	5.59ab	13.10a
BRA 051083	26.00ab	3.09c	6.29d
BRA 051108	27.66ab	6.76a	9.03b
BRA 051126	24.50ab	5.06abc	6.41d
BRA 051129	21.17b	5.85ab	8.88b
BRA 051130	25.67ab	3.73bc	8.64bc
BRA 051134	23.50	5.96ab	5.91d
BRA 051135	23.00ab	4.66abc	5.33d
BRA 051250	27.00ab	4.57abc	6.81cd
Average	24.84	5.17b	7.27a
F-test			
N rate (N)	**	**	
Genotype (G)	*	**	
N × G	NS	**	
CV (%) N rate		8.94	
CV (%) Genotype		11.77	

NS, nonsignificant.

*, **Significant at the 5% and 1% probability levels, respectively. Means followed by the same letter in the same column (separate for N rate and genotypes) are not significantly different at the 5% probability level by Tukey's test.

genotypes grown on a Brazilian Oxisol. Overall, root length was 5% higher at the higher N rate compared to lower N rate. However, 35% genotypes produced lower root length at the higher N rate compared to lower N rate. Fageria (1992) reported higher root length of rice at low N rate compared to high N rate in nutrient solution. Fageria (1992) also reported that at nutrient-deficient levels, root length is higher compared to high nutrient levels because of the tendency of plants to tap nutrients from deeper soil layers. Forde and Lorenzo (2001) reported that plants growing on concentrated nutrient solution developed a short, compact, and densely branched root system, while on dilute solutions or water the roots were long and more sparsely branched. Ye et al. (2010) also reported that the maximum root length of two rapeseed (*Brassica napus* L.) genotypes was higher at low N rate compared to high N

BRA 02654

300   mg N kg^{-1}

0

FIGURE 7.8   Root growth of lowland rice genotype BRA 02654 at low N and high N rates.

rate treatment. This general response to variations in the nutrient supply has been reported in a wide range of plant species (Wiersum, 1958).

Root dry weight was significantly influenced by N rate and genotype treatments (Table 7.1). The N × G interaction was also significant for root dry weight, suggesting that genotypes responded differently at the two N rates. At lower N rate, the root dry weight varied from 3.09 g plant^{-1} produced by genotype BRA 051083 to 6.76 g plant^{-1} produced by genotype BRA 051108, with an average value of 5.17 g plant^{-1}. At higher N rate, root dry weight varied from 2.67 g plant^{-1} produced by genotype BRS Jaçanã to 13.10 g plant^{-1} produced by genotype BRA 051077, with an average value of 7.27 g plant^{-1}. The application of 300 mg N kg^{-1} produced 41% higher dry weight compared to 0 mg N kg^{-1} of soil. Figure 7.8 shows the root growth of genotype BRA 02654 at low and high N rates. Noulas et al. (2010) reported that higher N rate improved root dry weight of wheat genotypes. These authors also reported that N fertilization altered the anatomy of root branches and that well-fertilized plants produce a greater proportion of fine root hairs compared to lower N fertilized plants. Drew et al. (1973) and Hackett (1972) also reported a proliferation of fine roots in nutrient-rich zones and in wetter soil profiles compared to dry and nutrient-poor soil profiles. Root systems with fine hairs absorb nutrients and water more efficiently than plants with a root system of fewer fine lateral roots (Barber and Silverbush, 1984).

Fageria (2010) studied the influence of N on root dry weights of 20 upland rice genotypes. Root dry weight varied significantly among genotypes. Overall, root dry weight was 97% higher at the higher N rate compared to the lower N rate. Fageria and Baligar (2005) and Fageria (2009) reported that N fertilization improved root dry weight in crop plants, including upland rice. The positive effect of N on root dry matter has been previously documented (Fageria, 2009). Ye et al. (2010) also reported that root dry weight of two rapeseed (*Brassica napus* L.) genotypes decreased significantly at the low N treatment compared to the high N treatment. Figures 7.9 through 7.11

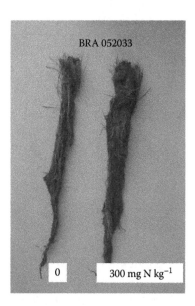

FIGURE 7.9 Root growth of upland rice genotype BRA 052033 at low and high N rates grown on a Brazilian Oxisol.

FIGURE 7.10 Root growth of upland rice genotype BRA 02535 at low and high N rates grown on a Brazilian Oxisol.

show the root growth of three upland rice genotypes at low and high N rates. The root growth of these genotypes was different at low as well as high N rates. However, it was improved with the addition of N fertilization. Thus, selecting genotypes for N-use efficiency may be an important aspect of improving root growth and, consequently, the yields of upland rice in Brazilian Oxisols (Fageria and Moreira, 2011).

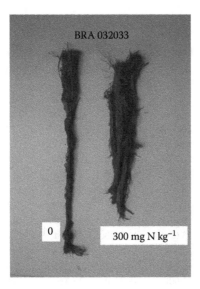

FIGURE 7.11 Root growth of upland rice genotype BRA 032033 at low and high N rates grown on a Brazilian Oxisol.

The influence of N and rhizobia on the root growth of dry bean is shown in Table 7.2. The nitrogen × genotype-significant interactions for the maximum root length and root dry weight of dry bean were observed, indicating that some genotypes were highly responsive to N treatments while others were not (Table 7.2). Maximum root length varied from 11.00 to 28.33 cm at 0 mg N kg^{-1}, 10.33 to 22.67 cm at 0 mg kg^{-1} + inoculation with rhizobia, 11.33 to 24.33 cm at inoculation + 50 mg N kg^{-1}, and 12.00 to 26.67 cm at 200 mg N kg^{-1} soil treatments. Average values for these treatments were 18.2 cm for 0 mg N kg^{-1}, 15.31 cm for inoculation treatment, 16.04 cm for inoculation + 50 mg N kg^{-1}, and 18.71 cm for 200 mg N kg^{-1}. Across four N levels, the values of root length varied from 13.33 to 19.66 cm. The variation in the root length of dry bean genotypes has been reported by Fageria and Moreira (2011) and Fageria (2009).

Root dry weight varied significantly among bean genotypes at different N treatments (Table 7.3). It was significantly higher at 200 mg N kg^{-1} soil treatment compared to other N treatments. The increase in root dry weight was about 20% at 200 mg N kg^{-1} treatment compared to control treatment. Figure 7.12 shows root growth of dry bean genotypes BRS Pontal as influenced by N and rhizobia treatments. Similarly, the influence of N on the root growth of dry bean genotype CNFC 10470 is presented in Figure 7.13. Variability in root growth among crop species and among the genotypes of same species is widely reported in the literature (O'Toole and Bland, 1987; Gregory, 1994; Fageria, 2009). This variability can be used in improving the yields of annual crops by incorporating vigorous root growth into desirable cultivars. Vigorous root growth is especially important when nutrient and water stress are significant (Gregory, 1994). Ludlow and Muchow (1990), in their review of traits that are likely to improve yields in water-limited environments, place

TABLE 7.2

Maximum Root Length (cm) of 15 Dry Bean Genotypes at Different N and Rhizobium Inoculation Treatments

Genotype	N_0	N_1	N_2	N_3	Average
Aporé	21.33bcd	13.00cd	17.00bcd	22.33a–c	18.41abcd
Pérola	25.33ab	11.67cd	11.33d	23.33ab	17bcd
BRSMG Talisma	19.67cd	14.00cd	20.33abc	12.00d	16.50cde
BRS Requinte	24.33abc	13.33cd	21.67ab	26.67a	21.50a
BRS Pontal	18.33de	11.67cd	24.33a	14.00vd	17.08bcd
BRS 9435 Cometa	18.33de	17.33abc	12.33d	21.00abcd	17.25bcd
BRS Estilo	21.33bcd	22.67a	17.00bcd	17.67abcd	19.66ab
CNFC 10408	18.67d	21.00ab	11.67d	16.33bcd	16.91bcd
CNFC 10470	13.67ef	17.33abc	11.67d	19.00abcd	15.41def
Diamante Negro	13.33f	21.00ab	15.67bcd	16.67bcd	16.67bcde
Corrente	12.00f	14.00cd	17.33bcd	21.33abc	16.17def
BRS Valente	11.00f	12.33cd	14.33cd	15.67bcd	13.33f
BRS Grafite	28.33a	15.33bcd	14.00cd	19.67abcd	19.33abc
BRS Marfim	13.67ef	10.33d	14.33cd	16.33bcd	13.67ef
BRS Agreste	13.67ef	14.67cd	17.67abcd	18.67abcd	16.16def
Average	18.20a	15.31b	16.04b	18.71a	
F-test					
N levels (N)	**				
Genotypes (G)	**				
N × G	**				
CVN (%)	17.94				
CVG (%)	13.07				

Source: Fageria, N.K. et al., *Commun. Soil Sci. Plant Anal.*, 43, in press.
**Significant at the 1% probability level. Means in the same column followed by the same letter are not significantly different at the 5% probability level by Tukey's test. Average values were compared in the same line for significant differences among N rates. N_0 = 0 mg N kg^{-1} (control); N_1 = 0 mg N kg^{-1} + inoculation with rhizobial strains; N_2 = inoculation with rhizobial strains + 50 mg N kg^{-1}; and N_3 = 200 mg N kg^{-1}.

a vigorous rooting system high in their list of properties to be sought. The genotypic variability in the root growth of annual crops has been used to identify superior genotypes for drought-prone environments (Hurd et al., 1972; Gregory, 1994). Gregory and Brown (1989) reviewed the role of root characters in moderating the effects of drought and concluded that roots may have a direct effect, by increasing the supply of water available to the crop, or an indirect effect by changing the rate at which the supply becomes available. Where crops are grown on deep soils and water is stored throughout the whole soil profile, the depth of rooting has a major influence on the potential supply of water (Gregory, 1994). Rain may replenish the upper soil

TABLE 7.3
Root Dry Weight (g/Plant) of 15 Dry Bean Genotypes at Different N and Rhizobium Inoculation Treatments

Genotype	N_0	N_1	N_2	N_3	Average
Aporé	0.44bcdef	0.23def	0.26ef	0.58bcde	0.37cde
Pérola	0.33def	0.37bcd	0.26ef	0.56bcde	0.38cde
BRSMG Talisma	0.56bc	0.19f	0.29de	0.54bcde	0.39bcde
BRS Requinte	0.52bcde	0.25cdef	0.42abc	1.13a	0.58a
BRS Pontal	0.49bcdef	0.32cdef	0.47ab	0.51bcde	0.45bc
BRS 9435 Cometa	0.55bcd	0.39bcde	0.25ef	0.83ab	0.50ab
BRS Estilo	0.52bcde	0.38abc	0.31cde	0.57bcde	0.44bcd
CNFC 10408	0.28f	0.49ab	0.29de	0.24e	0.33e
CNFC 10470	0.27f	0.21ef	0.17f	0.21e	0.21f
Diamante Negro	0.66ab	0.54a	0.42abc	0.69bcd	0.57a
Corrente	0.41cdef	0.29cdef	0.36bcde	0.42cde	0.37cde
BRS Valente	0.31ef	0.32cdef	0.29de	0.41cde	0.33de
BRS Grafite	0.80a	0.39abc	0.38bcd	0.78abc	0.58a
BRS Marfim	0.40cdef	0.40abc	0.50a	0.47bcde	0.44bcd
BRS Agreste	0.41cdef	0.28cdef	0.27def	0.35de	0.33e
Average	0.46b	0.33c	0.33c	0.55a	

F-test
N levels (N)   **
Genotypes (G)   **
N × G   **
CVN (%)   19.89
CVG (%)   18.72

Source: Fageria, N.K. et al., *Commun. Soil Sci. Plant Anal.*, 43, in press.

**Significant at the 1% probability level. Means in the same column followed by the same letter are not significantly different at the 5% probability level by Tukey's test. Average values were compared in the same line for significant differences among N rates. $N_0 = 0$ mg N kg^{-1} (control); $N_1 = 0$ mg N kg^{-1} + inoculation with rhizobial strains; $N_2 =$ inoculation with rhizobial strains + 50 mg N kg^{-1}; and $N_3 = 200$ mg N kg^{-1}.

during the season, but later growth and grain filling in many crops is accomplished during periods of low rainfall when soil moisture stored deep in the profile must be utilized. Sponchiado et al. (1989) reported that in dry bean, drought avoidance results from root growth and soil water extraction deep in the profile. Nitrogen fertilization may increase crop root growth by increasing soil N availability (Fageria and Moreira, 2011). Nitrogen also improves the production of lateral roots and root hairs, as well as increases rooting depth and root length density deep in the profile (Hansson and Andren, 1987). Hoad et al. (2001) reported that the surface application of nitrogen fertilizer increases root densities in the surface layers of the soil.

FIGURE 7.12 Root growth of dry bean genotype BRS Pontal at different N treatments. (Left to right) 0 mg N kg⁻¹, 0 mg N kg⁻¹ + inoculant, inoculant + 50 mg N kg⁻¹, and 200 mg N kg⁻¹. (From Fageria, N.K. et al., *Commun. Soil Sci. Plant Anal.*, 43, in press; Fageria, N.K. et al., *Commun. Soil Sci. Plant Anal.*, 43, in press.)

FIGURE 7.13 Growth of root system of dry bean genotype CNFC 10470 at 0 and 200 mg N kg⁻¹ soil. (From Fageria, N.K. et al., *Commun. Soil Sci. Plant Anal.*, 43, in press; Fageria, N.K. et al., *Commun. Soil Sci. Plant Anal.*, 43, in press.)

Data related to maximum root length of 10 tropical legume cover crops are presented in Table 7.4. The N × cover crops root length interaction was significant for root length, indicating that some crop species were highly responsive to the applied N and rhizobial inoculants, while others were not. In the control treatment (N_0), maximal root length varied from 22 cm produced by showy crotalaria to 33 cm produced by

TABLE 7.4

Maximal Root Length of 10 Tropical Cover Crops as Influenced by N and Bradyrhizobial Inoculants

Cover Crops	Maximal Root Length (cm)				
	N_0	N_1	N_2	N_3	Average
Crotalaria	29.00ab	24.00cde	24.67cd	20.67e	24.58b
Smooth crotalaria	33.00a	25.33bcd	30.67ab	28.67abc	29.42a
Showy crotalaria	22.00d	21.67de	29.33bc	26.33cd	24.83b
Calapo	27.67bc	29.33ab	34.00ab	32.33ab	30.83a
Pueraria	23.33cd	24.33cde	24.00d	28.00bc	24.92b
Pigeon pea	30.00ab	30.67a	34.33a	23.00de	29.50a
Lablab	26.67bcd	29.33ab	31.67ab	26.33cd	28.50a
Black velvet bean	25.33bcd	20.00e	18.33e	32.33ab	24.00b
Bengal bean	27.67bc	29.33ab	30.67ab	33.33a	30.25a
Jack bean	23.67cd	27.67abc	21.00de	20.00e	23.08b
Average	26.17a	26.83a	27.87a	27.87a	26.99

F-test

N rate (N)	NS
Cover crops (C)	**
N × C	**
CVN (%)	9.27
CVC (%)	6.75

Source: Fageria, N.K. et al., *Commun. Soil Sci. Plant Anal.*, 43, in press.
NS, nonsignificant.

**Significant at the 1% probability level. Means in the same column followed by the same letter are not significantly different at the 5% probability level by Tukey's test. Average values were compared in the same line for significant differences among N rates. $N_0 = 0$ mg N kg^{-1}; $N_1 = 0$ mg N kg^{-1} + Bradyrhizobial inoculants; $N_2 = 100$ mg N kg^{-1} + Bradyrhizobial inoculants; and $N_3 = 200$ mg N kg^{-1}.

smooth crotalaria, with an average value of 26.17 cm. At N_1 (0 mg N kg^{-1} + rhizobial inoculants) and N_2 treatments (100 mg N kg^{-1} + rhizobial inoculants), minimal root length was produced by black velvet bean and maximum root length was produced by pigeon pea. At N_3 (200 mg N kg^{-1}) treatment, the situation changed and a minimal root length of 20 cm was produced by jack bean and a maximum root length of 33.33 cm was produced by gray velvet bean. Across four N treatments, a minimal root length of 23.08 cm was produced by jack bean and a maximal root length of 30.83 cm was produced by calopogonio, with an average value of 26.99 cm. Variation in root length is genetically controlled and varies among plant species and it is also influenced by environmental factors (Comfort et al., 1988; Eghball et al., 1993; Costa et al., 2002; Fageria et al., 2006).

Similarly, root dry weight had a significant N × cover crop species interaction (Figure 7.14), indicating variation in root dry weight with the variation in N and

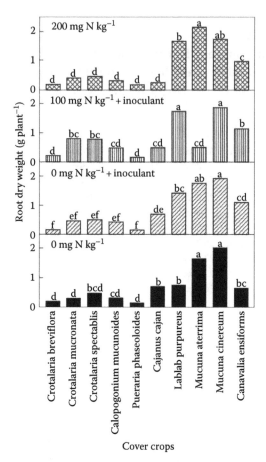

FIGURE 7.14 Root dry weights of 10 tropical legume cover crops as influenced by N and rhizobia inoculant. (Adapted from Fageria, N.K. et al., *Commun. Soil Sci. Plant Anal.*, 43, in press; Fageria, N.K. et al., *Commun. Soil Sci. Plant Anal.*, 43, in press.)

rhizobial inoculants. In the control treatment (N_0), shoot dry weight varied from 0.16 g plant^{-1} produced by pueraria (*Pueraria phaseoloides*) to 2.01 g plant^{-1} produced by gray velvet bean (*Mucuna cinereum*), with an average value of 0.72 g plant^{-1}. These two cover crops also produced minimal and maximal root dry weights at N_1 (0 mg N kg^{-1} + inoculants) and N_2 (100 mg N kg^{-1} + inoculant) treatments. However, at N_3 (200 mg N kg^{-1}) treatment, minimum root dry weight was produced by crotalaria and maximum root dry weight was produced by black velvet bean. Across four N levels, minimum root dry weight was produced by crotalaria and pueraria and maximum root dry weight was produced by gray velvet bean. Overall, gray velvet bean produced about 12-fold more root dry weight compared to minimum root-dry-weight-producing cover crops, crotalaria and pueraria. Root dry weight is an important trait in improving the organic matter content of the soil as well as in the absorption of water and nutrient (Sainju et al., 1998; Fageria et al., 2006). Vigorous root systems also assimilate large amounts of leaching nutrients such as N and provide to the succeeding economic crops (Kristensen and Thorup-Kristensen, 2004; Feaga et al., 2010). Root dry weight had significant positive association with shoot dry weight (Figure 7.15). Figures 7.16 through 7.18 show the root growth of showy crotalaria, calopogonio, and lablab, respectively. Root growth varied with N treatments, and overall, more vigorous root system was produced

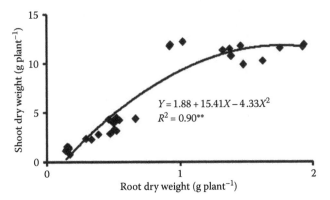

$$Y = 1.88 + 15.41X - 4.33X^2$$
$$R^2 = 0.90^{**}$$

FIGURE 7.15 Relationship between root dry weight and shoot dry weight of cover crop. Values are averages of 10 cover crops. **Significant at the 1% probability level. (From Fageria, N.K. et al., *Commun. Soil Sci. Plant Anal.*, 43, in press; Fageria, N.K. et al., *Commun. Soil Sci. Plant Anal.*, 43, in press.)

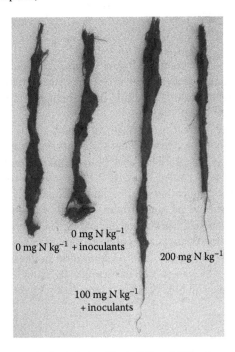

FIGURE 7.16 Showy crotalaria root growth at different N treatments. (Left to right) 0 mg N kg^{-1}, 0 mg N kg^{-1} + Bradyrhizobial inoculants, 100 mg N kg^{-1} + Bradyrhizobial inoculants, and 200 mg N kg^{-1}.

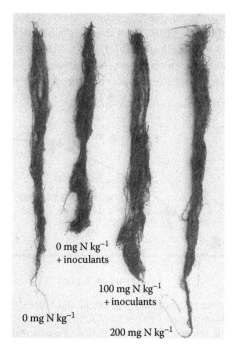

0 mg N kg^{-1}
+ inoculants

100 mg N kg^{-1}
+ inoculants

0 mg N kg^{-1}

200 mg N kg^{-1}

FIGURE 7.17 Calopogonio root growth at different N treatments. (Left to right) 0 mg N kg^{-1}, 0 mg N kg^{-1} + Bradyrhizobial inoculants, 100 mg N kg^{-1} + Bradyrhizobial inoculants, and 200 mg N kg^{-1}.

at N_2 (100 mg N kg^{-1} + inoculant) treatment compared to the other three N treatments in three cover crops. Baligar et al. (1998) reported that the root dry weight of legume crops was higher with the addition of N compared to treatment without N application.

Nitrogen sources also affect root growth in upland rice (Figure 7.19). Root dry weight increased in a quadratic exponential fashion with the application of N in the range of 0–400 mg kg^{-1} of soil. In the case of urea, maximum root dry weight was obtained with 281 mg N kg^{-1} of soil. Figure 7.20 shows how root growth of upland rice is affected by the application of urea and ammonium sulfate in a Brazilian Oxisol. Ammonium sulfate produced more vigorous root systems, especially at higher N rates, than urea, perhaps because upland rice is highly tolerant to soil acidity and ammonium sulfate reduces soil pH more than urea. Fageria (2009) reported that upland rice can tolerate up to 70% Al saturation in the soil. Another possible explanation is that ammonium sulfate has about 24% S, which may improve root growth if the extractable soil S level is lower than 10 mg kg^{-1}.

7.2.2 PHOSPHORUS

Phosphorus is one of the most yield-limiting nutrients in crop production, especially in highly weathered Oxisols and Ultisols (Fageria and Baligar, 2008). The deficiency of P in these soils is related to low natural level of this element as well

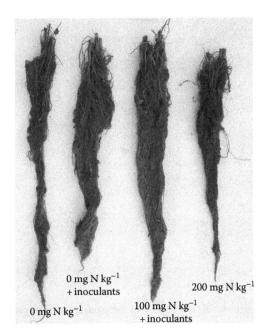

FIGURE 7.18 Lablab root growth at different N treatments. (Left to right) 0 mg N kg^{-1}, 0 mg N kg^{-1} + Bradyrhizobial inoculants, 100 mg N kg^{-1} + Bradyrhizobial inoculants, and 200 mg N kg$^-$.

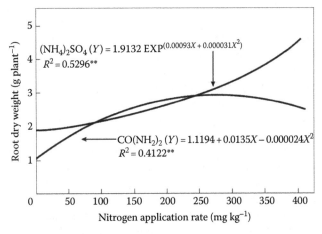

FIGURE 7.19 Relationship between nitrogen application rate by ammonium sulfate and urea and root dry weight of upland rice. **Significant at the 1% probability level. (From Fageria, N.K. et al., *J. Plant Nutr.*, 34, 361, 2011c.)

as high immobilization capacity by iron and aluminum oxides. Phosphorus plays an important role in many physiological and biochemical reactions in the plants. Its role in improving root growth is widely reported (Baligar et al., 1998; Fageria et al., 2011a; Fageria and Moreira, 2011). Phosphorus availability regulates many features of root architecture, including hypocotyls-borne rooting, basal root elongation,

$CO(NH_2)_2$ $(NH_4)_2SO_4$

FIGURE 7.20 Root growth of upland rice at two sources of nitrogen applied as 300 mg N kg^{-1} of soil.

basal root-growth angle, lateral rooting, and the density and length of root hairs (Bates and Lynch, 1996; Bonser et al., 1996; Liao et al., 2001; Ma et al., 2001; Miller et al., 2003; Vieira et al., 2008). Fageria et al. (2011b) studied the influence of P fertilization on the root growth of 20 upland rice genotypes grown on a Brazilian Oxisol (Table 7.5). Phosphorus level and genotype interactions for root dry weight and root length were significant, indicating different responses of genotypes at two P levels. The root dry weight of 20 upland rice genotypes at low P level varied from 2.00 to 5.68 g plant^{-1}, with an average value of 3.41 g plant^{-1}. At high P level, root dry weight varied from 2.43 to 8.55 g plant^{-1}, with an average value of 4.01 g plant^{-1}. The increase in dry weight at high P level was about 18% compared to low P level. Root length varied from 23.00 to 38.33 cm with an average value of 30.9 cm at low P level. At high P level, root length varied from 23.67 to 34.33 cm, with an average value of 28.20 cm. There was a 10% decrease in root length at high P level compared to low P level. Baligar et al. (1998) have reported increase in root dry weight of upland rice with the addition of P to Brazilian Oxisols. Similarly, these authors also reported decrease in root length at higher P level compared to lower P level. Higher P level roots had more fine hairs compared to lower P level. Hence, roots at higher P level had higher capacity to uptake nutrients and water compared to lower P level. Figure 7.21 shows the root growth of upland rice cultivar BRS Primavera at low and high P levels. Root growth was higher at 200 mg P kg^{-1} compared to 25 mg P kg^{-1}. Hence, selecting upland rice genotypes for better root geometry is possible, which may be helpful under drought conditions.

Fageria, Melo, Oliveira, and Coelho (in press) also studied the influence of P on the root growth of 30 dry bean genotypes (Table 7.6). Maximum root length and root

TABLE 7.5
Root Dry Weight and Root Length of 20 Upland Rice Genotypes as Influenced by P Levels

Genotype	Root Dry Weight (g Plant⁻¹)		Root Length (cm)	
	Low P (25 mg kg⁻¹)	High P (200 mg kg⁻¹)	Low P (25 mg kg⁻¹)	High P (200 mg kg⁻¹)
BRA 01506	3.92ab	3.22c	26.00ab	26.67a
BRA 01596	2.78ab	2.73c	35.67ab	28.00a
BRA 01600	2.81ab	3.03c	36.00ab	29.00a
BRA 02535	3.12ab	4.30c	28.67ab	33.33a
BRA 02601	4.42ab	3.20c	31.33ab	27.00a
BRA 032033	3.70ab	3.62c	23.00b	29.67a
BRA 032039	2.91ab	4.36c	27.67ab	27.00a
BRA 032048	3.96ab	3.91c	37.00a	33.33a
BRA 032051	2.00b	2.58c	36.00ab	30.67a
BRA 042094	2.82ab	3.92c	30.00ab	27.33a
BRA 042156	2.50b	2.91c	29.00ab	27.00a
BRA 042160	5.68a	8.32ab	32.67ab	33.00a
BRA 052015	3.91ab	2.98c	27.00ab	27.00a
BRA 052023	4.69ab	8.55a	29.67ab	34.33a
BRA 052033	2.23b	2.43c	27.00ab	29.00a
BRA 052034	3.18ab	3.99c	31.00ab	24.67a
BRA 052045	3.07ab	3.08c	38.33a	24.67a
BRA 052053	2.57ab	3.87c	28.33ab	23.67a
BRS Primavera	3.56ab	5.21bc	29.67ab	25.00a
BRS Sertaneja	4.36ab	3.92c	34.00ab	23.67a
Average	3.41	4.01	30.9	28.20
F-test				
P level (P)	NS		*	
Genotype (G)	**		**	
P × G	*		**	

Source: Fageria, N.K. et al., J. Plant Nutr., 34, in press.
NS, nonsignificant.
*, **Significant at the 5% and 1% probability levels, respectively. Means in the same column followed by the same letter are not significantly different at the 5% probability level by Tukey's test.

dry weight were significantly influenced by P level as well as genotype treatments (Table 7.6). Maximum root length varied from 8.00 to 29.67 cm, with an average value of 18.86 cm at low P level. At high P level, maximum root length varied from 17.00 to 30.67 cm, with an average value of 22.65 cm. There was a 20% increase in maximum root length at high P level compared to low P level. Root dry weight varied from 0.21 to 0.54 g plant⁻¹ at low P level. Similarly, at high P level root dry weight

FIGURE 7.21 Root growth of upland rice cultivar BRS Primavera at two P levels. The P levels are in mg kg^{-1}.

varied from 0.60 to 1.97 g plant^{-1}, with an average value of 1.27 g plant^{-1}. The average increase in root weight with the addition of P was 234% compared with control treatment. Figures 7.22 through 7.24 show that the root growth of three dry bean genotypes was more vigorous at high P level compared to low P level. Improvement in root dry weight with the addition of P in dry bean is reported by Fageria (2009). Fageria et al. (2010) reported that dry bean genotypes differ significantly in root dry weight at 25 and 200 mg P kg^{-1} of soil. These authors also reported that difference in root dry weight was about two times between lowest and highest dry weight producing genotypes at low P (25 mg P kg^{-1}) as well as at high P levels (200 mg P kg^{-1}).

Phosphorus is a key nutrient essential for root development in highly weathered tropical soils. Baligar et al. (1998) reported that P increased the root weight of wheat, dry bean, and cowpea in a quadratic fashion with increasing P rate from 0 to 200 mg kg^{-1} of soil. The regression equations related to P rates versus root dry weight were $Y = 0.4019 + 0.094X - 0.00031X^2$, $R^2 = 0.74*$ for wheat, $Y = 0.4813$ Exp. $0.019X - 0.000071X^2$, $R^2 = 0.63*$ for dry bean, and $Y = 0.7351 + 0.0232X - 0.000073X^2$, $R^2 = 0.80**$ for cowpea. Based on these regression equations, maximum root dry weight for wheat was achieved at 152 mg P kg^{-1}, whereas maximum root dry weight for common bean and cowpea were achieved at 134 and 159 mg P kg^{-1} of soil, respectively. These results indicate that increasing P levels increased root growth, but root growth was reduced at higher P levels, and the crops had different P requirements to achieve maximum growth. Overall, the root growth of cereals and legume crops was reduced if P was deficient. Most studies indicate that, within certain limits, both root and shoot growth varies similarly as P level increases. Above certain levels, further increases in P supply do not affect root or shoot growth (Troughton, 1962). Fageria et al. (2006) reported that root dry weight

TABLE 7.6
Maximum Root Length and Root Dry Weight of 30 Dry
Bean Genotypes at Two P Levels

Genotype	Maximum Root Length (cm)		Root Dry Weight (g Plant^{-1})	
	0 mg P kg^{-1}	200 mg P kg^{-1}	0 mg P kg^{-1}	200 mg P kg^{-1}
Aporé	21.00cdef	22.00bcdefgh	0.33ab	1.47abcd
Pérola	22.33bcd	25.00abcdefg	0.44ab	1.02abcd
BRSMG Talisma	20.33cdef	30.67a	0.32ab	1.29abcd
BRS Requinte	28.00ab	23.66abcdefgh	0.51ab	1.53abcd
BRS Pontal	20.67cdef	19.00efgh	0.54a	1.57abcd
BRS 9435 Cometa	19.67cdefg	26.33abcd	0.43ab	1.35abcd
BRS Estilo	22.00bcde	21.33cdefgh	0.43ab	1.00abcd
BRSMG Majestoso	23.33abcd	30.67a	0.26ab	1.33abcd
CNFC 10429	29.67a	29.00ab	0.36ab	1.12abcd
CNFC 10408	19.00cdefgh	20.00defgh	0.27ab	0.88bcd
CNFC 10467	19.33cdefg	28.00abc	0.39ab	1.49abcd
CNFC 10470	20.67cdef	26.00abcde	0.36ab	1.45abcd
Diamante Negro	18.67defgh	23.67abcdefgh	0.42ab	1.87ab
Corrente	15.67efghi	17.00h	0.30ab	0.74d
BRS Valente	25.33abc	25.67abcdef	0.47ab	1.97a
BRS Grafite	15.00fghi	20.33defgh	0.42ab	1.42abcd
BRS Campeiro	20.33cdef	20.33defgh	0.38ab	1.17abcd
BRS 7762 Supermo	13.33ghij	18.67fgh	0.43ab	1.31abcd
BRS Esplendor	21.67bcde	24.00abcdefgh	0.52ab	1.06abcd
CNFP 10104	23.67abcd	21.67cdefgh	0.49ab	1.49abcd
Bambuí	22.33bcd	19.33defgh	0.24ab	1.16abcd
BRS Marfim	13.67ghij	18.33gh	0.40ab	0.86bcd
BRS Agreste	13.67ghij	19.33defgh	0.35ab	1.81abc
BRS Pitamda	15.00fghi	24.00abcdefgh	0.21b	1.05abcd
BRS Verede	23.00bcd	25.67abcdef	0.41ab	1.96a
EMGOPA Ouro	8.00j	22.00bcdefgh	0.26ab	0.79cd
BRS Radiante	10.67ij	18.67fgh	0.23ab	0.60d
Jalo Precoce	13.33ghij	21.00cdefgh	0.25ab	1.03abcd
BRS Executivo	12.67hij	21.00cdefgh	0.38ab	1.20abcd
BRS Embaixador	13.67ghij	25.33abcdefg	0.30ab	1.13abcd
Average	18.86b	22.65a	0.38b	1.27a
F-test				
P Levels (P)	**		**	
Genotype (G)	**		**	
P X G	**		**	
CVP (%)	8.07		72.39	
CVG (%)	12.78		26.28	

Source: Fageria, N.K. et al., *Commun. Soil Sci. Plant Anal.*, 43, in press.

**Significant at the 1% probability level. Means followed by the same letter in the same column or same line (P levels) are not significantly different at the 5% probability level by Tukey's test.

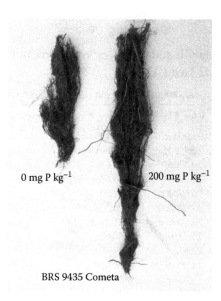

FIGURE 7.22 Root growth of dry bean genotype BRS 9435 Cometa at low and high P levels. (From Fageria, N.K. et al., *Commun. Soil Sci. Plant Anal.*, 43, in press; Fageria, N.K. et al., *Commun. Soil Sci. Plant Anal.*, 43, in press.)

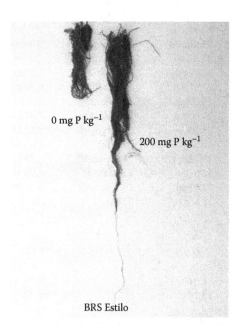

FIGURE 7.23 Root growth of dry bean genotype BRS Estilo at low and high P levels. (From Fageria, N.K. et al., *Commun. Soil Sci. Plant Anal.*, 43, in press; Fageria, N.K. et al., *Commun. Soil Sci. Plant Anal.*, 43, in press.)

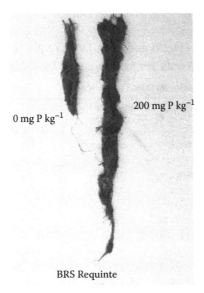

0 mg P kg^{-1}

200 mg P kg^{-1}

BRS Requinte

FIGURE 7.24 Root growth of dry bean genotype BRS Requinte at low and high P levels. (From Fageria, N.K. et al., *Commun. Soil Sci. Plant Anal.*, 43, in press; Fageria, N.K. et al., *Commun. Soil Sci. Plant Anal.*, 43, in press.)

was reduced by 62% in rice, 74% in common bean, 50% in corn, and 21% in soybean without added soil P, compared to adequate P in a Brazilian Oxisol. Gong et al. (2011) reported significant increase in the root dry weight of four corn genotypes with the addition of 1 mM P compared to 1 μM P in nutrient solution. These authors further reported that phosphorus deficiency (1 μM P) evidently enhanced the root-to-shoot ratio of four genotypes, compared with those in the presence of 1 mM P. However, root length was significantly higher in low P treatment compared to high P treatment. There were also significant differences among genotypes in root weight and root length.

7.2.3 Potassium

Potassium plays a vital role in many physiological and biochemical processes in plants. Soil is the major reservoir for plant-available K (Lorenz et al., 2010). The total soil K concentration typically ranges from 0.2% to 3.3% of the total soil mass. Up to 98% of the total K in the plow layer, however, is relatively unavailable for plant growth as it is fixed in minerals such as mica or feldspar (Sparks and Huang, 1985). The plant-available and temporarily dynamic pools, on the other hand, include exchangeable K, nonexchangeable K, and microbial biomass K (Lorenz et al., 2010). The exchangeable K pool in soils ranges from 0.1% to 3% of soil K, and includes K electrostatistically bound to clay mineral surfaces and organic matter, and K in the soil solution. The nonexchangeable K pool represents 1%–10% of soil K. Within this pool, K is bound primarily in the interlayers of micaceous clay minerals and is, therefore, slowly available (Sparks and Huang, 1985). Potassium is also taken up in

TABLE 7.7

Maximum Root Length and Root Dry Weight (g Plant⁻¹) of 12 Lowland Rice Genotypes as Influenced by K Fertilization

Genotype	Maximum Root Length (cm)		Root Dry Weight	
	0 mg K kg⁻¹	300 mg K kg⁻¹	0 mg K kg⁻¹	300 mg K kg⁻¹
BRS Tropical	23.33bc	33.00abc	0.44cd	5.48a
BRS Jaçanã	26.00b	39.00a	2.11a	6.18
BRS 02654	17.33cd	25.67cde	1.14bc	2.02
BRA 051077	36.00a	35.00ab	2.01a	6.57a
BRA 051083	8.66f	26.00cde	0.51cd	1.34b
BRA 051108	25.66b	31.33abc	1.62ab	6.58a
BRA 051126	10.66ef	21.00def	0.49cd	1.81b
BRA 051129	14.00def	14.00f	0.56cd	1.13b
BRA 051130	10.33ef	29.00bcd	0.48cd	1.36b
BRA 051134	11.67def	14.67f	0.49cd	1.04b
BRA 051135	15.00de	20.33ef	0.58cd	2.17b
BRS 051250	11.00ef	21.67def	0.34d	1.25b
Average	17.47b	25.89a	0.89b	3.08a
F-test				
K rate (K)	**		**	
Genotypes (G)	**		**	
K × G	**		**	
CV (%) (K)	13.00		5.00	
CV (%) (G)	11.47		20.12	

**Significant at the 1% probability level. Means followed by the same letter in the same column (in the same line for averages) are not significantly different at the 5% probability level by Tukey's test.

maximum amount compared to other essential nutrients by important food crops (Fageria, 2009). Under this situation, vigorous root growth of crop plants can be helpful in satisfying plant needs for K. This can be achieved with the addition of K fertilizers in the soils deficient or low in available K.

Data in Table 7.7 show the root length and root dry weight of 12 lowland rice genotypes as influenced by K fertilization, grown on a Brazilian Inceptisol. Root length and root dry weight were significantly increased with the addition of K fertilizer. In addition, K × genotype interactions were significant for these two growth parameters, indicating variation in root length as well as root dry weight at two K levels. Hence, it is possible to select lowland rice genotypes for root length and root dry weight at low as well as at high K levels. Root length at low K level varied from 8.66 cm produced by genotype BRA 051083 to 36 cm produced by genotype BRA 051077, with an average value of 17.47 cm. Root length at high K level varied from 14.67 cm produced by genotype BRA 051134 to 39 cm

FIGURE 7.25   Root growth of lowland rice cultivar BRS Tropical at 0 and 300 mg K kg^{-1} of soil.

produced by genotype BRS Jaçanã, with an average value of 25.89 cm. Overall, increase in root length was 48% with the addition of 300 mg K kg^{-1} compared to control treatment.

Root dry weight varied from 0.34 g plant^{-1} produced by genotype BRS 051250 to 2.11 g plant^{-1} produced by genotype BRS Jaçanã, with an average value of 0.89 g plant^{-1} at low K level. At high K level, root dry weight varied from 1.04 g plant^{-1} produced by genotype BRA 051134 to 6.58 g plant^{-1} produced by genotype BRA 051108, with an average value of 3.08 g plant^{-1}. Overall, increase in root dry weight at high K level was 246% compared to low level of K. Figures 7.25 through 7.27 show the root growth of three lowland rice genotypes at low and high K levels. The root growth of these genotypes was higher at high K level compared to low K level. There was also a difference among genotypes in root growth at low as well as at high K levels.

7.2.4   CALCIUM

Calcium is an essential macronutrient for crop growth and the development of plant root system. In calcium-deficient soils, roots become prone to infection by bacteria and fungi (Fageria, 2009). Adequate rate of Ca depends on crop species and genotypes within species (Fageria, 2009; Fageria et al., 2011a). Calcium requirements are high for legumes compared to cereals. A study conducted by the author at the National Rice and Bean Research Center of EMBRAPA, Brazil showed that calcium applied as lime improved the root growth of soybean (Figure 7.28). Gonzalez-Erico et al. (1979) evaluated the response of maize to deep incorporation of limestone on an Oxisol. They reported that

FIGURE 7.26 Root growth of lowland rice genotype BRA 051108 at low and high K levels.

FIGURE 7.27 Root growth of lowland rice genotype BRA 051250 at low and high K levels.

0   0.71   1.42 g lime kg⁻¹ of soil

FIGURE 7.28   Root growth of soybean at 0, 0.71, and 1.42 g lime kg⁻¹ of soil.

the incorporation of limestone to depths of 30 cm improved root growth and increased water utilization and grain yield of maize. Similar results were obtained for maize and cotton when limestone was incorporated to depths up to 45 cm (Doss et al., 1979).

7.2.5 Magnesium

Magnesium plays a significant role. Magnesium deficiency is common in highly weathered acid soils like Oxisols and Ultisols (Fageria and Baligar, 2008). Data in Table 7.8 show significant increase in the root dry weight of dry bean to increasing Mg concentration in the range of 0.30–6.22 cmol$_c$ kg⁻¹ of soil. The increase in root dry weight was quadratic in fashion with increasing Mg levels in the soil. Fageria and Souza (1991) determined the effects of Mg levels on the root weights of rice, common bean, and cowpea grown in Oxisols of central Brazil (Figure 7.29). Dry weights of rice roots were higher at the lowest Mg concentration compared to the highest soil Mg concentration. Initial exchangeable Mg levels of surface soils were 0.1 cmol$_c$ kg⁻¹. They increased to 0.3 cmol$_c$ kg⁻¹ within 3 days after liming and to 0.75 cmol$_c$ kg⁻¹ at harvest time (33 days after sowing). The lack of growth responses to applications of Mg indicated that this level of exchangeable Mg was adequate to meet Mg requirements of upland rice grown in this limed soil. Dry weight of roots of common bean increased with Mg application up to 3 cmol$_c$ kg⁻¹ of soil. Similarly, significant responses of cowpea root growth to soil Mg levels were observed, and

TABLE 7.8

Influence of Mg on Root Growth of Dry Bean Grown on Brazilian Oxisol

Mg Level in Soil (cmol$_c$ kg^{-1})	Root Dry Weight (g Plant^{-1})
0.30	0.70
1.05	0.81
1.15	0.83
1.33	0.74
3.52	1.00
6.22	0.56
R^2	0.86*

Source: Adapted from Fageria, N.K., *The Use of Nutrients in Crop Plants*, CRC Press, Boca Raton, FL, 2009.

*Significant at the 5% probability level.

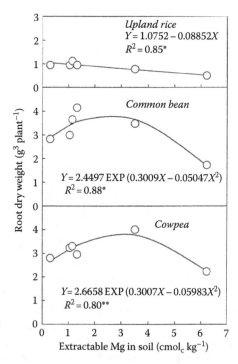

FIGURE 7.29 Relationship between extractable Mg in soil and root dry weight of five field crops. *Significant at the 5% probability level and **Significant at the 1% probability level. (From Fageria, N.K. and Souza, C.M.R., *Commun. Soil Sci. Plant Anal.*, 22, 1805, 1991.)

maximum root weight was achieved at $2.5 \, cmol_c$ Mg kg^{-1} of soil. Silva et al. (2001a) reported that low concentration of Mg (200 µmol L^{-1}) improved soybean root elongation in the presence of toxic Al^{3+} concentration, but this protective effect was not observed for wheat roots. In another study with soybean, Silva et al. (2001b,c) reported that Mg was more efficient than Ca in reducing Al accumulation at the root tip and improving root elongation when supplemented at micromolar concentrations. However, the two cations were equally effective in alleviating Al rhizotoxicity at millimolar concentrations (Silva et al., 2001b,c). Silva et al. (2009) also reported that the addition of 50 µmol L^{-1} Mg to solution containing Al increased Al tolerance in 15 soybean cultivars.

7.2.6 SULFUR

Sulfur plays many important roles in the growth and development of plants. Fageria and Gheyi (1999) summarized the important functions of sulfur in the plant. It is an important component of two amino acids, cysteine and methionine, which are essential for protein formation. Since animals cannot reduce sulfate, plants play a vital role in supplying essential S-containing amino acids to them. Sulfur plays an important role in enzyme activation. It promotes nodule formation in legumes. Sulfur is necessary in chlorophyll formation, although it is not a constituent of chlorophyll. Maturity of seeds and fruits is delayed in the absence of adequate sulfur. Sulfur is required by the plants in the formation of nitrogenase. It increases the crude protein content of forages, and it improves the quality of cereals for milling and baking. Sulfur increases the oil content of oilseed crops, and it increases winter hardiness in plants. It increases drought tolerance in plants, controls certain soilborne diseases, and helps in the formation of glucosides that give characteristic odors and flavors to onion, garlic, and mustard. Sulfur is necessary for the formation of vitamins and the synthesis of some hormones and glutathione, and it is involved in oxidation–reduction reactions. Sulfur improves tolerance to heavy metal toxicity in plants, and it is a component of sulfolipids. Organic sulfates may serve to enhance water solubility of organic compounds, which may be important in dealing with salinity stress, and fertilization with sulfate decreases fungal diseases in many crops. Few studies have assessed the impact of sulfur on root growth and function; however, the effects of sulfur on root growth may be similar to those of N. Zhao et al. (2008) reported that S application increased the root number and root dry weight of soybean compared to control treatment.

The author also studied the influence of S on the root growth of lowland rice genotypes grown on a Brazilian Inceptisol. Root length and root dry weight were significantly influenced by sulfur and genotype treatments and sulfur × genotype interactions were also significant for root length and root dry weight (Table 7.9). Root length varied from 18 to 24 cm, with an average value of 21.38 cm at lower S rate. At higher S rate, root length varied from 20.33 cm, with an average value of 25.22 cm. The increase in root dry weight was 18% at higher S rate compared to lower S rate. Root dry weight varied from 2.18 to 4.67 g $plant^{-1}$ at the lower S level, with an average value of 3.38 g $plant^{-1}$. At higher S rate, root dry weight varied from 2.51 to 4.49 g $plant^{-1}$, with an average value of 3.58 g $plant^{-1}$.

TABLE 7.9

Maximum Root Length and Root Dry Weight of 12 Lowland Rice Genotypes as Influenced by S and Genotype Treatments

Genotype	Root Length (cm)		Root Dry Weight (g Plant^{-1})	
	0 mg S kg^{-1}	80 mg S kg^{-1}	0 mg S kg^{-1}	80 mg S kg^{-1}
BRS Tropical	18.00e	20.33d	4.10abc	4.42ab
BRS Jaçanã	22.66ab	24.00bc	3.36abcd	4.49a
BRA 02654	24.00a	25.66bc	4.17ab	4.07abc
BRA 051077	22.33abc	25.33	4.67a	3.68abc
BRA 051083	20.00cde	22.66cd	3.54abcd	3.17abc
BRA 051108	22.33abc	25.00bc	3.73abcd	3.85abc
BRA 051126	19.66de	30.33a	3.30abcd	4.01abc
BRA 051129	20.00cde	23.66bc	3.47abcd	2.73bc
BRA 051130	22.00abcd	25.66bc	2.98bcd	3.74abc
BRA 051134	21.33bcd	30.66a	2.18d	2.51c
BRA 051135	21.66abcd	26.00b	2.52cd	3.32abc
BRA 051250	22.67ab	23.33bcd	2.59bcd	2.98abc
Average	21.38b	25.22a	3.38a	3.58a
F-test				
S level (S)	**		NS	
Genotype (G)	**		**	
S × G	**		*	
CV (%) (S)	3.57		22.75	
CV (%) (G)	4.38		15.62	

NS, nonsignificant.

*, **Significant at the 5% and 1% probability levels, respectively. Means followed by the same letter in the same column do not differ significantly at the 5% probability level by Tukey's test. For average values, means were compared across the same line.

Figures 7.30 through 7.32 show the root growth of three genotypes of lowland rice at low (0 mg S kg^{-1}) and high S (80 mg S kg^{-1}) fertilization. The root growth of all the three genotypes was more vigorous at higher S level compared with low S level. Overall, the improvement in root dry weight with the addition of S was 6% compared to control treatment. Significant differences in root length and root dry weight among lowland rice genotypes at lower as well as at higher S rates is an important genetic variation and also influenced by environmental conditions. Hence, it can be concluded that the use of S-efficient genotypes and at the same time the use of adequate S fertilizer can improve root length and root dry weight and the uptake of water and nutrients in favor of higher yield. Root length also had significant positive quadratic correlation ($r = 0.43^{**}$) with grain yield.

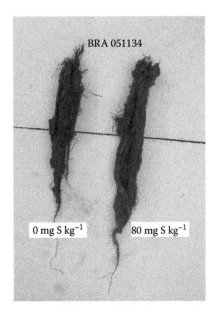

FIGURE 7.30   Root growth of lowland rice genotype BRA 051134 at 0 and 80 mg S kg^{-1}.

FIGURE 7.31   Root growth of lowland rice cultivar BRS Jaçanã at 0 and 80 mg S kg^{-1}.

FIGURE 7.32   Root growth of lowland rice genotype BRA 051130 at 0 and 80 mg S kg^{-1}.

Root growth variation among crop genotypes and improvement with the addition of nutrient is reported by Fageria et al. (2006) and Fageria (2009).

7.3   MICRONUTRIENTS

Micronutrients have also been called minor or trace elements, indicating that their concentrations in plant tissues are minor or in trace amounts relative to the macronutrients (Mortvedt, 2000). Even though micronutrients are required in small quantities by field crops, their influence is as large as that of macronutrients in crop production (Fageria et al., 2002). Micronutrients are normally constituents of prosthetic groups that catalyze redox processes by electron transfer, such as with the transition elements Cu, Fe, Mn, and Mo, and form enzyme–substrate complexes by coupling enzyme with substrate (Fe and Zn) or enhance enzyme reactions by influencing molecular configurations between enzyme and substrate (Zn; Fageria et al., 2002).

Micronutrient deficiencies in crop plants are widespread because of (1) increased micronutrient demands from intensive cropping practices and the adaptation of high-yielding cultivars which may have higher micronutrient demand; (2) enhanced production of crops on marginal soils that contain low levels of essential nutrients; (3) increased use of high analysis fertilizers with low amounts of micronutrients; (4) decreased use of animal manures, composts, and crop residues; (5) the use of many soils that are inherently low in micronutrient reserves; and (6) the involvement of natural and anthropogenic factors that limit adequate supplies and create element imbalances (Fageria et al., 2002). Data related to the influence of micronutrients are scarce as compared to macronutrients.

7.3.1 Zinc

The deficiency of Zn in crop production is spread worldwide (Alloway, 2008). Graham (2008) reported that half of the world's soils are intrinsically deficient in Zn. Zinc deficiency in annual crops is reported in Brazil (Fageria and Stone, 2008), Australia (Graham, 2008), India (Singh, 2008), China (Zou et al., 2008), Turkey (Cakmak, 2008), Europe (Sinclair and Edwards, 2008), the United States (Brown, 2008), and Africa (Waals and Laker, 2008). Micronutrient deficiencies are also a worldwide problem in human health (Welch, 2008). Zinc deficiency is the highest priority among micronutrients for agriculture to address (Graham, 2008; Fageria and Moreira, 2011).

The application of Zn at an adequate level in Zn-deficient soils improves the root growth of crop plants. Fageria (2009) and Fageria and Moreira (2011) reported that the root growth of upland rice, soybean, and dry bean improved with the application of Zn in Brazilian Oxisols. Similarly, the author studied the influence of Zn on root dry weight and maximum root length of 20 upland rice genotypes grown on a Brazilian Oxisol (Table 7.10). Maximum root length was significantly influenced by genotype treatment and it varied from 19.00 to 25.33 cm. Root dry weight was significantly influenced by Zn and genotype, and Zn × genotype interaction was also significant. Hence, there was a variation among genotypes in relation to root dry weight as the Zn level was changed. Zinc fertilization produced significantly higher root dry weight compared with control or without Zn fertilization treatment. This is an important conclusion because upland rice is subject to drought stress in Brazilian Oxisols (Fageria et al., 2006) and adequate Zn level may improve root growth and, consequently, result in higher uptake of water and nutrients. Figures 7.33 through 7.35 show that the root growth of three upland rice genotypes was higher at higher Zn level compared with low Zn level. Sadaghiani et al. (2011) reported that the root weight of two barley genotypes significantly increased with the addition of Zn in the range of 0–3.2 mg kg^{-1}. These authors also reported significant differences in the root weight of two genotypes. Root weight is genetically controlled and also influenced by environmental factors, especially mineral nutrition (Baligar et al., 1998; Fageria et al., 2006). Specific root length was significantly influenced by Zn and genotype treatments (Table 7.11). The Zn × genotype interaction was significant for this trait, indicating variation in specific root length among genotypes with the variation in Zn levels. Specific root length decreased at higher Zn level compared with low Zn level. This may be related to high root weight at high Zn level compared with low Zn level. Zhao et al. (2011) reported that the root dry weight of wheat increased significantly and quadratically with the addition of Zn in the range of 0–50 mg L^{-1} in nutrient solution.

7.3.2 Copper

The copper requirement of plants is lower than zinc, iron, and manganese. It activates many enzymatic reactions in the plant. Factors that influence the level of available Cu in soil are organic matter, clay type and content, oxide type and content, redox

TABLE 7.10

Maximum Root Length and Root Dry Weight of 20 Upland Rice Genotypes as Influenced by Zn Fertilization; Zn Levels were 0 and 20 mg kg^{-1} of Soil

Genotype	Root Length (cm)	Root Dry Weight (g Plant^{-1})	
		Zn_0	Zn_{20}
BRA 01506	21.83ab	2.61abcde	2.92bcdef
BRA 01596	19.83ab	1.60e	2.06f
BRA 01600	21.33ab	1.65de	2.34ef
BRA 025535	25.33a	2.33abcde	4.29abcd
BRA 02601	21.33ab	3.05ab	5.33a
BRA 032033	21.66ab	1.97bcde	4.67ab
BRA 032039	22.33ab	2.22bcde	4.36abc
BRA 032048	20.33ab	1.95bcde	4.15abcd
BRA 0320 51	21.50ab	1.93bcde	2.01f
BRA 042094	22.16ab	1.83cde	2.25ef
BRA 042156	20.33ab	1.72cde	1.78f
BRA 042160	21.16ab	2.73abcd	2.54def
BRA 052015	21.83ab	1.64de	2.02f
BRA 052023	21.00ab	3.39a	2.91bcdef
BRA 052033	19.33b	2.51abcde	2.70cdef
BRA 052034	21.50ab	2.21bcde	2.75cdef
BRA 052045	22.16ab	2.79abc	3.87abcde
BRA 052053	19.00b	1.69cde	1.45f
BRS Primavera	23.50ab	1.83cde	1.52f
BRS Sertaneja	22.33ab	1.69cde	2.75cdef
Average	21.48	2.17b	2.94a
F-test			
Zn level (Zn)	NS	**	
Genotype (G)	*	**	
Zn × G	NS	**	
CV Zn (%)	16.94	34.03	
CVG (%)	12.43	17.57	

Values of root length are across two Zn levels.

NS, nonsignificant.

*, **Significant at the 5% and 1% probability levels, respectively. Means followed by the same letter in the same column are not significantly different at 5% probability level. Means of the average values are compared at low and high Zn levels.

FIGURE 7.33 Root growth of upland rice genotype BRA 01600 at 0 and 20 mg Zn kg^{-1}.

FIGURE 7.34 Root growth of upland rice genotype BRA 02601 at 0 and 20 mg Zn kg^{-1}.

BRA 052045

0 mg Zn kg^{-1}

20 mg Zn kg^{-1}

FIGURE 7.35   Root growth of upland rice genotype BRA 052045 at 0 and 20 mg Zn kg^{-1}.

potential, and microorganisms, and the nature of other elements associated with Cu (Fageria et al., 2002). Copper uptake is metabolically mediated and strongly inhibited by other divalent transition metals, especially Zn (Fageria, 2009). Applications of relatively high levels of N and P fertilizers have induced Cu deficiency in plants grown in low-Cu soils. Even though N and Cu interact, no significant effects of NO_3-N or NH_4-N on Cu uptake have been noted (Kochian, 1991). Even though increased soil P induced Cu deficiency, this response was related to dilution effects from increased growth and the depressing effects of P on Cu absorption (Fageria et al., 2002). Copper toxicity has also been noted in P-deficient plants (Wallace, 1984). Plants grown in coarse-textured soils with low available P and Fe and high in Cu had induced Cu toxicity (Fageria et al., 2011). Copper application in adequate amount can improve the root growth of crop plants in Cu-deficient soils. The root dry weight of wheat and root length of dry bean were significantly increased by the application of copper fertilizers (Table 7.12). Similarly, application of Cu improved the root growth of soybean grown on a Brazilian Oxisol (Figure 7.36). Maximum root growth was achieved at Cu level of 2 mg kg^{-1} of soil. Copper application increased the shoot as well as root growth of cover crop species grown on Brazilian Oxisol (Figures 7.37 through 7.40).

7.3.3 BORON

Boron has been used as a fertilizer for more than 400 years, but it was not shown to be an essential element until the twentieth century (Fageria, 2009). Boron deficiency is widespread and has been reported in at least 80 countries on 132 crop species (Shorrocks, 1997). Fageria et al. (2007) conducted a field experiment with upland rice and common bean grown in rotation on a Brazilian Oxisol. Upland

TABLE 7.11
Specific Root Length of 20 Upland Genotypes as Influenced by Zn Fertilization

	Specific Root Length (cm g^{-1})	
Genotype	Zn$_0$	Zn$_{20}$
BRA 01506	7.91bc	8.12bcdef
BRA 01596	13.40ab	9.12bcdef
BRA 01600	16.01a	8.59bcdef
BRA 025535	10.15abc	6.46cdef
BRA 02601	7.17bc	4.03f
BRA 032033	11.17abc	4.76def
BRA 032039	10.53abc	5.22def
BRA 032048	11.70abc	4.59ef
BRA 032051	12.55abc	9.57bcdef
BRA 042094	12.10abc	10.03abcde
BRA 042156	11.49abc	11.83abc
BRA 042160	8.21bc	8.38bcdef
BRA 052015	14.18ab	10.38abcd
BRA 052023	5.55c	8.21bcdef
BRA 052033	7.70bc	7.81bcdef
BRA 052034	9.58abc	8.13bcdef
BRA 052045	8.12bc	5.71def
BRA 052053	12.41abc	12.19ab
BRS Primavera	13.09abc	15.68a
BRS Sertaneja	12.84abc	8.49bcdef
Average	10.79a	8.36b
F-test		
Zn level (Zn)	*	
Genotype (G)	**	
Zn × G	**	

*, **Significant at the 5% and 1% probability levels, respectively.
Means followed by the same letter in the same column are not
significantly different at 5% probability level. Means of the average values are compared at low and high Zn levels.

rice did not respond to B applications but the yield of common bean was significantly increased with B fertilization. Maximum grain yield was achieved at about 2 kg B ha^{-1}.

Fageria (2009) reported on the influence of B on the root growth of soybean and corn grown on Brazilian Oxisols. The application of B at the rate of 1 mg B kg^{-1} of soil improved the root growth of soybean compared to control treatment (Figure 7.41). However, B rate at 12 mg kg^{-1} decreased root growth, indicating toxicity of this element. Similarly, the root growth of corn was also improved with the addition of B

TABLE 7.12
Root Dry Weight of Wheat
and Root Length of Dry Bean
as Influenced by Cu Fertilization

Cu Rate (mg kg⁻¹)	Wheat (g 4 Plants⁻¹)[1]	Dry Bean (cm)[a]
0	0.53	25
2	0.60	30
4	0.50	27
8	0.48	28
16	0.47	28
32	0.47	24
64	0.43	30
96	0.17	14
R^2	0.88**	0.42**

Source: Adapted from Fageria, N.K., *J. Plant Nutr.*, 25, 613, 2002.

[a] Wheat and dry bean plants were harvested 5 weeks after sowing.

**Significant at the 1% probability level.

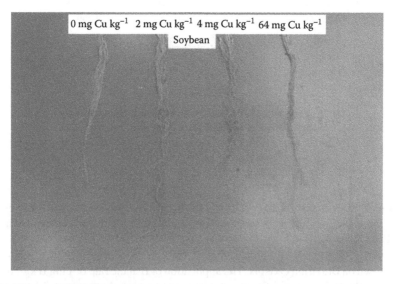

FIGURE 7.36 Influence of copper on root growth of soybean. (Left to right) 0, 2, 4, and 64 mg Cu kg⁻¹ of soil. (From Fageria, N.K., *The Use of Nutrients in Crop Plants*, CRC Press, Boca Raton, FL, 2009.)

FIGURE 7.37 Shoot growth of cover crop species *Crotalaria spectabilis* at 0, 5, 10, and 20 mg Cu kg^{-1} of soil.

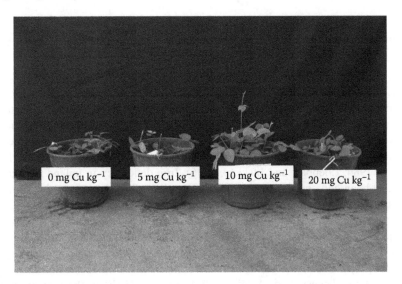

FIGURE 7.38 Shoot growth of cover crop species tropical kudzu at 0, 5, 10, and 20 mg Cu kg^{-1} of soil.

fertilizers (Figure 7.42). Maximum root growth was achieved at 3 mg B kg^{-1} of soil and higher B rate (12 mg kg^{-1}) decreased root growth.

7.3.4 Iron

Total Fe concentrations vary widely and range from <1% to >20% (depending on soil types and management practices); the median concentration is approximately

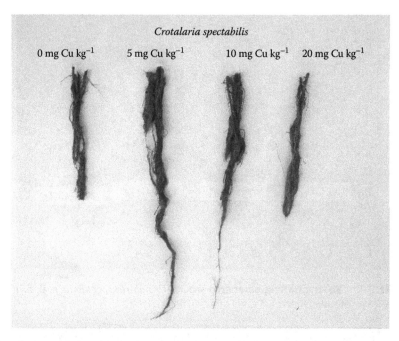

FIGURE 7.39 Root growth of cover crop species *Crotalaria spectabilis* at 0, 5, 10, and 20 mg Cu kg⁻¹ of soil.

3% (Loeppert and Inskeep, 1996). Under the earth's surface conditions, Fe may exist in either Fe^{2+} (ferrous) or Fe^{3+} (ferric) oxidation state. Although Fe is an abundant element in primary and secondary minerals in soils, its low availability frequently limits plant growth, especially in alkaline and calcareous soils, because of the low solubilities of the Fe-containing secondary minerals (Lindsay and Schwab, 1982; Lindsay, 1984; Loeppert and Inskeep, 1996). Iron deficiency in upland rice grown after dry bean and soybean in rotation on Brazilian Oxisols that received lime to raise pH higher than 6, which is considered optimum for dry bean and soybean, is frequently reported (Fageria, 2009; Fageria et al., 2001la). This deficiency of Fe in upland rice is not due to low level of Fe in these soils, but low availability by upland rice. Low availability of iron reduces root growth in crops plants (Fageria and Moreira, 2011). However, differences exist among crop species or genotypes within species (Fageria and Moreira, 2011). Figure 7.43 shows the root growth of upland rice as influenced by iron application rate in a Brazilian Oxisol. Root growth was at maximum at 50 mg Fe kg⁻¹ of soil and reduced when Fe rate was raised to 1600 mg kg⁻¹ of soil.

7.3.5 OTHER MICRONUTRIENTS

Other micronutrients that are essential for plant growth are manganese, molybdenum, chlorine, and nickel. Manganese is similar to Fe in its chemical behavior in soils and geological materials (Gambrell, 1996). Manganese deficiency as well

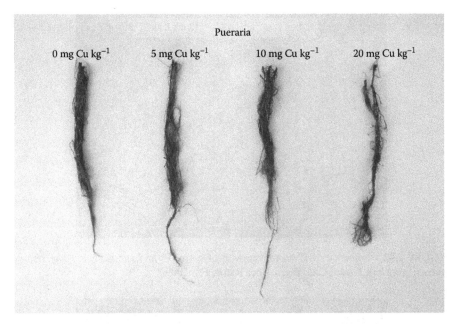

FIGURE 7.40   Root growth of cover crop species tropical kudzu at 0, 5, 10, and 20 mg Cu kg^{-1} of soil.

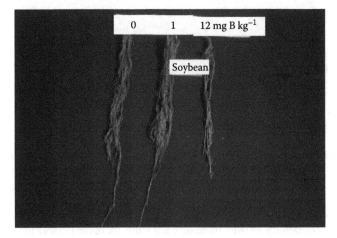

FIGURE 7.41   Influence of B on root growth of soybean. (From Fageria, N.K., *The Use of Nutrients in Crop Plants*, CRC Press, Boca Raton, FL, 2009.)

as toxicity is reported in crop plants (Foy et al., 1981; Fageria, 2009). Soil pH is one of the important factors affecting Mn availability to plants. Increasing soil pH decreases Mn availability to plants (Fageria et al., 2002). Table 7.13 shows the influence of increasing soil pH on the uptake of Mn by upland rice plants grown on a Brazilian Oxisol. Uptake was significantly and linearly decreased with increasing soil pH from 4.6 to 6.8.

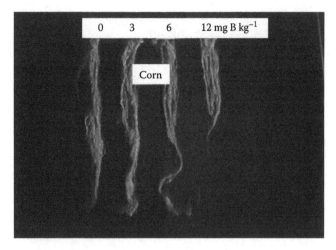

FIGURE 7.42 Influence of B on root growth of corn. (From Fageria, N.K., *The Use of Nutrients in Crop Plants*, CRC Press, Boca Raton, FL, 2009.)

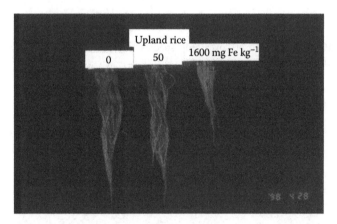

FIGURE 7.43 Root growth of upland rice at different Fe rates applied to Brazilian Oxisol. (From Fageria, N.K., *The Use of Nutrients in Crop Plants*, CRC Press, Boca Raton, FL, 2009.)

Molybdenum uptake is at minimum in crop plants compared to other micronutrients. The availability of Mo increases with the increase of soil pH contrary to other micronutrients. Hence, Mo deficiency in acid soils can be corrected with the addition of liming. Molybdenum is important for N_2 fixation in legume crops. It is an essential nutrient and functions as an electron transport agent in flavor enzymes, including xanthine oxidase, sulfite oxidase, and aldehyde oxidase (Hille, 1999; Boojar and Tavakkoli, 2011). Lack of this micronutrient may result in the accumulation of nitrate and the decrease of amino acids and levels of vitamin metabolism in plant tissues (Wang, 1991). The Mo requirements by plants varied among crop species and genotypes within species. The Mo sufficiency levels in the plant tissue of legumes (>0.5 mg Mo kg^{-1}) is 2–3 times higher compared to nonlegumes (Sims,

TABLE 7.13
Influence of Soil pH on Acquisition
of Mn by Upland Rice Grown in an
Oxisol of Brazil

Soil pH	Mn Uptake (μg Plant^{-1})
4.6	11,160
5.7	5,010
6.2	4,310
6.4	3,610
6.6	2,760
6.8	2,360
R^2	0.99**

Source: Adapted from Fageria, N.K., *Pesq. Agropec. Bras.*, 35, 2303, 2000.
**Significant at the 1% probability level.

1996). Chlorine is required by plants for water-splitting reaction in photosystem II, and nickel is a constituent of urease. Data on the influence of root growth of Mo, Cl, and Ni are not available and this aspect is not discussed here.

7.4 CONCLUSIONS

Roots are important organs that supply water, nutrients, hormones, and mechanical support (anchorage) to crop plants and, consequently, affect economic yields. In addition, roots improve soil organic matter (OM) by contributing to soil pools of organic carbon (C), nitrogen (N), and microbial biomass. Root-derived soil C is retained and forms more stable soil aggregates than shoot-derived soil C. Although roots normally contribute only 10%–20% of total plant weight, a well-developed root system is essential for healthy plant growth and development. Root growth is controlled genetically, but it is also influenced by environmental factors. The response of root growth to chemical fertilization is similar to that of shoot growth; however, the magnitude of the response may differ. In nutrient-deficient soils, root weight often increases in a quadratic manner with the addition of chemical fertilizers. Increasing nutrient supplies in the soil may also decrease root length but increase root weight in a quadratic fashion. Roots with adequate nutrient supplies may also have more root hairs than nutrient-deficient roots. This may result in greater uptake of water and nutrients by roots well supplied with essential plant nutrients, compared with roots grown in nutrient-deficient soils. Rooting pattern in crop plants is under multi- or polygenic control, and breeding programs can be used to improve root system properties for environments where drought is a problem. The use of crop species and cultivars tolerant to biotic and abiotic stresses, as well as the use of appropriate cultural practices, can improve plant root system function under favorable and unfavorable environmental conditions. Based on the review of the literature, it can be concluded that nutrients have profound effects on the root growth of crop plants. Furthermore, many of these effects are specific to particular nutrients and are strongly dependent on the genotype of the crop species.

Nitrogen sources also affect root growth in upland rice. Root dry weight increased in a quadratic exponential fashion with the application of N in the range of 0–400 mg kg^{-1} of soil. Ammonium sulfate produced more vigorous root systems, especially at higher N rates, than urea, perhaps because upland rice is highly tolerant to soil acidity and ammonium sulfate reduces soil pH more than urea. Brazilian rice cultivars (lowland as well as upland) are highly tolerant to soil acidity. Another possible explanation is that ammonium sulfate has about 24% S, which may improve root growth if the extractable soil S level is lower than 10 mg kg^{-1}.

REFERENCES

Alloway, B. J. 2008. Micronutrients and crop production: An introduction. In: *Micronutrient Deficiencies in Global Crop Production*, ed., B. J. Alloway, pp. 1–39. New York: Springer Verlag.

Baligar, V. C., N. K. Fageria, and M. Elrashidi. 1998. Toxicity and nutrient constraints on root growth. *Hort. Sci.* 33:960–965.

Barber, S. A. and M. Silverbush. 1984. Plant root morphology and nutrient uptake. In: *Roots, Nutrient and Water Influx and Plant Growth*, eds., S. A. Barber, D. R. Bouldin, D. M. Kral, and S. L. Hawkins, pp. 65–88. Madison, WI: American Society of Agronomy.

Barber, S. A., J. M. Walker, and E. H. Vasey. 1963. Mechanisms for movement of plant nutrients from the soil and fertilizer to the plant root. *J. Agric. Food Chem.* 11:204–207.

Barley, K. P. 1970. The configuration of the root system in relation to nutrient uptake. *Adv. Agron.* 22:159–301.

Bates, T. R. and J. P. Lynch. 1996. Stimulation of root hair elongation in *Arabidopsis thaliana* by low phosphorus availability. *Plant Cell Environ.* 19:529–538.

Bonser, A. M., J. Lynch, and S. Snapp. 1996. Effect of phosphorus deficiency on growth angle of basal roots in *Phaseolus vulgaris*. *New Phytol.* 132:281–288.

Boojar, M. M. A. and Z. Tavakkoli. 2011. New molybdenum hyperaccumulator among plant species growing on molybdenum mine-a biochemical study on tolerance mechanism against metal toxicity. *J. Plant Nutr.* 34:1532–1557.

Brown, P. H. 2008. Micronutrient use in agriculture in the United States of America: Current practices, trends and constraints. In: *Micronutrient Deficiencies in Global Crop Production*, ed., B. J. Alloway, pp. 267–286. New York: Springer Verlag.

Cakmak, I. 2008. Zinc deficiency in wheat in Turkey. In: *Micronutrient Deficiencies in Global Crop Production*, ed., B. J. Alloway, pp. 181–200. New York: Springer Verlag.

Chapin, F. S. 1980. The mineral nutrition of wild plants. *Annu. Rev. Ecol. Syst.* 11:233–260.

Comfort, S. D., G. L. Malzer, and R. H. Busch. 1988. Nitrogen fertilization of spring wheat genotypes: Influence on root growth and soil water depletion. *Agron. J.* 80:114–120.

Costa, C., L. M. Dwyer, X. Zhou, P. Dutilleul, L. M. Reid, and D. L. Smith. 2002. Root morphology of contrasting maize genotypes. *Agron. J.* 94:96–101.

Denton, M. D., C. Sasse, M. Tibbett, and M. H. Ryan. 2006. Root distributions of Australian herbaceous perennial legumes in response to phosphorus placement. *Funct. Plant Biol.* 33:1091–1102.

Doss, B. D., W. T. Dumas, and Z. F. Lund. 1979. Depth of lime incorporation for correction of subsoil acidity. *Agron. J.* 71:541–544.

Drew, M. C., L. R. Saker, and T. W. Ashley. 1973. Nutrient supply and the growth of the seminal root system in barley. I. The effect of nitrate concentration on the growth of axes and laterals. *J. Exp. Bot.* 24:1189–1202.

Eghball, B., J. R. Settimi, J. W. Maranville, and A. M. Parkhurst. 1993. Fractal analysis for morphological description of corn roots under nitrogen stress. *Agron. J.* 85:147–152.

Ericsson, T. 1995. Growth and shoot/root ration of seedlings in relation to nutrient availability. *Plant Soil* 169:205–214.

Fageria, N. K. 1992. *Maximizing Crop Yields.* New York: Marcel Dekker.

Fageria, N. K. 2000. Upland rice response to soil acidity in Cerrado soil. *Pesq. Agropec. Bras.* 35:2303–2307.

Fageria, N. K. 2002. Micronutrients influence on root growth of upland rice, common bean, corn, wheat, and soybean. *J. Plant Nutr.* 25:613–622.

Fageria, N. K. 2009. *The Use of Nutrients in Crop Plants.* Boca Raton, FL: CRC Press.

Fageria, N. K. 2010. Root growth of upland rice genotypes as influenced by nitrogen fertilization. Paper presented at the *19th World Soil Science Congress*, August 1–6, 2010, Brisbane, Australia.

Fageria, N. K. and V. C. Baligar. 2005. Enhancing nitrogen use efficiency in crop plants. *Adv. Agron.* 88:97–185.

Fageria, N. K. and V. C. Baligar. 2008. Ameliorating soil acidity of tropical Oxisols by liming for sustainable crop production. *Adv. Agron.* 99:345–399.

Fageria, N. K., V. C. Baligar, and R. B. Clark. 2002. Micronutrients in crop production. *Adv. Agron.* 77:185–268.

Fageria, N. K., V. C. Baligar, and R. B. Clark. 2006. *Physiology of Crop Production.* New York: The Haworth Press.

Fageria, N. K., V. C. Baligar, and C. A. Jones. 2011a. *Growth and Mineral Nutrition of Field Crops*, 3rd edn. Boca Raton, FL: CRC Press.

Fageria, N. K., V. C. Baligar, and Y. C. Li. 2008. The role of nutrient efficient plants in improving crop yields in the twenty first century. *J. Plant Nutr.* 31:1121–1157.

Fageria, N. K., V. C. Baligar, A. Moreira, and T. A. Portes. 2010. Dry bean genotypes evaluation for growth, yield components and phosphorus use efficiency. *J. Plant Nutr.* 33:2167–2181.

Fageria, N. K., V. C. Baligar, and R. W. Zobel. 2007. Yield, nutrient uptake, and soil chemical properties as influenced by liming and boron application in common bean in a no-tillage system. *Commun. Soil Sci. Plant Anal.* 38:1637–1653.

Fageria, N. K., E. P. B. Ferreira, V. C. Baligar, and A. M. Knupp. (in press). Growth of tropical cover crops as influenced by nitrogen fertilization and rhizobia. *Commun. Soil Sci. Plant Anal.* 43.

Fageria, N. K. and H. R. Gheyi. 1999. *Efficient Crop Production.* Campina Grande, Brazil: Federal University of Paraiba.

Fageria, N. K., L. C. Melo, J. P. Oliveira and A. M. Coelho. (in press). Yield and yield components of dry bean genotypes as influenced by phosphorus fetilization. *Commun. Soil Sci. Plant Anal.* 43.

Fageria, N. K. and A. Moreira. 2011. The role of mineral nutrition on root growth of crop plants. *Adv. Agron.* 110:251–331.

Fageria, N. K., A. Moreira, and A. M. Coelho. 2011b. Yield and yield components of upland rice as influenced by nitrogen sources. *J. Plant Nutr.* 34:361–370.

Fageria, N. K. and C. M. R. Souza. 1991. Upland rice, common bean, and cowpea response to magnesium application on an Oxisols. *Commun. Soil Sci. Plant Anal.* 22:1805–1816.

Fageria, N. K. and L. F. Stone. 2008. Micronutrient deficiency problems in South America. In: *Micronutrient Deficiencies in Global Crop Production*, ed., B. J. Alloway, pp. 245–266. New York: Springer Verlag.

Feaga, J. B., J. S. Selker, R. P. Dick, and D. D. Hemphill. 2010. Long-term nitrate leaching under vegetable production with cover crops in the Pacific Northwest. *Soil Sci. Soc. Am. J.* 74:186–195.

Forde, B. and H. Lorenzo. 2001. The nutritional control of root development. *Plant Soil* 232:51–68.

Foy, C. D., H. W. Webb, and J. E. Jones. 1981. Adaptation of cotton genotypes to an acid, manganese toxic soil. *Agron. J.* 73:107–111.

Gambrell, R. P. 1996. Manganese. In: *Methods of Soil Analysis, Part 3, Chemical Methods*, ed., J. M. Bartels, pp. 665–682. Madison, WI: SSSA and ASA.

Gong, Y., Z. Guo, L. He, and J. Li. 2011. Identification of maize genotypes with high tolerance or sensitivity to phosphorus deficiency. *J. Plant Nutr.* 34:1290–1302.

Gonzalez-Erico, E., E. J. Kamprath, G. C. Nederman, and W. V. Soares. 1979. Effect of depth of lime incorporation on the growth of corn on an Oxisols of central Brazil. *Soil Sci. Soc. Am. J.* 43:1155–1158.

Graham, R. D. 2008. Micronutrient deficiencies in crops and their global significance. In: *Micronutrient Deficiencies in Global Crop Production*, ed., B. J. Alloway, pp. 41–61. New York: Springer Verlag.

Gregory, P. J. 1994. Root growth and activity. In: *Physiology and Determination of Crop Yield*, ed., G. A. Peterson, pp. 65–93. Madison, WI: ASA, CSSA, and SSSA.

Gregory, P. J. and S. C. Brown. 1989. Root growth, water use and yield of crops in dry environments: What characteristics are desirable? *Aspec. Appl. Biol.* 22:235–243.

Hackett, C. 1972. A method of applying nutrients locally to roots under controlled conditions and some morphological effects of local applied nitrate on the branching of wheat roots. *Aust. J. Biol. Sci.* 25:1169–1180.

Hansson, A. C. and O. Andren. 1987. Root dynamics in barley, Lucerne, and meadow fescue investigated with a minirhizotron technique. *Plant Soil* 103:33–38.

Hille, R. 1999. Molybdenum enzymes. *Essays Biochem.* 34:125–137.

Hoad, S. P., G. Russell, M. E. Lucas, and I. J. Bingham. 2001. The management of wheat, barley, and oat root systems. *Adv. Agron.* 74:193–246.

Hodge, A. 2004. The plastic plant: Root responses to heterogeneous supplies of nutrients. *New Phytol.* 162:9–24.

Hurd, E. A., T. F. Townley-Smith, L. A. Patterson, and C. H. Owen. 1972. Techniques used in producing Wascana wheat. *Can. J. Plant Sci.* 52:689–691.

Kochian, L. V. 1991. Mechanisms of micronutrient uptake and translocation in plants. In: *Micronutrient in Agriculture*, eds., J. J. Mortvedt, F. R. Cox, L. M. Shuman, and R. M. Welch, pp. 229–296. Madison, WI: SSSA.

Kristensen, H. L. and K. Thorup-Kristensen. 2004. Root growth and nitrate uptake of three different catch crops in deep soil layers. *Soil Sci. Soc. Am. J.* 68:529–537.

Liao, H., G. Rubio, X. Yan, A. Cao, K. M. Brown, and J. P. lynch. 2001. Effect of phosphorus availability on basal root shallowness in common bean. *Plant Soil* 232:69–79.

Lindsay, W. L. 1984. Soil and plant relationships associated with iron deficiency with emphasis on nutrient interactions. *J. Plant Nutr.* 7:489–500.

Lindsay, W. L. and A. P. Schwab. 1982. The chemistry of iron in soils and its availability to plants. *J. Plant Nutr.* 5:821–840.

Loeppert, R. H. and W. P. Inskeep. 1996. Iron. In: *Methods of Soil Analysis, Part 3, Chemical Methods*, ed., J. M. Bartels, pp. 639–664. Madison, WI: SSSA and ASA.

Lorenz, N., K. Verdell, C. Ramsier, and R. P. Dick. 2010. A rapid assay to estimate soil microbial biomass potassium in agricultural soils. *Soil Sci. Soc. Am. J.* 74:512–516.

Ludlow, M. M. and R. C. Muchow. 1990. A critical evaluation of traits for improving crop yields in water-limited environments. *Adv. Agron.* 43:107–153.

Lynch, J. P. 2007. Roots of the second green revolution. *Aust. J. Bot.* 55:493–512.

Ma, Z., D. G. Bielenberg, K. M. Brown, and J. P. Lynch. 2001. Regulation of root hair density by phosphorus availability in *Arabidopsis thaliana*. *Plant Cell Environ.* 24:459–467.

Marschner, H. 1995. *Mineral Nutrition of Higher Plants*. New York: Academic Press.

Miller, C. R., I. Ochoa, K. L. Nielsen, D. Beck, and J. P. Lynch. 2003. Genetic variation for adventitious rooting in response to low phosphorus availability: Potential utility for phosphorus acquisition from stratified soil. *Funct. Plant Biol.* 30:973–985.

Mortvedt, J. J. 2000. Bioavailability of micronutrients. In: *Handbook of Soil Science*, ed., M. E. Sumner, pp. D71–D88, Boca Raton, FL: CRC Press.

Noulas, C., M. Liedgens, P. Stamp, I. Alexiou, and J. M. Herrera. 2010. Subsoil root growth of field grown spring wheat genotypes (*Triticum aestivum* L.) differing in nitrogen use efficiency parameters. *J. Plant Nutr.* 33:1887–1903.

Okajima, H. 2001. Historical significance of nutrient acquisition in plant nutrition research. In: *Plant Nutrition Acquisition: New Perspectives*, eds., N. Ae, J. Arihara, K. Okada, and A. Srinivasan, pp. 3–31. New York: Springer Verlag.

O'Toole, J. C. and W. L. Bland. 1987. Genotypic variation in crop plant root systems. *Adv. Agron.* 41:91–145.

Russelle, M. P. and J. F. S. Lamb. 2011. Divergent alfalfa root system architecture is maintained across environment and nutrient supply. *Agron. J.* 103:1115–1123.

Sadaghiani, M. R., B. Sadeghzadeh, E. Sepehr, and Z. Rengel. 2011. Root exudation and zinc uptake by barley genotypes differing in Zn efficiency. *J. Plant Nutr.* 34:1120–1132.

Sainju, U. M., B. P. Singh, and W. F. Whitehead. 1998. Cover crop root distribution and its effect on soil nitrogen cycling. *Agron. J.* 90:511–518.

Shorrocks, V. M. 1997. The occurrence and correction of boron deficiency. In: *Boron in Soils and Plants: Reviews*, eds., B. Dell, P. H. Brown, and R. W. Bell, pp. 121–148. Dordrecht, the Netherlands: Kluwer Academic Publishers.

Silva, I. R., T. F. C. Correa, R. F. Novais, T. J. Smyth, T. Rufty, E. F. Silva, F. O. Gebrim, and F. N. Nunes. 2009. Timing, location and crop species influence the magnitude of amelioration of aluminum toxicity by magnesium. *Rev. Bras. Ciênc. Solo* 33:65–76.

Silva, I. R., T. J. Smith, T. E. Carter, and T. W. Rufty. 2001a. Altered aluminum sensitivity in soybean genotypes in the presence of magnesium. *Plant Soil* 230:223–230.

Silva, I. R., T. J. Smith, D. W. Israel, T. E. Carter, T. E. Raper, and T. W. Rufty. 2001b. Differential aluminum tolerance in soybean: An evaluation of the role of organic acids. *Physiol. Plant* 112:200–210.

Silva, I. R., T. J. Smith, D. W. Israel, T. E. Raper, and T. W. Rufty. 2001c. Magnesium is more efficient than calcium in alleviating aluminum toxicity in soybean and its ameliorative effect is not explained by the Gouy-Chapman-Stern model. *Plant Cell Physiol.* 42:538–545.

Sims, J. L. 1996. Molybdenum. In: *Methods of Soil Analysis, Part 3, Chemical Methods*, ed., J. M. Bartels, pp. 723–737. Madison, WI: SSSA and ASA.

Sinclair, A. H. and A. C. Edwards. 2008. Micronutrient deficiency problems in agricultural crops in Europe. In: *Micronutrient Deficiencies in Global Crop Production*, ed., B. J. Alloway, pp. 225–266. New York: Springer Verlag.

Singh, M. V. 2008. Micronutrient deficiencies in crop and soils in India. In: *Micronutrient Deficiencies in Global Crop Production*, ed., B. J. Alloway, pp. 93–123. New York: Springer Verlag.

Sparks, D. and P. M. Huang. 1985. Physical chemistry of soil potassium. In: *Potassium in Agriculture*, eds., R. E. Munson, pp. 201–276. Madison, WI: ASA, CSSA, and SSSA.

Sponchiado, B. N., J. W. White, J. A. Castillo, and P. G. Jones. 1989. Root growth of four common bean cultivars in relation to drought tolerance in environments with contrasting soil types. *Exp. Agric.* 25:249–257.

Troughton, A. 1962. The roots of temperate cereals (wheat, barley, oats and rye). Mimeographed publication No. 2. Hurley, England: Commonwealth Bureau Pastures and Field Crops.

Vance, C. P., C. Uhde-Stone, and D. L. Allan. 2003. Phosphorus acquisition and use: Critical adaptation by plants for securing a nonrenewable resource. *New Phytol.* 157:423–447.

Vieira, R. F., J. E. S. Carneiro, and J. P. Lynch. 2008. Root traits of common bean genotypes used in breeding programs for disease resistance. *Peaq. Agropec. Bras.* 43:707–712.

Waals, J. H. V. and M. C. Laker. 2008. Micronutrient deficiencies in crops in Africa with emphasis on southern Africa. In: *Micronutrient Deficiencies in Global Crop Production*, ed., B. J. Alloway, pp. 201–224. New York: Springer Verlag.

Wallace, A. 1984. Effect of phosphorus deficiency and copper excess on vegetative growth of bush bean plants in solution culture at two different solution pH. *J. Plant Nutr.* 7:603–608.

Wang, K. 1991. *Trace Elements in Life Science*. Beijing, China: Chinese Measurement Press.

Welch, R. M. 2008. Linkages between trace elements in food crops and human health. In: *Micronutrient Deficiencies in Global Crop Production*, ed., B. J. Alloway, pp. 287–317. New York: Springer Verlag.

Wiersum, L. K. 1958. Density of root branching as affected by substrate and separate ions. *Acta Bot. Neerl.* 7:174–190.

Ye, X., J. Hong, L. Shi, and F. Xu. 2010. Adaptability mechanism of nitrogen efficient germplasm of natural variation to low nitrogen stress in *Brassica Napus*. *J. Plant Nutr.* 33:2028–2040.

Zhao, R. Q., X. H. Tian, W. H. Lu, W. J. Gale, X. C. Lu, and Y. X. Cao. 2011. Effect of zinc on cadmium toxicity in winter wheat. *J. Plant Nutr.* 34:1372–1385.

Zhao, Y., X. Xiao, D. Bi, and F. Hu. 2008. Effects of sulfur fertilization on soybean root and leaf traits, and soil microbial activity. *J. Plant Nutr.* 31:473–483.

Zou, C., X. Gao, R. Shi, X. Fan, and F. Zhang. 2008. Micronutrient deficiencies in crop production in China. In: *Micronutrient Deficiencies in Global Crop Production*, ed., B. J. Alloway, pp. 127–148. New York: Springer Verlag.

8 Ecophysiology of Major Root Crops

8.1 INTRODUCTION

Ecophysiology is defined as the influence of environmental factors on physiological processes of plants. Ecophysiology is one of the most important topics in understanding the influence of environmental factors on the growth and development of crop plants. Environmental factors that affect the growth and development of plants are climate (precipitation, temperature, and solar radiation) and soil (physical, chemical, and biological). Part of the roots of plants such as sugar beet (*Beta vulgaris* L.), cassava (*Manihot esculenta* Crantz), sweet potato (*Ipomoea batatas* L. Lam.), and carrot (*Daucus carota* L.) are specifically adapted to store products that are photosynthesized in the shoot. According to Gregory (2006), the products are synthesized aboveground and transported to the root in the phloem where they reside until needed to complete the life cycle. In biennials such as carrot and sugar beet, the storage organs are frequently harvested for human use before the life cycle is complete, but if allowed to mature, the stored materials are retranslocated to the shoot where they are used to produce flowers, fruits, and seeds. The development of storage roots is similar to that of nonstorage roots, except that parenchyma cells predominate in the secondary xylem and phloem of the storage roots (Gregory, 2006).

The importance of root crops as a food is enormous due to their higher starch content. Starch content in the roots of cassava and sweet potato ranges from 65% to 90% of the total dry matter—a result of a long period of starch deposition. The pattern of starch accumulation is specific to the species and is related to the particular pattern of differentiation of the organ (Preiss and Sivak, 1996). Root crops can be a good calorie supplement along with cereals. A large population of the tropics depends on root crops for calorie supplement. For example, roasted cassava flour is eaten with rice and bean by people of the northeastern region of Brazil every day. The use of root crop as a food is not restricted only to the tropics; China and Japan make extensive use of the sweet potato even though these countries lie mostly within temperate zones. Similarly, sugar beetroots are eaten by Europeans as a cooked salad or vegetable. Apparently, the world has a distorted view of root crops in relation to seed crops. There are a large number of root crops and it is not possible to discuss all the root crops in one chapter. Hence, the ecophysiology of major root crops, that is, sugar beet, cassava, sweet potato, and carrot will be discussed in this chapter because they are largely used as food crops in developed as well as developing countries.

8.2   SUGAR BEET

The sugar beet is a member of the Chenopodiaceae family. It has 11–13 species in Europe and Asia (Letschert, 1993; Mabberley, 1997). Some of the wild forms have been included in *B. vulgaris* or recognized as distinct species (Austin, 1991). Agronomically, the sugar beet is one of the four cultural types of *B. vulgaris* L.: sugar beet, red beet or garden beet, Swiss chard, and fodder beet (Ulrich et al., 1993). At a young stage of growth, the tops of all cultural types, especially Swiss chard, may serve as a leafy vegetable and in later stages of growth as livestock forage. The fleshy root and tops of the red beet are excellent vegetables (Ulrich et al., 1993). The sugar beet is a biennial plant that is agriculturally important because of its ability to store sucrose to high concentrations in its storage root. Ulrich et al. (1993) reported that the commercial production of beet sugar has been an outstanding achievement scientifically and economically as an alternative source of sugar when other supplies are insecure. Crystalline sugar was a scarce luxury in the Western world prior to the seventeenth century (Campbell, 1984). Originally, all sugar came from sugarcane grown in the tropics but at the present time, beet sugar accounts for nearly half the total world production of the refined product (Campbell, 1984). Sugar is extracted from the beet in factories, using a process similar to that for sugarcane.

The yield of sugar in the storage roots depends on the way photosynthate is partitioned within in the crop and is the product of the total amount of dry matter produced during growth, the proportion allocated to the storage root, and the proportion of the storage root dry matter accumulated as sucrose (Bell et al., 1996). In addition, the efficiency of the sugar extraction process is dependent on the concentration of solutes other than sucrose (K, Na, amino acids, and glycinebetaine), and the interrelationships among the accumulation of sucrose and these so-called impurities are important determinants of root quality. The propagation of sugar beet is always from seed. Regarding origin and domestication, there is no archeological record that exists for preclassical times; linguistic records, however, place leafy forms of the cultivated beet to the eighteenth century BC in Babylonia (Siemonsma and Piluek, 1993; Zohary and Hopf, 1993). Sugar beet is also used for ethanol production in Europe. The largest ethanol-producing factory from sugar beet is located in France. White varieties of sugar beet are used for ethanol production rather than red varieties. White varieties' roots are larger compared to red sugar beet varieties.

Europe is the largest producer of sugar beet followed by North and Central America (Simpson and Conner-Ogorzaly, 1995). Sugar beetroots contain 7%–10% carbohydrates (sucrose), 1.5%–2% protein, and small quantities of fat, ash, and fiber. Roots contain a lower mineral and vitamin content than most other vegetables. The red color is produced by betanins (red betacyanins). Geosmin causes the earthy smell (Austin et al., 1991). To produce good yields of sugar beet, growers need to plant early to a uniform stand, meet the water and fertilizer requirements of the crop, prevent weed competition, and control pests.

8.2.1   Climate and Soil Requirements

For the successful production of a crop, knowledge of its soil and climate requirements is fundamental. Each crop has a minimum and maximum temperature requirement from germination to maturity. The temperature values should be in the favorable range for maximizing yields. Sugar beet thrives best in cold climates but is also tolerant to high temperature. The optimum temperature reported for sugar beet is 24°C (Radke and Bauer, 1969). Adequacy of moisture is vital for achieving maximum economic yields. Sugar beet is mostly grown as an irrigated crop. The frequency of irrigation depends on temperature and soil type. Hot weather and sandy soils require more water compared to cold weather and clay soils. The irrigation frequency is 4–5 days during hot weather and 8–10 days during cold weather. Sugar beet does not tolerate waterlogging.

Sugar beet grows best on deep well-drained loam soil rich in mineral nutrients, including N, which is the key element for vigorous top and root growth. Soil pH and adequate soil fertility are important characteristics of soils in determining sugar beet yield. It is susceptible to soil acidity and the critical soil pH range is 6.0–6.5 (Fageria et al., 2011). Sugar beet grown on acid mineral soils with a pH of less than 5.0 is likely to show symptoms of Al toxicity. Sugar beet is tolerant to soil salinity. The salinity threshold (the maximum soil salinity that does not reduce yield below that obtained under nonsaline conditions) for sugar beet is reported to be 7 dS m^{-1} (Maas, 1993). For soil salinity exceeding the threshold of any given crop, the relative yield (Y_r) can be estimated with the following equation (Maas, 1993):

$$Y_r = 100 - b(EC_e - a)$$

where

a is the salinity threshold expressed in dS m^{-1}

b is the yield reduction, or slope, expressed in % per dS m^{-1}

EC_e is the mean electrical conductivity of saturated soil extracts taken from the root zone

The slope (% per dS m^{-1}) reported for sugar beet is 5.9 (Maas, 1993).

8.2.2 DISEASES AND INSECTS

Sugar beet is not as susceptible to diseases as some other crops. However, leaves of beets can be infested with downy mildew (*Peronospora parasitica*) and *Cercospora beticola*. *Phoma betae* and other fungi cause damping off. Beet mosaic virus also causes problems (Dusi and Peters, 1999). The larvae of beet web worms feed on leaves, and aphids and beet leaf miners also cause damage. Root knot nematodes (*Meloidogyne* spp.) affect the root system (Austin et al., 1991). The control of diseases and insects is recommended with appropriate fungicides, insecticides, and cultural practices to improve yields and the quality of sugar beets. The supply of essential nutrients in adequate rate and balance reduces disease and insect infestation. The effect of P on disease resistance seems to be most closely related to its effect on plant growth, particularly during the seedling stages. Phosphorus applications have a positive effect on seedling vigor, which increases the number of plants that survive the attack of various seedling diseases (Lorenz and Vittum, 1980). Draycott (1972) reported that sugar beet seedlings on severely P-deficient soils die from the attack of black leg (*Pythium* spp.) and other fungal infestations.

8.2.3 Nutrient Requirements

Sugar beet requires a large amount of nutrients for profitable yields. However, the nutrient requirements depend on soil types and yield levels. Nitrogen is the key element in the nutrition of the sugar beet plant when all other nutrients are present in ample supply and no other factor is limiting growth from the time of planting to harvest. Nitrogen deficiency is indicated under field conditions when petioles from recently matured leaves contain less than 500 mg of nitrate N (Ulrich et al., 1993). Sugar beet needs 90 kg N ha^{-1} for producing good yields. Usually, the recommended rates range from 60 to 80 kg P ha^{-1} for most vegetables grown on low fertility soils (Lorenz and Vittum, 1980). A sugar beet yield of about 57 Mg fresh root removed 32 kg P ha^{-1} in total plant roots and tops (Lorenz and Vittum, 1980). Draycott (1972) summarized P uptake by sugar beet as determined by various researchers in the United Kingdom and reported that total P uptake varied from 10 to 33 kg ha^{-1} with an overall average of 23 kg ha^{-1}.

Soil testing and plant tissue analysis are generally used to identify nutrition deficiency/sufficiency for achieving maximum economic yields. There is considerable variation in the deficiency levels of available soil P as recommended by various researchers. Lorenz and Bartz (1968), after reviewing much published data, concluded that levels above about 20 mg P kg^{-1} of bicarbonate-extractable P were adequate for vegetables grown on calcareous soils. On acid soils, when the Bray extractant of 0.025 N HCl is used, levels of less than 8.5 mg P ka^{-1} are considered low, 8.5–17 mg P kg^{-1} medium, and above 17 mg P kg^{-1} high. Reisenauer et al. (1976) set the critical range for cool-season vegetables at 12–20 mg P kg^{-1} by bicarbonate-extractable P and for warm-season vegetables at 5–9 mg P kg^{-1} of soil. Adequate levels of macro- and micronutrients in the sugar beet plant tissues are presented in Table 8.1. These values can be used as a guideline to identify nutrient sufficiency or deficiency in sugar beet plants.

8.2.4 Growth and Development

Botanically, the sugar beet is classified as a biennial plant that produces leaves in the first year until the plant becomes dormant during winter. The dry matter accumulation pattern in the sugar beet is slow in the beginning due to low leaf area and low interception of solar radiation; however, it increases in a linear fashion as the leaf area increases. Dry matter production becomes directly proportional to radiation receipts once the leaf cover of the ground is complete. A major part of the dry matter accumulated by sugar beet during growth period is in the root and majority of that is as sucrose. As the taproot enlarges, sucrose is stored within the parenchyma cells of the storage root at an equilibrium concentration characteristic of the cultivar for sugar production. Under high N conditions, most commercial cultivars have a sucrose concentration from 6% to 10%, which may reach 18%–20% for a sugar type. These equilibrium values increase moderately during the cool, sunny weather of autumn, and when combined with N depletion for 4–6 weeks prior to harvest, the sugar concentration often increases dramatically. Under these conditions, beets of medium size with N deficiency in a uniform stand will ripen rapidly, often to 18% sugar, to produce an easily processed quality crop (Ulrich et al., 1993).

TABLE 8.1
Sufficiency Level of Nutrients in Sugar Beet
Plant Tissues

Nutrient	Growth Stage	Plant Part	Sufficiency Level (g kg⁻¹ or mg kg⁻¹)
N	Prior to flowering	YMB	25
P	Prior to flowering	YMB	1.8
K	Prior to flowering	YMB	10
S	Prior to flowering	YMB	0.75
Ca	Prior to flowering	YMB	5
Mg	Prior to flowering	YMB	10
Cu	50–60 days	ML	7–15
Zn	Prior to flowering	YMB	9
Mn	Prior to flowering	YMB	10
Fe	Prior to flowering	YMB	55
B	Prior to flowering	YMB	27
Mo	U	YMB	0.2–20

Sources: Compiled from Reuter, D.J., Temperate and sub-tropical crops, in: *Plant Analysis: An Interpretation Manual*, eds., D.J. Reuter, and J.B. Robinson, pp. 38–99, Inkata Press, Melbourne, Victoria, Australia, 1986; Huett, D.O. et al., Vegetables, in: *Plant Analysis: An Interpretation Manual*, 2nd edn., eds., D.J. Reuter, and J.B. Robinson, pp. 385–502, CSIRO Publishing, Collingwood, Victoria, Australia, 1997.
Macronutrients concentration in g kg⁻¹ and micronutrients in mg kg⁻¹.
U, unknown; YMB, youngest (uppermost) mature leaf blade; ML, mature leaf.

The aerial part of the sugar beet has about 20% of the total dry matter accumulated by the crop during growth cycle (Bell et al., 1996). The distribution of dry matter in the root and aerial part is significantly influenced by environmental factors and cultural practices adopted. The main factors that affect dry matter partitioning in roots and shoots are genotypes, nitrogen fertilization, plant density, type of soil, and climatic conditions during crop growth. Solar radiation is also one of the important factors in determining dry matter production and distribution in roots and shoots (Scoot and Jaggard, 1993). Depending on these factors, the proportion of total dry matter allocated to the storage root ranges from 47% to 77%, and the proportion in the storage root present as sucrose from 72% to 79% (Bell et al., 1996).

8.3 CASSAVA

Cassava (*M. esculenta* Crantz) is a perennial shrub belonging to the family Euphorbiaceae, subfamily Crotonoideae, and class/subclass Angiospermae (Dicotyledones). Three subspecies are recognized. *M. esculenta* subspecies *esculenta* is domesticated and includes all cultivars known in cultivation. The wild *M. esculenta* subspecies *peruviana* occurs in eastern Peru and western Brazil. The wild *M. esculenta* subspecies *flabellifolia* shows a wider distribution and ranges from the central Brazilian state of Goiás northward to Venezuelan Amazonia. The large area of distribution of the two wild subspecies makes it difficult to assign a place of initial domestication (Allem, 1994). Plants are monoecious and pollination is by insects. Seeds, however, are not used in cultivation; propagation is by cuttings. Although, cassava is a perennial, it is usually harvested during the first or the second year.

Cassava is commonly known as tapioca, manioc, mandioca, yuca, and sagu, and in Africa more than a half dozen vernacular names prevail. The exact origin of cassava is not known, but apparently it was first domesticated somewhere in South America. It was taken to Africa by the Portuguese as early as 1558 and spread to Asia in the seventeenth century (Cock, 1984). The maximum production of cassava is in Africa, followed by Asia and South America. Cassava is Africa's second-most important food staple, after corn, in terms of calories consumed (Okechukwu and Dixon, 2009). Virtually all the cassava produced in Africa is for human consumption, feeding more than 50% of the 788 million people in sub-Saharan Africa (Okechukwu and Dixon, 2009).

Cassava is Africa's food insurance because it gives stable yields even in the face of more frequent droughts, low soil fertility, and low intensity management. In addition, it can remain in the soil until needed, spreading out the food supply over time, helping families through annual scarcities when seasonal harvests run out, and averting the tragic boom-and-bust cycle of oversupply followed by shortage (Dixon et al., 2003). In some African countries, particularly the Democratic Republic of Congo, Congo Republic, Gabon, Central African Republic, Angola, Sierra Leone, and Liberia, the leaves are eaten as green vegetable. Cassava leaves provide a cheap and rich source of protein, minerals, and vitamins (Hahn et al., 1989). Its production in Nigeria has grown at an annual rate of 4.6% from 1970 to 2006 and the crop is now cultivated commercially in large hectares in different parts of the country (Okechukwu and Dixon, 2009).

Although it is one of the world's most important food staples, cassava is known in North America and Europe almost solely as tapioca, an occasional dessert (Janick et al., 1974). The major cassava-producing countries are Brazil, Indonesia, Thailand, Zaire, Nigeria, India, and China. The average world yield is about 8.7 Mg of fresh root ha^{-1}, which is far below the potential yield of 80 Mg ha^{-1} produced under experimental conditions (Howeler, 1985). This large gap between potential and actual productivity is due to the fact that the crop is largely produced by subsistence farmers with low technology. Cassava's high-yield potential has been attributed to a high total crop dry weight in relation to foliage development, leaf area duration, canopy architecture, and a high ratio of dry weight of storage roots to total dry weight, the so-called harvest index (HI; Jose and Mayobre, 1982). Cassava roots contain about 35% starch, 102% protein, and 102% fiber. The important minerals are mostly phosphorus and iron. Vitamin C is usually 0.35% of fresh weight, and there are traces of niacin and vitamins A, B$_1$, and B$_2$ (Austin, 2002).

8.3.1 CLIMATE AND SOIL REQUIREMENTS

Although cassava is, by origin, a tropical crop, it is successfully grown in latitudes up to 25°S in South America, southern Africa, and in some trial commercial plantings in high latitudes of Australia (Harris, 1978). Cook and Rosas (1975) and Cock (1983) have suggested that cassava's ecological zone lies between latitudes 30°N and 30°S at elevations up to 2300 m above mean sea level. The cassava plant is sensitive to frost and cannot be grown successfully where the temperature is below 15°C (Cock, 1984). In areas with marked seasonal temperature changes, cassava is grown only when the annual mean temperature is greater than 20°C (Cock, 1983). Optimum temperature is reported to be in the range of 25°C–30°C (Austin, 2002). Cassava is grown in areas with as little as 750 mm rainfall per year since the crop has an extremely conservative pattern of water use. Reduced leaf area and stomatal closure markedly reduce crop growth rates during periods of stress (Connor et al., 1981).

Cassava can grow on infertile soils varying in texture from light to heavy, and in pH from 3.5 to 7.8, but it will not tolerate excess water or high salinity. The highest yields occur on well-drained, medium-to-heavy texture, fertile soils with a pH of about 5.5–7 (Howeler, 1981). Abruna et al. (1982) found that Cassava is highly tolerant to soil acidity. Yields of cassava grown on corozal clay (Aquic Tropudults) were affected only when the soil pH dropped below 4.3 and the exchangeable Al increased above 60% saturation. Cassava is more susceptible than other food crops to soil salinity and alkalinity, but large varietal differences in tolerance exist. Yield of this root crop are markedly reduced when the Na saturation is above 2%–5% and the electrical conductivity is above 0.5–0.7 dS m^{-1} (Cock and Howeler, 1978). It is classified as moderately sensitive to salinity (Maas, 1993).

8.3.2 DISEASES AND INSECTS

Cassava mosaic is the most important viral disease and whitefly (*Bemista tabaci*) is its primary vector. Bacterial blight (*Xanthomonas manihotis*) and stem rot (*Erwinia* spp.) are also problems (Austin, 2002). Fungal diseases include brown leaf spot (*Cercospora henningsii*), leaf spot (*Phyllosticia* spp.), white thread (*Fomes lignosus*), anthracnose (*Colletotrichum gloeosporioides*), and super elongation disease (*Sphaceloma* spp.). Insect pests include variegated grasshopper (*Zonocerus variegatus*), green spider mite (*Monomychellus tanajoa*), red spider mite (*Tetranychus telarius*), web mite (*Aoindomtilus albus*), and mealybug (*Phenacoccus manihoti*) (Austin, 2002). In South America termites also infest cassava crops. Using appropriate crop rotations can control cassava diseases and insects.

8.3.3 NUTRIENT REQUIREMENTS

Cassava is considered to be tolerant to low-fertility conditions and grows well on acid soils where other crops cannot be grown satisfactorily (Howeler, 1981). Although this crop has been traditionally grown without the use of fertilizers, it is now well established that the cassava plant responds well to fertilization, and in order to obtain high yield, adequate nutrients should be supplied (Howeler, 1981; Howeler and Cadavid, 1983). Howeler (1981) summarized the response to fertilization and liming in the three most extensive tropic soils (Oxisols, Ultisols, and Inceptisols). Phosphorus is generally the element most limiting to yield. Cassava extracts large amounts of K from soil and may cause the depletion of this element if grown continuously without adequate K fertilization. Compared to other crops, cassava has a low requirement for N, and high N application may lead to excessive top growth, a reduction in starch synthesis, and poor root thickening (Howeler, 1981). Cassava is acid tolerant, and the optimum pH is between 5.5 and 7.5. The crop often responds to a low rate of liming, but overliming may induce micronutrient deficiencies. The critical levels of various nutrients in the soil reported by Howeler (1981), based upon summarization of results from various studies, are as follows: P (7–9 mg kg^{-1}), K (0.06–0.15 cmol$_c$ kg^{-1}), Ca (0.25 cmol$_c$ kg^{-1}), Zn (1 mg kg^{-1}), Mn (5–9 mg kg^{-1}), and S (8 mg kg^{-1}) of soil. In the following sections, nutrient concentrations and the quantity of nutrients removed by cassava are discussed. This information may form the basis for an understanding of the nutritional requirements of this important root crop.

8.3.3.1   Nutrient Concentration

Nutrient concentrations in plants vary with the plant part analyzed, stage of plant growth, soil fertility, climatic conditions, and management practices. Modern cassava cultivars had their greatest growth rate and thus their greatest nutrient demands (Howeler, 1981; Howeler and Cadavid, 1983). Nutrient sufficiency levels in cassava tissue reported in the literature are summarized in Table 8.2. In general, it may be concluded that a fertilizer response is not likely when the uppermost mature leaf blades contain 50–60 g kg^{-1} N, 3–5 g kg^{-1} P, 12–20 g kg^{-1} K, 6–15 g kg^{-1} Ca, 2.5–5 g kg^{-1} Mg, and 3–4 g kg^{-1} S. Similarly, a fertilizer response is not expected when the Cu concentration in plant tissue is 7–15 mg kg^{-1}, Zn 40–100 mg kg^{-1}, Mn 50–250 mg kg^{-1}, Fe 60–200 mg kg^{-1}, and B 15–20 mg kg^{-1}. Irizarry and Rivera (1983), working in a Puerto Rican Ultisol, reported that 6 months after planting, optimum leaf concentrations were approximately 43, 1.2, 18, 14, and 4 g kg^{-1} for N, P, K, Ca, and Mg, respectively.

8.3.3.2   Nutrient Accumulation

The nutrient accumulation (concentration × dry matter) pattern can provide useful information about the nutrient requirements of a crop. Howeler and Cadavid (1983) studied nutrient accumulation in different plant parts of cassava in Colombia during different growth stages. Most nutrients accumulated initially in leaves and stems, but were translocated to roots in the latter part of the growth cycle. Only Ca, Mg, and Mn accumulated more in stems than in roots. Nutrient removal at the 12 month growth period was in the order of N > K > Ca > P > Mg > S > Fe > Mn > Zn > Cu > B. Irizarry and Rivera (1983) reported that in an Ultisol of Puerto Rico, 10 months after planting, cassava accumulated 204 kg N ha^{-1}, 13 kg P ha^{-1}, 222 kg K ha^{-1}, 86 kg Ca ha^{-1}, and 33 kg Mg ha^{-1}, with a total dry matter production of about 23 Mg ha^{-1} of leaves, stems, and roots. Howeler (1981) calculated nutrient removal by cassava roots based on various studies in the literature and concluded that on the average each ton of cassava roots removed 2.3 kg N, 0.5 kg P, 4.1 kg K, 0.6 kg Ca, and 0.3 kg Mg.

TABLE 8.2
Nutrient Sufficiency Levels in Cassava Plant

Nutrient	Growth Stage	Plant Part	Sufficiency Level (g kg⁻¹ or mg kg⁻¹)
N	Vegetative	YMB	50–60
N	3–4 months	YMB	51–58
N	3–5 months	YMB	56–65
P	28 DAS	WS	4.7–6.6
P	3–4 months	YMB	3.6–5.0
P	3–5 months	YMB	4.2–4.7
K	Vegetative	YMB	12–20
K	3–4 months	YMB	13–20
K	3–5 months	YMB	15–20
S	Vegetative	YMB	3–4
S	3–4 months	YMB	2.6–3.0
Ca	Vegetative	YMB	6–15
Ca	3–4 months	YMB	7.5–8.5
Mg	Vegetative	YMB	2.5–5
Mg	3–4 months	YMB	2.9–3.1
Cu	Vegetative	YMB	7–15
Cu	3–4 months	YMB	6–10
Zn	Vegetative	YMB	40–100
Zn	3–4 months	YMB	30–60
Mn	Vegetative	YMB	50–250
Mn	3–4 months	YMB	50–120
Fe	Vegetative	YMB	60–200
Fe	3–4 months	YMB	120–140
B	Vegetative	YMB	15–50
B	3–4 months	YMB	30–60

Source: Compiled from Huett, D.O. et al., Vegetables, in: *Plant Analysis: An Interpretation Manual*, 2nd edn., eds., D.J. Reuter and J.B. Robinson, pp. 385–502, CSIRO Publishing, Collingwood, Victoria, Australia, 1997.

Values of macronutrients are in g kg⁻¹ and micronutrients in mg kg⁻¹.

YMB, youngest (upper most) mature leaf blade; DAS, days after sowing; WS, whole shoot or tops.

8.3.4 GROWTH AND DEVELOPMENT

The basic constituents of the cassava plant are (1) nodal units that consist of a leaf blade, petiole, and internode, and (2) thickened roots that form mainly at the base of the stem cutting that is used as planting material. The weight of the dry stem internode, which varies with variety, averages from 0.5 to 3.0 g/internode for a mature plant, and leaf blades together with petioles have an area-to-weight ratio of about 135 cm² g⁻¹.

The plant generally shows strong apical dominance and does not generally produce leaves from the axillary buds (Cock et al., 1979).

Propagation of cassava is possible either through true seed or through cuttings. Planting through cuttings is the normal practice for commercial production, while planting through seed is practiced in breeding programs. Cutting length is normally 20 cm and preferably taken from the middle of the stems. The growth duration of the crop depends on environmental conditions. The period from planting to harvest is about 9–12 months in hot regions and 2 years in cooler or drier regions (Cock, 1984).

8.3.4.1   Roots

Roots are the main storage organs in cassava. Secondary thickening results in tuber development as swellings on adventitious roots a short distance from the stem. There are usually 5–10 tubers per plant. The tubers are cylindrical or tapering, 15–100 cm long, 3–15 cm across, and occasionally branched (Purseglove, 1987). Keating et al. (1982) reported that the final number of storage roots was generally reached within 90–135 days after planting and ranged from 10 to 14 storage roots per plant, independent of planting date and season. Campos and Sena (1974) studied the root system of cassava and found that most roots were concentrated in the top 30 cm, with some roots as far down as 140 cm; however, Connor et al. (1981) found roots at a depth of 250 cm. They also observed root characteristics and found that cassava has rather thicker roots than most species (0.37–0.67 mm diameter), but has very short roots (less than 1 km m^{-2} land surface).

8.3.4.2   Tops

Cassava plants vary in height and branching habit. Each nodal unit consists of a node that subtends a leaf on an internode. The total number of nodes per plant depends on the number of nodes per shoot and the number of shoots, or apices, per plant (Cock, 1984). Cassava leaves are spirally arranged (phyllotaxis 2/5) and are variable in size, color of stipules, petioles, midribs and laminae, in the number of lobes, in the depth of lobes, and in the shape and width of lobes (Purseglove, 1987). Sangoi and Kruse (1993) measured plant height during the crop growth cycle on the highlands of Santa Catarina, Brazil. During the first 60 days of plant growth, increase in the plant height was slow. Maximum plant height increased during 60–180 days after planting. Maximum plant height was reached at about 210 days after sowing, after which it was more or less constant.

8.3.4.3   Leaf Area Index

The leaf area index (LAI) is an important growth parameter that determines the photosynthetic capacity of a crop. It is affected by climatic factors, soil fertility levels, and cultural practices. The LAI normally increases as the crop grows, reaches a maximum value, and declines in the latter part of the crop growth due to leaf abscission. The LAI of cassava increases in the first 4–6 months after planting and then declines (Williams, 1972; CIAT, 1979). The maximum LAI values reported in the literature are between 6 and 8 (Enyi, 1972). According to Keating et al. (1982) substantial leaf abscission begins at LAI values of the order of 5–6. Crop growth is related to the LAI. Cassava is a C$_3$ plant that reaches a maximum growth rate of 120–150 g m^{-2} week^{-1} at LAI of about 4 under solar radiation of about

450 cal cm^{-2} day^{-1} (Cock, 1983). Cock (1983) studied relationship between LAI and crop growth rate. There is a linear relationship between the LAI and the growth rate of leaves and stems. The relationship between the LAI and the growth rate of roots was quadratic, with a maximum growth rate achieved at LAI of about 3, followed by a decline with further increase in the LAI. Total plant growth rate increased up to LAI of 4 and then remained constant. The maximum crop growth rates of cassava within the tropics are of the order of 120 g m^{-2} week^{-1} (Cock, 1984).

8.3.4.4 Dry Matter

A study conducted by Irizarry and Rivera (1983) in Ultisol of Puerto Rico showed that dry matter in the stems, roots, and whole plant increased steadily with plant age. Dry matter in the leaves declined sharply from 5 to 8 months after planting and thereafter remained stable. Total plant dry matter 10 months after planting was 23 Mg ha^{-1}, and yield of edible fresh roots was about 11 Mg ha^{-1}. Howeler and Cadavid (1983) reported in an experiment carried out in Colombia that cassava dry matter accumulation was slow during the first 2 months, increased rapidly during the next 4 months, and slowed down during the final 6 months as dry matter production was partly offset by leaf fall. At harvest (12 months) dry matter was present mainly in roots followed by stems, leaves, and petioles. Sangoi and Kruse (1993) determined the fresh weight of roots in a Brazilian Cambisol during the crop growth cycle. At the age of about 210 days, a maximum root weight of about 15 Mg ha^{-1} was achieved.

8.3.4.5 Harvest Index

HI is the measure of the distribution of dry matter to the economically useful plant parts. In cereals this is measured by taking into consideration the grain yield and total aboveground plant parts. In cassava, the roots are the major economic part, and root to total dry matter production is used as an index to quantify economically useful plant parts. Cock (1976) obtained HI values after 1 year ranging from less than 0.30–0.57 for a number of clones when a population of 20,000 plants ha^{-1} was used. In one trial with M col.22, HI values of more than 0.7 were obtained, but HI decreased as population increased above 12,000 plants ha^{-1} (Cock et al., 1977). HI can be used as one of the selection criteria for higher yield potential in cassava cultivars (Kawano, 1978). According to Iglesias et al. (1994), an HI value of 0.5–0.6 is the optimum level because at higher values of HI, root production potential is affected as a result of reduced photosynthetic area.

8.4 SWEET POTATO

Sweet potato (*I. batatas* L. Lam) is an important root crop for human consumption. The sweet potato is basically a starchy vegetable, subsidiary or complementary to the *Solanum* potato (Yen, 1984). It is known as *batata doce* in Portuguese. The economically important part of the plant is the tuberous root. It is eaten by the population of developed and developing countries as cooked and fried vegetable. Sweet potato is a dicotyledonous plant native to Central America (Hahn and Hozyo, 1984). The origin of the sweet potato is reported to be in northwestern South America (Austin, 2002). It belongs to the family Convolvulaceae. Columbus took the sweet potato back to

Europe along with the Caribbean name *age* or *aje*. The species was spread into the old world tropics in the early 1500s; it reached China by or before 1594 and Japan in 1674 (Austin, 2002). The sweet potato is a perennial crop; however, in agricultural systems it is treated as an annual crop with a growing period that is normally 3–7 months, depending on the environment and cultivar (Hahn and Hozyo, 1984). The pollination of sweet potato is entomophilous. Propagation of this crop is either by roots or by stems, although stems are most commonly used. Scientific interest in the sweet potato began late in the nineteenth century in the United States, with the cataloguing of regional collections (Yen, 1984).

Sweet potato contains 44%–78% starch, 8%–27% sugar, and 1%–11% protein (Austin, 2002). In addition, riboflavin, fair amounts of carotene, thiamine, calcium, iron, and vitamins A and C are also present. Carotene and vitamin A are antioxidants that are claimed to be associated with cancer prevention. Some studies have suggested that sweet potato roots also produce antimutagenic compounds (Yoshimoto et al., 1998). Asia is the largest sweet potato–producing continent. Among Asian countries, China and Japan produce a large amount of sweet potatoes. The African continent is second in the production of sweet potato (Simpson and Conner-Ogorzaly, 1995). An average yield of about 20 Mg fresh weight per hectare (equivalent to 6 Mg dry matter per hectare) in 4–5 months has been reported in a number of countries (Hahn and Hozyo, 1984). In Japan, the average yield of 21 Mg ha^{-1} fresh tubers is equivalent in food energy to 2.94×10^7 kcal ha^{-1}. This is 1.9 times the energy from a crop of rice with an average grain yield of 4500 kg ha^{-1} (Murata et al., 1976).

8.4.1 Climatic and Soil Requirements

The sweet potato is a warm-season crop. However, it is grown over a wide range of environments between the latitudes 40°N and 40°S of the equator and between sea level and 2300 m altitude (Hahn, 1977). The optimum temperature for the growth of sweet potato is around 30°C. The optimum temperature for photosynthesis is between 23°C and 33°C, which is relatively warm for a C$_3$ plant (Hahn and Hozyo, 1984). Its growth is significantly reduced in temperature below 10°C. Although the sweet potato is grown in areas with a relatively high rainfall, it cannot tolerate waterlogging and is usually grown on ridges or mounds. It has good drought tolerance (Hahn and Hozyo, 1984). It grows well on poor fertility soils but also responds well to the addition of fertilizers (Austin, 2002). Excess nitrogen may produce more vegetative growth and less root yield. The salinity threshold for sweet potato is reported to be 1.5 dS m^{-1} and slope 11 (% per dS m^{-1}). It is classified as moderately sensitive to salinity (Maas, 1993).

8.4.2 Diseases and Insects

Perhaps the most damaging pest for the sweet potato is the weevil (*Cyclas formicarius*), which has now spread around the world (Austin et al., 1991). Other insect pests are the beetle (*Euscepes* spp.), a vine borer (*Omphisa anastomosalis*), and the hawk moth (*Herse convolvuli*). In addition, there are several fungal diseases that affect the sweet potato adversely, including black rot (*Ceratocystis fimbriata*), scurf (*Monilochaetes cans*), fusarium wilt (*Fusarium oxysporum f. batatais*), and soft rot

(*Rhizopus stolonifer*) (Austin, 2002). Mosaic virus is a problem in sweet potatoes in several parts of the world, as is feathery mottle complex transmitted by whiteflies (*Bemisia and Traileurodes*). Internal cork, leaf spot, and russet crack are other viral problems. Nematodes are uncommon problems in sweet potato, but exceptions are sting nematodes (*Belonoliamus gracilis*), root lesion nematodes (*Prathlenchus* spp.), and root knot nematodes (*Meloidogyne* spp.). Detailed discussion about pest problems in sweet potatoes is given by Jansson and Raman (1991).

8.4.3 NUTRIENT REQUIREMENTS

Sweet potato plants produce well in poor soils; however, they respond well to fertilization. Fertilizer rate depends on type of soil, initial soil fertility, and cropping system adopted. Osaki et al. (1995a) studied the response of sweet potato to N, P, and K fertilization under field conditions. These authors concluded that the treatments were (1) the application of 50 kg N ha^{-1}, 100 kg P$_2$O$_5$ ha^{-1} (44 kg P ha^{-1}), and 100 kg K$_2$O ha^{-1}; (2) −N plots; (3) −P plots; and (4) −K plots. These authors concluded that under −N treatment, both biological and economic yields decreased compared to plots that received N treatments. Maximum biological yield (15.8 Mg ha^{-1}) and root yield (5 Mg ha^{-1}) were obtained in the plots that received adequate rates of N, P, and K fertilization. Under P deficiency, total dry weight was the same as that in the plots that received adequate rate of P, while tuber root dry weight was strictly restricted, suggesting that the construction of proteins or cellulose in leaves was stimulated, and the construction of starch in tuber roots was restricted in the −P plots. Under K deficiency, since N absorption was not restricted compared to the control treatment (adequate nutrient rates), and P absorption was restricted, the nutritional status of the plants showed an N excess, suggesting that N excess and disorder of phloem transport adversely affected sweet potato growth. Hartemink (2006) reported that 34.5 Mg ha^{-1} yield of sweet potato removed 175 kg N ha^{-1}, 34 kg P ha^{-1}, and 290 kg K ha^{-1}.

Adequate nutrient concentrations for sweet potatoes are given in Table 8.3. The values of these nutrient concentrations in specific plant parts are determined experimentally and used as guides to indicate how well plants are supplied with these nutrients at a certain time of sampling. These concentration values can be used as a tool to assist the agronomists and/or extension workers in evaluating nutrient disorders and in improving fertilizer practices. A basic concept is that the concentration of a nutrient within the plant at any particular moment is an integrated value of all the factors that have influenced the nutrient concentration up to the time of sampling. Plant analysis values (sufficient/deficient) are more stable across environmental conditions compared to soil analysis (Fageria et al., 2011). Hence, plant analysis of different regions or even different countries can be used as reference values for a particular crop, provided plant sampling or growth stage is well defined.

8.4.4 GROWTH AND DEVELOPMENT

The sweet potato is a C$_3$ plant and commercial cultivars of sweet potato are vegetatively propagated. The maximum height of sweet potato is reported to be about 50 cm by Osaki et al. (1995a). Similarly, the growth rate of sweet potato was reported to be

TABLE 8.3
Nutrient Sufficiency Level in Sweet Potato

Nutrient	Growth Stage	Plant Part	Sufficiency Level (g kg⁻¹ or mg kg⁻¹)
N	28 DAT	YMB	43–45
N	44 DAT	YMB	35.2–41.4
N	58 DAT	YMB	26.5–32.4
N	Harvest	WS	23–33
P	28 DAT	YMB	2.6–4.5
P	Mid-growth	YML	2–3
P	Harvest	Tuber	1.2–2.2
K	28 DAT	YMB	47–60
K	Mid-growth	YML	29–50
K	Harvest	WS	46–54
S	28 DAT	YMB	3.5–4.5
Ca	28 DAT	YMB	9–12
Ca	Mid-growth	YML	7.3–9.5
Ca	Harvest	WS	7.3–7.4
Mg	28 DAT	YMB	1.5–3.5
Mg	Mid-growth	YML	4–8
Mg	Harvest	Tubers	0.6
Zn	28 DAT	YMB	12–40
Zn	Harvest	WS	20–45
Mn	28 DAT	YMB	26–500
Mn	Mid-growth	YML	40–100
Mn	Harvest	WS	40–200
Fe	28 DAT	YML	45–80
B	28 DAT	YML	45–75
B	Harvest	WS	118

Source: Compiled from Huett, D.O. et al., Vegetables, in: *Plant Analysis: An Interpretation Manual*, 2nd edn., eds., D.J. Reuter and J.B. Robinson, pp. 385–502, CSIRO Publishing, Collingwood, Victoria, Australia, 1997.

Macronutrients in g kg⁻¹ and micronutrients in mg kg⁻¹.

DAT, days after transplanting; YMB, youngest (upper most) mature leaf blade; YML, youngest mature leaf; WS, whole shoot or tops.

$25.6\,g\,m^{-2}\,day^{-1}$ by these authors. Growth rate is related to the LAI of a crop species. Osaki et al. (1995a) reported that LAI of 7 for sweet potato reached about 100 days after transplanting. High productivity of root crops is attributed to the active root–shoot interaction because photosynthates of root crops translocate predominantly to underground organs (tubers, tuberous roots, and roots) (Osaki et al., 1995a).

Osaki et al. (1995b) studied the dry matter accumulation in shoot and root of sweet potato during the growth cycle of the crop. The dry matter accumulation in the tops as well as in the roots followed a sigmoid-type relationship with increasing plant age. Maximum biological yield of 15.8 Mg ha^{-1} was obtained when plants attended an age of 125 days after transplanting. Similarly, root weight of about 5.5 Mg ha^{-1} was achieved at 100 days after transplanting and decreased to about 5 Mg ha^{-1} at harvest (about 125 days after transplanting). This growth pattern was obtained in plots that received adequate amounts of N, P, and K fertilization. In plots that did not receive N, P, and K fertilization, total dry weight was reduced, except in the P-treated plot. Hence, it is suggested that the sweet potato is efficient in P uptake and utilization compared to N and K fertilization.

The optimum LAI of the sweet potato is small compared with cereals. In Japan where the average solar radiation was 430 g cal cm^{-2} day^{-1} the optimum LAI was reported to be 3.2, and this gave a maximum dry matter production of 120 g m^{-2} week^{-1} (Tsuno and Fujise, 1963). However, Kodama et al. (1970) and Enyi (1977) reported that the maximum crop growth rates of 150 and 163 g m^{-2} week^{-1} were attained when the LAI was 5.5 and 8.0, respectively. Osaki et al. (1995a) reported that photosynthetic rate increases more or less proportionally to the increase in the LAI of plants until a certain LAI value is reached and the photosynthetic rate may decrease by mutual shading of leaves when the LAI exceeds this value. The respiration rate increases almost proportionally to the increase in the LAI because it is not affected by mutual shading. For these reasons the dry matter production reaches a maximum value at the optimum LAI (Osaki et al., 1995a). The rate of respiration per unit of dry weight is much larger in the leaves and stems than in the tubers (Tsuno and Fujise, 1963). Leaves account for about 50% of the total respiration of the plant during its complete growth period, compared with only 20%–30% that is accounted for by the tubers (Tsuno and Fujise, 1964).

8.5 CARROT

The carrot (*D. carota* L.) is a biennial crop grown for its edible root. It is a genus of 22 species that belongs to the family Umbellifereae. Ten species occur in Europe and the remaining are spread through the Mediterranean, southwest and central Asia, tropical Africa, Australia, New Zealand, and the Americas (Heywood, 1983; Austin, 2002). The species has been divided into 13 wild and cultivated subspecies (Heywood, 1983). It originated in Europe and Asia and is consumed as a raw or cooked vegetable. It is also used as a feed for animals in developing countries. Carrots are well known for their high carotene content, which has been recently claimed to correlate with cancer prevention (Austin, 2002). The roots also contain 6%–9% carbohydrates (sugar), 1% protein, ash, and 5–10 mg vitamin C, 40 mg Ca, and 1 mg Fe per 100 g fresh weight (Austin, 2002). Terpenoids and other volatile compounds are part of the taste of raw carrots (Siemonsma and Piluek, 1993).

The propagation of carrot is from seeds. Hence, one of the most essential requirements of successful carrot growing is good seed, which should have high vitality and good breeding. It should be free from disease and insect injury, suitable for local conditions, and free from foreign matter such as weeds and dirt. The seeds, when sown, should germinate rapidly giving healthy, vigorous seedlings to ensure a good stand of thrifty quick-growing plants.

8.5.1 CLIMATE AND SOIL REQUIREMENTS

The carrot is a vegetable crop of temperate climate but is now grown in both temperate and tropical regions. A loam or sandy loam soil is preferred for carrot cultivation. However, it can be cultivated on clay loam soil if there is no compact soil layer within about 90 cm soil depth. It does not tolerate waterlogging. It requires a mellow bed for successful cultivation or higher yields. The benefits of a mellow seedbed for good growth of taproots are widely reported in the literature. Bowen (1981) reported that a mellow seedbed with its many well-aerated pores allows roots to grow unhindered in any and every direction and to place their absorbing surfaces in vital touch with the soil grains and soil moisture. In this way, nourishment in the seed provides the maximum root surfaces in the shortest time (Unger and Kaspar, 1994). Carrot is susceptible to soil salinity and the threshold value for this root crop is reported to be 1.0 dS m^{-1} and the slope (% dS m^{-1}) reported is 14 (Maas, 1993).

8.5.2 NUTRIENT REQUIREMENTS

The addition of fertilizer is determined by a soil fertility test and the yield level. However, as a vegetable crop, carrot requires heavy fertilization. A combination of organic and chemical fertilizers is the best strategy to supply essential nutrients to carrot crops. The use of about 20 Mg ha^{-1} well-decomposed farmyard manure or compost is very helpful to keep the soil in proper physical condition and achieve good yields. In addition, the use of chemical fertilizers containing N, P, and K in an appropriate balance will help in improving the yield and quality of the crop. Sometimes topdressing of N during crop growth cycle also helps in improving yields and the quality of carrot. Adequate concentrations of essential nutrients for maximum yields of carrot are given in Table 8.4.

8.5.3 DISEASES AND INSECTS

There are several diseases and insects that infest carrot. Among these are leaf blight (*Alternaria dauci* and *Cercospora carotae*) and root knot nematodes (*Meloidogyne hapla*). Plants are also attacked by powdery mildews, white rust, and bacterial blights. Root rot occurs from a variety of organisms, including *Botrytis cinerea*, *Fusarium* spp., *Sclerotinia sclerotiorum*, *Pythium violae*, and *Erwinia carotovora*. The most common insects are root fly (*Psila rosae*), lygus bug (*Lygus hesperus* and *Lagus elisus*), leafhoppers (*Macrosteles fascifrons*), carrot weevil (*Listronotus oregonensis*), and ary worm (*Spodoptera* spp.). Aphids are vectors of at least 14 viral diseases in carrots (Austin, 2002).

8.5.4 GROWTH AND DEVELOPMENT

The germination of carrot is epigeal, and, hence, it initially establishes a shoot with cotyledonary leaves. These are borne on a hypocotyl linking them to a taproot. The growth of shoot and root proceeds concurrently in carrot. This is in

TABLE 8.4
Nutrient Sufficiency Level in Carrot

Nutrient	Growth Stage	Plant Part	Sufficiency Level (g kg⁻¹ or mg kg⁻¹)
N	Mid-growth	YML	20–35
N	Peak harvest	Root	8.5–9.5
P	Mid-growth	WS	3–5
P	Peak harvest	YML	3–4
K	Mid-growth	WS	27–40
K	Peak harvest	Root	14–19
S	Peak harvest	WS	3.2–6.3
Ca	Mid-growth	YML	14–30
Ca	Peak harvest	Root	3.0–3.5
Mg	Mid-growth	YML	3.0–5.5
Mg	Peak harvest	Root	1.2–1.5
Cu	28 DAS	Whole plant	4
Cu	Mid-growth	YML	10–25
Zn	28 DAS	Whole plant	<30
Zn	Mid-growth	YML	20–50
Mn	28 DAS	Whole plant	<50
Mn	Mid-growth	WS	50–100
Fe	28 DAS	Whole plant	466–818
Fe	Mid-growth	YML	190–350
B	28 DAS	Whole plant	<30
B	Mid-growth	WS	30–80
Mo	28 DAS	Whole plant	<3.6
Mo	Mid-growth	WS	0.5–1.5
Cl[a]	Roots, 1–3 cm diam	WS	20–30
Cl[a]	Peak harvest	YML	30–36
Cl[a]	Peak harvest	Root	10–12

Source: Compiled from Huett, D.O. et al., Vegetables, in: *Plant Analysis: An Interpretation Manual*, 2nd edn., eds., D.J. Reuter, and J.B. Robinson, pp. 385–502, CSIRO Publishing, Collingwood, Victoria, Australia, 1997.

Macronutrients in g kg⁻¹ and micronutrients in mg kg⁻¹.

DAS, days after sowing; YML, youngest mature leaf; WS, whole shoot or tops.

[a] Values of chlorine are in g kg⁻¹.

contrast to some crops, such as potato, where an extensive leaf canopy is established before there is a switch to initiation and filling of the yield organs, with considerable monopoly of the supply by these sinks (Hole, 1996). The estimates of the photosynthetic rate expressed as carbohydrate equivalent ranged from 39.3 mg g⁻¹ shoot dry weight h⁻¹ in 14 day old plants (Hole and Dearman, 1994) to 9.4 mg g⁻¹ shoot dry weight h^{-1} in 37 day old plants (Steingrover, 1981). The proportion of assimilate allocated to the shoot and root was very similar in both the studies (i.e., 64% in the shoot and 36% in the root).

8.6   CONCLUSIONS

Root crops along with grain crops are an important source of calories for a large part of the world population. They are consumed as cooked vegetables, eaten as raw roots, and also consumed as roasted flour. Many root crops are also used as animal feed. The major root crops are sugar beet, cassava, sweet potato, carrot, radish, turnip, and yams. The ecophysiology of these crops includes soil and climatic requirements, nutrient requirements, principal diseases and insects that infect these crops, and growth and development. The optimum temperature, soil type, soil pH, and salinity threshold values of these crops are given. These climatic and soil properties can be manipulated in favor of higher yields by adopting appropriate soil and crop management practices. In addition, discussion about the growth and development of these crops is also given. The knowledge of growth and development can help in adopting appropriate fertilizers, irrigation, the control of diseases and insects, and weed management practices that can lead to higher yields.

REFERENCES

Abruna, F., J. Vicente-Chandler, and J. Badillo. 1982. Effect of soil acidity components on yield and foliar composition of tropical root crops. *Soil Sci. Soc. Am. J.* 46:1004–1007.

Allem, A. C. 1994. The origin of *Manihot esculenta* Crantz (*Euphorbiaceae*). *Genet. Resour. Crop Eval.* 41:133–150.

Austin, D. F. 2002. Roots as a source of food. In: *Plant Roots: The Hidden Half*, 3rd edn., eds., Y. Waisel, A. Eshel, and L. Kafkafi, pp. 1025–1043. New York: Marcel Dekker.

Austin, D. F., R. K. Jansson, and G. W. Wolfe. 1991. Convolvulaceae and *Cyclas*: A proposed hypothesis on the origins of the plant/insect relationship. *Trop. Agric.* 68:162–170.

Bell, C. I., G. F. J. Milford, and R. A. Leigh. 1996. Sugarbeet. In: *Photoassimilate Distribution in Plants and Crops*, eds., E. Zamski and A. A. Schaffer, pp. 691–707. New York: Marcel Dekker.

Bowen, H. D. 1981. Alleviating mechanical impedance. In: *Modifying the Root Environment to Reduce Crop Stress*, eds., A. F. Arkin and H. M. Taylor, pp. 21–57. St. Joseph, MN: American Society of Agricultural and Engineering.

Campbell, K. G. K. 1984. Sugarbeet. In: *Evolution of Crop Plants*, ed., N. W. Simmonds, pp. 25–28. London, U.K.: Longman.

Campos, H. R. and Z. F. Sena. 1974. Root depth of cassava (*Manihot esculenta* Crantz) as a function of plant age em diferentes idades. Cruz das Almos, Brazil: Universidade Federal da Bahia. Escola de Agronomia.

CIAT (Centro Internacional de Agricultura Tropical). 1979. Annual report of the CIAT, Cali, Colombia.

Cock, J. H. 1976. Characteristics of high yielding cassava varieties. *Exp. Agric.* 12:135–143.

Cock, J. H. 1983. Cassava. In: *Potential Productivity of Field Crops under Different Environments,* ed., IRRI, pp. 341–359. Los Baños, Philippines: IRRI.

Cock, J. H. 1984. Cassava. In: *The Physiology of Tropical Field Crops*, eds., P. R. Goldsworthy and N. M. Fisher, pp. 529–549. New York: John Wiley & Sons.

Cock, J. H., D. Franklin, G. Sandoval, and P. Juri. 1979. The ideal cassava plant for maximum yield. *Crop Sci.* 19:271–279.

Cock, J. H. and R. H. Howeler. 1978. The ability of cassava to grow on poor soils. In: *Crop Tolerance to Suboptimal Land Conditions*, ed., G. A. Jung, pp. 145–154. Madison, WI: ASA.

Cock, J. H., D. W. Wholey, and O. G. Casas. 1977. Effects of spacing on cassava (*Manihot ecsulenta* Crantz). *Exp. Agric.* 13:289–299.

Connor, D. J., J. H. Cock, and G. H. Parra. 1981. The response of cassava to water shortage. I. Growth and yield. *Field Crops Res.* 4:181–200.

Cook, J. H. and C. R. Rosas. 1975. Ecophysiology of cassava. In: *Symposium on Ecophysiology of Tropical Crops*, ed., CEPLAC, pp. 1–14. Ilheus-Itabuna, Brazil: Communication Division of CEPLAC.

Dixon, A. G. O., R. Bandyopadhya, D. Coyne, M. Ferguson, S. B. Ferris, R. Hanna, J. A. Hughes et al. 2003. Cassava: From poor farmers crop to pacesetter of African rural development. *Chron. Hortic.* 43:8–15.

Draycott, A. P. 1972. *Sugarbeet Nutrition*. New York: John Wiley & Sons.

Dusi, A. N. and D. Peters. 1999. Beet mosaic virus: Its vector and host relationships. *Phytopathol. Z.* 147:293–298.

Enyi, B. A. C. 1972. The effects of spacing on growth, development and yield of single and multi shoot plants of cassava (*Manihot esculenta* Crantz). I. Root tuber yield and attributes. *East Afr. Agric. For. J.* 38:23–26.

Enyi, B. A. C. 1977. Analysis of growth and tuber yield in sweet potato cultivars. *J. Agric. Sci. Camb.* 89:421–430.

Fageria, N. K., V. C. Baligar, and C. A. Jones. 2011. *Growth and Mineral Nutrition of Field Crops*, 3rd edn. Boca Raton, FL: CRC Press.

Gregory, P. 2006. *Plant Roots: Growth, Activity and Interaction with Soils*. Oxford, U.K: Blackwell Publishing.

Hahn, S. K. 1977. Sweet potato. In: *Ecophysiology of Tropical Crops*, eds., R. T. Alvim and T. T. Kozlowski, pp. 237–248. New York: Academic Press.

Hahn, S. K. and Y. Hozyo. 1984. Sweet potato. In: *The Physiology of Tropical Field Crops*, eds., P. R. Goldsworthy and N. M. Fisher, pp. 551–567. New York: John Wiley & Sons.

Hahn, S. K., J. C. G. Isoba, and T. Ikotun. 1989. Resistance breeding in root and storage root crops at the International Institute of Tropical Agriculture, Ibadan Nigeria. *Crop Prot.* 8:147–168.

Harris, N. V. 1978. The potential of cassava in coastal Queensland: Some observations at the Yandaran plantation. In: *Proceedings of the Conference on Alcohol Fuels*, ed., Institution of Chemical Engineers, pp. 76–79. Sydney, NSW, Australia: Institution of Chemical Engineers.

Hartemink, A. E. 2006. Assessing soil fertility decline in the tropics using soil chemical data. *Adv. Agron.* 89:179–225.

Heywood, V. H. 1983. Relationships and evolution of *the Dacus carota* complex. *Isr. J. Bot.* 35:51–65.

Hole, C. C. 1996. Carrots. In: *Photoassimilate Distribution in Plants and Crops*, eds., E. Zamski and A. A. Schaffer, pp. 671–690. New York: Marcel Dekker.

Hole, C. C. and J. Dearman. 1994. Sucrose uptake by phloem parenchyma of carrot storage root. *J. Exp. Bot.* 45:7–15.

Howeler, R. H. 1981. *Mineral Nutrition and Fertilization of Cassava*, 52 pp. Cali, Colombia: CIAT.

Howeler, R. H. 1985. Potassium nutrition of cassava. In: *Potassium in Agriculture*, ed., R. D. Munson, pp. 819–841. Madison, WI: ASA.

Howeler, R. H. and L. F. Cadavid. 1983. Accumulation and distribution of dry matter and nutrients during a 12-month growth cycle of cassava. *Field Crops Res.* 7:123–139.

Huett, D. O., N. A. Maier, L. A. Sparrow, and T. J. Piggott. 1997. Vegetables. In: *Plant Analysis: An Interpretation Manual*, 2nd edn., eds., D. J. Reuter and J. B. Robinson, pp. 385–502. Collingwood, Victoria, Australia: CSIRO Publishing.

Iglesias, C. A., F. Calle, C. Hershey, G. Jaramillo, and E. Mesa. 1994. Sensitivity of cassava (*Manihot esculenta* Crantz) clones to environmental changes. *Field Crops Res.* 36:213–220.

Irizarry, H. and E. Rivera. 1983. Nutrient and dry matter contents of intensively managed cassava grown on an Ultisol. *J. Agric. Univ. Puerto Rico* 67:213–220.

Janick, J., R. W. Schery, F. W. Woods, and V. W. Ruttan. 1974. *Plant Science: An Introduction to World Crops*, 2nd edn. San Francisco, CA: W. H. Freeman and Company.

Jansson, R. K. and K. V. Raman. 1991. *Sweet Potato Pest Management: A Global Perspective*. Boulder, CO: Westview Press.

Jose, J. J. S. and F. Mayobre. 1982. Quantitative growth relationships of cassava (*Manihot esculenta* Crantz): Crop development in a savanna wet season. *Ann. Bot.* 50:309–316.

Kawano, K. 1978. Genetic improvement of cassava (*Manihot esculenta* Crantz) for productivity. *Trop. Agric. Res. Ser.* 11:9–21.

Keating, R. A., J. P. Evenson, and S. Fukai. 1982. Environmental effects on growth and development of cassava (*Manihot esculenta* Crantz). I. Crop development. *Field Crop Res.* 5:271–281.

Kodama, S., K. Chuman, and M. Tanoue. 1970. On the growth differentials of sweet potato to different soil fertility. *Bull. Kyushu Agr. Exp. Stn. Jpn*, 15:493–514.

Letschert, J. P. W. 1993. *Beta* section *Beta*: Biogeographical patterns of variation, and taxonomy. Thesis. Wageningen, the Netherlands: Wageningen Agriculture University. Papers 91, pp. 1–55.

Lorenz, O. A. and J. F. Bartz. 1968. Fertilization for high yields and quality of vegetable crops. In: *Changing Patterns in Fertilizer Use*, ed., L. B. Nelson, pp. 327–352. Madison, WI: SSSA.

Lorenz, O. A. and M. T. Vittum. 1980. Phosphorus nutrition of vegetable crops and sugar beets. In: *The Role of Phosphorus in Agriculture*, ed., R. C. Dinauer, pp.737–762. Madison, WI: SSSA.

Maas, E. V. 1993. Testing crops for salinity tolerance, In: *Proceedings of a Workshop on Adaption of Plants to Soil Stresses*, ed., J. W. Maranville, pp. 234–247. Lincoln, NE: University of Nebraska.

Mabberley, D. J. 1997. *The Plant Book*, 2nd edn. Cambridge, U.K.: Cambridge University Press.

Murata, Y., A. Kumura, and L. Ishiyee. 1976. Photosynthesis and ecology of crop plants, pp. 204–233. Tokyo, Japan: The Cultural Association of Agriculture, Forestry and Fishery.

Okechukwu, R. U. and A. G. O. Dixon. 2009. Performance of improved cassava genotypes for early bulking, disease resistance, and culinary qualities an inland valley ecosystem. *Agron. J.* 101:1258–1265.

Osaki, M., M. Matsumoto, T. Shinano, and T. Tadano. 1995a. A root–shoot interaction hypothesis for high productivity of root crops. *Soil Sci. Plant Nutr.* 42:289–301.

Osaki, M., H. Ueda, T. Shinano, H. Matsui, and T. Tadano. 1995b. Accumulation of carbon and nitrogen compounds in sweet potato plants grown under deficiency of N, P, or K nutrients. *Soil Sci. Plant Nutr.* 41:557–566.

Preiss, J. and M. N. Sivak. 1996. Starch synthesis in sinks and sources. In: *Photoassimilate Distribution in Plants and Crops: Source-Sink Relationships*, eds., E. Zamski and A. A. Schafeer, pp. 63–96. New York: Marcel Dekker.

Purseglove, J. W. 1987. *Tropical Crops: Dicotyledons*. New York: Longman Scientific and Technical.

Radke, J. F. and R. E. Bauer. 1969. Growth of sugarbeets as affected by root temperatures. Part I. Greenhouse studies. *Agron. J.* 61:860–863.

Reisenauer, H. M., J. Quick, and R. E. Voss. 1976. Soil test interpretive guides in soil and plant-tissue testing in California. University of California Bulletin 1879, pp. 39–40. Berkeley, CA: University of California.

Reuter, D. J. 1986. Temperate and sub-tropical crops. In: *Plant Analysis: An Interpretation Manual*, eds., D. J. Reuter and J. B. Robinson, pp. 38–99. Melbourne, Victoria, Australia: Inkata Press.

Sangoi, L. and N. D. Kruse. 1993. Accumulation and distribution of dry matter of cassava on the highlands of Santa Catarina, Brazil. *Pesq. Agropec. Bras.* 28:1151–1164.

Scoot, R. K. and K. W. Jaggard. 1993. Crop physiology and agronomy. In: *The Sugarbeet Crop: Science into Practice*, eds., D. A. Cooke and R. K. Scoot, pp. 179–237. London, U.K.: Chapman and Hall.

Siemonsma, J. J. and K. Piluek. 1993. *Plant Resources of South-Easr Asia No. 8. Vegetables*. Wageningen, the Netherlands: Pudoc Scientific.

Simpson, B. B. and M. Conner-Ogorzaly. 1995. *Economic Botany. Plant in Our World*, 2nd edn. New York: McGraw Hill.

Steingrover, E. 1981. The relationship between cyanide-resistant root respiration and the storage of sugars in the taproot in *Daucus carota* L. *J. Exp. Bot.* 32:911–919.

Tsuno, Y. and K. Fujise. 1963. Studies on the dry matter production of sweet potato. II. Aspect of dry matter production in the field. *Crop Sci. Soc. Jpn.* 26:285–288.

Tsuno, Y. and K. Fujise. 1964. Studies on the dry matter production of sweet potato. III. The relationship between the dry matter production and the absorption of mineral nutrients. *Proc. Crop Sci. Jpn.* 27:293–300.

Ulrich, A., J. T. Moraghan, and E. D. Whitney. 1993. Sugarbeet. In: *Nutrient Deficiencies & Toxicities in Crop Plants*, ed., W. F. Bennett, pp. 91–98. St. Paul, MN: The American Phytopathological Society.

Unger, P. W. and T. C. Kaspar. 1994. Soil compaction and root growth: A review. *Agron. J.* 86:759–766.

Williams, C. N. 1972. Growth and productivity of tapioca (*Manihot utilissima*). III. Crop ratio, spacing and yield. *Exp. Agric.* 8:15–23.

Yen, D. E. 1984. Sweet potato. In: *Evolution of Crop Plants*, ed., N. W. Simmonds, pp. 42–44. London, U.K.: Longman.

Yoshimoto, M., S. Okuno, M. Yoshinaga, and O. Yamakawa. 1998. Antimutagenic activity of water extracts from sweet potato. *Trop. Agric.* 75:308–313.

Zohary, D. and M. Hopf. 1993. *Domestication of Plants in the Old World*. Oxford, U.K.: Clarendon Press.

9 Management Strategies for Maximizing Root Systems

9.1 INTRODUCTION

The environmental conditions that prevail in the rhizosphere during plant growth will determine root growth and development. The environmental conditions that are linked to climate are precipitation, temperature, and solar radiation, which can modify root growth and development. Soil properties that determine the rhizosphere environment and can affect root growth are physical, chemical, and biological processes operating during plant growth. However, climatic and soil properties are interrelated in changing or defining the rhizosphere environment. Hence, management strategies for improving root growth involve optimizing climatic variables like the availability of water, temperature, and solar radiation, and the physical, chemical, and biological properties of the soil. This strategy is known as the modification of soil to fit the plants. The use of crop genotypes with higher yield potential and better root geometry is another strategy. This strategy is known as the modification of plants to fit the soil. In addition, the root growth of crop plants can also be improved with the adoption of appropriate cultural practices. All these strategies involve soil and plant management and should be used together rather than in isolation to get best results in optimizing root systems. Nielsen (1974) reported that if conditions are favorable to growth, the root system should be structurally and morphologically able to use as much carbohydrates as the tops can supply and in turn supply as much water and nutrients as the tops can use. This chapter is meant to discuss soil, plant, and cultural practices that can be modified in favor of higher root system of crop plants.

9.2 OPTIMIZING SOIL PHYSICAL PROPERTIES

Physical properties are the characteristics, processes, or reactions of soil that are caused by physical forces and that can be described by or expressed in physical or chemical equations (Fageria et al., 2011). The physical properties of soil that can be modified in favor of higher root growth are temperature, moisture content, structure, and bulk density. Soil texture is also an important physical property that determines root growth. However, soil texture cannot be modified easily. The physical properties of soil are interrelated, and a change in one may cause a change in others, which may be favorable or adverse. Root growth is genetically determined under optimum conditions and may be further modified by the environment. There is also interaction between genetic variability and environmental factors for root growth. Hence, the creation of favorable physical conditions of soil for root growth is a very complex phenomenon.

9.2.1 TEMPERATURE

Temperature is one of the important factors affecting the distribution of plants on earth. Root zone temperature has a dominant influence on seed germination and vegetative and reproductive growth stages in crop plants. As with the shoot, temperature affects both the expansion of the root system through effects on development and growth and a range of metabolic processes that determine the activity of the root system (Gregory, 2006). Root zone temperature affects root function and metabolism (Cooper, 1973; Bland, 1993; McMichael and Burke, 1998). At optimum temperature, cell division is more rapid but of shorter duration than at lower temperatures. At cooler temperatures, roots are usually whiter, thicker in diameter, and less branched than at warmer temperatures (Ketellapper, 1960; Nielsen and Cunningham, 1964; Garwood, 1968), although there are exceptions (Bowen, 1970). Soil temperature regimes are defined in soil taxonomy using measurements taken at a depth of 50 cm (Rodriguez et al., 2010). Root temperature is generally lower than that of air, but seasonal fluctuations can occur with depth, depending on soil and aboveground factors (McMichael and Burke, 1998).

The impact of soil temperature on the function of root systems has been documented in a number of species (McMichael and Burke, 1998). At low temperatures, water and nutrient uptake by root systems may be reduced (Nielsen and Humphries, 1966; Nielsen, 1974). In general, root growth tends to increase with increasing temperature until an optimum is reached above which root growth reduces (Brar et al., 1970; Pearson et al., 1970; Cooper, 1973; Glinski and Lipiec, 1990). Higher root temperature can affect the overall enzymatic activity of root systems (Nielsen, 1974). The efficiency of extracellular enzymes increases with temperature, and microbes in warmer soils may invest fewer resources in their production, in order to incur lesser metabolic cost (Allison, 2005; Bell et al., 2010). Alternatively, warming can decrease the water content of the soil during the growing season, which can limit enzyme and substrate diffusion (Allison, 2005). Warming during the plant growth season can decrease microbial biomass, possibly due to decreased soil moisture or increased predation (Cole et al., 2002; Rinnan et al., 2007). Warming can also shift the balance toward higher fungal dominance over bacteria in microbial communities (Zhang et al., 2005), although the direct effects of warming on soil microbes are potentially confounded by the indirect effects of warming on plant productivity and species composition (Jonasson et al., 1999) or microbial consumers (Rinnan et al., 2008; Bell et al., 2010).

McMichael and Burke (1994) reported that root metabolism may become more temperature sensitive as the mobilization of reserves in the cotyledons declines during early seedling growth, indicating that temperature dependency is developmentally regulated. McMichael and Quisenberry (1993) observed that the optimum temperature for root growth in cotton was between 28°C and 35°C versus between 23°C and 25°C in sunflower. Genetic variability exists for root growth in response to changes in temperature, both, between and within species (McMichael and Burke, 1998). McMicheal and Quisenberry (1991) also reported that the genetic variability observed in root development in cotton genotypes was somewhat independent of their variability in shoot development, suggesting that breeding for more favorable root traits (i.e., deeper, more branched roots, less sensitive to low temperature) might be possible.

Crop species respond differently to temperature throughout their life cycles. Each species has a defined range of maximum and minimum temperatures within which growth occurs and an optimum temperature at which plant growth progresses at its fastest rate (Hatfield et al., 2011). Vegetative development usually has a higher optimum temperature than reproductive development. The progression of a crop through phonological phases is accelerated by increasing temperatures up to the species-dependent optimum temperature. Exposure to higher temperatures causes faster development in food crops, which does not translate into an optimum for maximum production because the shorter life cycle means shorter reproductive period and shorter radiation interception period (Hatfield et al., 2011). Data related to optimum temperature for maximum yield and root growth of several crop species are presented in Tables 9.1 and 9.2.

Soil temperature is affected by many factors, of which the more important are air temperature; intensity, quality, and duration of radiant energy; precipitation and evaporative potential of air; surface cover and color; and the thermal conductivity of the soil. Management practice that affects the quantity of incident radiation at the soil surface, the insulating effects of the soil surface, or the thermal conductivity of the soil will modify the influence of radiation on the temperature of the root zone. Hence, a well-watered soil with a straw mulch and a dense crop growing on it will have a lower root zone temperature than a soil in fallow (Nielsen, 1974). In addition, tillage operation may loosen the upper structure, incorporating organic residues. The higher porosity

TABLE 9.1
Optimum Soil Temperature for Maximum Yield of Important Field Crops

Crop	Optimal Temperature (°C)	Reference
Barley (*Hordeum vulgare* L.)	18	Power et al. (1970)
Oats (*Avena sativa* L.)	15–20	Case et al. (1964)
Wheat (*Triticum aestivum* L.)	20	Whitfield and Smika (1971)
Corn (*Z. mays* L.)	25–30	Dormaar and Ketcheson (1960)
Cotton (*Gossypium hirsutum* L.)	28–30	Pearson et al. (1970)
Potato (*Solanum tuberosum* L.)	20–23	Epstein (1966)
Rice (*Oryza sativa* L.)	25–30	Owen (1971)
Bean (*Phaseolus vulgaris* L.)	28	Mack et al. (1964)
Soybean (*Glycine max* L. Merr.)	30	Voorhees et al. (1981)
Sugar beet (*Beta vulgaris* L.)	24	Radke and Bauer (1969)
Sugarcane (*Saccharum officinarum* L.)	25–30	Hartt (1965)
Alfalfa (*Medicago sativa* L.)	28	Heinrichs and Nielsen (1966)

TABLE 9.2

Optimum Temperature for Root Growth
of Some Crop Species

Crop Species	Parameter Measured	Temperature (°C)
Sunflower	Root elongation rate	20
Tomato	Root elongation rate	30
Corn	Root elongation rate	30
Corn	Root mass	26
Oats	Root mass	5
Cotton	Root elongation rate	33
Soybean	Taproot extension rate	25
Rape	Root extension	23
Rice	Root growth	25–37
Lolium perenne	Root mass	17

Source: Adapted from McMichael, B.L. and Burke, J.J., *Hort. Sci.*, 33, 947, 1998.

makes the tilled zone more responsive to ambient air temperatures because both the heat conductivity of the cold from lower depths and the volumetric heat capacity are smaller than in the untilled soil. These effects pertain only to soils where the moisture content does not exceed field capacity. In addition, ridged soil can be warmer than unridged soil. There is strong interaction between root temperature and water supply and the availability of nutrients. The effect of unfavorable root temperatures can be partially mitigated by favorable supplies of water and nutrients (Nielsen, 1974).

9.2.2 MOISTURE CONTENT

Moisture availability is one of the most important factors determining crop production. The distribution of vegetation in an agroecological region is controlled more by the availability of water than by any other single factor. Water is required by plants for the translocation of mineral elements, for the manufacture of carbohydrates, and for the maintenance of the hydration of protoplasm. Crop yield can be reduced at both very low and very high levels of moisture. Excess moisture reduces soil aeration and, thus, the supply of O_2 available to roots. With poor aeration, the activities of beneficial microorganisms and water and nutrient uptake by plants can be seriously inhibited, although aquatic plants and rice are adapted to and function well even when soils are saturated. Soil moisture deficits can cause stomata in the leaf to close, reducing transpiration and helping in maintaining the hydration of protoplasm, but also reducing photosynthesis. Moisture stress also causes reduction in both cell division and cell elongation and, hence, in growth (Fageria et al., 2011).

The interaction between plant-root systems and soil, especially soil with moisture, is very important in many respects. For example, root water uptake from the soil plays an important role in the hydrological process of water flow through soil, plants, and air. In the field of crop science, many experimental studies indicate that variation in soil water conditions affects the root system architecture of various kinds of crops, and that the architecture affects absorption efficiency (Tsutsumi et al., 2003). It is also reported in the literature that there exists root hydrotropism, that is, the bending or growth of roots in the direction of increasing moisture in a moist gradient (Takahashi, 1994; Takano et al., 1995). Takahashi and Scoot (1993) investigated the behavior of root hydrotropism in pea (*Pisum sativum* L.) and corn (*Zea mays* L.) and concluded that the root system of these two crops bends in the direction of increasing moisture.

The storage terms relate only to the portion of soil moisture available to the plant. Field capacity and wilting coefficient or permanent wilting point are the practical upper and lower limits of water availability for crops. The upper limit of soil water availability to plants is often considered to be the water content after saturated soil has freely drained for 2–3 days or wetted soils have been subjected to pressures in the range of 5–30 kPa (kilopascals) or 0.05–0.3 bars (Unger et al., 1981). The lower value is generally applicable to light-textured soils and the higher value to heavy-textured soils. Root growth is better when water content in the soil is around field capacity.

9.2.3 STRUCTURE

Soil structure is the combination or arrangement of primary soil particles into secondary units or peds. The secondary units are characterized on the basis of size, shape, and grade (degree of distinctness) (Soil Science Society of America, 2008). Hence, the binding of soil particles into aggregates results in structure. Soil structure, in combination with texture, governs the porosity of the soil and, thus, affects aeration, water infiltration, root penetration, and microbiological activities of soil flora and fauna (Fageria et al., 2011). Primary factors in the development of soil structure are shrink-and-swell phenomena during wetting and drying. Pressure is also exerted by plant roots. Soil separates are bonded together by clays, iron, aluminum compounds, and organic substances such as humus, polyuronides, polysaccharides, and proteins (Baver et al., 1972). Soil structure plays an important role in plant growth and, consequently, in crop production. Soil must have a favorable structure for high productivity. A good soil structure provides adequate aeration and drainage, sufficient water storage capacity, good root growth, and access to nutrients (Russell, 1977). No one structure is completely ideal, however, because requirements differ among crop plants. Moreover, crops will generally grow satisfactorily over a range of structural conditions (Low, 1979).

Soil structure can be modified much more readily through cultural practices than can texture. The tillage of virgin soils modifies soil structure (MacDonald et al., 2010). When soils are brought under cultivation, organic carbon content decreases, which generally reduces the stability of aggregates, creating unfavorable conditions for root growth. In addition, cultivation also decreases microbial biomass carbon (Granatstein et al., 1987; Gupta and Germid, 1988; Follett and Schimel, 1989; Lupwayi et al., 1999; Grandy and Robertson, 2006; MacDonald et al., 2010), which may create an unfavorable structure for root growth. On the other hand, conversion to no-till production systems is observed to increase aggregate size, stability, and surface soil organic carbon, and improve soil structure (Cambardella and Elliott, 1993), which may create favorable conditions for plant root growth in the upper soil horizon.

Soil structure can be extremely important to root growth in fine-textured soils, but soil strength usually is more important than soil structure in sandy soils. Soil strength is defined as the ability or capacity of a particular soil in a particular condition to resist or endure an applied force (Gill and Vandenberg, 1967). Soil strength is easily affected by changes in the water content and bulk density of soil, although other factors including texture, mineralogy, cementation, cation composition, and organic matter content also affect it (Soil Science Society of America, 2008).

The addition of lime to acid soils can improve soil structure and, consequently, root growth. Chan and Heenan (1998) reported that Ca in liming materials helps in the formation of soil aggregates, hence improving soil structure. The lime-induced improvement in aggregate stability manifested through the effect of liming on dispersion and flocculation of soil particles (Bolan et al., 2003). Liming is often recommended for the successful colonization of earthworms in pasture soils. The lime-induced increase in earthworm activity may influence soil structure and macroporosity through the release of polysaccharide and the burrowing activity of earthworms (Springett and Syers, 1984).

Soil structure is related to soil aggregation, which can be defined as the process whereby primary soil particles (sand, silt, clay) are bound together, usually by natural forces and substances derived from root exudates and microbial activity (Soil Science Society of America, 2008). Soil aggregate stability is the ability of soil aggregates to resist a wide range of disruptive forces, the impact of raindrops being one of the most severe of such forces (Jimba and Lowery, 2010). The aggregation of soil is an essential function in the physicochemical and biological processes of soil, and has been shown to influence soil quality through the protection of existing soil organic matter (SOM), moisture-holding capacity, and soil nutrient retention (Tisdall and oades, 1982; Angers and Giroux, 1996; Angers and Caron, 1998; Jiao et al., 2006; Corbin et al., 2010). Factors affecting aggregate stability can be grouped as abiotic (clay minerals, sesquioxides, exchangeable cations), biotic (activities of plant roots, soil fauna, and microorganisms), and environmental (soil temperature and moisture) (Marquez et al., 2004). The concept of aggregate stability depends on both the forces that bind particles together and the nature and magnitude of the disruptive stress (Beare and Bruce, 1993). It has been documented that no-tillage soils improved soil aggregation, C sequestration, and aggregate stability (Mikha and Rice, 2004; Zhang et al., 2007; Mikha et al., 2010), which results in increased water infiltration and resistance to wind and water erosion (Zhang et al., 2007). Soil aggregation is one of the most important characteristics that mediates many chemical, physical, and biological properties of soil and improves soil quality and sustainability (Mikha et al., 2010).

Growing perennial warm-season grasses improves the structural properties of soil (Blanco-Canqui, 2010). Active and decaying root systems of stiff-stemmed grasses may improve the porosity of the soil and result in increased hydraulic conductivity within grass hedge areas (Rachman et al., 2004). The addition of farmyard manures to organic systems has been shown to enrich SOM directly and indirectly through improved soil properties such as increased numbers and the distribution of soil macroaggregates, microfauna, macro- and micronutrients, and improved crop yields (Jokela, 1992; Ghoshal and Singh, 1995; Edmeades, 2003; Mikha and Rice, 2004; Butler and Muir, 2006; Jiao et al., 2006). The addition of organic matter improves soil quality by decreasing bulk density, surface sealing, and crust formation, and by increasing aggregate stability, cation exchange capacity (CEC), nutrient cycling, and biological activity (Kay, 1990; Jokela et al., 2009). In addition, adopting a no-tillage system also increased total water stable soil aggregation (Mikha and Rice, 2004; Green et al., 2005; Zotarelli et al., 2007).

Legumes and grass cover crops have also been shown to improve the physical, chemical, and microbiological properties of soil that impact soil quality (Dabney et al., 2001; Magdoff and Weil, 2004). Deep-rooted cover crops are one possible solution to compaction problems, especially in no-till farming systems (Unger and Kaspar, 1994). The deep-growing taproots of the perennial "alfalfa" can increase infiltration rate on compacted no-till soils and the recolonization of root channels left by alfalfa has been shown to benefit corn root systems that follow (Rasse and Smucker, 1998). Stirzaker and White (1995) showed a doubling of yield for lettuce (*Letuca sativa* L.) grown on compacted soil following cover crop of subterranean clover (*Trifolium subterraneum* L.). They speculated that the yield increase was due to biopores made by the cover crop roots (Williams and Weil, 2004). In loam soils, cover crops increased the percentage of water stable aggregates (Dapaah and Vyn, 1998; Villamil et al., 2006) and decreased bulk density and penetrometer resistance. Jokela et al. (2009) reported that cover/companion crop treatments generally had more large macroaggregates and greater aggregate mean-weight diameters.

The release of organic compounds in the rhizosphere is recognized as a major energy input to the soil, providing an essential driving force for microbial-mediated processes like carbon mineralization–humification; nutrient metabolization, mineralization, and immobilization; denitrification; and the maintenance of soil structure (Bottner et al., 1999). Hence, planting crop species or genotypes within species that produce vigorous root systems is an important strategy in improving soil structure. In regard to the use and management of soils, SOM is important as a fuel for microbial activity; a source of macro- and micronutrients; a sink for heavy metals; a major contributor to the soil exchange capacity; a critical component for soil aggregation, structure, and infiltration; and the storehouse of all the soil organic carbon (Stolt et al., 2010). Hence, improving SOM is fundamental to improving soil structure and, consequently, the root growth of crop plants.

9.2.4 BULK DENSITY AND POROSITY

Bulk density is defined as the mass of dry soil per unit bulk volume. The value is expressed as megagram per cubic meter ($Mg\ m^{-3}$). This volume includes both solids and pores. Bulk density can be calculated by using the following formula:

$$\text{Bulk density} = \frac{\text{Weight of oven dry soil}}{\text{Volume of soil (solids + pores)}}$$

Soil bulk density is an important physical property of the soil that provides a measure of potential soil aeration conditions for root growth and defines the upper limit to the saturated water-holding capacity of the soil (Nemes et al., 2010). It can also be used as a measure of soil compaction, whether natural or human imposed.

Soil compaction is the process by which the soil grains are rearranged to decrease void space, thereby increasing bulk density (Soil Science Society of America, 2008). In addition, bulk density is a useful physical property of the soil in determining solute movement in the profile, water-holding capacity, and soil erosion risk. Average bulk density values of different soil orders are presented in Tables 9.3 and 9.4. As a rule,

TABLE 9.3
Average Values of Soil Bulk Density of Different Soil Orders

Soil Order[a]	Bulk Density (g cm⁻³)	Number of Samples Analyzed
Alfisols	1.51	7541
Andisols	1.01	593
Aridisols	1.41	3893
Entisols	1.41	1991
Histosols	1.36	17
Inceptisols	1.40	2982
Mollisols	1.38	9398
Spodosols	1.46	692
Ultisols	1.52	2514
Vertisols	1.31	1387

Source: Adapted from Nemes, A. et al., *Soil Sci. Soc. Am. J.*, 74, 1938, 2010.

[a] Data from soil orders of 48 U.S. states (excluding Alaska and Hawaii).

TABLE 9.4
Soil Bulk Density of Principal Soil Orders in Different Countries

Country	Location	Soil Order	Bulk Density (Mg m⁻³)
Bangladesh	Mymensingh	Haplaquepts	1.38
Brazil	Piracicaba	Ultisol	1.35
Chile	Santa Rosa	Andisol	1.08
China	Hangzhou	Inceptisol	1.06
Egypt	Minia	Entisol	1.68
Malaysia	Puchong	Ultisol	1.28
Morocco	Atlas	Inceptisol	1.39
Sri Lanka	Kundasala	Ultisol	1.41
Vietnam	Ba Ria Vung Tau	Ultisol	1.38

Source: Adapted from Dourado-Neto, D. et al., *Soil Sci. Soc. Am. J.*, 74, 139, 2010.

the higher the bulk density, the more compact the soil, the more poorly defined the structure, and the smaller the amount of pore space (Fageria et al., 2011). The bulk density of uncultivated soils varies from 0.93 to 1.43 Mg cm^{-3} and cultivated soils varies from 1.13 to 1.54 Mg m^{-3} (Brady and Weil, 2002). The growth of roots into moist soil is generally limited by bulk density ranging from 1.45 Mg m^{-3} in clays to 1.85 Mg m^{-3} in loamy sands (Brady and Weil, 2002). Bowen (1981) reported that as a general indication (with many exceptions), roots will be severely impeded if bulk density exceeds 1.55 Mg m^{-3} in clay loam soils, 1.65 Mg m^{-3} in silt loam soils, 1.80 in fine sandy loam soils, and 1.85 in loamy fine-sand soils. High soil strength and bulk density can confine crop root growth (Laboski et al., 1998) and result in a shallower root system (Oussible et al., 1992). When soil compaction suppresses total root length, shoot growth may also be reduced (Montagu et al., 2001). Crop cultivation increases bulk density of ploughed soil layer. Soil bulk density has been used as an important indicator of soil quality (Karlen et al., 1994; Werner, 1997) and has been shown to decrease with the incorporation of organic amendments such as plant residue and farmyard manure (Latif et al., 1992; Sharma and Gupta, 1998).

Porosity is defined as the volume of pores in a soil sample (nonsolid volume) divided by the bulk volume of the sample. Porosity can be calculated by using the following formula:

$$\text{Pore space}\,(\%) = 100 - \frac{\text{Bulk density}}{\text{Particle density}} \times 100$$

Particle density of most silicate-dominated mineral soils is assumed to be 2.65 Mg cm^{-3}. Bulk density and porosity are interrelated. As bulk density increases, total porosity decreases. Both high bulk density and low porosity restrict root growth (Singh and Sainju, 1998). Rosolem and Takahashi (1996) reported a 10% and 50% decrease in soybean root growth when the bulk density of an Oxisol was increased from 1.06 Mg cm^{-3} to 1.45 and 1.69 Mg cm^{-3}, respectively. Shierlaw and Alston (1984) reported that bulk density equal to or >1.2 Mg cm^{-3} reduced root growth of corn seedlings. Sainju and Good (1993) found a negative relationship between root length density and bulk density and a positive relationship between root length density and porosity. A higher soil bulk density reduces root length (Gerard et al., 1982; Freitas et al., 1999) but increases the diameter of the root (Chassot and Richner, 2002).

Soil should contain more than 10% air-filled pores to maintain aeration (Box, 1996; Singh and Sainju, 1998). Pores not filled with water are filled with air. Root growth is restricted by poor soil aeration. Gardner and Danielson (1964) found a high correlation ($r = 0.99$) between soil penetration by cotton roots and the percentage of aeration porosity. Root elongation is more sensitive to oxygen diffusion rate than to oxygen concentration (Singh and Sainju, 1998). When the oxygen diffusion rate in the soil falls below 58 mg cm^{-2} s^{-1}, root growth is restricted (Erickson and Van Doren, 1961). Compacted soil may have oxygen diffusion rate of <33 mg cm^{-2} s^{-1} (Erickson, 1982). Hence, in compacted soil, even air-filled pore space >10%

can limit root growth. Asady et al. (1985) reported that root penetration of compacted layers decreased linearly as air-filled porosity decreased from 30% to 0%. Bulk density or porosity can be altered by using appropriate crop rotation, tillage, or the addition of organic matter. Corbin et al. (2010) reported that bulk density decreased 14%–6% for the corn, soybean, wheat/alfalfa, corn, and the corn/alfalfa–alfalfa/corn rotations, respectively. Crop rotation provides functions that maintain or improve SOM content, manage deficient or excess plant nutrients, and provide erosion control (USDA, 2008).

Growing perennial warm-season grasses decreases soil bulk density. The lower soil bulk density under grasses is due to root perenniality, deep rooting system, and continued biomass input. The permanent surface cover and the extended roots provide resilience against both external (i.e., wheel traffic) and internal compactive forces (Clark et al., 1998). The high-biomass-producing and deep-rooted grass species can also reduce the soils' susceptibility to compaction by increasing SOM concentration (Thomas et al., 1996). Soils under grasses have more continuous macropores created by root channels and biological activity than under row crops (Blanco-Canqui, 2010). Rachman et al. (2004) observed that soil macroporosity under switchgrass was about two times greater than that under row crops. The higher proportion of macropores partly explains the lower bulk density under warm-season grasses (Blanco-Canqui, 2010). The higher proportion of macropores partly explains the lower bulk density under warm-season grasses. Growing warm-season grasses protects soil surface from raindrop impact, reduces surface sealing and crusting, maintains the integrity of open-ended macropores, and reduces the risks of soil compaction and consolidation (Blanco-Canqui, 2010). The abundant and dense network of grass roots forms a skeleton to enmesh soil particles and forms stable aggregates (Acosta-Martinez et al., 2004).

9.3 OPTIMIZING SOIL CHEMICAL PROPERTIES

Principal soil chemical properties that affect root growth are soil pH, base saturation, aluminum saturation, SOM content, soil fertility, salinity, and alkalinity. Most of these chemical properties of soil can be modified by adopting appropriate soil management practices in favor of higher root growth.

9.3.1 LIMING ACID SOILS

Historically, civilizations have tended to settle on high-base-status soils, but when populations increased they migrated to other environments with mainly acid soils (Sanchez, 1976). Acid soils cover about 30% of the total ice-free land area, equivalent to 3950 million ha of the earth. About 16.7% of Africa, 6.1% of Australia and New Zealand, 9.9% of Europe, 26.4% of Asia, and 40.9% of America have acid soils (Von Uexkull and Mutert, 1995). They cover a significant part of at least 48 developing countries located mainly in the tropical area, being more frequent in Oxisols and Ultisols in South America and Oxisols in Africa. In Asia, Gleisols, Podzoluvisols, and Ultizols are the most important acid soils (Narro et al., 2001). In tropical South

America, 85% of the soils are acidic, and ≈850 million ha of this area is underutilized (Fageria and Baligra, 2008).

Theoretically, soil acidity is quantified on the basis of hydrogen (H^+) and aluminum (Al^{3+}) concentrations in soils. For crop production, however, soil acidity is a complex of numerous factors involving nutrient/element deficiencies and toxicities, low activities of beneficial microorganisms, and reduced plant root growth that limits absorption of nutrients and water (Fageria and Baligar, 2003, 2008). Plant utilization of many nutrients becomes less efficient as soil acidity increases (Haynes and Ludecke, 1981; Black, 1993). In addition, acid soils have low water-holding capacity and are subject to compaction and water erosion (Fageria and Baligar, 2003, 2008). The situation is further complicated by various interactions among these factors (Foy, 1992). Detrimental effects from soil acidity vary with crop, rooting depth, and crop tolerance (Black, 1993; Tumusiime et al., 2011). The components of the soil acidity complex have been thoroughly discussed in various publications (Foy, 1984, 1992; Kamprath and Foy, 1985; Tang and Rengel, 2003; Fageria and Baligar, 2008).

Although low fertility is a characteristic of acid soils, these vast areas have a large proportion of favorable topography for agriculture, adequate temperatures for plant growth throughout the year, and sufficient moisture availability year round in 70% of the region and for 6–9 months in the remaining 30%. When chemical constraints are eliminated by liming and using adequate amounts of fertilizers, the productivity of Oxisols and Ultisols is among the highest in the world (Sanchez and Salinas, 1981).

The use of ammonical fertilizers in crop production has been shown to acidify soils (Mahler and Harder, 1984). Acidification due to N fertilization results from three factors (Tumusiime et al., 2011). Nitrogen fertilizers increase yields and, thus, increase the removal of bases in the harvested crop. The second effect comes from NO_3^- that is not taken up by the growing crops. Nitrate is very soluble and, if not taken up by plants, leaches to deep soil layers, taking with it base elements like Ca^{2+} and Mg^{2+} (Mahler and Harder, 1984). Another effect is the microbial oxidation of ammonical fertilizers (nitrification), a process that releases H into the soil (Adams, 1984). Soil acidity may result from parent materials that were acid and naturally low in the basic cations (Ca^{2+}, Mg^{2+}, K^+, and Na^+) or because these elements have been leached from soil profile by heavy rains (Kamprath and Foy, 1971). When precipitation exceeds evapotranspiration, leaching occurs. In highly leached soils, iron and aluminum oxides and some of the trace metal oxides, which are highly resistant to weathering, remain from the original parent material (Fageria et al., 1990; Bohn et al., 2001). Soil acidity may also develop from exposure to the air of mine spoils containing iron pyrite (FeS_2) or other sulfides. Soil acidity may also develop by the decomposition of plant residues or organic wastes into organic acids. This process is of particular importance in many forest soils.

The use of legume crops continuously or in rotation can increase soil acidity. In Australia and New Zealand, the continuous cultivation of legume crops decreased the pH of agricultural soils (Bolan and Hedley, 2003). Legume-based pastures also increased soil acidification (Williams, 1980; Loss et al., 1993). Williams (1980) reported that even the normal growth of clover pasture for 50 years decreased the pH of an Australian soil from 6.0 to 5.0 at a depth of 30 cm. Legumes also increase soil acidification in arable cropping systems (Burle et al., 1997). The reason for generating acidity is associated with higher absorption of basic cations by these crops and the release of H^+ ions by the root of legume crops to maintain ionic balance (Bolan and Hedley, 2003). According to Bolan and Hedley (2003), for different legume species about 0.2–0.7 mol of H^+ were released per mol N_2 fixed. These authors also state that the amount of H^+ ions released during N_2 fixation is really a function of carbon assimilation and, hence, depends mainly on the form and amount of amino acids and organic acids synthesized within the plants. Soil acidification is also caused by the release of protons (H^+) during the transformation and cycling of carbon, nitrogen, and sulfur in the soil–plant–animal system (Bolan and Hedley, 2003). Soil acidification is caused by acid precipitation, which is the result of industrial pollution (Foy, 1984; Ulrich et al., 1980).

Finally, crop fertilization with ammonia or ammonium fertilizers can result in soil acidification by a microbial-mediated reaction as follows (Fageria et al., 2010a):

$$(NH_4)_2SO_4 \Leftrightarrow 2NH_4^+ + SO_4^{2-}$$

$$NH_4^+ + 2O_2 \Leftrightarrow NO_3^- + 2H^+ + H_2O$$

$$CO(NH_2)_2 + 3H_2O \Leftrightarrow 2NH_4^+ + 2OH^- + CO_2$$

$$NH_4^+ + 2O_2 \Leftrightarrow NO_3^- + 2H^+ + H_2O$$

According to these equations, the NH_4^+ ions are oxidized to yield NO_3^- ions or nitrification, which results in the release of H^+ ions, leading to soil acidification. Another reason for soil acidification by ammonium sulfate and urea is the leaching of NO_3^- ions. During the leaching process, NO_3^- ions are accompanied by positively charged basic cations, like Ca^{2+}, Mg^{2+}, and K^+, in order to maintain the electric charge on the soil particles. After these basic cations are leached from soil particles, their sites are replaced by H^+ ions, and the acidification process is accelerated (Bolan and Hedley, 2003). The author conducted a field experiment for three consecutive years with lowland rice using ammonium sulfate and urea as N sources. After the harvest of the third rice crop, the soil pH was determined; the results are presented in Table 9.5. Soil pH was decreased linearly with increasing N rate by ammonium sulfate (0–210 kg N ha^{-1}) and urea (0–200 kg N ha^{-1}) (Table 9.5). The decrease in pH with ammonium sulfate was from 5.8 at 0 N ha^{-1} rate to 5.2 at 210 kg N ha^{-1}. This means that the decrease at the highest N rate applied with ammonium sulfate was 12% compared with the lowest N rate for the same fertilizer. Similarly, the decrease in pH with urea fertilizer was 5.7 at 0 rate ha^{-1} to 5.5 at 200 kg N ha^{-1}. This decrease corresponds to 4% at highest N rate compared with the lowest N rate. Hence, decrease in soil pH was higher with ammonium sulfate compared to urea. Ammonium sulfate fertilization accounted for about 40% variation in soil pH, and urea fertilization accounted for 26% variation in soil pH. Since both N fertilizer

TABLE 9.5

Influence of Nitrogen Rate Applied by Ammonium Sulfate and Urea on Soil pH after Harvest of Three Lowland Rice Crops

N Rate (kg ha⁻¹) $NH_4(SO_4)_2$	pH in H_2O	N Rate (kg ha⁻¹) Urea	pH in H_2O
0	5.8	0	5.7
30	5.8	50	5.6
60	5.5	100	5.7
90	5.4	150	5.5
120	5.4	200	5.5
150	5.4		
180	5.3		
210	5.2		
Average	5.5		5.6

Regression Analysis

N rate, ammonium sulfate (X) vs. pH (Y) = 5.7187 − 0.0026X,
  R^2 = 0.3969**

N rate, urea (X) vs. pH (Y) = 5.6927 − 0.00092X, R^2 = 0.2601*

Source: Fageria, N.K. et al., *Commun. Soil Sci. Plant Anal.*, 41, 1565, 2010c.

*, **Significant at the 5% and 1% probability levels, respectively.

sources are acidic, their acid equivalency (kg $CaCO_3$ required to neutralize acidity produced by 100 kg fertilizers) are ammonium sulfate 110 and urea 80 (Fageria, 1989). Decrease in soil pH has been reported by the use of ammonium sulfate and urea (Hetrick and Schwab, 1992). Stumpe and Vlek (1991) reported that decrease in pH of tropical acid soils (Oxisols, Ultisols, and Alfisols) due to the use of three N fertilizers was in the order of ammonium sulfate > urea > ammonium nitrate. These two fertilizer sources can acidify the soil by the following reactions (Fageria, 1989; Bolan and Hedley, 2003).

Increasing soil acidity following the application of commercial fertilizers, especially N fertilizers, also led to the development of infertile soils that did not respond well to increased fertilizer application for sustaining crop yields (Herrero et al., 2010) and reduced root growth (Fageria and Moreira, 2011). In such conditions, integrated crop–livestock system can be used as an option to improve soil quality and sustain crop yields (Franzluebbers, 2007; Maughan et al., 2009; Sainju et al., 2010). Improving soil quality will improve the root growth of crops planted on acid soils.

The amelioration of acid soils is commonly accomplished using liming materials such as limestone ($CaCO_3$), dolomitic lime ($CaCO_3 \cdot MgCO_3$), hydrated lime [$Ca(OH)_2$], burnt lime (CaO), or municipal and industrial lime wastes ($CaCO_3$ with impurities) (Fageria and Baligar, 2008; DeSutter and Cihacek, 2009). Liming is the most common and effective practice to achieve optimum yields of all crops grown on acid soils. Liming is the most widely used long-term method of soil acidity amelioration, and its success is well documented (Haynes, 1982; Conyers et al., 1991; Scott et al., 2001; Kaitibie et al., 2002; Fageria and Baligar, 2008). The application of lime at an appropriate rate brings several chemical and biological changes in the soil that are beneficial or helpful in improving crop yields on acid soils and improving root growth. Adequate liming eliminates soil acidity and toxicity of Al, Mn, and H; improves soil structure (aeration), the availability of Ca, P, Mo, and Mg, pH, and N_2 fixation; and reduces the availability of Mn, Zn, Cu, and Fe and the leaching loss of cations. These beneficial effects are responsible for better root growth of crop plants.

Liming raises soil pH, base saturation, Ca and Mg contents, and reduces aluminum concentration in Brazilian Oxisols (Fageria, 2000, 2001). The changes in these chemical properties of soil with the use of dolomitic lime $[CaMg(CO_3)_2]$ can be explained on the basis of the following equation (Fageria and Baligar, 2005, 2008):

$$CaMg(CO_3)_2 + 2H^+ \Leftrightarrow 2HCO_3^- + Ca^{2+} + Mg^{2+}$$

$$2HCO_3^- + 2H^+ \Leftrightarrow 2CO_2 + 2H_2O$$

$$CaMg(CO_3)_2 + 4H^+ \Leftrightarrow Ca^{2+} + Mg^{2+} + 2CO_2 + 2H_2O$$

The aforementioned equations show that the acidity neutralizing reactions of lime occur in two steps. In the first step, Ca and Mg react with H on the exchange complex and H is replaced by Ca^{2+} and Mg^{2+} on the exchange sites (negatively charged particles of clay or organic matter), forming HCO_3^-. In the second step, HCO_3^- reacts with H^+ to form CO_2 and H_2O to increase pH. Soil moisture and temperature and quantity and quality of liming material mainly determine the reaction rate of lime. To get maximum benefits from liming or for improving crop yields, liming materials should be applied in advance of crop sowing and thoroughly mixed into the soil to enhance its reaction with soil exchange acidity. Selected soil chemical properties that are influenced by applied lime in a Brazilian Oxisols are presented in Table 9.6. The data in Table 9.6 show that liming improves pH, Ca, and Mg content, base saturation, and reduces H^+ Al and acidity saturation. All these changes in the chemical properties of soil can improve root growth in limed acid soils.

Data presented in Table 9.7 show that the maximum root length (MRL) and root dry weight (RDW) of soybean and upland rice improved with the addition of lime in Brazilian Oxisols. The improvement in soybean root length as well as dry weight was quadratic with increasing lime rate in the range of 0–4.28 g kg^{-1} soil. Similarly, upland rice root length and dry weight were also increased in a quadratic fashion with the addition of lime. Figures 9.1 and 9.2 show the root growth of soybean and upland rice under different lime rates.

TABLE 9.6

Influence of Liming on Selected Soil Chemical Properties of Oxisols at 0–10 and 10–20 cm Depth

	Lime Rate (Mg ha⁻¹)				
Soil Property	**0**	**12**	**24**	**F-Test**	**CV (%)**
0–10 cm depth					
pH (1:2.5 soil water)	5.4c	6.7b	7.1a	**	2
Base saturation (%)	28.9c	72.0b	84.9a	**	19
H⁺ Al (cmol$_c$ kg⁻¹)	6.5a	2.0b	1.0c	**	14
Acidity saturation (%)	71.1a	27.8b	15.1c	**	14
Ca (cmol$_c$ kg⁻¹)	1.8c	3.6b	4.3a	**	9
Mg (cmol$_c$ kg⁻¹)	0.6b	1.3ᵃ	1.3a	**	12
10–20 cm depth					
pH	5.3c	6.2b	6.5a	**	3
Base saturation (%)	22.9c	51.2b	61.7a	**	24
H⁺ Al (cmol$_c$ kg⁻¹)	6.6a	3.8b	2.9c	**	13
Acidity saturation (%)	77.1a	49.0b	38.3c	**	12
Ca (cmol$_c$ kg⁻¹)	1.4c	2.7b	3.3a	**	15
Mg (cmol$_c$ kg⁻¹)	0.4b	1.0ᵃ	1.1a	**	15

Source: Fageria, N.K., *J. Plant Nutr.*, 29, 1219, 2006.

**Significant at the 1% probability level. Means followed by the same letter in the same line under different lime treatments are not statistically significant at the 5% probability level by Tukey's test.

9.3.2 BASE SATURATION

Base saturation is defined as the proportion of the CEC occupied by exchangeable bases. It is calculated as follows (Fageria et al., 2007; Fageria and Baligar, 2008):

$$\text{Base saturation (\%)} = \frac{\sum (\text{Ca, Mg, K, Na})}{\text{CEC at pH 7 or 8.2}} \times 100$$

where CEC is the sum of Ca, Mg, K, Na, H, and Al expressed in cmol$_c$ kg⁻¹.

In Brazil, Na⁺ is generally not determined because of a very low level of this element in Brazilian Oxisols (Raij, 1991). Hence, Na is not considered when calculating CEC or base saturation.

For crop production, base saturation levels in soil may be grouped into very low (lower than 25%), low (25%–50%), medium (50%–75%), and high (>75%) (Fageria and Baligar, 2008). Very low or low base saturation means a predominance of

TABLE 9.7

Influence of Liming on MRL (Maximum Root Length) and RDW (Root Dry Weight) of Soybean and Upland Rice at Physiological Maturity

Lime Rate (g kg⁻¹)	Soybean		Upland Rice	
	MRL Length (cm)	RDW (g Plant⁻¹)	MRL (cm)	RDW (g Plant⁻¹)
0	16.00	0.29	24.0	2.53
0.71	22.25	0.42	22.5	2.35
1.42	23.50	0.37	27.3	2.89
2.14	26.50	0.35	23.8	3.05
2.85	15.50	0.20	22.0	2.13
4.28	15.50	0.14	21.0	2.31
F-test	**	**	**	NS
CV (%)	10.65	12.82	8.08	26.54

Regression Analysis

Lime rate vs. MRL of soybean $(Y) = 17.6154 + 5.6909X - 1.5300X^2$, $R^2 = 0.4378$**

Lime rate vs. RDW of soybean $(Y) = 0.6938 + 0.3915X - 0.1169X^2$, $R^2 = 0.4954$**

Lime rate vs. MRL of upland rice $(Y) = 23.7836 + 1.0253X - 0.4141X^2$, $R^2 = 0.2592$*

Lime rate vs. RDW of upland rice $(Y) = 2.4817 + 0.2438X - 0.0714X^2$, $R^2 = 0.0583$NS

NS, nonsignificant.

*, **Significant at the 5% and 1% probability levels, respectively.

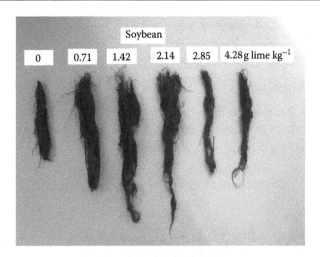

FIGURE 9.1   Soybean root growth at different lime rates.

adsorbed hydrogen and aluminum on the exchange complex. Deficiencies of calcium, magnesium, and potassium are likely to occur in soils with low CEC and very low to low percent base saturation.

For Brazilian Oxisols, the desired optimum base saturation for most cereals is in the range of 50%–60%, and for legumes it is in the range of 60%–70%

FIGURE 9.2   Upland rice root growth at different lime rate.

(Fageria et al., 1990). However, there may be exceptions, like upland rice, which is very tolerant to soil acidity and can produce good yield at base saturation lower than 50%. Specific optimal base saturation values for important annual crops grown on Brazilian Oxisols are given in Table 9.8. Nature of the soil alters the optimum base saturation required by any given crop specie. Bean yield had significant quadratic response in relation to base saturation (Figure 9.3). Maximum yield was obtained with base saturation of 73% at 0–10 cm soil depth, with base saturation of 62%

TABLE 9.8
Optimal Base Saturation for Important Annual Crops Grown on Brazilian Oxisols

Crop Species	Type of Experiment	Plant Part Measured	Base Sat. (%)	Reference
Common bean	Field	Grain yield	60	Fageria and Santos (2005)
Common bean	Field	Grain yield	69	Fageria and Stone (2004)
Upland rice	Field	Grain yield	40	Fageria and Baligar (2001)
Common bean	Field	Grain yield	70	Lopes et al. (1991)
Corn	Field	Grain yield	59	Fageria (2001)
Soybean	Field	Grain yield	63	Fageria (2001)
Upland rice	Field	Grain yield	50	Lopes et al. (1991)
Upland rice	Field	Grain yield	30	Sousa et al. (1996)
Common bean	Field	Grain yield	71	Fageria and Stone (2004)
Corn	Field	Grain yield	60	Raij et al. (1985)
Wheat	Field	Grain yield	60	Lopes et al. (1991)
Soybean	Field	Grain yield	60	Raij et al. (1985)
Cotton	Field	Grain yield	60	Raij et al. (1985)
Sugarcane	Field	Cane yield	50	Raij et al. (1985)
Soybean	Field	Grain yield	61	Gallo et al. (1986)

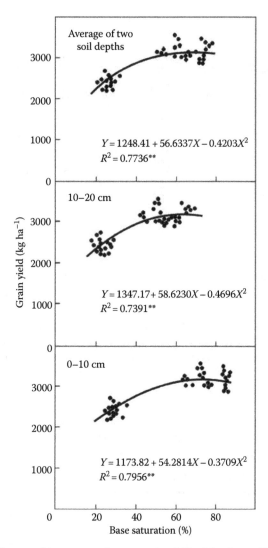

FIGURE 9.3 Influence of base saturation on grain yield of dry bean. **Significant at the 1% probability level. (From Fageria, N.K., *Commun. Soil Sci. Plant Anal.*, 39, 845, 2008.)

at 10–20 cm soil depth and at 67% base saturation when averaged across two soil depths. Hence, at topsoil layer higher base saturation was required compared with lower soil layer (Fageria, 2008). Optimum base saturation for grain yield is also optimum for root growth for a given crop in a given soil.

9.3.3 ALUMINUM SATURATION

Aluminum inhibits root growth and Al-injured roots are stubby and brittle. Root tips and lateral roots become thickened and may turn brown. The root system as a whole becomes corraloid, with many stubby lateral roots but reduced fine branching. Such roots are inefficient in absorbing both nutrients and water (Foy, 1992). Caires et al. (2008) reported that under unfavorable rainfall conditions, Al toxicity severely compromises root growth and yield of soybean. Aluminum toxicity in strongly acidic subsoil (pH 5 or below) is believed to be a major cause of shallow rooting and drought, particularly in the coastal plain of the southeastern United States (Foy, 1992). Excess Al can reduce water use by restricting both rooting depth and branching in acid subsoils (Ritchey et al., 1980).

Aluminum toxicity is a complex disorder that may be manifested as a deficiency of P, Ca, Mg, Fe, or as drought stress (Foy, 1992). Crops grown in soils with acceptable levels of basic cations do not show Al toxicity symptoms even when the levels of KCl-extractable Al are considered high (Kariuki et al., 2007). Hence, the mere presence of Al in the soil is not an indicator of Al toxicity (Johnson et al., 1997). A more reliable measure of the potential for Al toxicity is Al saturation (Kariuki et al., 2007). Aluminum saturation, or the proportion of aluminum among the cations, is calculated by using the following formula:

$$Al^{3+} \text{ saturation } (\%) = \frac{Al^{3+}}{ECEC} \times 100$$

where ECEC is effective cation exchange capacity in $cmol_c$ kg^{-1}, which is the sum of exchangeable Al^{3+}, Ca^{2+}, Mg^{2+}, and K^+ in $cmol_c$ kg^{-1}.

Critical Al saturation values for important plant species are given in Table 9.9. These values can be used as a reference guide to calculate lime rates for different crop species.

The effects of Al on root growth depend on the species of Al in the soil solution. The toxic effect of these species to root growth decreases in the following order: $Al^{3+} > Al(OH)^{2+} > Al(OH)_2^+ > Al(OH)_4^-$ (Alleoni et al., 2010). The species of aluminum ions present in soil solution vary with pH. In general, the net positive charge of the hydroxyl aluminum species decreases as the pH increases and then becomes negative in the alkaline pH range. The species of aluminum ions generates hydrogen ions through a series of hydrolysis reactions as shown in the following (Lindsay, 1979):

$$Al^{3+} + H_2O \Leftrightarrow Al(OH)^{2+} + H^+$$

$$Al^{3+} + 2H_2O \Leftrightarrow Al(OH)_2^+ + 2H^+$$

$$Al^{3+} + 3H_2O \Leftrightarrow Al(OH)_3^0 + 3H^+$$

$$Al^{3+} + 4H_2O \Leftrightarrow Al(OH)_4^- + 4H^+$$

$$Al^{3+} + 5H_2O \Leftrightarrow Al(OH)_5^{2-} + 5H^+$$

TABLE 9.9

Critical Soil Aluminum Saturation for Important Field Crops at 90%–95% of Maximum Yield

Crop	Type of Soil	Critical Al Saturation (%)
Cassava	Oxisols/Ultisols	80
Upland rice	Oxisols/Ultisols	70
Cowpea	Oxisols/Ultisols	55
Cowpea	Oxisols	42
Peanut	Oxisols/Ultisols	65
Peanut	Oxisols	54
Soybean	Oxisols	19
Soybean	Oxisols	27
Soybean	Oxisols/Ultisols	15
Soybean	Not given	<20
Corn	Oxisols	19
Corn	Oxisols/Ultisols	29
Corn	Oxisols/Ultisols	25
Corn	Oxisols	28
Mungbean	Oxisols/Ultisols	15
Mungbean	Oxisols/Ultisols	5
Coffee	Oxisols/Ultisols	60
Sorghum	Oxisols/Ultisols	20
Common bean	Oxisols/Ultisols	10
Common bean	Oxisols/Ultisols	8–10
Common bean	Oxisols/Ultisols	23
Cotton	Not given	<10

Source: Compiled from various sources by Fageria, N.K. et al., *Growth and Mineral Nutrition of Field Crops*, 3rd edn., CRC Press, Boca Raton, FL, 2011.

The exchangeable Al^{3+} precipitates as insoluble Al hydroxyl species as pH increases and is reported to decrease 1000-fold for each unit increase in pH (Lindsay, 1979). However, at pH values greater than 6.5, Al becomes increasingly soluble as negatively charged aluminates form (Haynes, 1984). The $Al(OH)^{2+}$ species is of minor importance and exists over only a narrow pH range. The Al^{3+} ion is predominant below pH 4.7, $Al(OH)_2^+$ between pH 4.7 and 6.5, $Al(OH)_3^0$ between pH 6.5 and 8.0, $Al(OH)_4^-$ above pH 8, and $Al(OH)_5^{2-}$ species occurs at pH values above those usually found in soils (Bohn et al., 2001; Fageria and Baligar, 2008).

Aluminum toxicity in acid soils can be reduced with liming. The rate of liming should be determined on the basis of crop response curves under each agroecological regions. However, in Brazil liming recommendations are made on the basis of Ca, Mg, and Al content in the soils as well as on base saturation values. The equation used for lime rate determination is (Fageria et al., 1990; Raij, 1991)

$$\text{Lime rate (Mg ha}^{-1}) = (2 \times \text{Al}^{3+}) + [2 - (\text{Ca}^{2+} + \text{Mg}^{+})]$$

where values of Al^{3+}, Ca^{2+}, and Mg^{2+} are expressed in cmol_c kg^{-1}. If values of Ca^{2+} and Mg^{2+} cations are more than $2\,\text{cmol}_c$ kg^{-1}, only Al multiplied by a factor of 2 is considered. This criterion was originally suggested by Kamprath (1970) for tropical soils and is still largely used for liming recommendation for Brazilian acid soils (Paula et al., 1987; Raij, 1991). Alvarez and Ribeiro (1999) recommended that the factor used to multiply Al should be varied according to soil texture. These authors suggested that in sandy soil with clay content of 0%–15%, a factor of 0–1 should be used; for medium-texture soils with clay content of 15%–35%, a factor of 1–2 should be used, for clayey soil with clay content of 35%–60%, a factor of 2–3 should be used; and for heavy clayey soil with clay content of 60%–100%, a factor of 3–4 should be used.

Similarly, the lime rate determined on the basis of base saturation method is calculated by using the following formula (Fageria et al., 1990; Fageria and Bligar, 2008):

$$\text{Lime rate (Mg ha}^{-1}) = \frac{\text{CEC}(B_2 - B_1)}{\text{TRNP}} \times \text{df}$$

where
  CEC is the cation exchange capacity or total exchangeable cations (Ca^{2+}, Mg^{2+}, K^+, H^+ + Al^{3+}) in cmol_c kg^{-1}
  B_2 is the desired optimum base saturation
  B_1 is the existing base saturation
  TRNP is the total relative neutralizing power of liming material
  df is the depth factor, 1 for 20 cm depth and 1.5 for 30 cm depth

For Brazilian Oxisols, the desired optimum base saturation for most cereals is in the range of 50%–60%, and for legumes it is in the range of 60%–70% (Fageria et al., 1990). However, there may be exceptions like upland rice, which is very tolerant to soil acidity and can produce good yield at base saturation lower than 50%.

The addition of organic matter can reduce Al toxicity in crop plants. Organic acids either secreted by plant roots or released from plant decomposing form Al-organic complexes, reducing the toxic Al in the soil solution. In addition, the use of Al-tolerant crop species or genotypes within species is another option for improving root growth in high-Al-content soils. This approach is very useful where lime is difficult to obtain and rather costly and Al-tolerant cultivars are available. It has been widely reported in the literature that differences in Al tolerance are found among plant species and cultivars within species (Foy, 1992; Kochian, 1995; Okada and Fischer, 2001; Fageria and Baligar, 2003; Yang et al., 2004).

9.3.4 Use of Gypsum

Gypsum ($CaSO_4 \cdot 2H_2O$), the common name for calcium sulfate and also known as phosphogypsum (PG), is an industrial by-product from phosphoric acid plants that is used as an ameliorant for acid soil infertility (Alva and Sumner, 1991). Table 9.10 reports gypsum production per country since a small amount of this product is used for agricultural purposes as a soil amendment and nutrient source. As an example, a little over 1 million MT of the 12.7 million MT of gypsum produced in the United States in 2008 was used in agriculture (Fixen, 2010). Alva and Sumner (1991) reported that PG contains 242 g kg^{-1} Ca, 0.37 g kg^{-1} Mg,

TABLE 9.10
Gypsum Production in Different Countries in 2008

Country	Gypsum Production (million MT)
China	40.7
United States	12.7
Iran	12.0
Spain	11.3
Thailand	8.8
Canada	7.3
Mexico	5.8
Japan	5.7
Italy	5.5
France	4.7
Australia	4.1
India	2.8
Russia	2.4
Egypt	2.0
Brazil	1.7
Germany	1.7
Poland	1.7
United Kingdom	1.7
Algeria	1.3
Uruguay	1.1
Austria	1.0

Source: Adapted from Fixen, P., World fertilizer nutrient reserves, in *Good Practices for Efficient Use of Fertilizers*, eds., Prochnow, L.I., Casarin, V., and Stipp, S.R., pp. 91–109, International Plant Nutrition Institute, Piracicaba, Brazil, 2010.

0.8 g kg^{-1} K, 2.8 g kg^{-1} P, 3.8 g kg^{-1} F, 189.3 mg kg^{-1} S, 1800 mg kg^{-1} Si, 711 mg kg^{-1} Al, 91 mg kg^{-1} Fe, and 65 mg kg^{-1} Mo. DeSutter and Cihacek (2009) gave the detailed chemical composition of commercial gypsum (Table 9.11). The high percentages of calcium (24.7%) and sulfur (20.7%) in commercial gypsum make this by-product an attractive fertilizer source. The form of calcium utilized by the plant is the Ca^{2+} and its function in the plant is to stimulate the development of roots and shoots, strength cell wall structure, activate some enzyme systems

TABLE 9.11
Chemical Properties of Commercial Gypsum

Chemical Property	Value	Nutrient Supplied by Application of 1 Mg Gypsum (kg ha^{-1})
Electrical conductivity (dS m^{-1})	7.5	—
pH (1:1 product/water ratio)	6.5	—
P (μg g^{-1})	13	0.013
K (μg g^{-1})	772	0.772
Ca (%)	24.7	247
S (%)	20.7	207
Mg (μg g^{-1})	1286	1.286
Al (μg g^{-1})	1516	1.516
B (μg g^{-1})	168	0.168
Cu (μg g^{-1})	<0.8	<0.0008
Fe (μg g^{-1})	906	0.906
Mn (μg g^{-1})	31	0.031
Mo (μg g^{-1})	0.8	0.0008
Zn (μg g^{-1})	6.1	0.0061
As (μg g^{-1})	<2.6	<0.0026
Ba (μg g^{-1})	68	0.068
Cd (μg g^{-1})	0.3	0.0003
Co (μg g^{-1})	0.7	0.0007
Cr (μg g^{-1})	4.8	0.0048
Li (μg g^{-1})	31	0.031
Ni (μg g^{-1})	2.3	0.0023
Pb (μg g^{-1})	1.2	0.0012
Sb (μg g^{-1})	7.0	0.007
Se (μg g^{-1})	<1.2	0.0012
Si (μg g^{-1})	1152	1.152
Sr (μg g^{-1})	1335	1.335
V (μg g^{-1})	5.0	0.005
Hg (μg g^{-1})	0.02	0.00002

Source: Adapted from DeSutter, T.M. and Cihacek, L.J., *Agron. J.*, 101, 817, 2009.

(Potash and Phosphate Institute, 1993), and it functions in cell wall division and membrane permeability (Westerman, 2005; DeSutter and Cihacek, 2009). However, gypsum has not been typically used as a sulfur source due to its low solubility (2.1 g L^{-1} for laboratory grade product) compared with ammonium sulfate (412 g L^{-1}) (DeSutter and Cihacek, 2009).

In addition, the application of gypsum brings several physical and chemical changes in soil that influence crop growth, including root system. Data presented in Table 9.12 show that gypsum improves Ca^{2+} content of soil and, thus, increases base saturation and reduces soil acidity effects on crop growth. The application of gypsum has been shown to increase the infiltration rate of soils (Keren and Shainberg, 1981), thereby resulting in a decrease in runoff and an increase in water use efficiency (Alva and Gascho, 1991). The application of gypsum has resulted in the alleviation of subsoil acidity in several studies conducted in Brazil, South Africa, and the southern United States (Shainberg et al., 1989). The application of gypsum or PG not only improves the Ca status of subsoils but also alleviates Al toxicity due to its precipitation and complexation reactions with SO$_4^{2-}$ and F$^-$ (Alva and Sumner, 1991). Although gypsum is not a liming material and has a minimal impact on soil pH (Sumner et al., 1986; Liu and Hue, 2001), it has been shown to ameliorate subsoil acidity by reducing the concentration of exchangeable Al^{3+} that can be toxic to some plants. Exchangeable Al^{3+} is most prevalent in soil when pH values are below 4.7 (Marion et al., 1976). Alva and Sumner (1991) cited several mechanisms of alleviating Al toxicity by PG. These mechanisms are as follows: (1) an increase in SO$_4^{2-}$ status of a soil solution following the application of PG favors ligand exchange reaction between SO$_4^{2-}$ and OH$^-$, resulting in the release of OH$^-$ ions that will precipitate Al, thereby decreasing Al toxicity; (2) the presence of Al and SO$_4^{2-}$ in acid media can result in the formation of basic Al sulfate minerals including jurbanite, basaluminite, or alunite, formation of these solid phases in turn results in the alleviation of Al toxicity; (3) specific sorption of SO$_4^{2-}$ results in

TABLE 9.12

Influence of Gypsum on Oxisol Chemical Properties after Harvesting Soybean Crop

Gypsum Rate (g kg^{-1})	pH in H$_2$O	H$^+$ Al (cmol$_c$ kg^{-1})	Ca (cmol$_c$ kg^{-1})	Base Sat. (%)	Soybean Grain Yield (g Plant^{-1})
0	4.93	3.50	0.57	24.83	1.28
0.28	4.80	2.37	0.90	36.54	6.82
0.57	5.07	2.67	1.23	39.14	7.89
1.14	5.03	2.50	1.60	44.20	7.98
1.71	5.27	2.40	2.27	52.64	7.79
2.28	5.07	2.40	22.80	57.02	7.60
R^2	0.11^{NS}	0.28^{NS}	0.98**	0.82**	0.71**

NS, nonsignificant.

**Significant at the 1% probability level.

an increase in negative charges, which results in the sorption of Al very strongly contributing to a decrease in Al toxicity; and (4) an increase in SO_4^{2-} concentration in solution has been shown to alleviate Al toxicity despite the fact that no precipitation of Al was evident in solution. Since Al readily forms ion pair with SO_4^{2-} ($Al^{3+} + CaSO_4 \Leftrightarrow AlSO_4^+ + Ca^{2+}$), a decrease in toxicity of Al in the presence of SO_4^{2-} was thought to be due to the less toxic nature of the $AlSO_4^+$ ion pair than the uncomplexed Al^{3+} ion.

Nobel et al. (1988) reported that ion pair formation ($AlSO_4^+$) is dependent on the solution pH; the magnitude of the alleviation of Al toxicity by $CaSO_4$ may be influenced by pH. These authors studied the influence of $CaSO_4$ on the root growth of soybean in nutrient solution (625–10.000 μM Ca) at two pH levels (4.2 and 4.8) and concluded that an increase in $CaSO_4$ in solution increased the root length by three- to twofold in solution at pH 4.2 and 4.8, respectively. The predicted activity of Al^{3+} decreased while that of $AlSO_4^+$ increased with an increase in added $CaSO_4$. The magnitude of the alleviation of Al toxicity by $CaSO_4$ was smaller at pH 4.8 than at pH 4.2, together with an increase in the formation of $Al(OH)_2^+$ at pH 4.8. Cameron et al. (1986) and Kinraide and Parker (1987) also reported that the $AlSO_4^+$ complex is less phytotoxic than the other Al species in solution.

Gypsum has been used to alleviate both physical and chemical growth-limiting factors for roots (Foy, 1992). Gypsum has been used extensively in the reclamation of sodic soils with infiltration problems. It is well known that the application of gypsum to sodic soils improves the soil physical conditions by promoting flocculation, enhancing aggregate stability, and increasing the infiltration rate (Lebron et al., 2002). Radcliffe et al. (1986) reported that gypsum increased subsoil root activities which, in turn, reduced mechanical impedance. Sumner and Carter (1987) applied gypsum to alfalfa crop and yield response was highly significant, which was attributed to a reduction in subsoil Al and an increase in Ca. Root examination showed that yield response was due to increased root growth in subsoil and increased resistance to drought. Farina and Channon (1988) concluded that the surface incorporation of gypsum (10 Mg ha^{-1}) is an economical, viable approach to the amelioration of acid subsoils of South Africa. Initial costs of the treatment were recovered within 3 years and benefits were expected to persist for many years thereafter. Four years after the surface application of gypsum, the treated plots had decreased levels of exchangeable Al and increased exchangeable Ca, Mg, and SO_4-S in subsoil, and grain yields were increased by 13.4 Mg ha^{-1}.

Gypsum ($CaSO_4 \cdot 2H_2O$) provides a more mobile source of Ca than $CaCO_3$ because the SO_4^{2-} ions can easily comigrate with Ca^{2+}. Surface applied or incorporated gypsum can, therefore, significantly improve root-growing conditions by increasing exchangeable Ca and reducing exchangeable Al, without neutralizing any acidity (Wendell and Ritchey, 1996; Willert and Stehouwer, 2003). The effect can be very long lasting. Toma et al. (1999) found a significant and deep-reaching Ca increase in the subsoil of an Ultisol 16 years after a one-time gypsum application. Miller (1987) found that spreading gypsum at the soil surface significantly increased infiltration and decreased runoff and erosion under rainfall conditions for three typical soils from southeast United States. Similarly, Yu et al. (2003) also reported that the application of gypsum at the rate of 4 Mg ha^{-1} improved infiltration rate and decreased soil erosion. Gypsum dissolution maintained high concentration of electrolytes in the soil solution at the soil surface during a rainstorm, thus preventing chemical dispersion of the clay particles and the formation of low infiltration seal (Keren and Shainberg, 1981; Shainberg et al., 1990). PG was more effective than mined gypsum in decreasing seal formation because of its higher rate of dissolution (Shainberg et al., 1990).

Roots of many crop cultivars grew well in acid soils of pH 4.3–4.6 treated with $CaSO_4 \cdot 2H_2O$ (Inoue et al., 1988; Nobel et al., 1988; Foy, 1992). The author studied the influence of gypsum on the root growth of wheat, soybean, and upland rice grown on a Brazilian Oxisol (Tables 9.13 and 9.14). Root length as well as RDW were significantly affected with the addition of gypsum. Both these parameters increased in a quadratic fashion with the addition of gypsum in the range of 0–2.32 g kg^{-1} soil, except the RDW of rice. The RDW of rice increased linearly with the increasing gypsum rate. The variation in RDW of three species was higher compared to variation in root length. Hence, it can be concluded that RDW was more sensitive to gypsum application compared to root length. Figures 9.4 and 9.5 show the influence of gypsum on the root growth of soybean and upland rice. The Cooperative Extension Services in most states (United States) that grow peanut recommend that peanut-seed-producing farmers apply supplemental Ca, usually in the form of gypsum, regardless of the soil Ca test value (Tillman et al., 2010).

TABLE 9.13
Influence of Gypsum on MRL and RDW of Wheat and Soybean

Gypsum Rate (g kg^{-1})	Wheat		Soybean	
	MRL (cm)	RDW (g Plant^{-1})	MRL (cm)	RDW (g Plant^{-1})
0	10.30	0.05	16.00	0.53
0.28	14.50	0.39	22.25	1.05
0.57	17.25	0.40	23.50	1.16
1.14	18.75	0.45	26.50	1.23
1.71	14.25	0.34	15.50	0.35
2.28	13.75	0.15	15.50	0.37
F-test	**	**	**	**
CV (%)	16.65	14.82	10.65	12.82

Regression Analysis

Gypsum rate vs. MRL of wheat $(Y) = 11.3184 + 10.8081X + 4.4155X^2$, $R^2 = 0.4609$**
Gypsum rate vs. RDW of wheat $(Y) = 0.1353 + 0.5993X - 0.2606X^2$, $R^2 = 0.6676$**
Gypsum rate vs. MRL of soybean $(Y) = 15.4256 + 11.2965X - 5.0707X^2$, $R^2 = 0.2071$NS
Gypsum rate vs. RDW of soybean $(Y) = 0.6063 + 0.8793X - 0.3671X^2$, $R^2 = 0.6311$**

NS, nonsignificant.
**Significant at the 1% probability level.

TABLE 9.14

Influence of Gypsum on Maximum Root Length (MRL) and Root Dry Weight (RDW) of Upland Rice

Gypsum Rate (g kg⁻¹)	MRL (cm)	RDW (g Plant⁻¹)
0	30	21
0.28	27	17
0.57	29	16
1.14	31	25
1.71	31	23
2.28	31	31
F-Test	**	**
CV (%)	3.20	13.62

Regression Analysis

Gypsum rate vs. MRL (Y)
  $= 28.2090 + 2.1303X - 0.4400X^2$, $R^2 = 0.2951*$
Gypsum rate vs. RDW (Y) $= 4.2671 + 1.2811X$,
  $R^2 = 0.5242**$

*, **Significant at the 5% and 1% probability levels, respectively.

FIGURE 9.4 Root growth of soybean at different gypsum levels.

The application rates of gypsum to ameliorate subsoil acidity have ranged from 2.5 to 35 Mg ha⁻¹ (Sumner et al., 1986; Smith et al., 1994; Toma et al., 1999; Farina et al., 2000). Raij (2010) recommended the gypsum rate for maximum yield of corn to be 8 Mg ha⁻¹, sugarcane 2–10 Mg ha⁻¹ (depending on soil type and fertility level), soybean and cotton 6 Mg ha⁻¹ in Brazilian soils. An unintended consequence of

FIGURE 9.5   Upland rice root growth at different gypsum rate.

applying high rates of gypsum to soil may be the exchange of Ca^{2+} for Mg^{2+} or K^+ and the subsequent leaching of Mg^{2+} and K^+ (Ernani et al., 2006; Jalali and Rowell, 2008). In some soils, the use of gypsum as a fertilizer source of SO_4^{2-} may be more desirable than using ammonium sulfate due to soil acidification concerns (Shainberg et al., 1989). Gypsum is also used to ameliorate sodic soils. Sodic soil remediation requires the displacement of Na^+ from the soil exchange sites with Ca^{2+}, where the most common Ca^{2+} source used for this purpose is gypsum (Shainberg et al., 1982; Frenkel et al., 1989).

9.3.5   SOIL ORGANIC MATTER CONTENT

Maintaining SOM at an adequate level is one of the most important factors in improving the root growth of crop plants. SOM is key to soil functioning in agroecosystems, both influencing and being influenced by environmental conditions and fluxes of matter and energy (Viaud et al., 2010). It is a key component of soil quality because it directly affects the chemical, physical, and biological properties of soil and plays a crucial role in sustaining soil fertility (Tiessen et al., 1994) and environmental quality (Lal, 2009). SOM is universally recognized to be among the most important factors responsible for soil fertility, crop production, and land protection from contamination, degradation, erosion, and desertification (Santos et al., 2010). Changes in SOM content, composition, or dynamics can greatly modify nutrient availability (Agboola and Corey, 1973), soil aggregation and structural stability (Tisdall and Oades, 1982), erodibility (Le Bissonnais and Arrouays, 1997; Lal, 2005; Lal and Pimentel, 2008), porosity (Emerson and McGarry, 2003), water-holding capacity (Haynes and Naidu, 1998), and biological activity (Fonte et al., 2009). In addition to having a direct impact on soil itself, SOM has a strong impact on the local and global C cycles (Smith et al., 1997); even small changes in the soil organic pool may change the global C cycle (Johnson et al., 2004). Lal (2004, 2007), Trumbore and Czimczik

(2008), and Glumac et al. (2010) reported that soil organic carbon is the largest terrestrial C pool and its fluctuation may have a significant influence on atmospheric CO_2 levels and global climate.

The amount of organic matter in a soil is a function of the rates of C gains and losses from the soil under specific land use as well as the quantity and quality of organic matter inputs (Paustian et al., 1997; Paul et al., 2001; Johnson et al., 2006; Collins et al., 2010). Reduced levels of SOM in agricultural soils have been attributed to erosion and decomposition caused by intensive cultivation, resulting in a loss of up to 60% of soil C reserve (Paustian et al., 1997). Organic matter content of well-drained mineral surface soils varied from 1% to 6% or 10–60 g kg^{-1} (Brady and Weil, 2002). SOM can be maintained and/or improved with the addition of crop residues, organic manures, conservation tillage, and crop rotation. The change from cropping of C_3 to C_4 plants results in a change in the C content of the soil and provides an excellent marker to determine the source of soil C, C turnover rates, and C sequestration (Gregorich et al., 1996; Qian and Doran, 1996; Collins et al., 1999, 2000, 2010; Clay et al., 2007).

9.3.6 SOIL FERTILITY

Soil infertility (natural element deficiencies or unavailability) is probably the single most important factor limiting crop yields worldwide. Soil fertility decreases due to intensive crop production without adopting adequate soil and crop management practices. In addition, soil erosion by wind and water is another important factor that is responsible for decreasing soil fertility of arable lands. Poesen et al. (2003) estimated that, on average, 44% of the total soil erosion worldwide was by gullies, whereas the NRCS (1997) estimated that 35% of the total soil loss was by gully erosion in the United States. Fox and Wilson (2010) reported that for more gentle slopes, gully erosion, which is a form of hillslope failure, can be a significant source of stream sediment.

Optimal soil fertility is an important factor in modern agriculture to obtain higher crop yields. Higher crop yields are associated with better root systems that can supply adequate nutrients and water to the plants. To maintain soil fertility at an optimal level, the basic principles of soil fertility should be followed. These are the use of appropriate sources, adoption of effective method of applications, and applying adequate rates of chemical and/or organic manures. In addition, the appropriate timing of fertilizer application is also important for supplying essential nutrients to plants during growth cycle. Furthermore, control of soil erosion is also helpful in maintaining soil fertility.

The use of organic manures is one of the options to improve organic matter and nutrient contents of arable lands. Land application of organic manures recycle nutrients and maintain soil fertility (Moore et al., 2005; Havlin et al., 2006; Grijalva et al., 2010). The quantity of organic manure application varies with animal production systems, bedding materials, storage, and processing methods (Sims and Wolf, 1994; Chadwick et al., 2000; Siddique and Robinson, 2003; Grijalva et al., 2010). Obour et al. (2010) reported that the application of manure in combination with inorganic fertilizers is an alternative that can be used to improve soil fertility and crop yields.

Evers (2002) reported that the combination of broiler litter and commercial N fertilizer increased P removal in an annual ryegrass (*Lolium multiflora* L.)–Bermuda grass (*Cynodon dactylon* L. Pers.) pasture. Similarly, Obour et al. (2009) reported that addition of inorganic N fertilizer with organic manure may also provide readily available N that can promote yields and P uptake.

Fertilizer sources may have different reactions in soil, which may affect the availability of nutrients, and should be given due consideration in evaluating their efficiency in crop production. Mulvaney (1994) showed that nitrification occurred in the order of urea $[CO(NH_2)_2]$ > diammonium phosphate $[(NH_4)_2HPO_4]$ > ammonium sulfate $[(NH_4)_2SO_4]$ > ammonium nitrate $[(NH_4NO_3]$ > monoammonium phosphate $[NH_4H_2PO_4]$. This difference in nitrification rate has been attributed, in part, to the increase in soil pH associated with the hydrolysis of urea and diammonium phosphate soon after application, in contrast to the decline in soil pH associated with the application of ammonium sulfate and monoammonium phosphate (Lindsay et al., 1962; Allred and Ohlrogge, 1964). The use of rapidly nitrifying fertilizer materials and application well before plant uptake (i.e., in the fall) increases the accumulation of NO_3^-, thereby increasing the risk of N loss from denitrification or leaching when soils are excessively wet in the spring before rapid N uptake by plants (Fernandez et al., 2010). Further, research has shown that the presence of ammonium (NH_4^+) ions with P results in increased P uptake (Raun et al., 1987).

There are several sources of chemical fertilizers that can be used to maintain soil fertility at an adequate level. The principle sources of N, P, K, and micronutrients are listed in Tables 9.15 through 9.18, respectively. When soil fertility is low, immobile nutrients like P and K should be applied in the furrow or band near the

TABLE 9.15
Major Nitrogen Fertilizers, Their Chemical Formulas, and N Contents

Common Name	Formula	N (%)
Ammonium sulfate	$(NH_4)_2SO_4$	21
Urea	$CO(NH_2)_2$	46
Anhydrous ammonia	NH_3	82
Ammonium chloride	NH_4Cl	26
Ammonium nitrate	NH_4NO_3	35
Potassium nitrate	KNO_3	14
Sodium nitrate	$NaNO_3$	16
Calcium nitrate	$Ca(NO_3)_2$	16
Calcium cyanamide	$CaCN_2$	21
Ammonium nitrate sulfate	$NH_4NO_3(NH_4)_2SO_4$	26
Nitrochalk	$NH_4NO_3 + CaCO_3$	21
Monoammonium phosphate	$NH_4H_2PO_4$	11
Urea ammonium nitrate	$CO(NH_2)_2 + NH_4NO_3$	32
Diammonium phosphate	$(NH_4)_2HPO_4$	18

TABLE 9.16
Major Phosphorus Fertilizers, Their P Content, and Solubility

Common Name	Chemical Composition	P_2O_5 Content (%)	Solubility
Simple superphosphate	$Ca(H_2PO_4)_2 + CaSO_4$	18–22	Water soluble
Triple superphosphate	$Ca(H_2PO_4)_2$	46–47	Water soluble
Monoammonium phosphate	$NH_4H_2PO_4$	48–50	Water soluble
Diammonium phosphate	$(NH_4)_2HPO_4$	54	Water soluble
Phosphoric acid	H_3PO_4	55	Water soluble
Thermophosphate (yoorin)	$[3MgO \cdot CaO \cdot P_2O_5 + 3(CaO \cdot SiO_2)]$	17–18	Citric acid soluble
Rock phosphates	Apatites	24–40	Citric acid soluble
Basic slag	$Ca_3P_2O_8 \cdot Cao + CaO \cdot SiO_2$	10–22	Citric acid soluble

TABLE 9.17
Principal Potassium Fertilizers, Their Potassium Content, and Solubility

Common Name	Formula	K_2O (%)	Solubility
Potassium chloride	KCl	60	Water soluble
Potassium sulfate	K_2SO_4	50	Water soluble
Potassium–magnesium sulfate	$K_2SO_4 \cdot MgSO_4$	23	Water soluble
Potassium nitrate	KNO_3	44	Water soluble
Kainit	$MgSO_4 + KCl + NaCl$	12	Water soluble
Potassium metaphosphate	KPO_3	40	Low water solubility

root growth for better absorption. This practice is especially useful in soils that have high P immobilization capacity, like Oxisols and Ultisols. The adequate rate of nutrient application depends on crop growth response curves and/or soil test calibration curves. In the case of a soil mobile nutrient like N, adequate rate should be determined based on N rate versus grain yield curve. Adequate rate is where maximum grain yield is obtained. Such curve for upland rice grown on a Brazilian Oxisol is shown in Figure 9.6 for N and Figure 9.7 for P. The responses of four genotypes to N and three genotypes to P were significant and quadratic. Usually, grain yields of crops increase in response to nutrient application and then level off when they reach their maximum (McSwiney and Robertson, 2005; Gagnon and Ziadi, 2010). When the optimal N or P application is exceeded, N or P accumulation may continue to increase without affecting yield (Isfan et al., 1995), and the recovery of added N or P is higher at low applications (Jokela and Randall, 1997; Tran et al., 1997).

TABLE 9.18

Principal Sources of Micronutrient Fertilizers to Correct Deficiencies

Micronutrient	Source	Formula	Element (%)	Solubility
Zinc	Zinc sulfate (monohydrate)	$ZnSO_4 \cdot H_2O$	36	Soluble
	Zinc sulfate (heptahydrate)	$ZnSO_4 \cdot 7H_2O$	23	Soluble
	Zinc chloride	$ZnCl_2$	48–50	Soluble
	Zinc oxide	ZnO	50–80	Insoluble
	Basic Zinc sulfate	$ZnSO_4 \cdot 4Zn(OH)_2$	55	Slightly soluble
	Zinc chelate	$Na_2ZnEDTA$	14	Soluble
	Zinc chelate	$NaZnEDTA$	9	Soluble
	Zinc frits	Fritted glass	4–9	Slightly soluble
Copper	Copper sulfate (monohydrate)	$CuSO_4 \cdot H_2O$	35	Soluble
	Copper sulfate (pentahydrate)	$CuSO_4 \cdot 5H_2O$	25	Soluble
	Copper chloride	$CuCl_2$	47	Soluble
	Cuprous oxide	Cu_2O	89	Insoluble
	Cupric oxide	CuO	75	Insoluble
	Copper chelate	$Na_2CuEDTA$	13	Soluble
	Copper chelate	$NaCuHEDTA$	9	Soluble
Boron	Boric acid	$H_3BO_3[B(OH)_3]$	17	Soluble
	Borax	$Na_2B_4O_7 \cdot 10H_2O$	11	Soluble
	Sodium borate (anhydrous)	$Na_2B_4O_7$	20	Soluble
	Sodium (penta borate)	$Na_2B_{10}O_{16} \cdot 10H_2O$	18	Soluble
	Boron frits	Fritted glass	1.5–2.5	Slightly soluble
Iron	Ferrous sulfate (monohydrate)	$FeSO_4 \cdot H_2O$	33	Soluble
	Ferrous sulfate (heptahydrate)	$FeSO_4 \cdot 7H_2O$	19	Soluble
	Ferrous ammonium sulfate	$(NH_4)_2SO_4 \cdot FeSO_4 \cdot 6H_2O$	14	Soluble
	Ferric sulfate	$Fe_2(SO_4)_3 \cdot H_2O$	23	Soluble
	Fe chelate	$NaFEEDTA$	5–14	Soluble
	Fe chelate	$NaFeHEDTA$	5–9	Soluble
	Fe chelate	$NaFeEDDHA$	6	Soluble
	Fe chelate	$NaFeDTPA$	10	Soluble
	Fe frits	Fritted glass	2–6	Slightly soluble
Molybdenum	Sodium molybdate	$Na_2MoO_{24} \cdot 2H_2O$	39	Soluble
	Ammonium molybdate	$(NH_4)_6Mo_7O_{24} \cdot 4H_2O$	54	Soluble
	Mo-trioxide	MoO_3	66	Slightly soluble
	Molybdic acid	$H_2MoO_4 \cdot H_2O$	53	Soluble
	Mo-frits	Fritted glass	0.1–0.4	Slightly soluble

TABLE 9.18 (continued)
Principal Sources of Micronutrient Fertilizers to Correct Deficiencies

Micronutrient	Source	Formula	Element (%)	Solubility
Manganese	Mn sulfate (anhydrous)	$MnSO_4$	23–28	Soluble
	Mn sulfate (tetrahydrate)	$MnSO_4 \cdot 4H_2O$	26–28	Soluble
	Mn chloride	$MnCl_2$	17	Soluble
	Mn carbonate	$MnCO_3$	31	Insoluble
	Mn oxide	MnO	41–68	Insoluble
	Mn chelate	$Na_2MnEDTA$	5–12	Soluble
	Mn frits	Fritted glass	2–10	Slightly soluble
Chlorine	K chloride	KCl	48	Soluble
	Zn chloride	$ZnCl_2$	52	Soluble
	Ca chloride	$CaCl_2$	64	Soluble
	Mn chloride	$MnCl_2$	44	Soluble
Nickel	Ni chloride	$NiCl_2 \cdot 6H_2O$	25	Soluble
	Ni nitrate	$Ni(NO_3)_2 \cdot 6H_2O$	20	Soluble
	Ni oxide	NiO		Insoluble

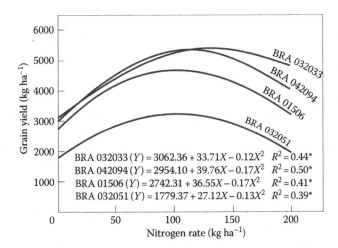

$$BRA\ 032033\ (Y) = 3062.36 + 33.71X - 0.12X^2 \quad R^2 = 0.44^*$$
$$BRA\ 042094\ (Y) = 2954.10 + 39.76X - 0.17X^2 \quad R^2 = 0.50^*$$
$$BRA\ 01506\ (Y) = 2742.31 + 36.55X - 0.17X^2 \quad R^2 = 0.41^*$$
$$BRA\ 032051\ (Y) = 1779.37 + 27.12X - 0.13X^2 \quad R^2 = 0.39^*$$

FIGURE 9.6 Relationship between N rate and grain yield of three upland rice genotypes grown on a Brazilian Oxisol. *Significant at the 5% probability level.

9.3.7 SALINITY

Salt-affected soils can be defined as those soils that have been adversely modified for the growth of most crop plants by the presence of soluble salts, with or without high amounts of exchangeable sodium (Soil Science Society of America, 2008). Common ions contributing to this problem are Ca^{2+}, Mg^{2+}, Cl^-, Na^+, SO_4^{2-}, HCO_3^-, and in some cases K^+ and NO_3^- (Bernstein, 1975). However, Diaz et al. (2011),

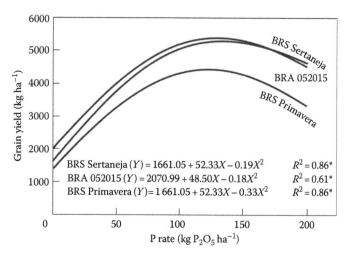

FIGURE 9.7 Relationship between P rate and grain yield of three upland rice genotypes grown on a Brazilian Oxisol. *Significant at the 5% probability level.

Tester and Davenport (2003), and Munns and Tester (2008) reported that Na^{2+} and Cl^- are the two key ions responsible for both osmotic and ion-specific damage that significantly reduce crop growth, yield, and productivity. The osmotic effects of salt stress can be observed immediately after salt application and are believed to continue during the time of exposure, resulting in an inhibition of cell expansion and cell division, as well as stomatal closure (Munns, 2002; Munns and Tester, 2008). The ionic stress results in premature senescence of older leaves and in toxicity symptoms (chlorosis, necrosis) in mature leaves (Munns, 2002; Tester and Davenport, 2003; Munns et al., 2006).

Salt-affected soils limit crop production around the world. Of the 1.5 billion ha arable land in the world, 75 million ha are affected by salinity (Munns, 2002; Zhou et al., 2010). Crop stress related to soil salinity is very common in irrigated agroecosystems. Globally, 20% of the irrigated land is affected by salinity (Munns, 2005). Civilizations have been destroyed by the encroachment of salinity on the soils; as a result vast areas of the land are rendered unfit for agriculture. Salt-affected soils normally occur in arid and semiarid regions where rainfall is insufficient to leach salts from the root zone. Salt problems, however, are not restricted to arid or semiarid regions. They can develop even in subhumid and humid regions under appropriate conditions (Bohn et al., 2001). In addition, these soils may also occur in coastal areas subject to tides. Salts generally originate from native soil and irrigation water. Roughly the land area of 263 million ha is irrigated worldwide and in most of that area salinity is a growing threat (Epstein and Bloom, 2005). The irrigated area represents about 20% of the total land used for crop production (Fageria, 1992). This represents about 19% of the total area of the world under crop production. The use of inappropriate levels of fertilizers with inadequate management practices can create saline conditions even in humid conditions.

Salinity at high level of salts (2%), primarily sodium chloride, usually poses mortal threats to germination, seedling growth, and establishment for profitable production of conventional crops (Ashraf and Foolad, 2005; Zhou et al., 2010). The predominant influence of salinity on plants is growth suppression. Typically, growth decreases linearly as salinity increases beyond a threshold salinity level (Figures 9.8 and 9.9). The effect is similar whether salinity is increased by raising the concentration of nutrients or by adding nonnutrient salts such as sodium chloride, sodium sulfate, or calcium chloride, which are common in saline soils (Hoffman, 1981). Figures 9.10 through 9.12 show the effects of salinity on lowland rice growth. At higher salinity levels (10 and 12 dS m^{-1}) none of the three cultivars of lowland rice survived. However, there was difference in the growth of these cultivars at low as well as at high salinity levels. Salt-affected plants usually appear normal although they are stunted in stature. Salt-affected leaves are smaller and have a dark blue–green color than normal leaves. Chlorosis is not a typical characteristic of salt-affected plants. Wilting is not a regular characteristic of salt-affected plants because it typically occurs when water availability decreases rather abruptly, as in a drying soil.

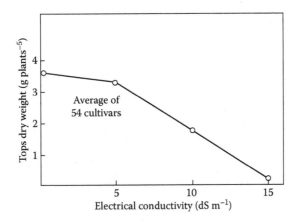

FIGURE 9.8 Dry weight of tops of lowland rice cultivars as influenced by salinity levels.

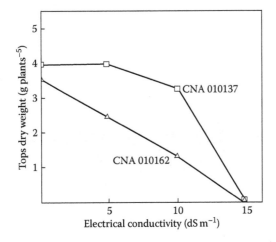

FIGURE 9.9 Influence of salinity on dry weight of tops of two lowland rice genotypes.

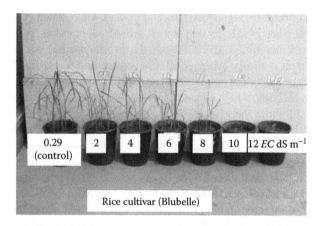

FIGURE 9.10 Growth of lowland rice cultivar Blubelle under different salinity levels.

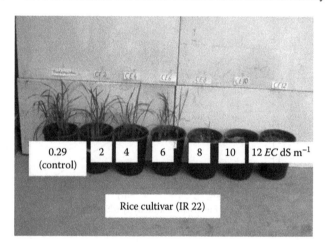

FIGURE 9.11 Growth of lowland rice cultivar IR 22 under different salinity levels.

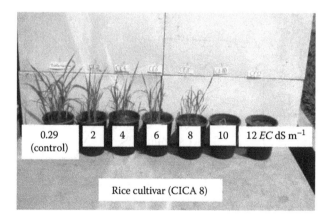

FIGURE 9.12 Growth of lowland rice cultivar CICA 8 under different salinity levels.

Under saline conditions moderately low water potentials are always present and water potential changes are usually gradual (Hoffman, 1981). Salt tolerance in plants depends mainly on the capability of roots for restricted or controlled uptake of sodium and chloride and the continued uptake of essential elements, particularly potassium and nitrate (Ashraf et al., 2004).

Salinity adversely affects photosynthesis and mineral uptake and accumulation in plants (Zhou et al., 2010). In the salt-affected environment, there is a preponderance of nonessential elements over essential elements. In salt-affected soils, plants must absorb the essential nutrients from a diluted source in the presence of highly concentrated nonessential nutrients. This requires extra energy, and plants sometimes are unable to fulfill their nutritional requirements. There are two main stresses imposed by salinity on plant growth. One is water stress, imposed by the increase in the osmotic potential of the rhizosphere as a result of high salt concentration. Another stress is the toxic effect of high concentration of ions. Hale and Orcutt (1987) reported that if the salt concentration is high enough to lower the water potential by 0.05–0.1 MPa, then the plant is under salt stress. If the salt concentration is not high enough, it is ion stress and may be caused by one particular species of ion (Hale and Orcutt., 1987).

The most common method of quantifying soil salinity is to measure the electrical conductivity of saturation extracts of the soil in the root zone. The lower limit of saturation extract electrical conductivity of salt-affected soils is conventionally set at 4 dS m^{-1} at 25°C (Soil Science Society of America, 2008). Root growth is significantly reduced in salt-affected soils. To improve root growth in salt-affected soils, salinity level should be reduced to the threshold level. Salinity threshold level varies from crop species to crop species and genotypes within species. Crops have most commonly been classified as tolerant, moderately tolerant, moderately sensitive, or sensitive, in terms of their response to increasing levels of salinity (Francois and Maas, 1999). However, the work of Steppuhn et al. (2005) and Brogan et al. (2011) suggested that a salinity tolerance index based on a nonlinear response function of crop growth to increasing salinity would most closely reflect agricultural crop response to root-zone salinity and be more useful in making comparisons between crops in terms of their relative salt tolerance. Salt tolerance of important food crops is given in Table 9.19. After these threshold values, crop yield decreases linearly with increasing salt concentrations (Maas and Hoffman, 1977).

Successful crop production on salt-affected soils depends on soil, water, and plant management practices. Management practices that can improve crop yields and, consequently, the root growth of crop plants grown on salt-affected soils are the use of soil amendments to reduce the effect of salts, the application of farmyard manures to create favorable plant growth environments, leaching salts from soil profile, and planting salt-tolerant crop species or genotypes within species (Ashraf et al., 2008). The incorporation of farmyard manure into saline soil is a traditional method to combat salt stress, especially during the germination and emergence stages. During the past decade, a technique of culturing corn and cotton seedlings in plastic matrices containing compost manure was successfully developed and widely adopted in double-crop cropping production regions in China (Zhou et al., 2010). The addition of fertilizers, especially potassium, may also help in reducing salinity effects and improving nutrient use efficiency. The maintenance of an internal positive turgor

TABLE 9.19

Salt Tolerance of Important Fiber, Food, and Special Crops at 100% Yield Potential

Crop Species	EC (dS m^{-1}) at 25°C	Slope (% per dS m^{-1})	Rating
Rice	3.0	12.0	Sensitive
Wheat	6.0	7.1	Moderately tolerant
Corn	1.7	12.0	Moderately sensitive
Barley	8.0	5.0	Tolerant
Sorghum	6.8	16.0	Moderately tolerant
Dry bean	1.0	19.0	Sensitive
Soybean	5.0	20.0	Moderately tolerant
Sugar beet	7.0	5.9	Tolerant
Potato	1.7	12.0	Moderately sensitive
Sugarcane	1.7	5.9	Moderately sensitive
Cotton	7.7	5.2	Tolerant
Cowpea	4.9	12.0	Moderately tolerant
Flax	1.7	12.0	Moderately sensitive
Peanut	3.2	29.0	Moderately sensitive
Sugar beet	7.0	5.9	Tolerant
Sunflower	—	—	Moderately tolerant
Potato	1.7	12.0	Moderately sensitive
Sweet potato	1.5	11.0	Moderately sensitive
Tomato	2.5	9.9	Moderately sensitive
Turnip	0.9	9.0	Moderately sensitive
Radish	1.2	13.0	Moderately sensitive
Onion	1.2	16.0	Sensitive
Guar	8.8	—	Moderately tolerant
Rye	11.4	10.8	Tolerant
Triticale	6.1	2.5	Tolerant

Source: Adapted from Maas, E.V., Testing crops for salinity tolerance, in: *A Workshop on Adaptation of Plants to Soil Stresses*, ed., Maranville, J.W., pp. 234–247, University of Nebraska, Lincoln, NE, 1993.

potential of plants exposed to saline conditions is an important factor for maintaining growth. This is accomplished by the uptake of ions, chiefly K^+, Na^+, and Cl^-, as well as by synthesizing organic metabolites (Yeo, 1983). Potassium is the most abundant cation in the cytoplasm and in glycophytes plays an important role in osmotic adjustment (Marschner, 1995; Munns and Tester, 2008). Thus, the application of high K^+ fertilization might enhance the capacity for osmotic adjustment of plants growing in saline habitats. Root growth is often much less affected by salinity than shoot growth, in common with the effect of dry soil, suggesting that the effect is probably due to factors associated with water stress than a salt-specific effect (Munns, 2002).

9.3.8 ALKALINITY

In alkaline soil, plant growth is reduced due to the excess amount of exchangeable sodium. The exchangeable sodium ratio is greater than 0.15, the conductivity of the soil solution at saturated water content is more than $4\,dS\,m^{-1}$ at 25°C, and pH is usually 8.5 or less in the saturated soil (Soil Science Society of America, 2008). There are about 558 million ha sodic soils worldwide (Lal et al., 1989). In Australia, approximately 23% of the total land area and 80% of the irrigated agricultural area (Rengasamy and Olsson, 1993) are affected by sodicity. A study by the Cooperative Research Center for Soil and Land Management in 1997 estimated that, in Australia, soil sodicity reduced farmers' incomes by $1.3 billion annually (Ghosh et al., 2010).

Alkali stress is the stress of alkaline salts and it may be more severe than salinity stress (Campbell and Nishio, 2000; Yang et al., 2007, 2010). When a salinized soil contains alkaline salt, thus raising soil pH, there is damage to plants from both salt and alkali stresses (Yang et al., 2010). The high pH environment surrounding the roots can cause the loss of the normal physiological functions of the roots and the destruction of root cell structure (Yang et al., 2008a–c). Alkaline stress can inhibit absorption of ions such as Cl^-, NO_3^-, and $H_2PO_4^-$, greatly affect the selective absorption of K, and disrupt the ionic balance (Yang et al., 2007, 2008b). Thus, plants in alkaline soil must cope with physiological drought and ion toxicity, and also maintain stable intercellular pH and regulate pH outside the roots (Yang et al., 2010).

Alkaline stress can be remediated by organic acid synthesis and secretion by the roots of alkalinity-tolerant plants. In addition, low molecular weight organic acids occur widely in the soil environment as natural products of root exudates, microbial secretions, and the decomposition of plant and animal residues. The most common low molecular weight organic acids identified to date in soils include oxalic, succinic, tartaric, fumaric, malic, and citric acids, among others (Zhang et al., 1997; Nardi et al., 2005; Gao et al., 2010). Organic acids are key components in mechanisms that some plants use to cope with drought (Timpa et al., 1986), Al^{3+} toxicity (Li et al., 2000), P deficiency (Koyama et al., 2000), Fe deficiency (Lopez-Millan et al., 2000), heavy metal stress (Lopez-Bucio et al., 2000), and plant microbial interactions at the root–soil interface (Lopez-Bucio et al., 2000). In recent years, reports have shown that some alkali-tolerant halophytes accumulate high concentration of organic acids under alkali stress (Yang et al., 2007, 2010). The organic acids secretion is important in protecting roots from high pH injury caused by alkali stress.

Traditionally, soil sodicity is managed by the use of gypsum and organic materials. The use of gypsum can replace the Na, which can be leached below the root zone by applying irrigation water of good quality (Ghosh et al., 2010). Typical organic materials used to ameliorate sodic soils include animal manure (Haynes and Naidu, 1998), sewage sludge (Albiach et al., 2001), and crop residue (De Neve and Hoffman, 2000). Tejada et al. (2006) reported significant decrease in exchangeable sodium due to the application of crushed cotton gin compost and poultry manure.

9.4 OPTIMIZING SOIL BIOLOGICAL PROPERTIES

Optimizing the activities of beneficial microbes in the rhizosphere is of fundamental importance in improving the physical, chemical, and biological properties of soil in favor of higher root growth. The important beneficial biological activities that are related to root growth are microbial biomass, including several bacteria, fungi and enzymes, and earthworms.

9.4.1 MICROBIAL BIOMASS

The composition of soil microbial communities plays a significant role in regulating soil processes (Balser and Firestone, 2005; Muruganandam et al., 2010). Soil microbial biomass is a sensitive indicator of soil quality and is influenced by many ecological factors such as the composition of plant community, SOM level, moisture, and temperature (Jenkinson and Ladd, 1981; Wardle, 1992). Soil microbial biomass C comprises only 1%–4% of C_{org} (Sparling, 1992) but due to its fast turnover time, the microbial biomass plays a key role in controlling nutrient cycling and energy flow (Jenkinson and Ladd, 1981; Li and Chen, 2004). The microbial metabolic quotient (qCO_2) has been used as a bioindicator of environmental stress on microbial communities (Anderson and Domsch, 1990, 1993), disturbances, and ecosystem development (Wardle and Ghani, 1995; Suman et al., 2006).

Microbial communities are important to soil quality and functioning because they control the potential for enzyme (i.e., hydrolases)-mediated substrate catalysis that drives biogeochemical cycles (Acosta-Martinez et al., 2010). In addition, the role of soil microbial biomass is fundamental in soil processes such as organic matter decomposition, maintenance of soil structure, nutrient cycling, and plant growth and root development. In addition, soil microorganisms contribute to key soil processes such as the degradation of agrochemicals and pollutants and the control of plant and animal pests (Stockdale and Brookes, 2006; Hummel et al., 2009; Lupwayi et al., 2011). Diverse soil microbial communities are more resilient in changing soil environments and are, therefore, important to the maintenance of soil function in agricultural systems (Lupwayi et al., 2001; Hummel et al., 2009). Because of their sensitivity to land management practices, particularly agricultural intensification, soil microbial communities are often used as indicators of soil health (Doran and Zeiss, 2000). Its role in these globally important biochemical processes has led to extensive research into its size and function in the soil ecosystem (Hood-Nowotny et al., 2010). Soil microbial biomass N is routinely measured in aerobic soils using the CH_3Cl fumigation–extraction method (Brooks et al., 1985a,b).

Increases in the soil microbial biomass have been associated with positive changes in soil quality and C sequestration under systems that are managed to support an extensive rooting system, protection of the soil surface by perennial grasses or the application of crop residues, and decreased tillage operations (Karlen et al., 1999; Moore et al., 2000; Acosta-Martinez et al., 2004, 2010; Sotomayor-Ramirez et al., 2009). Growing grasses (adopting pasture–crop rotation) can improve microbial biomass and the root growth of succeeding crops. Al-Kaisi and Grote (2007) reported that switchgrass (*Panicum virgatum* L.) had 200% more microbial biomass than either corn or soybean. Increased food supply (e.g., partly decomposed organic materials), adequate soil water content, and reduced fluctuations in soil temperature under grasses promote the proliferation of soil macro- and microorganisms. In a sandy loam in Alabama, earthworm numbers were greater under switchgrass than under an adjacent peanut–cotton cropping system (Katsvairo et al., 2007). The greater earthworm population under warm-season grasses in association with the extensive active and decaying switchgrass roots improves soil aggregation, macroporosity, and fluxes of water, air, and heat. Earthworms under trees and grass buffers are often more numerous than under croplands because of the abundant food source and cover (Blanco-Canqui, 2010). Roots and earthworm burrows under grasses can penetrate consolidated and compacted soil matrix layers and change the pore structure, thereby increasing water infiltration and storage at lower depths (Katsvairo et al., 2007). All these changes in the soil profile improve root growth and development.

The organic C additions from manure have increased soil microbial biomass (Fraser et al., 1988; Peacock et al., 2001; Lupwayi et al., 2005). However, effects on microbiomass vary from manure to manure. In a long-term (60 years) study, microbial biomass was higher in NPK fertilizer treatments than in those with residual manure. This response was attributed to higher turnover rates of available organic C in the fertilized treatment (Belay et al., 2002). In Saskatchewan, where urea and anhydrous NH_3 were applied to wheat at increasing rates up to 180 kg N ha^{-1}, soil populations of bacteria and fungi increased, but populations of actinomycetes decreased (Biederbeck et al., 1996). In China, Zhang et al., (2008) observed an optimum N rate of 160 kg ha^{-1} for soil microbial biomass and functional diversity in native grassland.

Using cover crops in cropping systems is another important strategy to improve microbial biomass. In studies by Galvez et al. (1995), Mendes et al. (1999), and Sainju et al. (2003), microbial biomass or other measures of microbial activity were increased by legume or cereal cover crops. Similarly, Kabir and Koide (2000, 2002) reported a linear relationship between arbuscular mycorrhizal fungi, which was increased by wheat and rye cover crops.

9.4.2 BIOPORES

Pores or channels created by living organisms in the soil are designated as biopores. Earthworm population is one of the important biological activities in the soil that create biopores. Earthworms contribute to biopores through fecal excretion (casts), burrowing, feeding, and digestion (Tian et al., 2000). Casts are nutrient rich and are an intimate mixture of soil, water, and microbial cells (Linden et al., 1994). Earthworm burrows provide pathways for root exploration (Logsdon and Linden, 1992). Earthworms are known to accelerate the decomposition of plant residues in the soil (Tian et al., 1995) and play a role in converting plant residue into SOM (Lee, 1985; Lavelle, 1988). Tian et al. (2000) reported that increase in earthworm populations by fallow led to an increase in leaf-litter decomposition, SOM, available P, and extractable cations and pH. These authors also reported that earthworms decrease in soil bulk density which thus decreases penetrometer resistance in fallow plots. Lavella and Martin (1992) discussed the short-term and long-term effects of earthworms on SOM dynamics in tropical soils. Temperature, moisture, and food supply are the major components of earthworm habitats (Edwards and Bohlen, 1996). Including periods of fallow in the crop rotation improves the earthworm habitat because of lower soil temperature, higher soil moisture, and better food supply, leading to a potential increase in earthworm populations (Tian et al., 2000).

9.4.3 CONTROL OF WEEDS

Walz (2004) and Corbin et al. (2010) reported that weed management has been identified by organic farmers as the greatest barrier to long-term success during the transition to certified production systems, with soil fertility and quality being the next barrier. Similarly, sorghum producers in the United States consider weed control to be the main problem that limits sorghum yield. A density of 12 redroot pigweed (*Amaranthus retroflexus* L.) plants per meter of row has been shown to reduce sorghum grain yield by 46% (Knezevic et al., 1997). Weeds compete with crop plants for water, nutrients, and solar radiation and reduce root growth and, consequently, crop yields (Tollenaar et al., 1994; Evans et al., 2003; Oerke, 2006; Bastiaans, 2008; Page et al., 2010). Yield losses from interspecific competition are most severe when weeds emerge at or near the time of crop emergence (O'Donovan et al., 1985; Kropff and Spitters, 1991; Bosnic and Swanton, 1997). For example, Bosnic and Swanton (1997) reported that seedlings of barnyard grass (*Echinochloa crusgalli* L. Beauv.) emerging before the third leaf tip stage of corn reduced yields by 26%–35%, whereas yield losses were only 6% if a similar density of seedlings emerged after the fourth leaf tip stage. The effect of interspecific competition in corn has been described by the critical period for weed control, which begins between the third and sixth leaf tip stage and can last up to the 13th leaf tip (Hall et al., 1992; Halford et al., 2001; Page et al., 2010). Although the critical period for weed control is defined by yield losses observed at maturity, the effects of interspecific competition can be traced to reductions in the crop biomass accumulation and leaf area expansion that occurred while the weeds were growing with the crop (Cox et al., 2006). The fact that these reductions in crop growth and development result in yield loss, even after weed removal, indicates that the crop is not able to recover or compensate for durations of interspecific competition occurring early in the growing season (Page et al., 2010).

Hence, weed control is an important strategy in crop production to improve root growth and yields. Weeds can be controlled manually, especially in developing countries with small holdings and labor availability and also by the use of herbicides. Several slow-release herbicides are now available for weed control. The slow-release formulations of herbicides are designed to maintain the threshold concentration of the active ingredient for weed control in the soil by providing release at the required rate. The consequent lowering of the required amounts of herbicides is both environmentally friendly and economically advantageous (Undabeytia et al., 2003, 2004, 2010). Integrated weed management (IWM) strategies is the best approach to control weeds in cropping systems. Using competitive crops and cultivars are important components of IWM (Callaway, 1992; Watson et al., 2006; Travlos et al., 2011). In addition, propane flaming could be an alternative to hand weeding or mechanical cultivation in cropping systems (Ulloa et al., 2010, 2011).

In recent years, genetic transformation has become an important approach for the introduction of novel agronomic traits into crops (Barro et al., 2002). In weed management, the launch of the commercial glyphosate-resistant (GR) (Roundup Ready) soybean cultivars in 1996 initiated a new era in row crop weed management (Dill, 2005). Since then, many crops including corn, cotton, canola, and wheat have been transformed (Hu et al., 2003; Dill, 2005; Larson et al., 2007, 2009). Genetically modified (GM) corn and soybean dominate the North American agricultural landscape and are becoming increasingly important as grain and biofuels (Krupke et al., 2009). Roughly, 90% of soybean (26.2 million ha) and 52% of corn (19.7 million ha) grown in the United States in 2007 were GR, allowing producers to apply glyphosate-only herbicide treatments to prevent crop yield loss due to weed competition (NASS, 2007; Krupke et al., 2009).

In addition, GR soybean and alfalfa cultivars have become commercially available and offer soybean and alfalfa producers new options for weed management in cropping systems (Bradley et al., 2010; Loecker et al., 2010). Glyphosate [N-(phosphonomethyl) glycine] is a nonselective herbicide that controls many grass and broadleaf weeds. Initial research with glyphosate in GR alfalfa revealed that excellent weed control could be achieved without significant alfalfa injury or yield loss (Miller and Alford, 2002; McCordick et al., 2008; Bradley et al., 2010). GR soybean production comprises the vast majority of all soybean production in the United States as well as other world markets (Duke and Powels, 2008). The overwhelming adoption of GR soybean varieties has greatly simplified weed management, assisted the adoption of conservation tillage, and increased crop rotation options (Cedeira and Duke, 2006).

Another option of weed control is the inherent ability of many crops to suppress weeds through a combination of high early vigor (competition) and allelopathic activity to further reduce weed interference (Bertholdsson, 2005; Ferreira and Reinhardt, 2010). Belz (2007) reported that allelopathy can be an important component of crop/weed interference. Crop allelopathy controls weeds by the release of allelochemicals from intact roots of living plants and/or through the decomposition of phytotoxic plant residues (Qasem and Hill, 1989; Wallace and Bellinder, 1992; Weston, 1996; Batish et al., 2002; Belz, 2004; Khanh et al., 2005; Ferreira and Reinhardt, 2010). Jones et al. (1999) reported that allelopathy is likely to be most beneficial where other options have become limiting due to herbicide resistance and high control costs.

Crop allelopathy controls weeds by the release of allelochemicals from intact roots of living plants and/or through the decomposition of phytotoxic plant residues (Weston, 1996; Batish et al., 2002; Belz, 2004; Khanh et al., 2005; Ferreira and Reinhardt, 2010). The incidence of growth inhibition of certain weeds and the induction of phytotoxic symptoms by plants and their residues is well documented for many crops, including all major grain crops such as rice, rye, barley, sorghum, and wheat (Belz, 2004).

Weeds can also be controlled by using high biomass cover crops and organic mulches. Applied in sufficient quantities, high biomass residues, either grown as cover crops or applied as mulches, have been shown to suppress weeds, control erosion, and conserve moisture (Rathore et al., 1998; Mulvaney et al., 2010).

9.4.4 CONTROL OF DISEASES AND INSECTS

Diseases and insects are responsible for large losses of agricultural crops. Fageria (1992) reported that average worldwide losses for the main agricultural crops were 11.8% caused by diseases and 12.2% caused by insect pests. The average combined losses caused by diseases, pests, and weeds are put at 33.7%. Although it is impossible to quantify these losses accurately, the foregoing estimates emphasize the enormous damage that is caused by diseases and pests. Diseases and insects reduce the nutrient and water use efficiency, thereby reducing the root growth of crops. Some diseases and insects directly infect root growth and reduce normal function such as absorption of water and nutrients.

Foliar diseases are major constraints to wheat production on the Canadian prairies. For example, annual yield losses due to leaf spot diseases may be as much as 20% (Gilbert and Tekauz, 1993). Blast disease in upland rice in the central part of Brazil is responsible for large losses in yield of this crop. Similarly, wheat is the most important food crop grown worldwide, being widely cultivated and used as a dietary staple by nearly 35% of the world's population (Zeng et al., 2007; Provance-Bowley et al., 2010). Wheat losses attributed to powdery mildew (*Erysiphe graminis* DC.) can reach 40% by affecting root growth and grain fill and by reducing test weights (Lipps and Madden, 1989; Agrios, 2007). Economic losses vary, depending on the current market price, but at expected yield losses >10% fungicides are generally recommended (Provance-Bowley et al., 2010).

Yellow dwarf viruses are transmitted to cereals by aphids and can cause yield losses of up to 31% in naturally infected wheat crops (Cowger et al., 2010). Soybean aphid feeding injury can reduce soybean photosynthetic rates by up to 50% in infested leaflets. Feeding injury affects biochemical pathways for restoring chlorophyll to a low energy, light-receptive stage (Macedo et al., 2003). Riedell et al. (2009a) presented evidence that soybean aphids are capable of reducing total nodule volume plant^{-1} by 34%, nodule leghemoglobin content by 31%, plant nitrogen fixation rate by 80%, and shoot ureide-n concentration by 20%. Similarly, Catangui et al. (2009) reported that on average, the calculated maximum possible yield loss was 75% for soybean aphid infestation starting at the V5 (five node) stage and 48% for soybean aphid infestations starting at the R2 stage.

Root lesion nematodes (*Pratylenchus neglectus* and *Pratylenchus thronnei*) are widely distributed and substantially reduce grain yields in wheat in the wheat-producing regions of the world (Smiley, 2009). Plants with impaired root branching and cortical degradation caused by lesion nematodes become less capable of extracting nutrients and water from soil and of yielding, thus growing as unhealthy plants (Trudgill, 1991; Thompson et al., 1995; Smiley and Machado, 2009). These nematodes tend to cause the greatest amount of yield reduction in the lowest rainfall environments because affected plants in higher rainfall or irrigated environments are more able to compensate for damage caused by a given preplant population (Smiley, 2009). These nematodes remain mobile and may move into and out of roots and may deposit eggs in soil as well as within root tissue. Wheat and barley cultivars vary in their ability to allow these nematodes to reproduce (Hollaway et al., 2000; Thompson et al., 2008). Cultivars that greatly suppress reproduction are classified as resistant, and those that allow high rates of reproduction are classified as susceptible (Thompson et al., 1999).

The pea leaf weevil (*Sitona lineatus* L.) (Coleoptra:Curculionidae) is a serious pest of field pea (*P. sativum* L.) and faba bean (*Vicia faba* L.) in Europe, Asia, Africa, and North America (Vankosky et al., 2011). Larvae of leaf weevil can destroy root nodules as a result of their consumption of *Rhizobium leguminosarum* biovar *viciae* Frank bacteria, compromising the ability of plants to fix N (Hoebeke and Wheeler, 1985). Cantot (1986) observed that 12 larvae destroyed 90% of root nodules on a single plant. Nodule damage by larvae is expected to have a greater impact on yield than foliar damage by adult weevils (Dore and Meynard, 1995; Corre-Hellou and Crozat, 2005).

Reducing the infestation of diseases and insects by adopting appropriate soil and crop management practices can improve the root growth of plants. The use of fungicides and insecticides in appropriate dose and time during crop growth can minimize diseases and insect infestations. Vankosky et al. (2011) reported that the most efficient method of preventing yield losses due to *Sitona lineatus* L. infestation is to inoculate field peas. Elbert et al. (2008) and Hamon et al. (1990) reported that systemic insecticides are generally compatible with biological control and that natural enemies of pea leaf weevil are available for such programs. In general, problems caused by insect pests may be more severe in monoculture cropping systems than in polyculture cropping systems, which is most often because of a greater abundance of natural enemy species in diversified systems compared with monocultural systems (Cai et al., 2007; Lai et al., 2011). Diversified vegetation supports greater arthropod diversity, which contributes toward ecosystem equilibrium (You et al., 2004). The use of disease- and insect-resistant genotypes is another option to improve crop growth and the root systems of crop plants. The use of biotechnology tools may aid the control of diseases and insects and improvement in the root system. The application of such technology in wheat has resulted in the production of transgenic plants with resistance to fungi (Bliffeld et al., 1999; Clausen et al., 2000), insects, and viral diseases (Altpeter et al., 1999).

GM plants of various crops are also used to control insects. For example, the introduction of transgenic Bt (*Bacillus thuringiensis*) cotton has dramatically reduced problems with late-season insect infestations. Similarly, corn hybrids are GM plants that produce insecticidal toxins in their tissues and thereby resist feeding by specific insect pests (Seydou et al., 2000; Stanger and Lauer, 2006; Krupke et al., 2009). These hybrids are also combined with other transgenic traits such as glyphosate herbicide tolerance. The adoption of corn hybrids that express insect-resistant traits (frequently known as "Bt corn") has increased dramatically in the United States in recent years, that is, from 29% (9.5 million ha) in 2004 to 49%

(18.6 million ha) in 2007 (NASS, 2007). Although Bt corn for Lepidopteran pests such as the European corn borer (*Ostrinia nubilalis* Hubner) has been available for over a decade, the recent increase in adoption is mainly due to the release of transgenic traits targeting the WCR, the key insect of corn in North America (Krupke et al., 2009).

However, developing cultivars with high level of resistance that is effective over multiple locations and stable over time is challenging because of the complexity of both the pathogen and the patterns of the inheritance of resistance (Urrea et al., 2010). In addition, to be commercially viable, cultivars must also possess desirable agronomic characteristics and some sources of resistance lack preferred yield and quality traits (Singh and Reddy, 1996). Hence, developing a disease- or insect-resistant crop cultivar requires joint efforts by scientists of various disciplines and a time-consuming breeding process.

9.5 OPTIMIZING CULTURAL PRACTICES

Optimizing cultural practices like deep plowing, conservation tillage, crop rotation, and using stress-tolerant plant species/genotypes can improve the root growth of crop plants. These cultural practices change the physical, chemical, and biological properties of soil, which are responsible for better root growth.

9.5.1 DEEP PLOWING

Deep plowing improves soil conditions for root growth by breaking compacted layers that the root cannot readily penetrate. When depths of root-restricting hardpans are relatively shallow (<0.25 m), chisel plowing can be effective for disrupting compacted layers (Fageria and Moreira, 2011).

9.5.2 CONSERVATION TILLAGE

Conservation tillage is an important soil management practice that can bring many benefits in favor of better root growth. This practice increases water infiltration and improves soil structure compared to conventional tillage (Truman et al., 2005; Afyuni and Wagger, 2006). Soil compaction has been minimized in reduced tillage systems while bulk density and other physical properties have been improved (Naderman et al., 2004). Improvement in organic matter content in surface soil is an important attribute of conservation tillage. Organic matter improves the physical, chemical, and biological properties of soil, which can improve root growth.

According to Godwin (1990), the most important role of tillage is to facilitate root development and function. So, it is not surprising that greater root penetration resistance exists under no-till or conservation tillage, especially in the upper 10 cm (Ferreras et al., 2000). Since crops have either a shallow fibrous or a deep taproot system, it is envisaged that the type of tillage system used may have differential effects on crop performance. Cereal crops, which have a fibrous root system, tend to have an extensive mass of adventitious roots occupying a large volume of soil around the plant's base, while the taproot system of noncereal crops generally penetrates deeper into the soil than the fibrous root system (Yau et al., 2010).

9.5.3 Crop Rotation

Appropriate crop rotations are a critical component of crop production systems, which can affect root growth by controlling diseases, insects, and weeds (Lamb et al., 1993; Karlen et al., 1994; Cox and Sholar, 1995; Jordan et al., 2002). In addition, the use of legume crops in crop rotation contributes to biological nitrogen fixation and improves yield and, consequently, the root growth of succeeding crops (Rochester et al., 2001; Drake et al., 2010). Deep-rooted legume crops such as alfalfa scavenge deep residual soil N and, thus, increase N availability to subsequent shallow-rooted crops (Karlen et al., 1994; Riedell et al., 2009). Crop rotation has been shown to improve soil structure, increase SOM levels (Bremer et al., 2008), increase water use efficiency (Tanaka et al., 2005), enhance mycorrhizal association (Johnson et al., 1992), improve grain quality (Kaye et al., 2007), and reduce grain yield variability (Varvel, 2000).

In addition to yield increase, crop rotations have positive effects on the physical, chemical, and biological properties of soil (Mady Kaye et al., 2007; Wright et al., 2008; Adeli et al., 2009) due to higher C inputs and the diversity of plant residues returned to soils in comparison with continuous monoculture cotton (Adeli et al., 2009). The combination of crop rotation and manure application may support crop yield and improve soil quality (Berzsenyi et al., 2000; Bronick and Lal, 2005).

9.5.4 Using Stress-Tolerant Crop Species/Genotypes

Abiotic and biotic stresses are responsible for decreasing the root growth of crop plants. Abiotic stresses are drought, extreme soil temperatures, unfavorable physical properties of soil, nutrient deficiency/elemental toxicities, and salinity. Important biotic stresses are diseases, insects, weeds, and harmful soil microorganisms that reduce the root growth of crop plants. To improve the root growth of crop plants under biotic and abiotic stresses by using crop species and/or genotypes within species that have greater tolerance to root-limiting factors is a very attractive strategy. Plant species and genotypes within species vary widely in tolerance to environmental stresses (Fageria et al., 2006). Stress-tolerant genotypes are being identified and bred worldwide to solve some of the more difficult problems of soil fertility, such as subsoil acidity, salinity, low plant availability of Fe in calcareous soils, and low P availability in acid soils (Foy, 1992; Fageria et al., 2008; Fageria, 2009). Many investigators have reported that plants display a wide array of adaptive responses to low P availability to enhance P mobility in the soil and increase its uptake (Faye et al., 2006; Fageria et al., 2008). A well-known adaptive response is the alteration

TABLE 9.20
Nitrogen Use Efficiency in 19 Upland Rice Genotypes

Genotype	AE (mg mg⁻¹)	PE (mg mg⁻¹)	APE (mg mg⁻¹)	AR (%)	UE (mg mg⁻¹)
CRO 7505	23.5ab	100.2a	55.4a	43.5a	42.5ab
CNAs 8993	26.7a	101.4a	59.0a	47.8a	45.5ab
CNAs 8812	24.2ab	80.5a	41.3a	58.2a	47.0ab
CNAs 8938	21.8ab	87.8a	46.3a	48.7a	40.9ab
CNAs 8960	21.7ab	79.3a	46.1a	48.3a	37.3ab
CNAs 8989	24.8aab	72.3a	42.3a	59.3a	42.4ab
CNAs 8824	17.0ab	73.7a	36.7a	47.0a	33.5b
CNAs 8957	22.4ab	72.3a	46.3a	50.7a	35.2ab
CRO 97422	19.8ab	85.7a	47.1a	45.0a	36.6ab
CNAs 8817	17.9ab	72.7a	33.6a	53.2a	38.3ab
CNAs 8934	20.3ab	83.3a	40.5a	50.2a	41.8ab
CNAs 9852	22.9ab	81.7a	48.9a	47.7a	38.4ab
CNAs 8950	21.0ab	88.5a	49.9a	44.7a	37.6ab
CNA 8540	22.2ab	89.7a	45.2a	53.5a	44.1ab
CNA 8711	19.4ab	79.2a	40.0a	48.1a	38.0ab
CNA 8170	12.8b	92.8a	28.0a	45.4a	42.0ab
BRS Primavera	22.6ab	89.8a	51.2a	45.8a	39.9ab
BRS Canastra	21.7ab	125.7a	57.7a	40.6a	47.4ab
BRS Carisma	24.2ab	89.6a	42.8a	56.2a	50.5a
Average	21.4	86.6	45.2	49.2	41.0
F-test					
Genotype	*	NS	NS	NS	**
CV (%)	17	23	26	24	12

Source: Fageria, N.K. et al., *J. Plant Nutr.*, 33, 1696, 2010b.

AE, agronomical efficiency; PE, physiological efficiency; APE, agrophysiological efficiency; ARE, apparent recovery efficiency; UE, utilization efficiency.

NS, nonsignificant.

*, **Significant at the 5% and 1% probability levels, respectively. Means followed by the same letter in the same column are not significantly different at the 5% probability level by Tukey's test.

of root morphology and architecture to increase P acquisition from the soil at minimum metabolic cost (Neumann et al., 1999; Jonathan and Kathleen, 2001). Results presented in Table 9.20 show N use efficiency of upland rice genotypes. Tables 9.21 and 9.22 show P use efficiency in upland rice and dry bean, respectively. Data in Table 9.23 show variations in K use efficiency of upland rice genotypes. Similarly, data in Table 9.24 show variations in acidity tolerance in upland rice genotypes. Results presented in Table 9.25 show variations in acidity tolerance among crop, pasture, and plantation crop species.

Genetic improvements in combination with adjustments in cropping systems and/or cultural practices will have a significant impact not only in developing

TABLE 9.21

Phosphorus Use Efficiency of Five Upland Rice Genotypes

Genotype	P Use Efficiency (kg Grain kg⁻¹ P Applied)
BRA 01596	19.30ab
BRA 01600	15.88b
BRA 02535	15.34b
BRA 02601	26.51a
BRA 032051	7.31c
Average	16.87

Source: Fageria, N.K. et al., *Growth and Mineral Nutrition of Field Crops*, 3rd edn., CRC Press, Boca Raton, FL, 2011.

Means within same column, followed by the same letter, do not differ significantly at the 5% probability level by Tukey's test.

countries, characterized by low input use, but also in developed countries, where they will contribute to making agricultural systems more environmentally sound.

9.6 BREEDING CROP VARIETIES FOR BETTER ROOT SYSTEMS

Breeding crop varieties for better root systems is a very attractive strategy. Large differences in root length and RDW among crop species and genotypes within species have been reported (Fageria et al., 2011; Fageria and Moreira, 2011). Figures 9.13 and 9.14 show that the root growth of two lowland rice cultivars differs at low and high N rates. Gregory (2006) reported that although root characteristics are determined genetically, their full expression is dependent on the environment, and many studies have indicated substantial genotype × environment interactions. Breeders often select for a characteristic (i.e., drought resistance) in the agroclimatic zone of interest but soils vary within regions so that what is thought to be a response to drought may also include responses to other environmental factors such as soil acidity, low nutrient availability, root diseases, etc., unless the soil is also characterized. The full exploitation of root traits depends then on appropriate targeting for soil properties in much the same way that appropriate phenology is currently linked with agroclimatic analysis (Gregory, 2006). O'Toole and Bland (1987) suggested the following steps for breeding for better root systems: (1) defining the problem well, (2) evaluation of root system parameters, (3) determination of existence, level, and nature of genetic variation, (4) hybridization and selection, and (5) field evaluation of resultant genotypes for validation.

TABLE 9.22
Classification of Dry Bean Genotypes to P Use Efficiency Based on Grain Yield Efficiency Index (GYEI)

Genotype	P Use Efficiency Index	Classification
CNFC 10467	1.20ab	E
CNFP 8000	2.06a	E
CNFC 10455	0.61ab	ME
CNFP 10035	1.97ab	E
CNFC 10410	1.88ab	E
CNFP 10076	1.22ab	E
CNFC 10432	0.99ab	ME
CNFP 10093	0.68ab	ME
CNFC 10408	0.88ab	ME
CNFP 10103	0.33ab	IE
CNFC 9461	1.42ab	E
CNFP 10104	1.90ab	E
CNFC 10429	0.92ab	ME
CNFP 10109	1.26ab	E
CNFC 10431	0.93ab	ME
CNFP 10120	0.32ab	IE
CNFC 10438	0.26b	IE
CNFP 10206	0.84ab	ME
CNFC 10444	0.43ab	IE
CNFC 10470	0.85ab	ME

E, efficient; ME, moderately efficient; IE, inefficient. Means in the same column followed by the same letter are not significantly different at the 5% probability level by Tukey's test.

The use of biotechnology is an important factor in identifying genes that is responsible for better root growth in crop species or genotypes within species. This technology makes it possible to investigate the inheritance of root traits that are either single gene or polygenic in nature, and select for chromosomal regions associated with the characters of interest (Gregory, 2006). This technology is quite new and not much progress has been achieved in improving the root systems of crop plants. However, significant progress has been made in studies with rice and to a lesser extent with corn (Champoux et al., 1995; Tuberosa et al., 2002; Babu et al., 2003; Zheng et al., 2003). The marked QTL × environment interactions show that the genetic control of root growth is complex and environmental conditions may have a large effect on the morphology of root systems.

TABLE 9.23
Potassium Use Efficiency in Six Upland Rice Genotypes

Genotype	AE (mg mg⁻¹)	PE (mg mg⁻¹)	APE (mg mg⁻¹)	ER (%)	EU (mg mg⁻¹)
BRS Bonança	4.11	17.63	6.57	35.50	8.39
BRS Primavera	11.33	45.61	29.47	38.55	17.53
BRSMG Curinga[a]	9.78	—	—	4.87	11.89
BRA 032033	10.00	57.35	36.62	28.70	16.08
BRA 01596	10.44	49.81	33.12	31.13	16.58
BRA 02582	7.75	90.07	34.20	25.41	19.56
Average	8.90	52.09	27.99	27.36	15.01

Source: Fageria, N.K. et al., *J. Plant Nutr.*, 41, 2676, 2010d.

AE, agronomic efficiency; PE, physiological efficiency; APE, agrophysiological efficiency; ER, recovery efficiency; EU, utilization efficiency.

[a] Due to negative values, PE and APE values were not presented.

TABLE 9.24
Grain Yield and Panicle Number of Six Upland Rice Genotypes at Two Soil Acidity Levels in Brazilian Oxisols

	Grain Yield (g Pot⁻¹)		Panicle Number (Pot⁻¹)	
Genotype	High Acidity (pH 4.5)	Low Acidity (pH 6.4)	High Acidity (pH 4.5)	Low Acidity (pH 6.4)
CRO 97505	74.3	52.0	38.0	28.3
CNAs 8983	55.2	42.9	29.0	25.7
Primavera	53.0	47.2	25.0	21.7
Canastra	51.6	38.9	32.0	26.3
Bonança	48.8	36.5	26.3	20.7
Carisma	50.8	17.5	43.3	17.7
Average	66.7	47.0	38.7	28.1

Source: Compiled from Fageria, N.K. et al., Response of upland rice genotypes to soil acidity, in: *The Red Soils of China: Their Nature, Management and Utilization*, eds., Wilson, M.J., He, Z., and Yang, X., pp. 219–237, Kluwer Academic Publishers, Dordrecht, the Netherlands, 2004.

To be useful to breeding programmers, the stability of QTL expression across soil environments is essential and this remains a challenge (Gregory, 2006). To achieve success for breeding a better root system in crop species and/or genotypes of the same species, a strong collaboration among soil scientists, physiologists, and breeders is required.

TABLE 9.25
Some Important Crop Species, Pasture Species, and Plantation Crops Tolerant to Soil Acidity in the Tropics

Annual Crop Species	Pasture Species	Plantation Crops
Rice	*Brachiaria*	Banana
Peanut	*Andropogon*	Oil palm
Cowpea	*Panicum*	Rubber
Potato	*Digitaria*	Coconut
Cassava	*Napier grass*	Cashew nut
Pigeon pea	*Jaraguagrass*	Coffee
Millet	*Centrosema*	Guarana
Kudzu	*Stylosanthes*	Tea
Mucuna		Leucaena
Crotolaria		Brazilian nut
		Eucalyptus
		Papaya

Source: Adapted from Sanchez, P. and Salinas, J., *Adv. Agron.*, 34, 279, 1981; Fageria, N.K. et al., *Growth and Mineral Nutrition of Field Crops*, 3rd edn., CRC Press, Boca Raton, FL, 2011; Brady, N.C. and Weil, R.R., *The Nature and Properties of Soils*, 13th edn., Prentice Hall, Upper Saddle River, NJ, 2002.

FIGURE 9.13   Root growth of lowland rice cultivars BRS Jaçanã at low and high N rates.

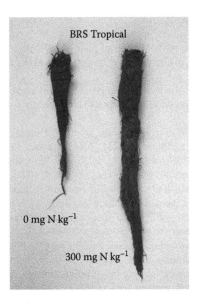

FIGURE 9.14 Root growth of lowland rice cultivar BRS Tropical at low and high N rates.

9.7 CONCLUSIONS

Roots serve the combined functions of anchorage, water and nutrient uptake, and hormone supply to plants. Hence, healthy functioning root systems are important to obtain maximum economic yield of crop plants. Root growth is genetically determined, but environmental factors significantly modify plant root systems. A number of physical, chemical, and biological soil factors, such as temperature, moisture content, air, pH, nutrients, salinity, and microbial activity define soil environment. Extreme values of these climatic and soil properties are detrimental to root growth. Many of these properties are highly variable in space and time, and there also exists interaction among most of these environmental variables. Hence, to obtain the optimum values of these environmental factors for root growth is very difficult and complex. However, adopting appropriate soil and crop management practices can bring these variables to near optimum values in favor of better root system. Soil management practices, particularly tillage (deep ploughing and conservation tillage), liming acid soils and the use of mulching can change the soil environment considerably in favor of better root growth. In addition, there is genetic variability between, as well as within, species in their response of roots to environmental conditions. Roots of different crop species as well as of cultivars within species differ considerably in their ability to penetrate through hard soil layers. Hence, the use of genetic variability in favor of better root system is another important strategy in improving root systems in different cropping systems. Since optimum root growth is crucial for maximizing crop yield, the discussion provided in this chapter may be very useful for scientists of different disciplines.

REFERENCES

Acosta-Martinez, V., G. Burow, T. M. Zobeck, and V. G. Allen. 2010. Soil microbial communities and function in alternative systems to continuous cotton. *Am. Soc. Soil Sci. J.* 74:1181–1192.

Acosta-Martinez, V., T. M. Zobeck, and V. Allen. 2004. Soil microbial, chemical and physical properties in continuous cotton and integrated crop-livestock systems. *Soil Sci. Soc. Am. J.* 68:1875–1884.

Adams, F. 1984. Crop response to lime in the southern United States. In: *Soil Acidity and Liming*, 2nd edn., ed., F. Adams, pp. 211–266. Madison, WI: ASA, CSSA, and SSSA.

Adeli, A., H. Tewolde, K. R. Sistani, and D. E. Rowe. 2009. Broiler litter fertilization and cropping system impacts on soil properties. *Agron. J.* 101:1304–1310.

Afyuni, M. and M. G. Wagger. 2006. Soil physical properties and bromide movement in relation to tillage system. *Commun. Soil Sci. Plant Anal.* 37:541–556.

Agboola, A. A. and R. B. Corey. 1973. The relationship between soil pH, organic matter, and nine elements in the maize tissue. *Soil Sci.* 115:367–375.

Agrios, G. E. 2007. *Plant Pathology*, 5th edn. San Diego, CA: Elsevier Academic Press.

Albiach, R., R. Cancer, F. Pomares, and F. Ingelmo. 2001. Organic matter components, aggregate stability and biological activity in a horticultural soil fertilized with different rates of two sewage sludges during ten years. *Bioresour. Technol.* 77:109–114.

Al-Kaisi, M. M. and J. B. Grote. 2007. Cropping systems effects on improving soil carbon stocks of exposed subsoil. *Soil Sci. Soc. Am. J.* 71:1381–1388.

Alleoni, L. R. F., M. A. Cambri, E. F. Caires, and F. J. Garbuio. 2010. Acidity and aluminum speciation as affected by surface liming in tropical no-till soils. *Soil Sci. Soc. Am. J.* 74:1010–1017.

Allison, S. D. 2005. Cheaters, diffusion and nutrients constrain decomposition by microbial enzymes in spatially structural environment. *Ecol. Lett.* 8:626–635.

Allred, S. E. and A. J. Ohlrogge. 1964. Principles of nutrient uptake from fertilizer bands. VI. Germination and emergence of corn as affected by ammonia and ammonium phosphate. *Agron. J.* 56:309–313.

Altpeter, F., I. Diaz, H. McAuslane, K. Gaddour, P. Carbonero, and I. K. Vasil. 1999. Increased insect resistance in transgenic wheat stably expressing trypsin inhibitor CMe. *Mol. Breed.* 5:53–63.

Alva, A. K. and G. J. Gascho. 1991. Differential leaching of cations and sulfate in gypsum amended soils. *Commun. Soil Sci. Plant Anal.* 22:1195–1206.

Alva, A. K. and M. E. Sumner. 1991. Characterization of phytotoxic aluminum in soil solutions from phosphogypsum amended soils. *Water Air Soil Pollut.* 57–58:121–130.

Alvarez, V. H. and A. C. Ribeiro. 1999. Liming. In: *Recommendations for Using Amendments and Fertilizer*, 5th Approximation, eds., A. C. Ribeiro, P. T. G. Guimarães, and V. H. Alvarez, pp. 43–60. Viçosa, Brazil: Soil Fertility Commission for State of Minas Gerais.

Anderson, T. H. and K. H. Domsch. 1990. Application of ecological quotients (qCO$_2$ and qD) on microbial biomass from soils of different cropping histories. *Soil Biol. Biochem.* 22:251–255.

Anderson, T. H. and K. H. Domsch. 1993. The metabolic quotients for CO$_2$ (qCO$_2$) as a specific activity parameter to assess the effects of environmental conditions, such as pH, on the microbial biomass of forest soils. *Soil Biol. Biochem.* 25:393–395.

Angers, D. A. and J. Caron. 1998. Plant-induced changes in soil structure: Process and feedbacks. *Biochemistry* 42:55–72.

Angers, D. A. and M. Giroux. 1996. Recently deposited organic matter in soil water-stable aggregates. *Soil Sci. Soc. Am. J.* 60:1547–1551.

Asady, G. H., A. J. M. Smucker, and M. W. Adams. 1985. Seedlings test for the quantitative measurement of root tolerances to compacted soils. *Crop Sci.* 25:802–806.

Ashraf, M., H. R. Athar, P. J. C. Harris, and T. R. Kwon. 2008. Some prospective strategies for improving crop salt tolerance. *Adv. Agron.* 88:223–271.

Ashraf, M. and M. R. Foolad. 2005. Pre-sowing seed treatment a shotgun approach to improve germination, plant growth, and crop yield under saline and non-saline conditions. *Adv. Agron.* 88:223–271.

Ashraf, M., S. Hasnain, O. Berge, and T. Mahmood. 2004. Inoculation wheat seedlings with exopolysaccharides-producing bacteria restricts sodium uptake and stimulates plant growth under salt stress. *Biol. Fertil. Soils* 40:157–162.

Babu, R. C., B. D. Nguyen, V. Chamarerk, P. Shanmugasundaram, P. Chezian, P. Jeyaprakash et al. 2003. Genetic analysis of drought resistance in rice by molecular markers: Association between secondary traits and field performance. *Crop Sci.* 43:1457–1469.

Balser, T. C. and M. K. Firestone. 2005. Linking microbial community composition and soil processes in a California annual grassland and mixed conifer forest. *Biogeochemistry* 73:395–415.

Barro, F., P. Barcelo, P. A. Lazzeri, P. R. Shewry, A. Martin, and J. Ballesteros. 2002. Field evaluation and agronomic performance of transgenic wheat. *Theor. Appl. Genet.* 105:980–984.

Bastiaans, L. 2008. Focus on ecological weed management: Are we addressing the right issues? *Weed Res.* 42:177–193.

Batish, D. R., P. Thung, H. P. Singh, and R. K. Kohli. 2002. Ohytotoxicity of sunflower residues against some summer season crops. *J. Agron. Crop Sci.* 188:19–24.

Baver, L. D., W. H. Gardner, and W. R. Gardner. 1972. *Soil Physics.* New York: Wiley Interscience.

Beare, M. H. and R. R. Bruce. 1993. A comparison of methods for measuring water-stable aggregates: Implications for determining environmental effects on soil structure. *Geoderma* 56:87–104.

Belay, A., A. S. Claassen, and F. C. Wehner. 2002. Soil nutrient contents, microbial properties and maize yield under long term based crop rotation and fertilization: A comparison of residual effect of manure and NPK fertilizers. *South Afr. J. Plant Soil* 19:104–110.

Bell, T. H., J. N. Klironomos, and H. A. L. Henry. 2010. Seasonal responses of extracellular enzyme activity and microbial biomass and nitrogen addition. *Soil Sci. Soc. Am. J.* 74:820–828.

Belz, R. G. 2004. Evaluation of allelophatic traits in *Triticum* L. *Spp.* and *Seale cereale* L. PhD thesis. Stuttgart, Germany: University of Hohenheim.

Belz, R. G. 2007. Allelopathy in crop/weed interactions-An update. *Pest Manag. Sci.* 63:308–326.

Bernstein, L. 1975. Effects of salinity and sodicity on plant growth. *Annu. Rev. Phytopathol.* 13:295–312.

Bertholdsson, N. O. 2005. Early vigor and allelopathy: Two useful traits for enhanced barley and wheat competitiveness against weeds. *Weed Res.* 45:94–102.

Berzsenyi, Z., B. Gyoreffy, and D. Lap. 2000. Effect of crop rotation and fertilization on maize and wheat yields and yield stability in a long term experiment. *Eur. J. Agron.* 13:225–244.

Biederbeck, V. C., C. A. Campbell, H. Ukrainetz, D. Curtin, and O. T. Bourman. 1996. Soil microbial and biochemical properties after ten years of fertilization with urea and anhydrous ammonia. *Can. J. Soil Sci.* 76:7–14.

Black, A. C. 1993. *Soil Fertility Evaluation and Control.* Boca Raton, FL: CRC Press.

Blanco-Canqui, H. 2010. Energy crops and their implications on soil and environment. *Agron. J.* 102:403–419.

Bland, W. L. 1993. Cotton and soybean root system growth in three soil temperature regimes. *Agron. J.* 85:906–911.

Bliffeld, M., J. Mundy, I. Potrykus, and J. Futterer. 1999. Genetic engineering of wheat for increased resistance to powdery mildew disease. *Theor. Appl. Genet.* 98:6–7.

Bohn, H. L., B. L. McNeal, and G. E. O'Connor. 2001. *Soil Chemistry*, 3rd edn. New York: Wiley Interscience.

Bolan, N. S., D. C. Adriano, and D. Curtin. 2003. Soil acidification and liming interactions with nutrient and heavy metal transformation and bioavailability. *Adv. Agron.* 78:215–272.

Bolan, N. S. and M. J. Hedley. 2003. Role of carbon, nitrogen, and sulfur cycles in soil acidification. In: *Handbook of Soil Acidity*, ed., Z. Rengel, pp. 29–56. New York: Marcel Dekker.

Bosnic, A. C. and C. J. Swanton. 1997. Influence of barnyardgrass (*Echinochola crusgalli*) time of emergence and density on corn (*Zea mays* L.). *Weed Sci.* 45:276–282.

Bottner, P., M. Pansu, and Z. Sallih. 1999. Modelling the effect of active roots on soil organic matter turnover. *Plant Soil* 216:15–25.

Bowen, G. D. 1970. Effects of soil temperature on root growth and on phosphate uptake along *Pinus radiata* roots. *Aust. J. Soil Res.* 8:31–42.

Bowen, H. D. 1981. Alleviating mechanical impedance. In: *Modifying the Root Environment to Reduce Crop Stress*, eds., G. F. Arkin and H. M. Taylor, pp. 21–57. St. Paul, MN: American Society of Agricultural Engineers.

Box, J. E., Jr. 1996. Modern methods of root investigation. In: *Plant Roots: The Hidden Half*, 2nd edn., eds., Y. Waisel, A. Eshel, and V. Kafkafi, pp. 193–237. New York: Marcel Dekker.

Bradley, K., R. Kallenbach, and C. A. Roberts. 2010. Influence of seedling rate and herbicide treatments on weed control, yield and quality of spring seeded glyphosate resistant alfalfa. *Agron. J.* 102:751–758.

Brady, N. C. and R. R. Weil. 2002. *The Nature and Properties of Soils*, 13th edn. Upper Saddle River, NJ: Prentice Hall.

Brar, G. S., J. F. Gomez, B. L. McMichael, A. G. Matches, and H. M. Taylor. 1970. Germination of twenty forage legumes as influenced by temperature. *Agron. J.* 83:173–175.

Bremer, E., H. H. Janzen, B. H. Ellert, and R. H. McKenzie. 2008. Soil organic carbon after twelve years of various crop rotations in an Aridic Boroll. *Soil Sci. Soc. Am. J.* 72:970–974.

Brogan, R. J., J. C. Froese, and M. H. Entz. 2011. Dry bean response to salinity in southern Manitoba. *Agron. J.* 103:663–672.

Bronick, C. J. and R. Lal. 2005. Manuring and rotation effects on soil organic carbon concentration for different aggregate size fractions on two soils in northeastern Ohio. *Soil Till. Res.* 81:239–252.

Brookes, P. C., A. Landman, G. Pruden, and D. S. Jenkinson. 1985a. Chloroform fumigation and the release of soil nitrogen: The effects of fumigation time and temperature. *Soil Biol. Biochem.* 17:831–835.

Brookes, P. C., A. Landman, G. Pruden, and D. S. Jenkinson. 1985b. Chloroform fumigation and the release of soil nitrogen: A rapid direct extraction method to measure microbial biomass nitrogen in soil. *Soil Biol. Biochem.* 17:837–842.

Burle, M. L., J. Mielniczuk, and S. Focchi. 1997. Effect of cropping systems on soil chemical characteristics, with emphasis on soil acidification. *Plant Soil* 190:309–316.

Butler, T. J. and J. P. Muir. 2006. Dairy manure compost improves soil and increases tall wheatgrass yield. *Agron. J.* 98:1090–1096.

Cai, H. J., Z. S. Li, and M. S. You. 2007. Impact of habitat diversification on arthropod communities: A study in the fields of Chinese cabbage, *Brassica chinensis. Insect Sci.* 14:241–249.

Caires, E. F., G. Barth, F. J. Garbuio, and S. Churka. 2008. Soil acidity, liming and soybean performance under no-till. *Sci. Agric.* 65:532–540.

Callaway, M. B. 1992. A compendium of crop varietal tolerance to weeds. *Am. J. Altern. Agric.* 7:169–180.

Cambardella, C. A. and E. T. Elliott. 1993. Carbon and nitrogen distribution in aggregates from cultivated and native grassland soils. *Soil Sci. Soc. Am. J.* 57:1071–1076.

Cameron, R. S., G. S. P. Ritchie, and A. D. Robson. 1986. Relative toxicities of inorganic aluminum complexes to barley. *Soil Sci. Soc. Am. J.* 50:1231–1236.

Campbell, S. A. and J. N. Nishio. 2000. Iron deficiency studies of sugar beet using an improved sodium bicarbonates-buffered hydroponic growth system. *J. Plant Nutr.* 23:741–757.

Cantot, P. 1986. Quantification population of *Sitona lineatus* L. and its attack on *Pisum sativum. Agronomie* 6:481–486.

Case, V. W., N. C. Brady, and D. J. Lathwell. 1964. The influence of soil temperature and phosphorus fertilizers of different water solubilities on the yield and phosphorus content of oats. *Soil Sci. Soc. Am. Proc.* 28:409–412.

Catangui, M. A., E. A. Beckendorf, and W. E. Riedell. 2009. Soybean aphid population dynamics, soybean yield loss, and development of stage-specific economic injury levels. *Agron. J.* 101:1080–1092.

Cedeira, A. L. and S. O. Duke. 2006. The current status and environmental impacts of glyphosate resistant crops: A review. *J. Environ. Qual.* 35:1633–1658.

Chadwick, D. R., F. John, B. F. Pain, B. J. Chambers, and J. Williams. 2000. Plant uptake of nitrogen from the organic nitrogen fraction of animal manures: A laboratory experiment. *J. Agric. Sci.* 134:159–168.

Champoux, M. C., G. Wang, S. Sarkarung, D. J. Mackill, J. C. O'Toole, N. Huang, and S. R. McCouch. 1995. Locating genes associated with root morphology and drought avoidance in rice via linkage to molecular markers. *Theor. Appl. Genet.* 90:969–981.

Chan, K. Y. and D. P. Heenan. 1998. Effect of lime (CaCO$_3$) application on soil structural stability of a red earth. *Aust. J. Soil Res.* 36:73–86.

Chassot, A. and W. Richner. 2002. Root characteristics and phosphorus uptake of maize seedlings in a bilayered soil. *Agron. J.* 94:118–127.

Clark, R. B., E. E. Alberts, R. W. Zobel, T. R. Sinclair, M. S. Miller, W. D. Kemper, and C. D. Foy. 1998. Eastern gamagrass (*Tripsacum dactyloides*) root penetration into and chemical properties of claypan soils. *Plant Soil* 200:33–45.

Clausen, M., R. Krauter, G. Schachermayr, I. Potrykus, and C. Sautter. 2000. Antifungal activity of a virally encoded gene in transgenic wheat. *Nat. Biotechnol.* 18:446–449.

Clay, D. E., C. E. Clapp, C. Reese, Z. Liu, C. G. Carlson, H. Woodard, and A. Bly. 2007. Carbon-13 fractionation of relic soil organic carbon during mineralization effects calculated half-lives. *Soil Sci. Soc. Am. J.* 71:1003–1009.

Cole, L., R. D. Bardgett, P. Ineson, and P. J. Hobbs. 2002. Enchytraeid worm (Oligochaeta) influences on microbial community structure, nutrient dynamics and plant growth in blanket peat subjected to warming. *Soil Biol. Biochem.* 34:83–92.

Collins, H. P., R. L. Blevins, L. G. Bundy, D. R. Christensen, W. A. Dick, D. R. Huggins, and E. A. Paul. 1999. Soil carbon dynamics in corn-based agroecosystems: Results from carbon-13 natural abundance. *Soil Sci. Soc. Am. J.* 63:584–591.

Collins, H. P., E. T. Elliott, K. Paustian, L. G. Bundy, W. A. Dick, D. R. Huggins, A. J. M. Smucker, and E. A. Paul. 2000. Soil carbon pools and fluxes in long-term Corn Belt agroecosystems. *Soil Biol. Biochem.* 32:157–168.

Collins, H. P., J. L. Smith, S. Fransen, A. K. Alva, C. E. Kruger, and D. M. Granatsein. 2010. Carbon sequestration under irrigated switchgrass (*Panicum virgatum* L.) production. *Soil Sci. Soc. Am. J.* 74:2049–2058.

Conyers, M. K., G. J. Poile, and B. R. Cullis. 1991. Lime responses by barley as related to available soil aluminum and manganese. *Aust. J. Agric. Res.* 42:379–390.

Cooper, A. J. 1973. *Root Temperature and Plant Growth*. Slough, U.K.: Commonwealth Agricultural Bureau.

Corbin, A. T., K. D. Thelen, G. P. Robertson, and R. H. Leep. 2010. Influence of cropping systems on soil aggregate and weed seedbank dynamics during the organic transition period. *Agron. J.* 102:1632–1640.

Corre-Hellou, G. and Y. Crozat. 2005. N$_2$ fixation and N supply in organic pea (*Pisum sativum* L.) cropping systems as affected by weeds and pea weevil (*Sitona lineatus* L.). *Eur. J. Agron.* 22:449–458.

Cowger, C., R. Weisz, J. M. Anderson, and J. R. Horton. 2010. Maize debris increases barley yellow dwarf virus severity in North Carolina winter wheat. *Agron. J.* 102:688–695.

Cox, W. J., R. R. Hahn, and P. J. Stachowski. 2006. Time of weed removal with glyphosate affects corn growth and yield components. *Agron. J.* 98:349–353.

Cox, F. R. and J. R. Sholar. 1995. Site selection, land preparation, and management of soil fertility. In: *Peanut Health Management*, eds., H. A. Melouk and F. M. Shokes, pp. 7–10. St. Paul, MN: The American Phytopathological Society.

Dabney, S. M., J. A. Delgado, and D. W. Reeves. 2001. Using winter cover crops to improve soil and water quality. *Commun. Soil Sci. Plant Anal.* 32:1221–1250.

Dapaah, H. K. and T. J. Vyn. 1998. Nitrogen fertilization and cover crops effects on soil structural stability and corn performance. *Commun. Soil Sci. Plant Anal.* 29:2557–2569.

De Neve, S. and G. Hoffman. 2000. Influence of soil compaction on carbon and nitrogen mineralization of soil organic matter and crop residues. *Biol. Fertil. Soils* 30:544–549.

DeSutter, T. M. and L. J. Cihacek. 2009. Potential agricultural uses of flue gas desulfurization gypsum in the Northern Great Plains. *Agron. J.* 101:817–825.

Diaz, M., E. Bastias, P. Pacheco, L. Tapia, M. C. Martinez-Ballesta, and M. Carvajal. 2011. Characterization of the physiological response of the highly-tolerant tomato cv. Poncho Negro to salinity and excess boron. *J. Plant Nutr.* 34:1254–1267.

Dill, G. M. 2005. Glyphosate-resistance crops: History, status and future. *Pest Manag. Sci.* 61:219–224.

Doran, J. W. and M. R. Zeiss. 2000. Soil health and sustainability: Managing the biotic component of soil quality. *Appl. Soil Ecol.* 15:3–11.

Dore, T. and J. M. Meynard. 1995. On-farm analysis of attacks by the pea weevil (*Sitona lineatus* L.) and the resulting damage to pea crops. *J. Appl. Entomol.* 119:49–54.

Dormaar, J. F. and J. W. Ketcheson. 1960. The effect of nitrogen form and soil temperature on the growth and phosphorus uptake of corn grown in the greenhouse. *Can. J. Soil Sci.* 40:177–184.

Dourado-Neto, D., D. Powlson, R. A. Bakar, O. O. S. Bacchi, M. V. Basanta, P. T. Cong et al. 2010. Multiseason recoveries of organic and inorganic nitrogen-15 in tropical cropping systems. *Soil Sci. Soc. Am. J.* 74:139–152.

Drake, W. L., D. L. Jordan, M. Schroeder-Moreno, P. D. Johnson, J. L. Heitman, Y. J. Cardoza et al. 2010. Crop response following tall fescue sod and agronomic crops. *Agron. J.* 102:1692–1699.

Duke, S. O. and S. B. Powels. 2008. Glyphosate resistant weeds and crops. *Pest Manag. Sci.* 64:317–318.

Edmeades, D. C. 2003. The long-term effects of manures and fertilizers on soil productivity and quality: A review. *Nutr. Cycl. Agroecosyst.* 66:165–180.

Edwards, C. A. and P. J. Bohlen. 1996. *Biology and Ecology of Earthworms*, 3rd edn. London, U.K.: Chapman and Hall.

Elbert, A., M. Haas, B. Springer, W. Thielert, and R. Nauen. 2008. Applied aspects of neonicotinoid uses in crop protection. *Pest Manag. Sci.* 64:1099–1105.

Emerson, W. W. and D. McGarry. 2003. Organic carbon and soil porosity. *Aust. J. Soil Res.* 41:107–118.

Epstein, E. 1966. Effect of soil temperature at different growth stages on growth and development of potato plants. *Agron. J.* 58:169–171.

Epstein, E. and A. J. Bloom. 2005. *Mineral Nutrition of Plants: Principal and Perspectives*, 2nd edn. Sunderland, MA: Sinauer Associates.

Erickson, A. E. 1982. Tillage effects on soil aeration. In: *Predicting Tillage Effects on Soil Physical Properties and Processes*, eds., P. W. Unger and D. M. Van Doren Jr., pp. 91–104. Madison, WI: ASA.

Erickson, A. E. and D. W. Van Doren. 1961. The relation of plant growth and yield to soil oxygen availability. *Transacyions of Seventh International Congress of Soil Science*, Vol. 4, pp. 428–432. Amsterdam, the Netherlands: Elsevier.

Ernani, P. R., D. J. Miquelluti, S. M. V. Fontoura, J. Kaminski, and J. A. Almeida. 2006. Downward movement of soil cations in highly weathered soils caused by addition of gypsum. *Commun. Soil Sci. Plant Anal.* 37:571–586.

Evans, S. P., S. Z. Knezevic, J. L. Lindquist, C. A. Shapiro, and E. E. Blankenship. 2003. Nitrogen application influences the critical period of weed control in corn. *Weed Sci.* 51:408–417.

Evers, G. W. 2002. Ryegrass-bermudagrass production and nutrient uptake when combining nitrogen fertilizer with broiler litter. *Agron. J.* 94:905–910.

Fageria, N. K. 1989. *Tropical Soils and Physiological Aspects of Crop Yield.* Brasilia, Brazil: EMBRAPA-National Rice and Bean Research Center.

Fageria, N. K. 1992. *Maximizing Crop Yields.* New York: Marcel Dekker.

Fageria, N. K. 2000. Upland rice response to soil acidity in cerrado soil. *Pesq. Agropec. Bras.* 35:2303–2307.

Fageria, N. K. 2001. Effect of liming on upland rice, common bean, corn, and soybean production in cerrado soil. *Pesq. Agropec. Bras.* 36:1419–1424.

Fageria, N. K. 2006. Liming and copper fertilization in dry bean production on an Oxisol in no-tillage system. *J. Plant Nutr.* 29:1219–1228.

Fageria, N. K. 2008. Optimum soil acidity indices for dry bean production on an Oxisol in no-tillage system. *Commun. Soil Sci. Plant Anal.* 39:845–857.

Fageria, N. K. 2009. *The Use of Nutrients in Crop Plants.* Boca Raton, FL: CRC Press.

Fageria, N. K. and V. C. Baligar. 2001. Improving nutrient use efficiency of annual crops in Brazilian acid soils for sustainable crop production. *Commun. Soil Sci. Plant Anal.* 32:1303–1319.

Fageria, N. K. and V. C. Baligar. 2003. Fertility management of tropical acid soils for sustainable crop production. In: *Handbook of Soil Acidity*, ed., Z. Rengel, pp. 359–385. New York: Marcel Dekker.

Fageria, N. K. and V. C. Baligar. 2005. Enhancing nitrogen use efficiency in crop plants. *Adv. Agron.* 88:97–185.

Fageria, N. K. and V. C. Baligar. 2008. Ameliorating soil acidity of tropical Oxisols by liming for sustainable crop production. *Adv. Agron.* 99:345–399.

Fageria, N. K., V. C. Baligar, and R. B. Clark. 2006. *Physiology of Crop Production.* New York: The Haworth Press.

Fageria, N. K., V. C. Baligar, and D. G. Edwards. 1990. Soil-plant nutrient relationships at low pH stress. In: *Crops as Enhancers of Nutrient Use*, eds., V. C. Baligar and R. R. Duncan, pp. 475–507. San Diego, CA: Academic Press.

Fageria, N. K., V. C. Baligar, and C. A. Jones. 2011. *Growth and Mineral Nutrition of Field Crops*, 3rd edn. Boca Raton, FL: CRC Press.

Fageria, N. K., V. C. Baligar, and Y. C. Li. 2008. The role of nutrient efficient plants in improving crop yields in the twenty first century. *J. Plant Nutr.* 31:1121–1157.

Fageria, N. K., V. C. Baligar, A. Moreira, and T. A. Portes. 2010a. Dry bean genotypes evaluation for growth, yield components and phosphorus use efficiency. *J. Plant Nutr.* 33:2167–2181.

Fageria, N. K., V. C. Baligar, and R. W. Zobel. 2007. Yield, nutrient uptake, and soil chemical properties as influenced by liming and boron application in common bean in a no-tillage system. *Commun. Soil. Sci. Plant Anal.* 38:1637–1653.

Fageria, N. K., E. M. Castro, and V. C. Baligar. 2004. Response of upland rice genotypes to soil acidity. In: *The Red Soils of China: Their Nature, Management and Utilization*, eds., M. J. Wilson, Z. He, and X. Yang, pp. 219–237. Dordrecht, the Netherlands: Kluwer Academic Publishers.

Fageria, N. K., O. P. Morais, and A. B. Santos. 2010b. Nitrogen use efficiency in upland rice. *J. Plant Nutr.* 33:1696–1711.

Fageria, N. K. and A. Moreira. 2011. The role of mineral nutrition on root growth of crop plants. *Adv. Agron.* 110:251–331.

Fageria, N. K. and A. B. Santos. 2005. Influence of base saturation and micronutrient rates on their concentration in the soil and bean productivity in cerrado soil in no-tillage system. Paper presented at the *Eighth National Bean Congress*, October 18–20, 2005. Goiânia, Brazil.

Fageria, N. K., A. B. Santos, and M. F. Moraes. 2010c. Influence of urea and ammonium sulfate on soil acidity indices in lowland rice production. *Commun. Soil Sci. Plant Anal.* 41:1565–1575.

Fageria, N. K., A. B. Santos, and M. F. Moraes. 2010d. Yield, potassium uptake, and use efficiency in upland rice genotypes. *J. Plant Nutr.* 41:2676–2684.

Fageria, N. K. and L. F. Stone. 2004. Yield of common bean in no-tillage system with application of lime and zinc. *Pesq. Agropec. Bras.* 39:73–78.

Farina, M. P. W. and P. Channon. 1988. Acid subsoil amelioration. II. Gypsum effects on growth and subsoil chemical properties. *Soil Sci. Soc. Am. J.* 52:175–180.

Farina, M. P. W., P. Chanon, and G. R. Thibaud. 2000. A comparison of strategies for ameliorating subsoil acidity: I. Long term growth effects. *Soil Sci. Soc. Am. J.* 64:646–651.

Faye, I., O. Diouf, A. Guisse, M. Sene, and N. Diallo. 2006. Characterizing root responses to low phosphorus in pearl millet (*Pennisetum glaucum* L. R. Br.). *Agron. J.* 98:1187–1194.

Fernandez, F. G., R. G. Hoeft, G. W. Randall, J. Vetsch, K. Greer, E. D. Nafziger, and M. B. Villamil. 2010. Apparent nitrogen recovery from fall-applied ammoniated phosphates and ammonium sulfate fertilizers. *Agron. J.* 102:1674–1681.

Ferreira, M. I. and C. F. Reinhardt. 2010. Field assessment of crop residues for allelopathic effects on both crops and weeds. *Agron. J.* 102:1593–1600.

Ferreras, L. A., J. L. Costa, F. O. Garcia, and C. Pecorari. 2000. Effect of tillage on some soil physical properties of a structural degraded Petrocalci Paleudoll of the Southern Pampa of Argentina. *Soil Till. Res.* 54:31–39.

Fixen, P. 2010. World fertilizer nutrient reserves. In: *Good Practices for Efficient Use of Fertilizers*, eds., L. I. Prochnow, V. Casarin, and S. R. Stipp, pp. 91–109. Piracicaba, Brazil: International Plant Nutrition Institute.

Follett, R. F. and D. S. Schimel. 1989. Effect of tillage practices on microbial biomass dynamics. *Soil Sci. Soc. Am. J.* 53:1091–1096.

Fonte, S. J., T. Winsome, and J. Six. 2009. Earthworm populations in relation to soil organic matter dynamics and management in California tomato cropping systems. *Appl. Soil Ecol.* 41:206–214.

Fox, G. A. and G. V. Wilson. 2010. The role of subsurface flow in hillslope and stream bank: A review. *Soil Sci. Soc. Am. J.* 74:717–733.

Foy, C. D. 1984. Physiological effects of hydrogen, aluminum and manganese toxicity in acid soils. In: *Soil Acidity and Liming*, 2nd edn., ed., F. Adams, pp. 57–97. Madison, WI: ASA, CSSA, and SSSA.

Foy, C. D. 1992. Soil chemical factors limiting plant root growth. *Adv. Soil Sci.* 19:97–149.

Francois, L. E. and E. V. Maas. 1999. Crop response and management of salt affected soils. In: *Handbook of Plant and Crop Stress*, ed., M. Pessarakli, pp. 169–201. New York: Marcel Dekker.

Franzluebbers, A. J. 2007. Integrated crop-livestock systems in the southern USA. *Agron. J.* 99:361–372.

Fraser, D. G., J. W. Doran, W. W. Sahs, and G. W. Lesoing. 1988. Soil microbial populations and activities under conventional and organic management. *J. Environ. Qual.* 17:585–590.

Freitas, P. L., R. W. Zobel, and V. A. Snyder. 1999. Corn root growth in soil columns with artificially constructed aggregates. *Crop Sci.* 39:725–730.

Frenkel, H., Z. Gerstl, and N. Alperovitch. 1989. Exchange induced dissolution of gypsum and the reclamation of sodic soils. *J. Soil Sci.* 40:599–611.

Gagnon, B. and N. Ziadi. 2010. Grain corn and soil nitrogen responses to side-dress nitrogen sources and applications. *Agron. J.* 102:1014–1022.

Gallo, P. B., H. A. A. Mascarenhas, J. A. Quaggio, and O. C. Bataglia. 1986. Differential responses of soybean and sorghum to liming. *Rev. Bras. Ciênc. Solo* 10:253–258.

Galvez, L., D. D. Douds Jr., P. Wagoner, L. R. Longnecker, L. F. Drinkwater, and R. R. Janke. 1995. An overwintering cover crop increases inoculum of VAM fungi in agriculture soil. *Am. J. Altern. Agric.* 10:152–156.

Gao, Y., L. Ren, W. Ling, F. Kang, X. Zhu, and B. Sun. 2010. Effects of low-molecular weight organic acids on sorption-desorption of phenanthrene in soils. *Soil Sci. Soc. Am. J.* 74:51–59.

Gardner, H. R. and R. E. Danielson. 1964. Penetration of wax layers by cotton roots as affected by some soil physical conditions. *Soil Sci. Soc. Am. Proc.* 28:457–460.

Garwood, E. A. 1968. Some effects of soil water conditions and temperature on the roots of grasses and clovers. II. Effects of variation in the soil water content and soil temperature on root growth. *J. Br. Grassland Soc.* 23:117–127.

Gerard, C. J., P. Sexton, and G. Shaw. 1982. Physical factors influencing soil strength and root growth. *Agron. J.* 74:875–879.

Ghosh, S., P. Lockwood, N. Hulugalle, H. Daniel, P. Kristiansen, and K. Dodd. 2010. Changes in properties of sodic Australian vertisols with application of organic waste products. *Soil Sci. Soc. Am. J.* 74:153–160.

Ghoshal, N. and K. P. Singh. 1995. Effects of farmyard manure and inorganic fertilizer on the dynamics of soil microbial biomass in a tropical dryland agroecosystems. *Biol. Fertil. Soils* 19:231–238.

Gilbert, J. and A. Tekauz. 1993. Reaction of Canadian spring wheats to *Septoria nodorum* and the relationship between disease severity and yield components. *Plant Dis.* 77:398–402.

Gill, W. R. and G. E. Vandenberg. 1967. *Soil Dynamics in Tillage and Traction*, USDA Handbook, No. 316. Washington, DC: US Government Printing Press.

Glinski, J. and J. Lipiec. 1990. *Soil Conditions and Plant Roots*. Boca Raton, FL: CRC Press.

Glumac, N. G., W. K. Dong, and W. M. Jarrell. 2010. Quantitative analysis of soil organic carbon using laser-induced breakdown spectroscopy: An improved method. *Soil Sci. Soc. Am. J.* 74:1922–1928.

Godwin, R. J. 1990. Agricultural engineering in development: Tillage for crop production in areas of low rainfall. Rome, Italy: FAO.

Granatstein, D. M., D. F. Bezdicek, V. L. Cochran, L. F. Elliott, and J. Hammel. 1987. Long term tillage and rotation effects on soil microbial biomass, carbon and nitrogen. *Biol. Fertil. Soils* 5:265–270.

Grandy, A. S. and G. P. Robertson. 2006. Aggregation and organic matter protection following tillage of a previously uncultivated soil. *Soil Sci. Soc. Am. J.* 70:1398–1406.

Green, V. S., M. A. Cavigelli, T. H. Dao, and D. C. Flanagan. 2005. Soil physical properties and aggregate-associated C, N, and P distributions in organic and conventional cropping systems. *Soil Sci.* 170:822–831.

Gregorich, E. G., C. M. Monreal, M. Schnitzer, and H. R. Schulten. 1996. Transformation of plant residues into soil organic matter: Chemical characterization of plant tissue, isolated soil fractions, and whole soils. *Soil Sci.* 161:680–693.

Gregory, P. 2006. *Plant Roots: Growth, Activity and Interactions with Soils*. Oxford, U.K.: Blackwell Publishing.

Grijalva, D. F. M., C. R. Crozier, T. J. Smyth, and D. H. Hardy. 2010. Nitrogen, phosphorus, and liming effects of poultry layer manures in coastal plain and piedmont soils. *Agron. J.* 102:1329–1339.

Gupta, V. V. S. R. and J. J. Germida. 1988. Distribution of microbial biomass and its activity in different soil aggregate size classes as affected by cultivation. *Soil Biol. Biochem.* 20:777–786.

Hale, M. G. and D. M. Orcutt. 1987. *The Physiology of Plants under Stress*. New York: John Wiley & Sons.

Halford, C., A. S. Hamill, J. Zhang, and C. Doucet. 2001. Critical period of weed control in no-till soybean (*Glycine max*) and corn (*Zea mays*). *Weed Technol.* 15:737–744.

Hall, M. R., C. J. Swanton, and G. W. Anderson. 1992. The critical period of weed control in grain corn (*Zea mays*). *Weed Sci.* 40:441–447.

Hamon, N., R. Bardner, L. Allen-Williams, and J. B. Lee. 1990. Carabid populations in field beans and their effect on the population dynamics of *Sitona lineatus* L. *Ann. Appl. Biol.* 117:51–62.

Hartt, C. E. 1965. The effects of temperature upon translocation of C^{14} in sugarcane. *Plant Physiol.* 40:74–81.

Hatfield, J. L., K. J. Boote, B. A. Kimball, L. H. Zisks, R. C. Izaurralde, D. Ort, A. M. Thompson, and D. Wolfe. 2011. Climate impacts on agriculture: Implications for crop production. *Agron. J.* 103:351–370.

Havlin, J. L., J. D. Beaton, S. L. Tisdale, and W. L. Nelson. 2006. *Soil Fertility and Fertilizers: An Introduction to Nutrient Management*, 7th edn. Englewood Cliffs, NJ: Prentice-Hall.

Haynes, R. J. 1982. Effects of liming on phosphate availability in acid soils. A critical review. *Plant Soil* 68:289–308.

Haynes, R. J. 1984. Lime and phosphate in the soil-plant system. *Adv. Agron.* 37:249–315.

Haynes, R. J. and T. E. Ludecke. 1981. Effect of lime and phosphorus applications on concentrations of available nutrients and on P, Al, and Mn uptake by two pasture legumes in an acid soil. *Plant Soil* 62:117–128.

Haynes, S. and R. Naidu. 1998. Influence of lime, fertilizer and manure applications on soil organic matter content and soil physical conditions: A review. *Nutr. Cycl. Agroecosyst.* 51:123–137.

Heinrichs, D. H. and K. F. Nielsen. 1966. Growth response of alfalfa varieties of diverse genetic origin to different root zone temperatures. *Can. J. Plant Sci.* 46:291–298.

Herrero, M., P. K. Thorton, A. M. Notenbaert, S. Wood, S. Msangi, H. A. Freeman et al. 2010. Smart investments in sustainable food productions: Revisiting mixed crop-livestock systems. *Science* 327:822–825.

Hetrick, J. A. and A. P. Schwab. 1992. Changes in aluminum and phosphorus solubility are in response to long term fertilization. *Soil Sci. Soc. Am. J.* 56:755–761.

Hoebeke, E. R. and A. G. Wheeler Jr. 1985. *Sitona lineatus* L., the pea leaf weevil: First records in eastern North America. *Proc. Entomol. Soc. Wash.* 87:216–220.

Hood-Nowotny, R., N. H. N. Umana, E. Inselbacher, P. O. Oswald-Lachouani, and W. Wanek. 2010. Alternative methods for measuring inorganic, organic, and total dissolved nitrogen in soil. *Soil Sci. Soc. Am. J.* 74:1018–1027.

Hoffman, G. J. 1981. Alleviating salinity stress. *ASEA* 4:305–346.

Hollaway, G. J., S. P. Taylor, R. F. Eastwood, and C. H. Hunt. 2000. Effect of field crops on density of *Pratylenchus* in south eastern Australia. Part 2. *P. thornei. J. Nematol.* 32:600–608.

Hu, T., S. Metz, C. Chay, H. P. Zhou, N. Biest, G. Chen et al. 2003. *Agrobacterium*-mediated large scale transformation of wheat (*Triticum aestivum* L.) using glyphosate selection. *Plant Cell Rep.* 21:1010–1019.

Hummel, J. D., L. M. Dosdall, G. W. Clayton, T. K. Turkington, N. Z. Lupwayi, K. N. Harker, and J. T. O'Donovan. 2009. Canola-wheat intercrops for improved agronomic performance and integrated pest management. *Agron. J.* 101:1190–1197.

Inoue, K., H. Yohota, and Y. Yamada. 1988. Effect of calcium in the medium on root growth under low pH conditions. *Soil Sci. Plant Nutr.* 34:359–374.

Isfan, D., J. Zizka, A. Davignon, and M. Deschenes. 1995. Relationships between nitrogen rate, plant nitrogen concentration, yield and residual soil nitrate-nitrogen in silage corn. *Commun. Soil Sci. Plant Anal.* 26:2531–2557.

Jalali, M. and D. L. Rowell. 2008. Potassium leaching in undisturbed soil cores following surface application of gypsum. *Environ. Geol.* 57:41–48.

Jenkinson, D. S. and J. N. Ladd. 1981. Microbial biomass in soil: Measurement and turnover. In: *Soil Biochemistry*, Vol. 5, eds., E. A. Paul and J. N. Ladd, pp. 415–471. New York: Marcel Dekker.

Jiao, Y., J. K. Whalen, and W. H. Hendershot. 2006. No-tillage and manure applications increase aggregation and improve nutrient retention in a sandy-loam soil. *Geoderma* 134:24–33.

Jimba, S. C. and B. Lowery. 2010. Automation of the water drop method for soil aggregate stability analysis. *Soil Sci. Soc. Am. J.* 74:38–41.

Johnson, J. M. F., R. R. Allmaras, and D. C. Reicosky. 2006. Estimating source carbon from crop residues, roots and rhizodeposits using the national grain-yield database. *Agron. J.* 622–636.

Johnson, J. P., B. F. Carver, and V. C. Baligar. 1997. Productivity in Great Plains acid soils of wheat genotypes selected for Al tolerance. *Plant Soil* 188:101–106.

Johnson, N. C., P. J. Copeland, R. K. Crookston, and F. L. Pfleger. 1992. Mycorrhizae: Possible explanation for the yield decrease with continuous corn and soybean. *Agron. J.* 84:387–390.

Johnson, C. A., P. Groffman, D. D. Breshears, Z. G. Cardon, W. Currie, and W. Emanuel. 2004. Carbon cycling in soil. *Front. Ecol. Environ.* 2:522–528.

Jokela, W. E. 1992. Nitrogen fertilizer and dairy manure effects on corn yield and soil nitrate. *Soil Sci. Soc. Am. J.* 5:148–154.

Jokela, W. E., J. H. Grabber, D. L. Karlen, T. C. Balser, and D. E. Palmquist. 2009. Cover crop and liquid manure effects on soil quality indicators in a corn silage system. *Agron. J.* 101:727–737.

Jokela, W. E. and G. W. Randall. 1997. Fate of fertilizer nitrogen as affected by time and rate of application on corn. *Soil Sci. Soc. Am. J.* 61:1695–1703.

Jonasson, S., A. Michelsen, I. K. Schmidt, and E. V. Nielsen. 1999. Responses in microbes and plants to changed temperature, nutrient, and light regimes in the Arctic. *Ecology* 80:1828–1843.

Jonathan, P. L. and M. B. Kathleen. 2001. Topsoil foraging-an architectural adaptation of plants to low P availability. *Plant Soil* 237:225–237.

Jones, E., R. S. Jessop, B. M. Sindel, and A. Hoult. 1999. Utilising crop residues to control weeds. In: *Proceedings of 12th Australian Weed Conference*, ed., A. Bishop, pp. 373–376. Devonport, TAS, Australia: Tasmania Weed Society.

Jordan, D. L., J. E. Bailey, J. S. Barnes, C. R. Bogle, S. G. Bullen, A. B. Brown, K. L. Edmisten, E. J. Dunphy, and P. D. Johnson. 2002. Yield and economic return of ten peanut-based cropping systems. *Agron. J.* 94:1289–1294.

Kabir, Z. and R. T. Koide. 2000. The effect of dandelion or a cover crop on mycorrhiza inoculum potential, soil aggregation and yield of maize. *Agric. Ecosyst. Environ.* 78:167–174.

Kabir, Z. and R. T. Koide. 2002. Effect of autumn and winter mycorrhizal cover crops on soil properties, nutrient uptake and yield of sweet corn in Pennsylvania, USA. *Plant Soil* 238:205–215.

Kaitibie, S., F. M. Epplin, E. G. Krenzer Jr., and H. Zhang. 2002. Economics of lime and phosphorus application for dual-purpose winter wheat production in low-pH soils. *Agron. J.* 94:1139–1145.

Kamprath, E. J. 1970. Exchangeable aluminum as a criterion for liming leached mineral soils. *Soil Sci. Soc. Am. Proc.* 34:252–254.

Kamprath, E. J. and C. D. Foy. 1971. Lime-fertilizers-plant interactions in acid soils. In: *Fertilizer Technology and Use*, 2nd edn., eds., R. W. Olsen, T. J. Army, J. J. Hanway, and V. J. Kilmer, pp. 105–151. Madison, WI: SSSA.

Kamprath, E. J. and C. D. Foy. 1985. Lime-fertilize-plant interactions in acid soils. In: *Fertilizer Technology and Use*, 3rd edn., ed., O. P. Engelstad, pp. 91–151. Madison, WI: SSSA.

Kariuki, S. K., H. Zhang, J. L. Schroder, J. Edwards, M. Payton, B. F. Carver, W. R. Raun, and E. G. Krenzer. 2007. Hard red winter wheat cultivar responses to a pH and aluminum concentration gradient. *Agron. J.* 99:88–98.

Karlen, D. L., M. J. Rosek, J. C. Gardner, D. L. Allan, M. J. Alms, D. F. Bezdicek, M. Flock, D. R. Huggins, B. S. Miller, and M. L. Staben. 1999. Conservation reserve program effects on soil quality indicators. *J. Soil Water Conserv.* 54:439–444.

Karlen, D. L., N. C. Wollenhaupt, D. C. Erbach, E. C. Berry, J. B. Swan, N. S. Eash, and J. L. Jordahl. 1994. Long-term tillage effects on soil quality. *Soil Till. Res.* 32:313–327.

Katsvairo, T. W., D. L. Wright, J. J. Marois, D. L. Hartzog, K. B. Balkcom, P. P. Wiatrak, and J. R. Rich. 2007. Cotton roots, earthworms, and infiltration characteristics in sod-peanut-cotton cropping systems. *Agron. J.* 99:390–398.

Kay, B. D. 1990. Rates of change of soil structure under different cropping systems. *Adv. Soil Sci.* 12:1–52.

Kaye, N. M., S. C. Mason, D. S. Jackson, and T. D. Galusha. 2007. Crop rotation and soil amendments alters sorghum grain quality. *Crop Sci.* 47:722–729.

Keren, R. and I. Shainberg. 1981. Effect of dilution rate on the efficiency of industrial and mined gypsum in improving infiltration of sodic soil. *Soil Sci. Soc. Am. J.* 45:103–107.

Ketellapper, H. J. 1960. The effect of soil temperature on the growth of *Phalaris tuberso* L. *Physiol. Plant* 13:641–647.

Khanh, T. D., M. I. Chung, T. D. Xuan, and S. Tawata. 2005. The exploitation of crop allelopathy in sustainable agricultural production. *J. Agron. Crop Sci.* 191:172–184.

Kinraide, T. B. and D. R. Parker. 1987. Non-phytotoxicity of the aluminum sulfate ion, $AlSO_4^+$. *Physiol. Planta.* 71:207–212.

Knezevic, S. Z., M. J. Horak, and R. L. Vanderlip. 1997. Relative time of redroot pigweed (*Amaranthus retroflexus* L.) emergence is critical in pigweed-sorghum (*Sorghum bicolor* L. Moench) competition. *Weed Sci.* 45:502–508.

Kochian, L. V. 1995. Cellular mechanisms of aluminum toxicity and resistance in plants. *Annu. Rev. Plant Physiol. Plant Mol. Biol.* 46:237–260.

Koyama, H., A. Kawamura, T. Kihara, T. Hara, E. Takita, and D. Shibata. 2000. Over-expression of mitochondrial citrate synthase in *Arabidopsis thaliana* improved growth on a phosphorus-limited soil. *Plant Cell Physiol.* 41:1030–1037.

Kropff, M. J. and C. J. T. Spitters. 1991. A simple model of crop loss by weed competition from early observations on relative leaf area of the weed. *Weed Res.* 31:97–105.

Krupke, C., P. Marquardt, W. Johnson, S. Weller, and S. P. Conley. 2009. Volunteer corn presents new challenges for insect resistant management. *Agron. J.* 101:797–799.

Laboski, C. A. M., R. H. Dowdy, R. R. Allmaras, and J. A. Lamb. 1998. Soil strength and water content influences on corn root distribution in a sandy soil. *Plant Soil* 203:239–247.

Lai, R., M. You, L. A. P. Lotz, and L. Vasseur. 2011. Response of green peach aphids and other arthropods to garlic intercropped with tobacco. *Agron. J.* 103:856–863.

Lal, R. 2004. Soil carbon sequestration impacts on global climate change and food security. *Science* 304:1623–1627.

Lal, R. 2005. Soil erosion and carbon dynamics. *Soil Till. Res.* 81:137–142.

Lal, R. 2007. Farming carbon. *Soil Till. Res.* 96:1–5.

Lal, R. 2009. Challenges and opportunities in soil organic matter research. *Eur. J. Soil Sci.* 60:158–169.

Lal, R., G. F. Hall, and F. P. Miller. 1989. Soil degradation. I. Basic processes. *Land Degrad. Rehabil.* 1:51–69.

Lal, R. and D. Pimentel. 2008. Soil erosion: A carbon sink or source? *Science* 319:1040–1042.

Lamb, M. C., J. I. Davidson, and C. L. Butts. 1993. Peanut yield decline in the southeast and economically reasonable solutions. *Peanut Sci.* 20:36–40.

Larson, J. A., C. O. Gwathmey, D. F. Mooney, L. E. Steckel, and R. K. Roberts. 2009. Does skip-row planting configuration improve cotton return? *Agron. J.* 101:738–746.

Larson, J. A., R. K. Roberts, and C. O. Gwathmey. 2007. Herbicide resistant technology price effects on the plant population density decision for ultra-narrow-row cotton. *J. Agric. Res. Econ.* 32:383–401.

Latif, M. A., G. R. Mehuys, A. F. Mackenzie, and M. A. Faris. 1992. Effects of legumes on soil physical quality in a maize crop. *Plant Soil* 140:15–23.

Lavella, P. and A. Martin. 1992. Small-scale and large-scale effects of endogenic earthworms on soil organic matter dynamics in soils of the humid tropics. *Soil Biol. Biochem.* 24:1491–1498.

Lavelle, P. 1988. Earthworm activities and the soil system. *Biol. Fertil. Soils* 6:237–251.

Lavelle, P. and A. Martin. 1992. Small-scale and large-scale effects of endogenic earthworms on soil organic matter dynamics in soils of humid tropics. *Soil Biol. Biochem.* 24:1491–1498.

Lee, K. E. 1985. *Earthworms: Their Ecology and Relationships with Soil and Land Use.* Sydney, New South Wales, Australia: Academic Press.

Le Bissonnais, Y. and D. Arrouays. 1997. Aggregate stability and assessment of soil crustability and erodibility: 2. Application to humic loamy soils with various organic carbon contents. *Eur. J. Soil Sci.* 48:39–48.

Lebron, I., D. L. Suarez, and T. Yoshida. 2002. Gypsum effect on the aggregate size and geometry of three sodic soils under reclamation. *Soil Sci. Soc. Am. J.* 66:92–98.

Li, X. and Z. Chen. 2004. Soil microbial biomass C and N along a climatic transect in the Mongolian steppe. *Biol. Fertil. Soils* 39:344–351.

Li, X. F., J. F. Ma, and H. Matsumoto. 2000. Pattern of aluminum induced secretion of organic acids differs between rye and wheat. *Plant Physiol.* 123:1537–1544.

Linden, D. R., P. F. Hendrix, D. C. Coleman, and P. C. J. Vliet. 1994. Faunal indicators of soil quality. In: *Defining Soil Quality for a Sustainable Environment*, eds., J. W. Doran, D. C. Coleman, D. F. Bezdicek, and B. A. Stewart, pp. 91–106. Madison, WI: SSSA.

Lindsay, W. L. 1979. *Chemical Equilibrium in Soils.* New York: John Wiley & Sons.

Lindsay, W. L., A. W. Frazier, and H. F. Stephenson. 1962. Identification of reaction products from phosphate fertilizers in soils. *Soil Sci. Soc. Am. Proc.* 26:446–452.

Lipps, P. E. and L. V. Madden. 1989. Assessment of methods of determining powdery mildew severity in relation to grain yield of winter wheat cultivars in Ohio. *Phytopathology* 79:462–470.

Liu, J. and N. V. Hue. 2001. Amending subsoil acidity by surface application of gypsum, lime, and composts. *Commun. Soil Sci. Plant Anal.* 32:2117–2132.

Loecker, J., N. O. Nelson, W. B. Gordon, L. D. Maddux, K. A. Janssen, and W. T. Schapaugh. 2010. Manganese response in conventional and glyphosate resistant soybean. *Agron. J.* 102:606–611.

Logsdon, S. L. and D. R. Linden. 1992. Interaction of earthworms with soil physical conditions and plant growth. *Soil Sci.* 154:330–337.

Lopes, A. S., M. C. Silva, and L. R. G. Guilherme. 1991. Soil acidity and liming. Technical bulletin 1, São Paulo, Brazil: National Association for Diffusion of Fertilizers and Agricultural Amendments.

Lopez-Bucio, J., M. F. Nieto-Jacobo, V. Ramirez-Rodriguez, and L. Herrera-Estrella. 2000. Organic acid metabolism in plants: From adaptive physiology to transgenic varieties for cultivation in extreme soils. *Plant Sci.* 160:1–13.

Lopez-Millan, A. F., F. Morales, S. Andaluz, Y. Gogorcena, A. Abadia, J. D. L. Rivas, and J. Abadia. 2000. Responses of sugar beet roots to iron deficiency: Changes in carbon assimilation and oxygen use. *Plant Physiol.* 124:885–898.

Loss, S. P., G. S. P. Ritchie, and A. D. Robson. 1993. Effect of lupins and pasture on soil acidification and fertility in Western Australia. *Aust. J. Exp. Agric.* 33:457–464.

Low, A. J. 1979. Soil structure. In: *The Encyclopedia of Soil Science*, Part 1, eds., R. W. Fairbridge and C. W. Finkl Jr., pp. 508–514. Stroudsburg, PA: Dowden, Hutchinson and Ross.

Lupwayi, N. Z., G. W. Clayton, J. T. O'Donovan, and C. A. Grant. 2011. Soil microbial response to nitrogen rate and placement and barley seedling rate under no till. *Agron. J.* 103:1064–1071.

Lupwayi, N. Z., T. Lea, J. L. Beaudoin, and G. W. Clayton. 2005. Soil microbial biomass, functional diversity and crop yields following application of cattle manure, hog manure and inorganic fertilizers. *Can. J. Soil Sci.* 85:193–201.

Lupwayi, N. Z., M. A. Montreal, G. W. Clayton, C. A. Grant, A. M. Johnston, and W. A. Rice. 2001. Soil microbial biomass and diversity respond to tillage and sulphur fertilizers. *Can. J. Soil Sci.* 81:577–589.

Lupwayi, N. Z., W. A. Rice, and G. W. Clayton. 1999. Soil microbial biomass and carbon dioxide flux under wheat as influenced by tillage and crop rotation. *Can. J. Soil Sci.* 79:273–280.

Maas, E. V. 1993. Testing crops for salinity tolerance. In: *A Workshop on Adaptation of Plants to Soil Stresses*, ed., J. W. Maranville, pp. 234–247. Lincoln, NE: University of Nebraska.

Maas, E. V. and G. J. Hoffman. 1977. Crop salt tolerance-current assessment. *J. Irrig. Drain. Div. Proc. Am. Soc. Civil Eng.* 103:115–134.

MacDonald, J. D., D. A. Angers, P. Rochette, M. H. Chantigny, I. Royer, and M. O. Gasser. 2010. Plowing a poorly drained grassland reduces soil respiration. *Soil Sci. Soc. Am. J.* 74:2067–2076.

Mack, H. J., S. C. Fang, and S. B. Butts. 1964. Effects of soil temperature and phosphorus fertilization on snap beans and peas. *Proc. Am. Soc. Hort. Sci.* 84:332–338.

Macedo, T. B., C. S. Bastos, L. G. Higley, K. R. Ostlie, and S. Madhavan. 2003. Photosynthetic responses of soybean to soybean aphid (Homoptera: Aphididae) injury. *J. Econ. Entomol.* 96:188–193.

Mady Kaye, N., S. C. Mason, D. S. Jackson, and T. D. Galusha. 2007. Crop rotation and soil amendment alters sorghum grain quality. *Crop Sci.* 47:722–727.

Magdoff, F. and R. R. Weil. 2004. Soil organic matter management strategies. In: *Soil Organic Matter in Sustainable Agriculture*, ed., R. R. Weil, pp. 45–65. Boca Raton, FL: CRC Press.

Mahler, R. L. and R. W. Harder. 1984. The influence of tillage methods, cropping sequence and N rates on the acidification of a northern Idaho soil. *Soil Sci.* 137:52–60.

Marion, G. M., D. M. Hendricks, G. R. Dutt, and W. H. Fuller. 1976. Aluminum and silicate solution in soils. *Soil Sci.* 127:76–85.

Marquez, C. O., V. J. Garcia, C. A. Cambardella, R. C. Schultz, and T. M. Isenhart. 2004. Aggregate-size stability distribution and soil stability. *Soil Sci. Soc. Am. J.* 68:725–735.

Marschner, H. 1995. *Mineral Nutrition of Higher Plants*, 2nd edn. New York: Academic Press.

Maughan, M. W., J. P. C. Flores, I. Anghinoni, G. Bollero, F. G. Fernandez, and B. G. Tracy. 2009. Soil quality and corn yield under crop-livestock integration in Illinois. *Agron. J.* 101:1503–1510.

McCordick, S. A., D. E. Hillger, R. H. Leep, and J. J. Kells. 2008. Establishment systems for glyphosate resistant alfalfa. *Weed Technol.* 22:22–29.

McMichael, B. L. and J. J. Burk. 1994. Metabolic activity of cotton roots in response to temperature. *Environ. Exp. Bot.* 31:461–470.

McMichael, B. L. and J. J. Burke. 1998. Soil temperature and root growth. *Hort. Sci.* 33:947–951.

McMichael, B. L. and J. E. Quisenberry. 1991. Genetic variation for root-shoot relationships among cotton germplasm. *Environ. Exp. Bot.* 31:461–470.

McMichael, B. L. and J. E. Quisenberry. 1993. The impact of the soil environment on the growth of root systems. *Environ. Exp. Bot.* 33:53–61.

McSwiney, C. P. and G. P. Robertson. 2005. Nonlinear response of N_2O flux to incremental fertilizer addition in a continuous maize (*Zea mays* L.) cropping system. *Global Change Biol.* 11:1712–1719.

Mendes, I. C., A. K. Bandick, R. P. Dick, and P. J. Bottomley. 1999. Microbial biomass and activities in soil aggregates affected by winter cover crops. *Soil Sci. Soc. Am. J.* 63:873–881.

Mikha, M. M., J. G. Benjamin, M. F. Vigil, and D. C. Nielson. 2010. Cropping intensity impacts on soil aggregation and carbon sequestration in the central Great Plains. *Soil Sci. Soc. Am. J.* 74:1712–1719.

Mikha, M. M. and C. W. Rice. 2004. Tillage and manure effects on soil and aggregate associated carbon and nitrogen. *Soil Sci. Soc. Am. J.* 68:809–816.

Miller, W. P. 1987. Infiltration and soil loss of three gypsum amended Ultisols under simulated rainfall. *Soil Sci. Soc. Am. J.* 51:1314–1320.

Miller, S. D. and C. M. Alford. 2002. Weed control and glyphosate tolerant alfalfa response to glyphosate rate and application timing. *Proc. North Cent. Weed Sci. Soc.* 57:201.

Montagu, K. D., J. P. Conroy, and B. J. Atwell. 2001. The position of localized soil compaction determines root and subsequent shoot growth responses. *J. Exp. Bot.* 52:2127–2133.

Moore, A. D., D. W. Israel, and R. L. Mikkelsen. 2005. Nitrogen availability of anaerobic swine lagoon sludge: Sludge source effects. *Bioresour. Technol.* 96:323–329.

Moore, J. M., S. Klose, and M. A. Tabatabai. 2000. Soil microbial biomass carbon and nitrogen as affected by cropping systems. *Biol. Fertil. Soils* 31:200–210.

Mulvaney, R. L. 1994. Nitrification of different nitrogen fertilizers. In: *Illinois Fertilizer Conference Proceedings*, ed., R. G. Hoeft, pp. 24–26. Peoria, IL: University of Illinois.

Mulvaney, M. J., C. W. Wood, K. S. Balkcom, D. A. Shannon, and J. M. Kemble. 2010. Carbon and nitrogen mineralization and persistence of organic residues under conservation and conventional tillage. *Agron. J.* 102:1425–1433.

Munns, R. 2002. Comparative physiology of salt and water stress. *Plant Cell Environ.* 25:239–250.

Munns, R. 2005. Gene and salt tolerance: Bringing them together. *New Phytol.* 167:645–663.

Munns, R., R. James, and A. Lauchli. 2006. Approaches to increasing the salt tolerance of wheat and other cereals. *J. Exp. Bot.* 57:1025–1043.

Munns, R. and M. Tester. 2008. Mechanisms of salinity tolerance. *Annu. Rev. Plant Biol.* 59:651–681.

Muruganandam, S., D. W. Israel, and W. P. Robarge. 2010. Nitrogen transformations and microbial communities in soil aggregates from three tillage systems. *Soil Sci. Soc. Am. J.* 74:120–129.

Naderman, G. C., B. G. Brock, G. B. Reddy, and C. W. Raczkowski. 2004. Six years of continuous conservation tillage at the Center for Environmental Farming Systems (CEFS). Part I. Impacts on soil bulk density and carbon content for differing soils and crop rotations. *Annual Meeting of the Soil Science Society of North Carolina*, January 21, 2004, Raleigh, NC.

Nardi, S., M. Tosoni, D. Pizzeghello, M. R. Provenzano, A. Cilenti, A. Sturaro, R. Rella, and A. Vianello. 2005. Chemical characteristics and biological activity of organic substances extracted from soils by root exudates. *Soil Sci. Soc. Am. J.* 69:2012–2019.

Narro, L., S. Pandey, C. D. Leon, F. Salazar, and M. P. Arias. 2001. Implications of soil acidity tolerant maize cultivars to increase production in developing countries. In: *Plant Nutrient Acquisition: New Perspectives*, eds., N. Ae, J. Arihara, K. Okada, and A. Srinivasan, pp. 447–463. Tokyo, Japan: Springer Verlag.

NASS. 2007. Acreage. http://usda.mannlib.cornell.edu/usda/nass/Acre//2000s/2007/Acre-06-29-2007.pdf (posted on June 29, 2007; cited on October 20, 2008; verified on March 30, 2009). Washington, DC: USDA-NASS.

Nemes, A., B. Quebedeaux, and D. J. Timlin. 2010. Ensemble approach to provide uncertainty estimate of soil bulk density. *Soil Sci. Soc. Am. J.* 74:1938–1945.

Neumann, G., A. Massonneau, E. Martoonia, and V. Romheld. 1999. Physiological adaptations to P deficiency during proteoid root development in white lupin. *Planta* 208:373–382.

Nielsen, K. F. 1974. Roots and root temperatures. In: *The Plant Root and Its Environment*, ed., E. W. Carson, pp. 293–333. Charlottesville, VA: University Press of Virginia.

Nielsen, K. F. and R. K. Cunningham. 1964. The effects of soil temperature, form and level of N on growth and chemical composition of Italian ryegrass. *Soil Sci. Soc. Am. Proc.* 28:213–218.

Nielsen, K. F., R. L. Halstead, A. J. MacLean, R. M. Holmes, and S. J. Bourget. 1960. The influence of soil temperature on the growth and mineral composition of oats. *Can. J. Soil Sci.* 40:255–264.

Nielsen, K. F. and E. C. Humphries. 1966. Effects of root temperature on plant growth. *Soil Fertil.* 29:1–7.

Nobel, A. D., M. E. Sumner, and A. K. Alva. 1988. The pH dependency of aluminum phytotoxicity alleviation by calcium sulfate. *Soil Sci. Soc. Am. J.* 52:1398–1402.

NRCS. 1997. *Americas Private Land: A Geography of Hope.* Washington, DC: US Government Printing Press.

Obour, A. K., M. L. Silveira, M. B. Adjei, J. M. Vendramini, and J. E. Rechcigl. 2009. Cattle manure application strategies effects on bahiagrass yield, nutritive value, and phosphorus recovery. *Agron. J.* 101:1099–1107.

Obour, A. K., M. L. Silveira, J. M. B. Vendramini, M. B. Adjei, and L. E. Sollenberger. 2010. Evaluating cattle manure application strategies on phosphorus and nitrogen losses from a Florida Spodosol. *Agron. J.* 102:1511–1520.

O'Donovan, J. T., A. E. De St. Remy, P. A. O'Sullivan, D. A. Dew, and A. K. Sharma. 1985. Influence of relative time of emergence of wild oat (*Avena fatua*) on yield loss in barley (*Hordeum vulgare*) and wheat (*Triticum aestivum*). *Weed Sci.* 33:498–503.

Oerke, E. C. 2006. Crop losses to pests. *J. Agric. Sci.* 144:31–43.

Okada, K. and A. J. Fischer. 2001. Adaptation mechanisms of upland rice genotypes to highly weathered acid soils of South American savannas. In: *Plant Nutrient Acquisition: New Perspectives*, eds., N. Ae, J. Arihara, K. Okada, and A. Srinivasan, pp. 185–200. Tokyo, Japan: Springer Verlag.

O'Toole, J. C. and W. L. Bland. 1987. Genotypic variation in crop plant root systems. *Adv. Agron.* 41:91–145.

Oussible, M., R. K. Crookston, and W. E. Larson. 1992. Subsurface compaction reduces the root and shoot growth and grain yield of wheat. *Agron. J.* 84:34–38.

Owen, P. C. 1971. The effects of temperature on the growth and development of rice. *Field Crop Abstr.* 24:1–8.

Page, E. R., M. Tollenaar, E. A. Lee, L. Lukens, and C. J. Swanton. 2010. Timing, effect, and recovery from intraspecific competition in maize. *Agron. J.* 102:1007–1013.

Paul, E. A., H. P. Collins, and S. Leavitt. 2001. Dynamics of resistant soil C measured by naturally occurring ^{14}C abundance. *Geoderma* 104:239–256.

Paula, M. B., F. D. Nogueira, H. Andrade, and J. E. Pitts. 1987. Effect of liming on dry matter yield of wheat in pots of low humic gley soil. *Plant Soil* 97:85–91.

Paustian, K. H., P. Collins, and E. A. Paul. 1997. Management controls on soil carbon. In: *Soil Organic Matter in Temperate Agroecosystems: Long-Term Experiments in North America*, ed., E. A. Paul, pp. 15–49. Boca Raton, FL: CRC Press.

Peacock, A. D., M. D. Mullen, D. B. Ringelberg, D. D. Tyler, D. B. Hedrick, P. M. Gale, and D. C. White. 2001. Soil microbial community responses to daily manure or ammonium nitrate application. *Soil Biol. Biochem.* 33:1011–1019.

Pearson, R. W., L. F. Ratliff, and H. M. Taylor. 1970. Effect of soil temperature, strength, and pH on cotton seedling root elongation. *Agron. J.* 62:243–246.

Poesen, J., J. Nachtergaele, G. Versteaeten, and C. Valetin. 2003. Gully erosion and environmental change: Importance and research needs. *Catena* 69:197–205.

Potash and Phosphate Institute. 1993. *Soil Fertility Manual*, 14th edn. Norcross, GA: Potash and Phosphate Institute.

Power, J. F., D. L. Grunes, G. A. Reichman, and W. O. Willis. 1970. Effect of soil temperature on rate of barley development and nutrition. *Agron. J.* 62:567–571.

Provance-Bowley, M. C., J. R. Heckman, and E. F. Durner. 2010. Calcium silicate suppresses powdery mildew and increases yield of field grown wheat. *Soil Sci. Soc. Am. J.* 74:1652–1661.

Qasem, J. R. and T. A. Hill. 1989. Possible role of allelopathy in the competition between tomato, *Senecio vulgaris* L. and *Chenopodium album* L. *Weed Res.* 29:349–356.

Qian, J. H. and J. W. Doran. 1996. Available C released from crop roots during growth by carbon-13 natural abundance. *Soil Sci. Soc. Am. J.* 60:828–831.

Rachman, A., S. H. Anderson, C. J. Gantzer, and E. E. Alberts. 2004. Soil hydraulic properties influenced by stiff-stemmed grass hedge systems. *Soil Sci. Soc. Am. J.* 68:1386–1393.

Radcliffe, D. E., R. L. Clark, and M. E. Sumner. 1986. Effect of gypsum and deep rooting perennial on subsoil mechanical impedance. *Soil Sci. Soc. Am. J.* 50:1566–1570.

Radke, J. F. and R. E. Bauer. 1969. Growth of sugarbeets as affected by root temperatures. Part I. Greenhouse studies. *Agron. J.* 61:860–863.

Raij, B. V. 1991. *Soil Fertility and Fertilization.* São Paulo, Brazil: Agronomy Editor Ceres.

Raij, B. V. 2010. Fertility of Brazilian soils: Contribution of Agronomic Institute of Campinas. *Agron. Inf.* 132:1–13.

Raij, B. V., N. M. Silva, O. C. Bataglia, J. A. Quaggio, R. Hiroce, H. Cantarella, R. Bellinazzi Jr., A. R. Dechen, and P. E. Trani. 1985. Fertilizer and lime recommendations for the State of São Paulo, Brazil. Technical bulletin 100. Campinas, Brazil: Agronomy Institute.

Rasse, D. P. and A. J. M. Smucker. 1998. Root recolonization of previous root channels in corn and alfalfa rotations. *Plant Soil* 204:203–212.

Rathore, A. L., A. R. Pal, and K. K. Sahu. 1998. Tillage and mulching effects on water use, root growth and yield of rainfed mustard and chickpea grown after lowland rice. *J. Sci. Food Agric.* 78:149–161.

Raun, W. R., D. H. Sander, and R. A. Olson. 1987. Phosphorus-fertilizer carrier and their placement for minimum till corn under sprinkler irrigation. *Soil Sci. Soc. Am. J.* 51:1055–1062.

Rengasamy, P. and K. A. Olsson. 1993. Irrigation and sodicity. *Aust. J. Soil Res.* 31:821–837.

Riedell, W. E., M. A. Catangui, and E. A. Beckendorf. 2009a. Nitrogen fixation, ureide, and nitrate accumulation responses to soybean aphid injury in *Glycine max. J. Plant Nutr.* 32:1674–1686.

Riedell, W. E., J. L. Pikul Jr., A. A. Jaradat, and T. E. Schumacher. 2009b. Crop rotation and nitrogen input on soil fertility, maize mineral nutrition, yield and seed composition. *Agron. J.* 101:870–879.

Rinnan, R., A. Michelsen, E. Baath, and S. Jonasson. 2008. Fifteen years of climate change manipulations alter soil microbial communities in a subarctic heath ecosystems. *Global Change Biol.* 13:28–39.

Rinnan, R., A. Michelsen, and S. Jonasson. 2007. Effects of litter addition and warming on soil carbon, nutrient pools and microbial communities in a subarctic heath ecosystem. *Appl. Soil Ecol.* 39:271–281.

Ritchey, K. D., D. M. G. Souza, E. Lobato, and O. Correa. 1980. Calcium leaching to increase rooting depth in Brazilian savannah Oxisol. *Agron. J.* 72:40–44.

Rochester, I. J., M. B. Peoples, N. R. Hulugalle, R. R. Gault, and G. A. Constable. 2001. Using legumes to enhance nitrogen fertility and improve soil condition in cotton cropping systems. *Field Crop. Res.* 70:27–41.

Rodriguez, M., J. Neris, M. Tejedor, and C. Jimenez. 2010. Soil temperature regimes from different latitudes on a subtropical island (Tenerife, Spain). *Soil Sci. Soc. Am. J.* 74:1662–1669.

Rosolem, C. and M. Takahashi. 1996. Soybean root growth and nutrient uptake as affected by liming and soil strength (Abstract), p. 64. *The 5th Symposium of the International Society of Root Research,* July 14–18, 1996. Clemson, SC: Clemson University.

Russell, R. S. 1977. *Plant Root Systems.* New York: McGraw-Hill.

Sainju, U. M. and R. E. Good. 1993. Vertical root distribution in relation to soil properties in New Jersey Pinelands forests. *Plant Soil* 150:87–97.

Sainju, U. M., A. W. Lenssen, H. B. Goosey, E. Snyder, and P. G. Hatfield. 2010. Dryland soil carbon and nitrogen influenced by sheep grazing in the wheat-fallow system. *Agron. J.* 102:1553–1561.

Sainju, U. M., W. F. Whitehead, and B. R. Singh. 2003. Cover crops and nitrogen fertilization effects on soil aggregation and carbon and nitrogen pools. *Can. J. Soil Sci.* 83:155–165.

Sanchez, P. 1976. *Properties and Management of Acid Soils in the Tropics*. New York: John Wiley & Sons.

Sanchez, P. and J. Salinas. 1981. Low input technology for managing Oxisols and Ultisols in tropical America. *Adv. Agron.* 34:279–406.

Santos, L. M., D. M. Bastos, P. Milori, M. L. Simões, W. T. L. Silva, E. R. Pereira-Filho, W. J. Melo, and L. M. Neto. 2010. Characterization by fluorescence of organic matter from Oxisols under sewage sludge applications. *Soil Sci. Soc. Am. J.* 74:94–104.

Scott, B. J., J. A. Fisher, and B. R. Cullins. 2001. Aluminum tolerance and lime increase wheat yield on the acidic soils of central and southern New South Wales. *Aust. J. Exp. Agric.* 41:523–532.

Seydou, B. T., R. E. Carlson, C. D. Pilcher, and M. E. Rice. 2000. Bt and non-Bt maize growth and development as affected by temperature and drought stress. *Agron. J.* 92:1027–1035.

Shainberg, I., R. Keren, and H. Frenkel. 1982. Response of sodic soils to gypsum and calcium chloride application. *Soil Sci. Soc. Am. J.* 46:113–117.

Shainberg, I., M. E. Sumner, W. P. Miller, M. P. W. Farina, M. A. Pavan, and M. V. Fey. 1989. Use of gypsum on soils: A review. *Adv. Soil Sci.* 9:1–111.

Shainberg, I., D. Warrington, and P. Rengasamy. 1990. Water quality and PAM interactions in reducing surface sealing. *Soil Sci.* 149:301–307.

Sharma, M. P. and J. P. Gupta. 1998. Effect of organic materials on grain yield and soil properties in maize (*Zea mays* L.)-wheat (*Triticum aestivum* L.) cropping system. *Indian J. Agric. Sci.* 68:715–717.

Shierlaw, J. and A. M. Alston. 1984. Effect of soil compaction on root growth and uptake of phosphorus. *Plant Soil.* 77:15–28.

Siddique, M. T. and J. S. Robinson. 2003. Phosphorus forms in manure and compost and their release during simulated rainfall. *J. Environ. Qual.* 29:1462–1469.

Sims, J. T. and D. C. Wolf. 1994. Poultry waste management: Agricultural and environmental issues. *Adv. Agron.* 52:1–83.

Singh, K. V. and M. V. Reddy. 1996. Improving chickpea yield by incorporating resistance to ascochyta blight. *Theor. Genet.* 92:509–515.

Singh, B. P. and U. M. Sainju. 1998. Soil physical and morphological properties and root growth. *Hort. Sci.* 33:966–971.

Smiley, R. W. 2009. Root-lesion nematodes reduce yield of intolerant wheat and barley. *Agron. J.* 101:1322–1335.

Smiley, R. W. and S. Machado. 2009. *Pratylenchus neglectus* reduces yield of winter wheat in dryland cropping systems. *Plant Dis.* 93:263–271.

Smith, C. J., M. B. Peoles, G. Keerthisinghe, T. R. James, D. L. Garden, and S. S. Tuomi. 1994. Effect of surface applications of lime, gypsum, and phosphogypsum on the alleviating of surface and subsurface acidity in a soil under pasture. *Aust. J. Soil Res.* 32:995–1008.

Smith, P., D. S. Powlson, M. J. Glendining, and J. U. Smith. 1997. Potential for carbon sequestration in European soils: Preliminary estimates for five scenarios using results from long term experiments. *Global Change Biol.* 3:67–79.

Soil Science Society of America. 2008. *Glossary of Soil Science Terms*. Madison, WI: SSSA.

Sotomayor-Ramirez, D., Y. Espinoza, and V. Acosta-Martinez. 2009. Land use effects on microbial biomass C, β-glucosides and β-glucosaminidase activities, and availability, storage, and age of organic C in soil. *Biol. Fertil. Soils* 45:487–497.

Sousa, D. M. G., L. N. Miranda, and E. Lobato. 1996. Evaluation methods of lime requirements in cerrado soils. Cerrado Center of EMBRAPA technical circular 33, Planaltina, Brazil: Cerrado Center of EMBRAPA.

Sparling, G. P. 1992. Ratio of microbial biomass carbon to soil organic carbon as a sensitive indicator of changes in soil organic matter. *Aust. J. Soil Res.* 30:195–207.

Springett, J. A. and J. K. Syers. 1984. Effect of pH and calcium content of soil on earthworm cast production in the laboratory. *Soil Biol. Biochem.* 16:185–189.

Stanger, T. F. and J. G. Lauer. 2006. Optimum plant population of Bt and non-Bt corn in Wisconsin. *Agron. J.* 98:914–921.

Steppuhn, H., M. T. V. Genuchten, and C. M. Grieve. 2005. Root-zone salinity: I. Selecting a product-yield index and response function for crop tolerance. *Crop Sci.* 45:209–220.

Stirzaker, R. J. and I. White. 1995. Ameliorating of soil compaction by a cover crop for no-tillage lettuce production. *Aust. J. Agric. Res.* 46:553–568.

Stockdale, E. A. and P. C. Brookes. 2006. Detection and quantification of the soil microbial biomass-Impacts on the management of agricultural soils. *J. Agric. Sci.* 144:285–302.

Stolt, M. H., P. J. Drohan, and M. J. Richardson. 2010. Insights and approaches for mapping soil organic carbon as a dynamic soil property. *Soil Sci. Soc. Am. J.* 74:1685–1689.

Stumpe, J. M. and P. L. G. Vlek. 1991. Acidification induced by nitrogen sources in columns of selected tropical soils. *Soil Sci. Soc. Am. J.* 55:145–151.

Suman, A., M. Lal, A. K. Singh, and A. Gaur. 2006. Microbial biomass turnover in Indian subtropical soils under different sugarcane intercropping systems. *Agron. J.* 98:698–704.

Sumner, M. E. and E. Carter. 1987. Amelioration of subsoil acidity. *Commun. Soil Sci. Plant Anal.* 19:1309–1318.

Sumner, M. E., H. Shahandeh, J. Bouton, and J. Hammel. 1986. Amelioration of an acid soil profile through deep liming and surface application of gypsum. *Soil Sci. Soc. Am. J.* 50:1254–1258.

Takahashi, H. 1994. Hydrotropism and its interaction with gravitropism in roots. *Plant Soil* 165:301–308.

Takahashi, H. and T. K. Scoot. 1993. Intensity of hydrostimulation for the induction of root hydrotropism and its sensing by the root cap. *Plant Cell Environ.* 16:99–103.

Takano, M., H. Takahashi, T. Hirasawa, and H. Suge. 1995. Hydrotropism in roots: Sensing of a gradient in water potential by the root cap. *Planta* 197:410–413.

Tanaka, D. L., R. L. Anderson, and S. C. Rao. 2005. Crop sequencing to improve use of precipitation and synergize crop growth. *Agron. J.* 97:385–390.

Tang, C. and Z. Rengel. 2003. Role of plant cation/anion uptake ratio in soil acidification. In: *Handbook of Soil Acidity*, ed., Z. Rengel, pp. 57–81. New York: Marcel Dekker.

Tejada, M., C. Garcia, J. L. Gonzalez, and M. T. Hernandez. 2006. Use of organic amendments as a strategy for saline soil remediation: Influence on the physical, chemical and biological properties of soil. *Soil Biol. Biochem.* 38:1413–1421.

Tester, M. and R. Davenport. 2003. Na^{2+} tolerance and Na^{2+} transport in higher plants. *Ann. Bot.* 9:503–527.

Thomas, G. W., G. R. Hazler, and R. L. Blevins. 1996. The effects of organic matter and tillage on maximum compactibility of soils using the proctor test. *Soil Sci.* 161:502–508.

Thompson, J. P., P. S. Brennan, T. G. Sheedy, and N. P. Seymour. 1999. Progress in breeding wheat for tolerance and resistance to root lesion nematodes (*Pratylenchus thornei*). *Aust. Plant Pathol.* 28:45–52.

Thompson, J. P., J. Mackenzie, and R. Amos. 1995. Root-lesion nematode (*Pratylenchus thornei*) limits response of wheat but not barley to stored soil moisture in the hermitage long term tillage experiment. *Aust. J. Exp. Agric.* 35:1049–1055.

Thompson, J. P., K. J. Owen, G. R. Stirling, and M. J. Bell. 2008. Root-lesion nematodes (*Pratylenchus thornei* and *P. neglectus*): A review of recent progress in managing a significant pest of grain crops in northern Australia. *Aust. Plant Pathol.* 37:235–242.

Tian, G., L. Brussaard, and B. T. Kang. 1995. Breakdown of plant residue with contrasting chemical compositions: Effect of earthworms and millipedes. *Soil Biol. Biochem.* 27:277–280.

Tian, G., J. A. Olimah, G. O. Adeoye, and B. T. Kang. 2000. Regeneration of earthworm populations in a degraded soil by natural and planted fallows under humid tropical conditions. *Soil Sci. Soc. Am. J.* 64:222–228.

Tiessen, H., E. Cuevas, and P. Chacon. 1994. The role of soil organic matter in sustaining soil fertility. *Nature* 371:783–785.

Tillman, B. L., M. W. Gomillion, G. Person, and C. L. Mackowiak. 2010. Variation in response to calcium fertilization among four runner-type peanut cultivars. *Agron. J.* 102:469–474.

Timpa, J. D., J. J. Burke, J. E. Quiseberry, and C. W. Wendt. 1986. Effect of water stress on organic acid and carbohydrate composition of cotton plant. *Plant Physiol.* 82:724–730.

Tisdall, J. M. and J. M. Oades. 1982. Organic matter and water-stable aggregates in soils. *J. Soil Sci.* 33:141–163.

Tollenaar, M., S. P. Nissanka, A. Aguilera, S. F. Weise, and C. J. Swanton. 1994. Effect of weed interference and soil nitrogen on four maize hybrids. *Agron. J.* 86:596–601.

Toma, M., M. E. Sumner, G. Weeks, and M. Saigusa. 1999. Long-term effects of gypsum on crop yield and subsoil chemical properties. *Soil Sci. Soc. Am. J.* 39:891–895.

Tran, T. S., M. Giroux, and M. P. Cescas. 1997. Effect of N rate and application methods on ^{15}N-labelled fertilizer use by corn. *Can. J. Soil Sci.* 77:9–19.

Travlos, I. S., G. Economou, and P. J. Kanatas. 2011. Corn and barnyardgrass competition as influenced by relative time of weed emergence and corn hybrid. *Agron. J.* 103:1–6.

Trudgill, D. L. 1991. Resistance to and tolerance of plant parasitic nematodes in plants. *Annu. Rev. Phytopathol.* 29:167–192.

Truman, C. C., J. N. Shaw, and D. W. Reeves. 2005. Tillage effects on rainfall partitioning and sediment yield from an Ultisol in central Alabama. *J. Soil Water Conserv.* 60:89–98.

Trumbore, S. E. and C. I. Czimczik. 2008. An uncertain future for soil carbon. *Science* 321:1455–1456.

Tsutsumi, D., K. Kosugi, and T. Mizuyama. 2003. Root-system development and water-extraction model considering hydrotropism. *Soil Sci. Soc. Am. J.* 67:387–401.

Tuberosa, R., M. C. Sanguineti, P. Landi, M. M. Giuliani, S. Salvi, and S. Conti. 2002. Identification of QTLs for root characteristics in maize grown in hydrophonic and analysis of their overlap with QTLs for grain yield in the field at two water regimes. *Plant Mol. Biol.* 48:697–712.

Tumusiime, E., B. W. Brotsen, J. Mosali, and J. T. Biermacher. 2011. How much does considering the cost of lime affect the recommended level of nitrogen? *Agron. J.* 103:404–412.

Ulloa, S. M., A. Datta, and S. Z. Knezevic. 2010. Growth stage influenced differential response of foxtail and pigweed species to broadcast flaming. *Weed Technol.* 24:319–325.

Ulloa, S. M., A. Datta, and S. Z. Knezevic. 2011. Growth stage influenced sorghum response to broadcast flaming: Effects on yield and its components. *Agron. J.* 103:7–12.

Ulrich, B., R. Mayer, and P. K. Khanna. 1980. Chemical changes due to acid precipitation in a loess derived soil in central Europe. *Soil Sci.* 130:193–199.

Undabeytia, T., G. Mishael, S. Nir, B. Papahadjopoulos-Sternberg, B. Rubin, E. Morillo, and C. Maqueda. 2003. A novel system for reducing leaching from formulations of anionic herbicides: Clay-liposomes. *Environ. Sci. Technol.* 37:4475–4480.

Undabeytia, T., S. Nir, and M. J. Gomara. 2004. Clay-vesicle interactions: Fluorescence measurements and structural implications for slow release formulations of herbicides. *Langmuir* 20:6605–6610.

Undabeytia, T., F. Sopena, T. Sanchez-Verdejo, J. Villaverde, S. Nir, E. Morillo, and C. Maqueda. 2010. Performance of slow-release formulations of alachlor. *Soil Sci. Soc. Am. J.* 74:898–905.

Unger, P. W., H. V. Eck, and J. T. Musick. 1981. Alleviating plant water stress. In: *Modifying the Root Environment to Reduce Crop Stress*, eds., G. F. Arkin and H. Taylor, pp. 61–96. St. Joseph, MI: American Society of Agricultural and Engineering.

Unger, P. W. and T. C. Kaspar. 1994. Soil compaction and root growth: A review. *Agron. J.* 86:759–766.

Urrea, C. A., R. M. Harveson, A. E. Koehler, P. Burgener, and D. D. Baltensperger. 2010. Evaluating the agronomic potential of chickpea germplasm for western Nebraska. *Agron. J.* 102:1179–1185.

USDA. 2008. Electronic code of federal regulations. Available at http://ecfr.gpoaccess.gov/ cgr/t/text-idx?c=ecfr&tpl=%2Findex.tpl (Verified on September 10, 2010). Washington, DC: USDA/AMS/TMP/NOP.

Vankosky, M. A., H. A. Carcamo, R. H. Mckenzie, and L. M. Dosdall. 2011. Integrated management of *Sitona lineatus* with nitrogen fertilizer, rhizobium, and thiamethoxam insecticide. *Agron. J.* 103:565–572.

Varvel, G. E. 2000. Crop rotation and nitrogen effects on normalized grain yields in a long-term study. *Agron. J.* 86:204–208.

Viaud, V., D. A. Angers, and C. Walter. 2010. Toward landscape-scale modeling of soil organic matter dynamics in agroecosystems. *Soil Sci. Soc. Am. J.* 74:1847–1860.

Villamil, M. B., G. A. Bollero, R. G. Darmody, F. W. Simmons, and D. G. Bullock. 2006. No-till corn/soybean systems including winter cover crops: Effects on soil properties. *Soil Sci. Soc. Am. J.* 70:1936–1944.

Von Uexkull, H. and E. Mutert. 1995. Blobal extent, development and economic impact of acid soils. *Plant Soil* 171:1–15.

Voorhees, W. B., R. R. Allmaras, and C. E. Johnson. 1981. Alleviating temperature stress. In: *Modifying the Root Environment to Reduce Crop Stress*, eds., G. F. Arkin and H. Taylor, pp. 217–266. St. Joseph, MI: American Society of Agricultural Engineering.

Wallace, R. W. and R. R. Bellinder. 1992. Alternative tillage and herbicide options successful weed control in vegetables. *Hort. Sci.* 27:745–749.

Walz, E. 2004. Final results of the fourth national organic farmers survey: Sustainable organic farms in a changing organic marketplace. Available at http://ofrf.org/publications/ pubs/4thsurvey_results.pdf (Verified on September 10, 2010). Santa Cruz, CA: Organic Farming Research Foundation.

Wardle, D. A. 1992. A comparative assessment of factors which influence microbial biomass carbon and nitrogen levels in soil. *Biol. Rev. Cambr. Philos. Soc.* 67:321–358.

Wardle, D. A. and A. Ghani. 1995. A critique of the microbial metabolic quotient (qCO_2) as a bioindicator of disturbance and ecosystem development. *Soil Biol. Biochem.* 27:1601–1610.

Watson, P. R., D. A. Derksen, and R. C. Van Acker. 2006. The ability of 29 barley cultivars to compete and withstand competition. *Weed Sci.* 54:783–792.

Wendell, R. R. and K. D. Ritchey. 1996. High-calcium flue gas desulfurization products reduce aluminum toxicity in an Appalachian soil. *J. Environ. Qual.* 25:1401–1410.

Werner, M. R. 1997. Soil quality characteristics during conversion to organic orchard management. *Appl. Soil Ecol.* 5:151–167.

Westermann, D. T. 2005. Nutritional requirements of potatoes. *Am. J. Potato Res.* 82:301–307.

Weston, L. A. 1996. Utilization of allelopathy for weed management in agroecosystems. *Agron. J.* 88:860–866.

Whitfield, C. J. and D. E. Smika. 1971. Soil temperature and residue effects on growth components and nutrient uptake of four wheat varieties. *Agron. J.* 63:297–300.

Willert, F. J. V. and R. C. Stehouwer. 2003. Compost, limestone, and gypsum effects on calcium and aluminum transport in acidic minespoil. *Soil Sci. Soc. Am. J.* 67:778–786.

Williams, C. H. 1980. Soil acidification under clover pasture. *Aust. J. Exp. Agric. Anim. Husban.* 20:561–567.

Williams, S. M. and R. R. Weil. 2004. Crop cover root channels may alleviate soil compaction effects on soybean crop. *Soil Sci. Soc. Am. J.* 68:1403–1409.

Wright, A. L., F. M. Hons, R. G. Lemon, M. L. McFarland, and R. L. Nicholas. 2008. Microbial activity and soil c sequestration for reduced and conventional tillage cotton. *Appl. Soil Ecol.* 38:168–173.

Yang, C., J. Chong, C. Kim, C. Li, D. Shi, and D. Wang. 2007. Osmotic adjustment and ion balance traits of an alkali resistant halophyte *Kochia sieversiana* during adaptation to salt and alkali conditions. *Plant Soil* 294:263–276.

Yang, C., W. Guo, and D. Shi. 2010. Physiological roles of organic acids in alkali-tolerance of the alkali-tolerant halophyte *Chloris virgate. Agron. J.* 102:1081–1089.

Yang, C., A. Jianaer, C. Li, C. Shi, and D. Wang. 2008a. Comparison of the effects of salt-stress and alkali-stress on the photosynthetic production and energy storage of an alkali-resistant halophyte *Chloris virgata. Photosynthetica* 46:273–278.

Yang, C., D. Shi, and D. Wang. 2008b. Comparative effects of salt stress and alkali stress on growth, osmotic adjustment and ionic balance of an alkali resistant halophyte *Suaeda glauca* (Bge.). *Plant Growth Regul.* 56:179–190.

Yang, C., P. Wang, C. Li, D. Shi, and D. Wang. 2008c. Comparison of effects of salt and alkali stresses on the growth and photosynthesis of wheat. *Photosynthetica* 46:107–114.

Yang, X., W. Wang, Z. Ye, Z. He, and V. C. Baigar. 2004. Physiological and genetic aspects of crop plant adaptation to elemental stresses in acid soils. In: *The Red Soils of China: Their Nature, Management and Utilization*, eds., M. J. Wilson, Z. He, and X. Yang, pp. 171–218. Dordrecht, the Netherlands: Kluwer Academic Publishers.

Yau, S. K., M. Sidahmed, and M. Haidar. 2010. Conservation versus conventional tillage on performance of three different crops. *Agron. J.* 102:269–276.

Yeo, A. R. 1983. Salinity resistance: Physiologies and prices. *Physiol. Plantram* 58:214–222.

You, M. S., Y. F. Liu, and Y. M. Hou. 2004. Agricultural biodiversity and integrated pest management. *Ecology* 24:117–122.

Yu, J., T. Lei, I. Shainberg, A. I. Mamedov, and G. J. Levy. 2003. Infiltration and erosion in soils treated with dry PAM and gypsum. *Soil Sci. Soc. Am. J.* 67:630–636.

Zeng, Z., T. Fu, Y. Tang, Y. Chen, and Z. Ren. 2007. Identification and chromosomal locations of novel genes for resistance to powdery mildew and stripe rust in a wheat line 101–3. *Euphytica* 156:89–94.

Zhang, G. S., K. Y. Chan, A. Oates, D. P. Heenan, and G. B. Huang. 2007. Relationship between soil structure and runoff/soil loss after 24 years of conventional tillage. *Soil Till. Res.* 92:122–128.

Zhang, F. S., J. Ma, and Y. P. Cao. 1997. Phosphorus deficiency enhances root exudation of low-molecular weight organic acids and utilization of sparingly soluble inorganic phosphate by radish (*Raghanus satiuvs* L.) and rape (*Brassica napus* L.) plants. *Plant Soil* 196:261–264.

Zhang, W., K. M. Parker, Y. Luo, S. Wan, L. L. Wallace, and S. Hu. 2005. Soil microbial response to experimental warming and clipping in a tallgrass prairie. *Global Change Biol.* 11:266–277.

Zhang, N., S. Wan, L. Li, J. Bi, M. Zhao, and K. Ma. 2008. Impacts of urea addition on soil microbial community in a semi-arid temperate steppe in northern China. *Plant Soil* 311:19–28.

Zheng, B. S., L. Yang, W. P. Zhang, C. Z. Mao, Y. R. Wu, K. K. Yi, F. Y. Liu, and P. Wu. 2003. Mapping QTLs and candidate genes for rice root traits under different water supply conditions and comparative analysis across three populations. *Theor. Appl. Genet.* 107:1505–1515.

Zhou, C., Y. Xia, B. L. Ma, C. Feng, and P. Qin. 2010. Culture of seashore mallow under different salinity levels using plastic nutrient-rich matrices and transplantation. *Agron. J.* 102:395–402.

Zotarelli, L., B. J. R. Alves, S. Urquiaga, R. M. Boddey, and J. Six. 2007. Impact of tillage and crop rotation on light fraction and intra-aggregate soil organic matter in two Oxisols. *Soil Till. Res.* 95:196–206.

Index

A

Aluminum saturation
  aluminum toxicity, 387–388
  base saturation method, 389
  calculation, 387
  hydrolysis reactions, 387
  lime rate determination, 389
  for maximum yield field crops, 387–388
  root systems, 386

B

Base saturation
  for annual crops, 385
  calculation, 383
  dry bean grain yield, 385–386
  optimum base saturation, 384–385
Border strip irrigation method, 147
Brazilian Oxisols
  addition of lime, 382
  addition of P, 316
  addition of Zn, 331
  base saturation, 384–385, 389
  N fertilization, 57–58
  P fertilization, 57–58
  upland rice yield, 204–205

C

Carrot
  climate and soil requirements, 362
  diseases and insects, 362
  growth and development, 362–364
  nutrient requirements, 362–363
  Umbellifereae, 361
Cassava
  cassava-producing countries, 352
  climate and soil requirements, 353
  common names, 352
  diseases and insects, 353
  fertilization, 353
  growth and development
    constituents, 355
    dry matter, 357
    harvest index, 357
    leaf area index, 356–357
    roots, 356
    tops, 356

nutrient accumulation, 354
nutrient concentration, 354–355
pH, 354
subspecies, 351–352
Cation exchange capacity (CEC), 8, 34, 383, 389
Central pivot irrigation method, 146
Check basin irrigation method, 146
Conservation tillage
  climate change mitigation, 139
  definition, 137
  macroaggregates, 138
  minimum tillage/zero tillage, 137
  SOM content, 138
Crop growth stages
  biomass production, 155
  canola vegetative and seed-filling stages, 158–159
  drought stress effects, 156
  grain yield reduction, 156
  grain yield response determination, 154
  moisture-sensitive growth stages, 158
  photosynthesis, 156
  pod-filling growth stages, 157
  shoot dry weight accumulation and grain yield, 154–155
  water stress, 154
  yield reduction, 157–158

D

Dinitrogen fixation
  angiosperms, 210–211
  in food and fiber production, 209
  nitrogen-fixing bacteria and host plants, 210
  Rhizobium and *Bradyrhizobium,* 210
  rhizosphere pH, 211
Drip irrigation method, 146
Drought
  biological and economic yields, 151
  CO_2 assimilation, 150
  crop growth stages (*see* Crop growth stages)
  crop yield and water stress, 152–154
  dryness, 151
  reduction management strategies
    categories, 160
    deep-rooted crops, 163
    deficit irrigation, 161
    genetic variation, 163
    irrigation management, 160–161

9 780367 381042